Quark Confinement and the Hadron Spectrum

Quark

Proceedings
of the
International
Conference on

Confinement
and the
Hadron
Spectrum

Como, Italy   20 – 24 June 1994

Editors

N. Brambilla & G. M. Prosperi

Universita, Degli Studi Di Milano, Italy

World Scientific
Singapore • New Jersey • London • Hong Kong

Published by

World Scientific Publishing Co. Pte. Ltd.

P O Box 128, Farrer Road, Singapore 9128

USA office: Suite 1B, 1060 Main Street, River Edge, NJ 07661

UK office: 57 Shelton Street, Covent Garden, London WC2H 9HE

Proceedings of the International Conference on
QUARK CONFINEMENT AND THE HADRON SPECTRUM

ISBN 981-02-2085-5

Photography by Dr. A. Vairo

Printed in Singapore by Uto-Print

FOREWORD

Quantum Chromodynamics has been proposed originally as a possible field theoretic version of the naive quark model. In this sense its main motivation was an understanding of the spectrum and of the properties of hadrons. In practice, due to asymptotic freedom, a great insight has been attained in high energy phenomena, like deep inelastic scattering or jets, while many difficulties remain in the problem of bound states.

Obviously the main difficulty is to understand and manage the phenomenon of confinement. Many approaches to the problem have been attempted: lattice theory and numerical simulation, $\frac{1}{N}$ expansions, sum rules, potential and Bethe-Salpeter formalism, dominance of classical configurations, dual QCD, bag model, vacuum structure. Essentially one needs to develop some kind of strong coupling technique. Indeed strong coupling limit is supposed to dominate long range interactions, as the weak coupling dominates short range forces. Then, however, the problem is to connect these two regions. As a matter of fact only the brute force lattice simulations can claim to rest on first principles alone. All other methods combine theory, conjectures and phenomenological assumptions. Usually, in fitting the spectrum the strong coupling constant (or the scale constant) and the string tension are treated as independent parameters and possibly only *a posteriori* one checks if the results are consistent with lattice simulations.

In spite of this, the problem of "quark confinement and the hadron spectrum" remains central in QCD and the organizers of the present conference feel that true, though slow, progresses have been achieved on the subject in the last years. For this reason they thought that to attempt to gather together people working from various point of view could be very useful.

It seems to me that the conference was very successful in its purpose. Many topics have been covered, both in the general talks of the plenary sessions and in the more specific contributed papers of the parallel sessions. The discussions were very stimulating and new interesting collaborations have been established. I hope that these proceedings can reflect the climate of the meeting and provide a good update to the subject.

The proceedings are divided in five parts corresponding to the sections in which the conference was organized

- Plenary talks

- Sec. A (Convener A. Di Giacomo, Pisa): Vacuum structure and non perturbative effects; flux tube configuratios and strings; lattice theory and numerical simulations.

- Sec. B (Convener F. F. Schoeberl, Wien): Interquark potential and quarkonia; Bethe-Salpeter equation, relativistic equations.

- Sec. C (Convener N. Toernqvist, Helsinki): Quark pair creation effects, mixing, decays; exotic states, glueballs; effective theories.

- Sec. D : Experimental and phenomenological talks and miscellanea.

The nice final talk of Yuval Ne'eman closes the first part of the book. The talk concerned the origins and the story of QCD and a summary of the different approaches and lines of research present at the workshop.

Unfortunately the Editors did not received in time some interesting contributions. In this case a short summary is reported. For the papers originally accepted in the poster session only the abstracts are reproduced.

Among the social events during the conference the Wednesday evening concert and the boat trip on the lake were particularly appreciated. The concert was held in the ancient magnificent church of san Fedele, the maestro Marco Doni at the organ and the Rumanian soprano Gayaneh Tapacian played for us "baroque music". The trip was particularly enjoyable, the weather was nice and the partecipants could admire the beauty of the lake celebrated by many writers and poets since the roman times. On the way back they enjoyed the conference dinner at "Isola Comacina", an island with an interesting history related to the fights between the Italian city–states and the German emperor Friedrich Barbarossa in the middle age.

On behalf of the Organizing Committee, before closing this introduction, I would like to acknowledge all participants that made this conference so live and interesting. Special thanks are due to the staff of "Centro Volta" for the accurate logistic organization, to the entire Advisory Committee for the numerous suggestions, to Prof. W. Lucha for the help in the manuscript revision, and in particular to the Scientific Secretary Dr. Nora Brambilla, without whose enthusiasm this meeting would have not be possible.

Last but not least I would like to acknowledge the financial support from various agencies:

- The University of Milan

- The "Istituto Nazionale di Fisica Nucleare", Rome

- The Human Capital Program of the European Union, Brussels

- The Soros Foundation

Milano, November 1994

G. M. Prosperi
Chairman of the Conference

INTERNATIONAL ADVISORY COMMITTEE

W. Buchmueller (DESY)

N. Cabibbo (Roma)

M. Creutz (BNL)

A. Di Giacomo (Pisa)

B. Durand (Madison)

F. Gliozzi (Torino)

Gribov (Lund)

D. Gromes (Heidelberg)

M. Luescher (DESY)

M. G. Olsson (Madison)

F. Schoeberl (Wien)

Yu. A. Simonov (Moscow)

F. Zachariasen (Caltech)

N. Isgur (CEBAF)

J. Rosner (Chicago)

N.A. Toernqvist (Helsinki)

ORGANIZING COMMITTEE

G.M. Prosperi (Chairman)

N. Brambilla (Scientific Secretary)

Univ. Milano

G. Marchesini (Milano)

L. Fulcher (B.G.S. Ohio)

G. Casati (Como)

TABLE OF CONTENTS

PARALLEL SESSION CONTRIBUTIONS

SECTION A

Vacuum structure and nonperturbative effects; flux tube configurations and strings; lattice theory and numerical simulations.

SECTION B

Interquark potential and quarkonia; Bethe-Salpeter equation, relativistic equations

SECTION C   316

Quark pair creation effects, mixing, decays; exotic states, glueballs; heavy quark effective theory

SECTION D   359

Phenomenology and experiments; miscellanea

POSTER SESSION 377

PLENARY SESSION PRESENTATIONS

CHIRAL SYMMETRY AND LATTICE GAUGE THEORY*

MICHAEL CREUTZ

Physics Department, Brookhaven National Laboratory
Upton, NY 11973, USA

ABSTRACT

I review the problem of formulating chiral symmetry in lattice gauge theory. I discuss recent approaches involving an infinite tower of additional heavy states to absorb Fermion doublers. For hadronic physics this provides a natural scheme for taking quark masses to zero without requiring a precise tuning of parameters. A mirror Fermion variation provides a possible way of extending the picture to chirally coupled light Fermions.

1. Introduction

Lattice gauge theory is now entering its third decade. Throughout this period Fermions have presented challenging obstacles. One of these issues lies in Monte Carlo algorithms, which remain awkward when quarks are included. Indeed, in the presence of a background Fermion density representing a chemical potential, no truly viable algorithms are known.

In this talk, however, I will concentrate on the other "Fermion problem," that of doubling and chiral symmetry. This discussion is essentially an abridged version of [1]. For a more general review of the subject see Ref. [3].

Why do we care about chiral symmetry when the lattice gives us a first principles scheme for calculating hadronic physics? The reason is partly historical and partly aesthetic. Indeed, chiral symmetry has long played an essential role in particle theory. The pion is made of the same quarks as the rho meson, yet its mass is considerably less. The canonical explanation says that if quarks were massless, then the pions would be Goldstone bosons arising from the spontaneous breaking of an underlying chiral symmetry. Naively, pure gauge interactions are helicity conserving, and thus both the number of left and right handed massless quarks are separately conserved. Through confinement into the physical states of baryons and mesons, this symmetry is spontaneously broken. A host of predictions from current algebra are based on this picture [2].

These issues are complicated by the presence of "anomalies." Ultraviolet divergences make it impossible, even in perturbation theory, to conserve simultaneously all axial vector currents associated with chiral symmetry, and the vector currents coupled to gluons. As current conservation is crucial to our understanding of gauge symmetries, we must conserve the vector currents, implying that the axial symmetry cannot be exact.

One consequence of the anomaly is that there is one less Goldstone boson than naive counting would suggest. For two flavors of light quarks, of the four ways to form pseudoscalar mesons there are three light pions while the eta remains heavier. With $SU(3)$ flavor symmetry, it is the η' which is anomalously heavy compared to the other pseudoscalars.

Chiral issues arise in an even more fundamental way with the weak interactions. Here parity violation seems to maximally differentiate between left and right handed Fermions. While lattice methods have been dominantly applied to the strong interactions, there are reasons to desire a lattice formulation of the weak interactions as well. In particular, the lattice is the best founded non-perturbative regulator, and thus provides an elegant framework for the definition of a quantum field theory. Even though the smallness of the electromagnetic coupling makes non-perturbative effects quite small in the electroweak theory, at least in principle we would like a rigorous formulation. While exceptionally small, some interesting non-perturbative phenomena are directly related to the anomaly, such as the prediction that baryons can decay through the instanton mechanism of 't Hooft [4].

2. Massless Fermions and the anomaly

Chiral symmetry is intimately tied with Lorentz invariance. A massive particle of spin s has $2s+1$ distinct spin states which mix under a general Lorentz transformation. The helicity of a massless particle, on the other hand, is frame invariant. Indeed, for free particles one can write down local fields which create or destroy just a single helicity state. For spin 1/2 Fermions coupled minimally to gauge fields, their helicity remains naively conserved.

In one space dimension the roles of left and right handed helicities are replaced by left and right moving particles. Since an observer cannot go faster than light, he can never overtake a massless particle and a right mover will be so in all Lorentz frames.

The fact that Lorentz invariance is crucial here provides another warning that chiral symmetry on the lattice will be difficult. Indeed, lattice formulations inherently violate the usual space time symmetries. Chiral issues should only be expected to be useful for states of low energy which do not see the underlying lattice structure.

While separate phase rotations of left and right handed massless Fermions give a formal symmetry of a continuum gauge theory, this is broken by the anomalies mentioned above. In particular, there is the famous triangle diagram where a virtual Fermion loop couples an axial vector current to two vector currents. In two space time dimensions the analogous problem arises with a simple bubble diagram connecting a vector and an axial vector current.

The fact that the anomaly must exist can be intuitively argued in analogy with band theory in solid state physics. With massive Fermions the vacuum has a Fermi level midway in a gap between the filled Dirac sea and a continuum of positive energy particle states. This represents an insulator. As the mass is taken to zero, the gap closes and the vacuum becomes a conductor. External gauge fields applied to this conductor can induce currents. For a specific example, consider a one space dimensional world compactified into a ring. A changing magnetic field through this ring will induce currents, changing the relative number of left and right moving particles.

Without the anomaly, transformers would not work.

In this problem, physics should be periodic in the amount of flux through the ring. This is a two dimensional analog of the periodicity of four dimensional non-abelian gauge theories as one passes through topologically non-trivial configurations [8]. The latter case with the standard model gives rise to a non-conservation of the baryon current [4].

With the one dimensional ring, the strength of the flux characterizes the phase that a charged particle acquires in running around the ring. As this net phase adiabatically increases, the individual Fermionic energy levels shift monotonically. As one adds another unit of flux through the ring, one filled right moving level from the Dirac sea moves, say, up to positive energy, while one empty left moving level drops into the sea, leaving a hole. This induces a net current carried by a right moving particle and a left moving antiparticle. This way of visualizing how the anomaly works was nicely discussed some time ago [9].

3. The doubling problem

The essence of the lattice doubling problem already appears with the simplest Fermion Hamiltonian in one space dimension

$$H = iK \sum_j a_{j+1}^\dagger a_j - a_j^\dagger a_{j+1}. \tag{1}$$

Here j is an integer labeling the sites of an infinite chain and the a_j are Fermion annihilation operators satisfying standard anticommutation relations

$$\left[a_j, a_k^\dagger\right]_+ \equiv a_j a_k^\dagger + a_k^\dagger a_j = \delta_{j,k}. \tag{2}$$

The bare vacuum $|0\rangle$ satisfies $a_j|0\rangle = 0$. This vacuum is not the physical one, which contains a filled Dirac sea. I refer to K as the hopping parameter. Energy eigenstates in the single Fermion sector

$$|\chi\rangle = \sum_j \chi_j a_j^\dagger |0\rangle \tag{3}$$

can be easily found in momentum space

$$\chi_j = e^{iqj}\chi_0. \tag{4}$$

where $0 \le q < 2\pi$. The result is

$$E(q) = 2K\sin(q). \tag{5}$$

The physical vacuum has all negative energy states filled to form a Dirac sea. Particles are represented by excitations on this vacuum.

If I consider a Fermionic wave packet produced from a superposition of states carrying small momentum q, then, since the group velocity dE/dq is positive in this region, the packet will move to the right. On the other hand, a wave packet produced from momenta in the vicinity of $q \sim \pi$ will be left moving. The essence of the Nielsen Ninomiya theorem [10] is that we must have both types of excitation. The periodicity in q requires the dispersion relation to have an equal number of zeros with positive and negative slopes.

The recent attempts to circumvent this result add to the spectrum an infinite number of additional states at high energy[11]. The idea is to have a mode with $E = 2K \sin(q)$ still exist at small q, but then become absorbed in an infinite band of states before q reaches π. If the band is truly infinite, then the extra state does not have to reappear as the momentum increases to 2π. In the domain wall picture, this infinite tower of states is represented by a flow into an extra dimension [5] [6].

4. The Wilson approach

In this section I review Wilson's scheme for adding a non-chirally symmetric term to remove the doublers appearing in a naive lattice transcription of the Dirac equation. I do this in some detail because the general behavior of the Wilson-Fermion Hamiltonian will be central to the later construction of surface modes. To keep the discussion simple, I work in one dimension with a two component spinor

$$\psi = \begin{pmatrix} a \\ b \end{pmatrix}. \tag{6}$$

The most naive lattice Hamiltonian begins with the simple hopping case of Eq. (1) and adds in the lower components and a mass term to mix the upper and lower components

$$H = iK \sum_j a_{j+1}^\dagger a_j - a_j^\dagger a_{j+1} - b_{j+1}^\dagger b_j + b_j^\dagger b_{j+1}$$
$$+ M \sum_j a_j^\dagger b_j + b_j^\dagger a_j. \tag{7}$$

Introducing Dirac matrices

$$\gamma_0 = \begin{pmatrix} 0 & 1 \\ 1 & 0 \end{pmatrix}, \quad \gamma_1 = \begin{pmatrix} 0 & -1 \\ 1 & 0 \end{pmatrix} \tag{8}$$

and defining $\overline{\psi} = \psi^\dagger \gamma_0$, I can write the Hamiltonian more conventionally as

$$H = \sum_j iK(\overline{\psi}_{j+1}\gamma_1\psi_j - \overline{\psi}_j\gamma_1\psi_{j+1}) + M \sum_j \overline{\psi}_j\psi_j. \tag{9}$$

As before, the single particle states are easily found by Fourier transformation and satisfy

$$E^2 = 4K^2 \sin^2(q) + M^2 \tag{10}$$

Again, the negative energy sea is to be filled.

Naive chiral symmetry is implemented through distinct phase rotations for the upper and lower components of ψ. The mass term mixes these components and opens up a gap in the spectrum. The doublers at $q \sim \pi$, however, are still with us.

To remove the degenerate doublers, I make the mixing of the upper and lower components momentum dependent. A simple way of doing this was proposed by Wilson [12]. For this I add one more term to the Hamiltonian

$$
\begin{aligned}
H = {} & iK \sum_j a^\dagger_{j+1} a_j - a^\dagger_j a_{j+1} - b^\dagger_{j+1} b_j + b^\dagger_j b_{j+1} \\
& + M \sum_j a^\dagger_j b_j + b^\dagger_j a_j \\
& - rK \sum_j a^\dagger_j b_{j+1} + b^\dagger_j a_{j+1} + b^\dagger_{j+1} a_j + a^\dagger_{j+1} b_j \\
= {} & \sum_j K(\overline{\psi}_{j+1}(i\gamma_1 - r)\psi_j - \overline{\psi}_j(i\gamma_1 + r)\psi_{j+1}) + \sum_j M\overline{\psi}_j\psi_j .
\end{aligned}
\tag{11}
$$

Now the spectrum satisfies

$$
E^2 = 4K^2 \sin^2(q) + (M - 2rK \cos(q))^2 .
\tag{12}
$$

Note how the doublers at $q \sim \pi$ are increased in energy relative to the states at $q \sim 0$. The physical particle mass is now $m = M - 2rK$ while the doubler is at $M + 2rK$.

The hopping parameter has a critical value at

$$
K_{crit} = \frac{M}{2r}
\tag{13}
$$

At this point the gap in the spectrum closes and one species of Fermion becomes massless. The Wilson term, proportional to r, still mixes the a and b type particles; so, there is no exact chiral symmetry. Nevertheless, in the continuum limit this represents a candidate for a chirally symmetric theory. Beforehand, as discussed in Ref. [13], chiral symmetry does not provide a good order parameter.

A difficulty with this approach is that gauge interactions will renormalize the parameters. To obtain massless pions one must finely tune K to K_{crit}, an a priori unknown function of the gauge coupling. Despite the awkwardness of such tuning, this is how numerical simulations with Wilson quarks generally proceed. The hopping parameter is adjusted to get the pion mass right, and one hopes for the remaining predictions of current algebra to reappear in the continuum limit.

5. Supercritical K and surface modes

The case of K exceeding the critical value $M/2r$ is rarely discussed but quite interesting nevertheless. Aoki and Gocksch [13] have argued that as one passes through this point with gauge fields present, there occurs a spontaneous breaking of parity. Restricting ourselves to the free Fermion case for the time being, interesting things happen here for supercritical K as well. As the band closes and reopens with increasing K, the positive energy particles and the negative energy Dirac sea couple strongly. A similar situation was studied some time ago by Shockley [14], who observed that if the system is finite with open walls, then two discrete levels leave the bands and emerge bound to the ends of the system.

As the volume of the system goes to infinity, particle-hole symmetry forces these surface levels to go to exactly zero energy. In a finite box, the wave functions have exponential tails away from the walls, mixing the states and in general giving them a small residual energy.

A general result [1] is that there exists such a state bound to any interface separating a region with $K > K_{crit}$ from a region with $K < K_{crit}$. In Ref. [5], Kaplan uses $M = 2Kr + m\epsilon(x)$. I prefer to consider here the simpler approach of Shamir [15] and take $K = 0$ on one side, giving modes on an open surface.

In the later discussion of the anomaly in terms of currents into an extra dimension, it will always be a flow into a region of supercritical hopping. This should be contrasted with the continuum discussion of Ref. [16], where the flow is symmetric about the defect. This symmetry appears, however, to be regulator dependent [17]. For example, with a Pauli-Villars regulator, the relative sign of the Fermion to the regulator masses controls the direction of flow.

Following the usual procedure of filling half the states for the Dirac sea, we see that there is an ambiguity with the last Fermion, which could go into either of the degenerate surface modes. If I imagine coupling the Fermions to, say, a $U(1)$ gauge field, then this last Fermion will be a source of a background electric field which will run to the hole state on the opposite wall. This is the physical origin of the parity breaking proposed in Ref. [13]. In the continuum limit the vacuum should be equivalent to that of the massive Schwinger model with a half unit of background electric flux. The physics of this model in the continuum was extensively discussed in Ref. [18].

6. Extra dimensions

As the system size goes to infinity, particle-hole symmetry naturally forces the the surface modes to zero energy. This behavior forms the basis for a lattice approach to chiral Fermions. The picture of Kaplan [5] is to reinterpret the coordinate labeled by j in the above discussion as an extra dimension beyond the usual ones of space and time. Our physical world then exists on a four dimensional interface, with the light quarks and leptons being the above surface modes.

To be concrete, consider adding D space dimensions to the above Hamiltonian, where for the following D will either be 1 or 3. For simplicity I will take L^D space sites and use antiperiodic boundary conditions for each of these dimensions. The extra dimension, which I refer to as the fifth, has L_5 sites and open boundaries. I take the same hopping and Wilson parameters in each of the dimensions, including the fifth, although this is not essential.

The Dirac matrices γ_μ satisfy the usual

$$[\gamma_\mu, \gamma_\nu]_+ = 2g_{\mu\nu}. \tag{14}$$

I define $\gamma_5 = i\gamma_0\gamma_1\gamma_2\gamma_3$ for $D = 3$ and $\gamma_5 = \gamma_0\gamma_1$ for $D = 1$. I take γ_0 and γ_5 to be Hermitian, while the spatial γ matrices are anti-Hermitian. The Hamiltonian I am led to is then

$$
\begin{aligned}
H = \sum_{\mathbf{n},j} \Big(& K\overline{\psi}_{\mathbf{n},j+1}(\gamma_5 - r)\psi_{\mathbf{n},j} - K\overline{\psi}_{\mathbf{n},j}(\gamma_5 + r)\psi_{\mathbf{n},j+1} \\
& + \sum_{a=1}^{D}(K\overline{\psi}_{\mathbf{n}+\mathbf{e}_a,j}(i\gamma_a - r)\psi_{\mathbf{n},j} - K\overline{\psi}_{\mathbf{n},j}(i\gamma_a + r)\psi_{\mathbf{n}+\mathbf{e}_a,j}) \\
& + M\overline{\psi}_{\mathbf{n},j}\psi_{\mathbf{n},j} \Big).
\end{aligned} \tag{15}
$$

Here $\mathbf{n}$ denotes the spatial sites, j the extra coordinate, and $\mathbf{e}_a$ is the unit vector in the positive a'th direction.

The use of antiperiodic boundary conditions makes it simple to go to momentum space for the spatial coordinates. Denoting the components of the momentum by q_a, I write

$$\psi_{\mathbf{q},j} = \frac{1}{L^{D/2}} \sum_{\mathbf{n}} e^{-i\mathbf{q}\cdot\mathbf{n}}\psi_{\mathbf{n},j}. \tag{16}$$

Each component of the momentum takes discrete values from the set $(2k + 1)\pi/L$ where k runs from, say, 0 to $L - 1$. This makes the Hamiltonian block diagonal, with each value for $\mathbf{q}$ representing a separate block. In this way the Hamiltonian reduces to

$$
\begin{aligned}
H = \sum_{\mathbf{q},j} \Big(& K\overline{\psi}_{\mathbf{q},j+1}(\gamma_5 - r)\psi_{\mathbf{q},j} - K\overline{\psi}_{\mathbf{q},j}(\gamma_5 + r)\psi_{\mathbf{q},j+1} \\
& + \sum_a 2K \sin(q_a)\overline{\psi}_{\mathbf{q},j}\gamma_a\psi_{\mathbf{q},j} \\
& + (M - 2Kr \sum_a \cos(q_a))\overline{\psi}_{\mathbf{q},j}\psi_{\mathbf{q},j} \Big).
\end{aligned} \tag{17}
$$

Modes bound to the surface in the fifth direction exist whenever K exceeds the critical value

$$K_{crit} = M/2r - K \sum_a \cos(q_a). \tag{18}$$

Note how this critical value now depends on the spatial momentum. Appropriately choosing M, I can have the surface states exist for small q, but have them disappear when any component of $q_a \sim \pi$. This avoids the doublers [19]. Specifically, when the hopping is direction independent, I want (assuming K, r, and M are all positive)

$$(D - 1)K < M/2r < (D + 1)K. \tag{19}$$

The above discussion shows that on a single surface I have an elegant lattice theory for a low energy chiral Fermion. I would now like to add gauge fields. Here I adopt the attitude that I do not want a lot of new degrees of freedom, and follow Ref. [11] in regarding the extra dimension as a flavor space. In particular, I do not put gauge fields in the fifth dimension, and the physical gauge fields are independent of this dimension.

While this approach has the advantage of preserving an exact gauge invariance and not introducing lots of unwanted fields, it has the disadvantage that both walls are coupled equally to the gauge field. Thus, even when the size of the fifth dimension approaches infinity, the opposite chirality Fermions do not decouple. The main thing that has been accomplished so far is to find a theory of Fermions coupled in a vectorlike manner, without any doublers, and with a natural way to take the Fermion masses to zero.

7. The anomaly and rotating eigenvalues

One of the nice features of this formulation is how the chiral anomaly appears as a flow of Fermionic states in the extra dimension. The basic scenario was discussed in a somewhat different context by Callan and Harvey [16]. They consider a vector theory, whose mass term has a domain wall shape in an extra dimension, and show that it has a chiral zeromode living on the wall. The anomalous gauge current generated by this state has to be cancelled in the underlying $2n+1$ dimensional theory since that world is anomaly free. Indeed, the massive modes contribute to the low energy effective action a piece representing the flow of charge into (or out of) the wall from the extra dimension. When calculated far from the wall, it cancels the anomalous contribution. In the $U(1)$ case in $2+1$ dimensions this was recently explicitly checked on the lattice with both Kaplan's and Shamir's formulations [20]. Indeed the cancellation is valid even close to the wall [21]. Therefore, what on the interface looks like an anomaly is the flow of charge into the extra dimension and the role of the heavy modes is to carry that charge.

The above picture was studied in some detail in Ref. [22]. Since opposite chirality partners live on opposite walls, the charge has to be transported through the extra dimension. In the adiabatic limit of slowly varying gauge fields, the time evolution is a continuous change of one particle states. As one passes through an "instanton" configuration the low energy states at the lattice ends change energy without substantially changing their position in the extra dimension. The same is true for the very high energy states, residing deep in the lattice interior.

However, the surface states with energies close to the cutoff are very sensitive to the applied field. When the energy of such a level rises towards the bottom of the band of plane waves flowing in the extra coordinate, the wave function penetrates increasingly deeply into this dimension. At the same time, another level from the interior lowers its energy and flows towards the opposite wall. This is also true for levels with corresponding negative energies; they just move in the opposite direction.

In this way we see how the heavy modes right at the cutoff carry the charge on and off the surfaces. With a gauge field applied to the physical vacuum with all negative energy states filled, these "flying states" are responsible for what appears to be the gauge anomaly on the surfaces.

8. Weak interactions, mirror Fermion model

With an exact gauge invariance and a finite size for the extra dimension, the surface models are inherently vectorlike. The Fermions always appear with both chiralities, albeit separated in the extra dimension. However, experimentally we know that only left handed neutrinos couple to the weak bosons. In this section I discuss one way to break the symmetries between these states, resulting in a theory with only one light gauged chiral state. Here I keep the underlying gauge symmetry exact, but do require that the chiral gauge symmetry be spontaneously broken, just as observed in the standard model. The picture also contains heavy mirror Fermions. If anomalies are not cancelled amongst the light species, these heavy states must survive in the continuum limit. It remains an open question when anomalies are properly cancelled whether it might be possible to drive the heavy mirror states to arbitrarily large mass.

I start by considering two separate species ψ_1 and ψ_2 in the surface mode picture. However, I treat these in an unsymmetric way. For ψ_1 I use the previous Hamiltonian. For ψ_2 I change the sign of all terms proportional to γ_5. On a given wall, the surface modes associated with ψ_1 and ψ_2 will then have opposite chirality.

Now I introduce the gauge fields. Since I want to eventually couple only one-handed neutrinos to the vector bosons, consider gauging ψ_1 but not ψ_2. Indeed, at this stage ψ_2 represents a totally decoupled right handed Fermion on one wall. I still have a mirror situation on the opposite wall, consisting of a right handed gauged state and a left handed decoupled Fermion.

The next ingredient is to spontaneously break the gauge symmetry, as in the standard model, by introducing a Higgs field ϕ with a non-vanishing expectation value. I can use this field to generate masses as in the standard model by coupling ψ_1 and ψ_2 with a term of the form $\bar{\psi}_1\psi_2\phi$.

The new feature is to allow the coupling of the Higgs field to depend on the extra coordinate. In particular, let it be small or vanishing on one wall and large on the other. The surface modes are then light on one wall and heavy on the other.

This model is closely related to the proposal in [23], where the gauge field is suddenly shut off in the interior of the fifth dimension, and gauge invariance is restored via a Higgs field. Folding that lattice in half around this shut off point reduces it to the picture presented here.

As in other mirror Fermion models [24], triviality arguments suggest that there might exist bounds on the mass of the heavy particles. This is certainly expected to be the case where the light Fermions alone give an anomalous gauge theory, in which case I expect the mirror particles cannot become much heavier than the vector mesons, i.e. the W.

It is conceivable that the restrictions on the mirror Fermion masses are weaker when anomalies cancel amongst the light states. In this case there is no perturbative need for the heavy states, and perhaps they can be driven to infinite mass in the continuum limit. This is a rather speculative desire, but if possible would give a candidate for a lattice discretization of the standard model.

Unfortunately, the model as it stands does not lead to baryon number violation [25]. The anomaly will involve a tunnelling of baryons from one wall to the opposite, where they become mirror baryons. Even if these extra particles are heavy, the decay can only occur through mixing with the ordinary particle states. In this sense, the mirror particles still show their presence in low energy physics. This further hints that the mirror Fermions might not be removable in the continuum limit.

Another speculative proposal is to use the right handed mirror states in some way as observed particles. Indeed, the world has left handed leptons and right handed antibaryons. Any simple extension of this idea to a realistic model must unify these particles [26]. On the other hand, the fact that the anomalies are canceled between different representations of the $SU(3)$ of strong interactions may preclude such options.

9. Conclusions

The use of Shockley surface states may provide the basis for a theory of chiral Fermions. For strong interaction physics this yields an elegant formulation where the massless limit for the quarks is quite natural.

In this picture the anomaly appears as a flow into the extra dimension. For anomaly free currents, the net flow in this dimension cancels, and I expect the predictions of current algebra to arise naturally. On the other hand, the symmetries for singlet axial currents are strongly broken by this flow. This presumably precludes the need for a corresponding Goldstone boson and solves the $U(1)$ problem.

Several questions remain before we have a theory of the weak interactions on the lattice, where the gauge fields are to be coupled to chiral currents. One approach leads to a theory with mirror Fermions on the opposing walls of the system. In a spontaneously broken theory these extra states can be given different masses. Whether they can be driven to infinite mass in the continuum limit presumably depends on whether all necessary chiral anomalies have been cancelled.

The difficulties in formulating chiral theories with a fundamental non-perturbative cutoff, such as the lattice, hints that there may be a deeper hidden message. Perhaps mirror fermions must exist at a few times the W mass and we should be looking for them. Note also that spontaneous breaking of the gauge theory is central to that approach, hinting that perhaps the only consistent chiral theories are spontaneously broken.

* This manuscript has been authored under contract number DE-AC02-76CH00016 with the U.S. Department of Energy. Accordingly, the U.S. Government retains a non-exclusive, royalty-free license to publish or reproduce the published form of this contribution, or allow others to do so, for U.S. Government purposes.

10. References

1. M. Creutz and I. Horvath, *Phys. Rev.* **D50** (1994) 2297.

2. S. Treiman, R. Jackiw, B. Zumino, and E. Witten, *Current algebra and anomalies,* (World Scientific, 1985, QC793.3.A4C87).

3. D. Petcher, *Nucl. Phys.* **B (Proc. Suppl.) 30** (1993) 50.

4. G. 't Hooft, *Phys. Rev. Lett.* **37**, 8 (1976); *Phys. Rev.* D14 (1976) 3432.

5. D. Kaplan, *Phys. Lett.* **B288** (1992) 342; M. Golterman, K. Jansen, D. Kaplan, *Phys. Lett.* **B301** (1993) 219.

6. K. Jansen, *Phys. Lett.* **B288** (1992) 348.

7. M. Creutz and I. Horvath, *Nucl. Phys.* **B (Proc. Suppl.) 34** (1994) 583.

8. M. Creutz, I. Muzinich, and T. Tudron, *Phys. Rev.* **D17** (1979) 531.

9. J. Ambjorn, J. Greensite, and C. Peterson, *Nucl. Phys.* **B221** (1983) 381; B. Holstein, *Am. J. Phys.* **61** (1993) 142.

10. H. Nielsen and M. Ninomiya, *Nucl. Phys.* **B185** (1981) 20; **B193** (1981) 173.

11. R. Narayanan and H. Neuberger, *Phys. Lett.* **B302** (1993) 62; *Nucl. Phys.* **B412** (1993) 574; *Phys. Rev. Lett.* **71** (1993) 3251.

12. K. Wilson, in *New Phenomena in Subnuclear Physics*, edited by A. Zichichi (Plenum Press, N. Y., 1977).

13. S. Aoki, *Nucl. Phys.* **B314** (1989) 79; S. Aoki and A. Gocksch, *Phys. Rev.* **D45** (1992) 3845.

14. W. Shockley, *Phys. Rev.* **56** (1939) 317; see also F. Seitz, *The Modern Theory of Solids*, (McGraw-Hill, 1940), p. 323-4; W. G. Pollard, *Phys. Rev.* **56** (1939) 324.

15. Y. Shamir, *Nucl. Phys.* **B406** (1993) 90.

16. C. Callan and J. Harvey, *Nucl. Phys.* **B250** (1985) 427.

17. A. Coste and M. Lüscher, *Nucl. Phys.* **B323** (1989) 631.

18. S. Coleman, *Annals Phys.* **101** (1976) 239.

19. K. Jansen and M. Schmaltz, *Phys. Lett.* **B296** (1992) 374.

20. S. Aoki and H. Hirose, *Phys. Rev.* **D49** (1994) 2604.

21. S. Chandrasekharan, *Phys. Rev.* **D49** (1994) 1980.

22. Y. Shamir, *Nucl. Phys.* **B417** (1994) 167.

23. M. Golterman, K. Jansen, D. Petcher, and J. Vink, *Phys. Rev.* **D49** (1994) 1606.

24. I. Montvay, *Nucl. Phys.* **B (Proc. Suppl.) 30** (1993) 621; *Phys. Lett.* **199B** (1987) 89.

25. J. Distler and S. Rey, Princeton preprint PUPT-1386 (1993).

26. S. Frolov and A. Slavnov, *Nucl. Phys.* **B411** (1994) 647; S. Aoki and Y. Kikukawa, *Mod. Phys. Lett.* **A8** (1993) 3517.

MONOPOLE CONDENSATION, DUAL MEISSNER EFFECT AND COLOUR CONFIMENT.

LUIGI DEL DEBBIO

Dip Fisica Università and INFN, Piazza Torricelli 2
Pisa, 56100 Italy

ADRIANO DI GIACOMO

Dip Fisica Università and INFN, Piazza Torricelli 2
Pisa, 56100 Italy

GIAMPIERO PAFFUTI

Dip Fisica Università and INFN, Piazza Torricelli 2
Pisa, 56100 Italy

PIERBIAGIO PIERI

Dip Fisica Università and INFN, Piazza Torricelli 2
Pisa, 56100 Italy

ABSTRACT

A disorder parameter is constructed to detect dual superconductivity in the ground state of gauge theories. The construction is tested on compact $U(1)$ on lattice. An unambiguous demonstration is then obtained that the ground state of $SU(2)$ gauge theory is a dual superconductor, thus confirming a popular mechanism for colour confinement.

1. Introduction

Understanding the mechanism by which colour is confined in QCD is a fundamental problem. A physically appealing possibility is dual superconductivity of the ground state[1,2]. Dual means that the role of electric and magnetic fields (and charges) is interchanged with respect to usual superconductors. The chromoelectric field between a q-$\bar{q}$ pair is channeled by Meissner effect into an Abrikosov[3] flux tube with energy proportional to the distance

$$V(R) = \sigma R \tag{1}$$

σ is the string tension. These flux tubes behave as strings[4,5]. The existence and relevance of strings in hadronic physics was known before QCD[6]. Flux tubes have been visualized by numerical simulations on the lattice[7,8], and their excitations have been detected [9]. Lattice is the ideal tool to study QCD at large distances.

In this talk I will present an analysis of the way in which the superconductivity of the QCD vacuum can be investigated on the lattice. I will present numerical results demonstrating unambiguously that confinement is indeed produced by dual superconductivity[10,11].

In Sect.2 I introduce some basic concepts of superconductivity. In Sect.3 I review the ideas about the mechanism of confinement by dual superconductivity. An extended version of this material is contained in Ref.12. In particular I show that the problem is in any case reduced to $U(1)$ by the so called abelian projection.

In Sect.4 I present a method to detect dual superconductivity in a $U(1)$ gauge theory. A disorder parameter is constructed, whose non-vanishing detects condensation of monopoles in the vacuum. In Sect.5 the tool developed in Sect.4 is used to test dual superconductivity of QCD vacuum: the result is a clear evidence that QCD vacuum is a dual superconductor. Sect.5 contains some concluding remarks and a survey of problems under study.

2. Basic superconductivity

A relativistic version of a superconductor is the abelian Higgs model[13]:

$$\mathcal{L} = -\frac{1}{4} F_{\mu\nu} F^{\mu\nu} + (D_\mu \varphi)^\dagger (D^\mu \varphi) - V(\varphi). \tag{2}$$

The notation is standard: $D_\mu = \partial_\mu - iqA_\mu$ and

$$V(\varphi) = \frac{\lambda}{2} \left(\varphi^\dagger \varphi - \frac{\mu^2}{2} \right)^2 \tag{3}$$

If $\mu^2 > 0$ the shape of the potential versus $\mathrm{Re}\,\varphi$ and $\mathrm{Im}\,\varphi$ is as in Fig.1. The ground state can be any point $\bar{\varphi}$ in the valley of the minima in Fig.1 and is not $U(1)$ invariant: $U(1)$ is broken to the group of rotations by $2\pi/q$. More precisely the breaking is à la Higgs. A convenient parametrization for the field is $\varphi = \psi \, e^{i\theta}$ with $\psi \geq 0$. In terms of ψ, θ

$$D_\mu \varphi = [\partial_\mu \psi + i(\partial_\mu \theta - qA_\mu)\psi] \, e^{i\theta} \tag{4}$$

or, defining

$$\tilde{A}_\mu = A_\mu - \frac{1}{q}\partial_\mu\theta \quad ; \quad \tilde{F}_{\mu\nu} = \partial_\mu \tilde{A}_\nu - \partial_\nu \tilde{A}_\mu = F_{\mu\nu} \tag{5}$$

$$D_\mu \varphi = \left[\partial_\mu \psi - iq\tilde{A}_\mu \psi \right] e^{i\theta} \tag{6}$$

and

$$\mathcal{L} = -\frac{1}{4} \tilde{F}_{\mu\nu} \tilde{F}^{\mu\nu} + q^2 \psi^2 \tilde{A}_\mu \tilde{A}^\mu + \partial_\mu \psi \partial^\mu \psi - V(\psi) \tag{7}$$

Notice that $\tilde{A}_\mu$ is gauge invariant. In Eq.(7) all quantities are gauge invariant. The ground state configuration is $\tilde{A}_\mu = 0$, $\psi = \bar{\psi} = \mu/\sqrt{2}\lambda$, $\theta = \bar{\theta}$. The photon $\tilde{A}_\mu$ acquires a mass, and correspondingly there is a finite penetration depth of the fields. This is nothing but Meissner effect. For the same reason around a magnetic flux tube and far enough from it $\tilde{A}_\mu = 0$, $\oint \tilde{A}_i \, dx^i = 0$ or, by use of Eq.(5) $\oint A_i \, dx^i = 2\pi n/q$ which is flux quantization. The equation of motion for the field $\tilde{A}_\mu$ is, neglecting quantum fluctuations of the field φ:

$$\partial_\mu \tilde{F}^{\mu\nu} + q^2 \bar{\psi}^2 \tilde{A}^\nu = 0 \tag{8}$$

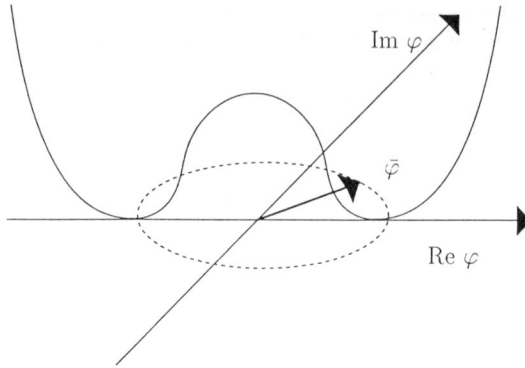

Fig. 1. Potential of the Higgs field.

The existence of static solutions with $\tilde{A}_i \neq 0$ means

$$\operatorname{curl} \vec{H} = q^2 \bar{\psi}^2 \vec{A} \tag{9}$$

or the existence of permanent currents (London equation)

$$\vec{j} = q^2 \bar{\psi}^2 \vec{A} \tag{10}$$

at zero electric field. Since $\vec{j} = \sigma \vec{E}$, $\sigma = \infty$.

The system has two scales of length, the inverse mass of the Higgs, $\mu^{-1} = \lambda_\varphi$, which is the correlation length of the field ψ and the penetration depth, or inverse mass of the photon $\lambda_A = 2\sqrt{\lambda}/\mu q$. If $\lambda_\varphi > \lambda_A$ the superconductivity is called of first kind, if, viceversa, $\lambda_A > \lambda_\varphi$ it is called of second kind. In the usual non relativistic treatment of superconductors $\bar{\varphi}$ is the Landau - Ginzburg order parameter, which is the v.e.v. of the operator creating a Cooper pair. In conclusion superconductivity is nothing but the spontaneous breaking of an $U(1)$ symmetry coupled to a gauge field. That breaking is signalled by the non vanishing of the v.e.v. of a charged field φ. Viceversa if there is a $U(1)$ symmetry coupled to a gauge field and the v.e.v. of any charged operator is different from zero, $U(1)$ is spontaneously broken, and under very general assumptions the system behaves as a superconductor.

To detect dual superconductivity of QCD vacuum we have to show that there exists at least one operator, with non trivial monopole charge, and non zero v.e.v.

3. A review of the confinement mechanism by dual superconductivity of the vacuum

3.1. Monopoles in gauge theories

The Georgi - Glashow model has monopoles as solitons[14,15]. The lagrangian is

$$\mathcal{L} = -\frac{1}{4}\vec{G}_{\mu\nu}\vec{G}_{\mu\nu} + \frac{1}{2}\overrightarrow{D_\mu\varphi}\overrightarrow{D_\mu\varphi} - \frac{\lambda}{4}\left[\vec{\varphi}^2 - \frac{\mu^2}{2\lambda}\right]^2 . \tag{11}$$

The gauge group is $SO(3)$, and $\vec{\varphi}$ transform as a vector. At large distances the monopole configuration is a hedgehog

$$\vec{\varphi}(\vec{r}) \underset{r \to \infty}{\simeq} \varphi_0 \frac{\vec{r}}{r} \quad ; \quad \vec{A}_0 = 0 \quad , \quad A_i^a = \frac{1}{g} \varepsilon_{iak} \frac{r_k}{r^2} \tag{12}$$

$\varphi_0 = \mu/\sqrt{2\lambda}$ is the v.e.v. of the Higgs field. Following 't Hooft a gauge invariant field strength can be defined

$$F_{\mu\nu} = \hat{\varphi}\vec{G}_{\mu\nu} - \frac{1}{g}\vec{\varphi}\left(\overrightarrow{D_\mu\varphi} \wedge \overrightarrow{D_\nu\varphi}\right) \tag{13}$$

$[\hat{\varphi} = \vec{\varphi}/|\vec{\varphi}|, \vec{G}_{\mu\nu}$ and $\overrightarrow{D_\mu\varphi}$ are gauge covariants].

In the hedgehog configuration, at large distances $F_{\mu\nu}$ is the field of a Dirac magnetic monopole

$$\vec{E} \underset{r \to \infty}{\simeq} = 0 \qquad \vec{H} \underset{r \to \infty}{\simeq} = \frac{1}{g}\frac{\vec{r}}{r^3}. \tag{14}$$

If we define $B_\mu = \hat{\varphi} \cdot \vec{A}_\mu$, B_μ is not gauge invariant, because $\vec{A}_\mu$ is not gauge covariant. The identity holds[17]

$$F_{\mu\nu} = (\partial_\mu B_\nu - \partial_\nu B_\mu) - \frac{1}{g}\hat{\varphi} \cdot (\partial_\mu\hat{\varphi} \wedge \partial_\nu\hat{\varphi}). \tag{15}$$

While $F_{\mu\nu}$ is gauge invariant, each of the two terms in the r.h.s. of Eq.(15) is not. In the unitary gauge, $\hat{\varphi} = (0, 0, 1)$, the second term in Eq.(15) vanishes and

$$F_{\mu\nu} = (\partial_\mu B_\nu - \partial_\nu B_\mu). \tag{16}$$

For that choice of the gauge $F_{\mu\nu}$ coincides with the abelian curl of the 3-d component of the gauge field. The gauge transformation which brings $F_{\mu\nu}$ to the form (16) is called an "abelian projection". The field $F_{\mu\nu}$ is the gauge field of a $U(1)$ symmetry, which is a symmetry of the system, being the residual gauge invariance after gauge fixing. However this $U(1)$ is not a subgroup of the original gauge group: $F_{\mu\nu}$ is a singlet. After abelian projection the 't Hooft - Polyakov monopoles become Dirac monopoles, and their monopole charge is a colour singlet.

Condensation of such monopoles in the ground state to produce dual superconductivity would give the correct physics. Their charge being a singlet the colour gauge group would be left unbroken, and so would be colour charge conjugation, since $F_{\mu\nu}$ is even as is evident from Eq.(13).

3.2. Monopoles in QCD

We shall confine our discussion to $SU(2)$ gauge group. Indeed for $SU(N)$ $N > 2$ the problem can be always reduced to $SU(2)$.

In QCD there are no Higgs fields. However configurations with the topology of monopoles do exist[1]. Let $\varphi(x) = \vec{\varphi}(x)\vec{\sigma}$ be any operator in the adjoint representation: we can define as abelian projection the gauge transformation U_φ which makes $\varphi(x)$ diagonal

$$U_\varphi^\dagger(x)\varphi(x)U_\varphi(x) = \vec{\varphi}'(x)\vec{\sigma} \qquad \vec{\varphi}\prime = (0,0,1). \tag{17}$$

In the points where $\varphi(x) = 0$ U_φ is singular, and the transformed field configuration has the topology of a monopole. The equation

$$\vec{\varphi}(x) = 0. \tag{18}$$

defines the world line of such monopoles.

't Hooft suggests that, although different choices define monopoles with different locations, and in different number, all of them are physically equivalent, in that they all define monopoles condensing in the vacuum and producing superconductivity[1]. A popular abelian projection is the so called maximal abelian, which is again of the type discussed above, even if the operator φ is defined implicitely by the condition[19]

$$\sum_{n,\mu} \mathrm{Tr} \left[\sigma_3 U_\mu(n)\sigma_3 U_\mu^\dagger(n) \right] = \text{max.} \tag{19}$$

Much attention has been devoted to the analysis on the lattice of the maximal abelian projection, mainly by determining quantities related to the density of monopoles[20], but also by looking for phenomenological evidence of dual superconductivity (eg. London equation)[21].

All these issues can be explored on the lattice if a suitable operator is constructed with non trivial monopole charge, whose v.e.v. can signal the spontaneous breaking of whatever $U(1)$ has been selected by the abelian projection.

4. Detecting dual superconductivity in $U(1)$ gauge theory

Let $A_i(\vec{x}, t)$ be the canonical variables of a $U(1)$ gauge theory, $\Pi_i(\vec{x}, t) = F_{0i}(\vec{x}, t)$ their conjugate momenta. Then the operator

$$\mu(\vec{y}, t) \equiv \exp \left(i \int d^3x \, \Pi_i(\vec{x}, t) b_i(\vec{x}, \vec{y}) \right) \tag{20}$$

creates a monopole if $b_i(\vec{x}, \vec{y})$ is the vector potential produced in the point $\vec{x}$ by a monopole sitting in $\vec{y}$.

Putting the string along the z axis

$$b_i = \frac{m}{2e} \frac{\varepsilon_{3ij} r_j}{r(r - r_3)} \qquad \vec{r} = \vec{x} - \vec{y} \tag{21}$$

m is the monopole charge in units of $1/e$. In the Schrödinger picture, when acting on a state $|A_i(\vec{x}, t)\rangle$ $\mu(\vec{y}, t)$ simply translates the field by $b_i(\vec{x}, \vec{y})$

$$\mu|A_i(\vec{x}, t)\rangle = |A_i(\vec{x}, t) + b_i(\vec{x}, \vec{y})\rangle \tag{22}$$

Eq.(22) is the field theoretic analog of the well known formula $e^{ipa}|x\rangle = |x + a\rangle$ In terms of commutators

$$\begin{aligned}
[A_i(\vec{x}, t), \mu(\vec{y}, t)] &= b_i(\vec{x}, \vec{y})\mu(\vec{y}, t) \\
[\Pi_i(\vec{x}, t), \mu(\vec{y}, t)] &= 0
\end{aligned} \tag{23}$$

which are simple consequences of the canonical commutation relations. μ has monopole charge m. The above construction is a rediscovery of a procedure which has been used in the literature in different languages and contexts[23]. In the same language a translation of the vector field by $\vec{g}(\vec{x})$, with $curl\, g(x) = 0$

$$\gamma(t) = \exp\left(i \int d^3x\, \Pi_i(\vec{x}, t)g_i(\vec{x})\right) \tag{24}$$

defines a gauge transformation. In the euclidean region Eq.(20) becomes

$$\mu(\vec{y}, y_4) \equiv \exp\left(-\int d^3x\, F_{4i}(\vec{x}, y_4)b_i(\vec{x}, \vec{y})\right). \tag{25}$$

We want to use $\langle\mu\rangle$ as a disorder parameter for dual superconductivity

$$\langle\mu\rangle = \frac{1}{Z}\int d[A]\, e^{-S}\mu \tag{26}$$

As a simple test we start computing $\langle\mu\rangle$ for a system of free photons, where $S = \frac{1}{4}F_{\mu\nu}F^{\mu\nu}$. Rescaling the fields $F \to F/e$ and b_i by extracting $1/e$ from Eq.(21),

$$\langle\mu\rangle = \frac{1}{Z}\int \mathcal{D}A \exp\left[-\frac{\beta}{4}\int F_{\mu\nu}F_{\mu\nu} - \beta\int d^3x\, F_{4i}(\vec{x}, y_4)b_i(\vec{x}, \vec{y})\right] \tag{27}$$

where $\beta = 1/e^2$. The integral in Eq.(27) is gaussian, and can be easily computed giving

$$\langle\mu\rangle = \exp\left\{\frac{\beta}{2}\langle F_{4i}(k)F_{4j}(-k)\rangle b_i(-\vec{k})b_j(\vec{k})\right\} \tag{28}$$

with

$$\langle F_{4i}(k)F_{4j}(-k)\rangle = \delta_{ij} - \frac{|\vec{k}|^2\delta_{ij} - k_i k_j}{k_0^2 + |\vec{k}|^2} \tag{29}$$

or

$$\langle\mu\rangle = \exp\left\{\frac{\beta}{2}\left[\int \frac{d^4k}{(2\pi)^4}|\vec{b}(\vec{k})|^2 - \frac{1}{2}\int \frac{d^3k}{(2\pi)^3}|\vec{k}||\vec{b}(\vec{k})|^2\right]\right\}. \tag{30}$$

The same construction for $\langle\gamma\rangle$ (Eq.(24)) gives

$$\langle\mu\rangle = \exp\left\{\frac{\beta}{2}\int \frac{d^4k}{(2\pi)^4}|\vec{g}(\vec{k})|^2\right\}. \tag{31}$$

By gauge invariance one should have $\langle\gamma\rangle = 1$.

In fact the factor corresponding to the first term at the exponent in Eq.(30), as well as the rhs of Eq.(31), disappear if a correct definition is given of the Feynman integral.

This can again be seen in a simple system with one degree of freedom, going back to the very definition of Feynman integral

$$\langle x', t'|e^{ip(t)a}|x_0, t_0\rangle = \prod_i^n dx_i \langle x'|e^{-iH\delta}|x_n\rangle\langle x_n|e^{-iH\delta}|x_{n-1}\rangle \cdots$$

$$\langle x_{j+1}|e^{-iH\delta}e^{ipa}|x_j\rangle \cdots \langle x_1|e^{-iH\delta}|x_0\rangle$$

with $H = \frac{p^2}{2} + V(x)$. For the matrix elements containing only $e^{-iH\delta}$

$$\langle x_{k+1}|e^{-iH\delta}|x_k\rangle = e^{-i\delta V(x_k)} \int \frac{dp_k}{(2\pi)} e^{-i\frac{p_k^2}{2}\delta - ip_k(x_k - x_{k+1})}$$

$$= e^{-i\delta V(x_k)} e^{i\delta \frac{(x_{k+1} - x_k)^2}{2\delta^2}} = e^{i\mathcal{L}(x_k, \dot{x}_k)\delta}$$

For the matrix element containing the translation instead

$$\langle x_{j+1}|e^{-iH\delta}e^{ipa}|x_j\rangle = e^{i\delta \mathcal{L}(x_j, \dot{x}_j)}e^{-im\dot{x}_j a\delta}e^{i\delta a^2} \tag{32}$$

Our definition of $\langle \mu \rangle$, Eq.(26) neglects the analogous of last factor in Eq.(32), wich exactly cancels the first term in the exponent of Eq.(30) and the exponent in Eq.(31), giving $\langle \gamma \rangle = 1$. In order to operate this cancellation, without doing any assumption on the form of the lagrangian we shall redefine $\langle \mu \rangle$ as

$$\langle \bar{\mu} \rangle = \frac{\langle \mu \rangle}{\langle \gamma \rangle} \tag{33}$$

with the condition

$$\int |\vec{b}(\vec{k})|^2 \frac{d^4k}{(2\pi)^4} = \int |\vec{g}(\vec{k})|^2 \frac{d^4k}{(2\pi)^4}. \tag{34}$$

For free photons then we get

$$\langle \bar{\mu} \rangle = \exp\left[-\frac{\beta}{2} \frac{d^3k}{(2\pi)^3} |\vec{k}| \, |\vec{b}(\vec{k})|^2 \right] \simeq \exp\left[-\beta C \log\left(\frac{V}{a^3}\right) \right] \tag{35}$$

with a an U.V. cutoff, $C > 0$.

In the limit $a \to 0$, or at fixed a as $V \to \infty$, $\langle \bar{\mu} \rangle \to 0$. For a gas of free photons there is no condensation of monopoles. Finally we write for our parameter $\langle \bar{\mu} \rangle$

$$\langle \bar{\mu} \rangle = \frac{\int dA \exp\left[-\beta(S + S_b)\right]}{\int dA \exp\left[-\beta(S + S_g)\right]} \tag{36}$$

where we have put

$$S_b = \int d^3x \, F_{4i}(\vec{x}, y_4) b_i(\vec{x}, \vec{y}) \qquad S_g = \int d^3x \, F_{4i}(\vec{x}, y_4) g_i(\vec{x}, \vec{y}) \qquad (37)$$

To transfer our construction on the lattice we simply make the replacement

$$F_{4i} = \mathrm{Im}\Pi_{4i} \qquad (38)$$

where $\Pi_{\mu\nu}$ is the parallel transport along the elementary square of the lattice (plaquette).

A construction for a disorder parameter for compact $U(1)$ on the lattice exists in the literature[24], for the Villain form of the action. A rigorous proof is given there that monopoles do condense in the vacuum for $\beta < \beta_c \simeq 1.0$. We have shown[10] that our construction is identical to ref.24 for the case of the Villain action. However contrary to ref.24 our construction does not rely on the specific form of the action, and can then be used to test QCD vacuum, where the form of the effective $U(1)$ action is unknown.

Computing $\langle \bar{\mu} \rangle$ directly on the lattice by Eq.(36) presents two kinds of difficulties

(i) $\langle \bar{\mu} \rangle$ is the average of the exponential of an extensive quantity, like the partition function. It is known that such quantities can have a non gaussian distribution of fluctuations.

(ii) The boundary conditions: periodic boundary conditions do not allow to put a monopole on the lattice; with the construction of ref.[24], with Villain action, antiperiodic b.c. are necessary for single monopole configurations, periodic for monopole antimonopole pairs[25].

This is of course an artefact. If the theory has a mass gap, as happens for compact $U(1)$ in the confined phase, boundary conditions should be unimportant for a sound construction of the disorder parameter.

To overcome these two difficulties, instead of measuring $\langle \bar{\mu} \rangle$ directly, we measure

$$\rho_b(\beta) = \frac{d}{d\beta} \log \langle \bar{\mu} \rangle \qquad (39)$$

From Eq.(36)

$$\rho_b(\beta) = \langle S + S_g \rangle_{S+S_g} - \langle S + S_b \rangle_{S+S_b} \qquad (40)$$

Since $\langle \bar{\mu} \rangle = 1$ at $\beta = 0$, $\langle \bar{\mu} \rangle$ can be reconstructed as

$$\langle \bar{\mu} \rangle = \exp \left[\int_0^\beta \rho_b(x) \, dx \right]. \qquad (41)$$

The advantage of Eq.(40) is that it measures a kind of internal energy, which is an extensive quantity, well measurable numerically, and sensitive to the bulk properties of the system, but not to the boundary conditions.

A similar construction can be repeated for a configuration containing a monopole and a antimonopole, when measuring their correlation

$$\langle \mu(x)\bar{\mu}(y)\rangle \underset{|x-y|\to\infty}{\simeq} \langle \mu\rangle^2. \tag{42}$$

Eq.(42) expresses the cluster property. In our formalism

$$\mu(x)\bar{\mu}(y) = \exp\left\{-\beta \int d^3z F_{4i}(\vec{z},x_4)b_i(\vec{z},\vec{x}) + \beta \int d^3z F_{4i}(\vec{z},y_4)b_i(\vec{z},\vec{y})\right\} \tag{43}$$

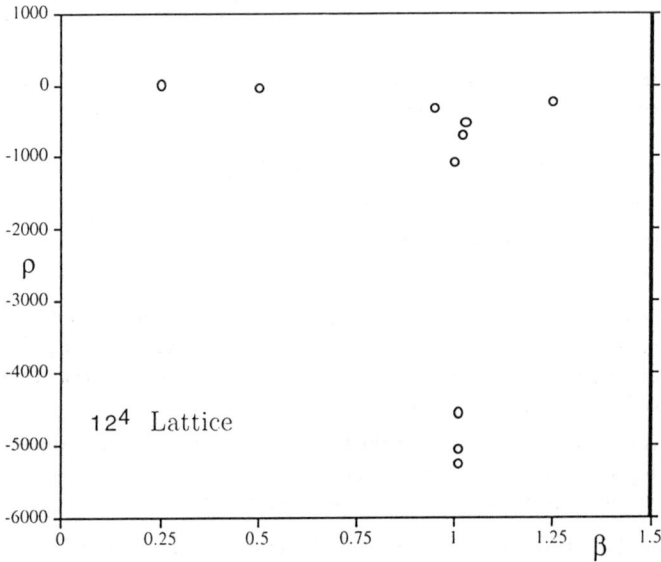

Fig. 2. $\rho_b(\beta)$ for $U(1)$, Wilson action.

Again we can consider

$$\frac{\langle \mu(x)\bar{\mu}(y)\rangle}{\langle\gamma\rangle^2} \underset{|x-y|\to\infty}{\simeq} \langle\bar{\mu}\rangle^2 \tag{44}$$

or better

$$\rho_{b\bar{b}} = \frac{d}{d\beta}\log\frac{\langle\mu(x)\bar{\mu}(y)\rangle}{\langle\gamma\rangle^2} = 2\langle S + S_g\rangle_{S+S_g} - \langle S + S_{b\bar{b}}\rangle_{S+S_{b\bar{b}}} - \langle S\rangle_S \tag{45}$$

where

$$S_{b\bar{b}} = \int d^3z \left[F_{4i}(\vec{z},x_4)b_i(\vec{z},\vec{x}) - F_{4i}(\vec{z},y_4)b_i(\vec{z},\vec{y})\right] \tag{46}$$

and

$$\frac{\langle \mu(x)\bar{\mu}(y)\rangle}{\langle \gamma \rangle^2} = \exp\left[\int_0^\beta \rho_{b\bar{b}}(x)dx\right]. \tag{47}$$

At large distances the cluster property Eq.(42) means

$$\rho_{b\bar{b}} = 2\rho_b \tag{48}$$

For the $U(1)$ system, with Wilson action we expect $\langle \bar{\mu} \rangle$ to be zero in the sense of (Eq.35) above β_c, where we have free photons, and to be different from zero below β_c. A sudden drop is expected at β_c, which will correspond to a large negative peak in ρ, while ρ should be finite below β_c, to ensure that $\langle \bar{\mu} \rangle \neq 0$. The observed behaviour is shown in Fig.2. $\rho(\beta)$ is pratically zero below β_c with a strong negative peak at β_c. Above β_c it agrees with Eq.(35). Roughly this means $\langle \bar{\mu} \rangle = \theta(\beta_c - \beta)$. We have shown that the signal is pratically independent of the choice of b_i (different orientation of the string, Wu - Yang form) and is zero if instead of $\vec{b}$ a pure gauge configuration like $\vec{r}/r^2$ is used[10]. The cluster property Eq.(48) is satisfied at distances of a few lattices spacings[10].

In conclusion we have a disorder parameter for compact $U(1)$ on the lattice, which can detect condensation of monopoles whatever the form of the action is. It coincides with rigorous definition for the Villain action. It is ready to be used for QCD.

5. QCD vacuum is a dual superconductor

For $SU(2)$ gauge theory we have measured[11] $\langle \mu \rangle$ and $\langle \mu(x)\bar{\mu}(y)\rangle$, for the monopoles defined by different abelian projections, to detect monopole condensation. Asymmetric lattices are used: $N_S^3 \times N_T$, $N_T \ll N_S$, to obtain finite temperature. The deconfining phase transition takes place at $\beta = 4/g^2$ such that

$$a(\beta)T = N_T \tag{49}$$

and it is well studied and known. The expectation is, according to the mechanism under study, that monopoles are condensed for $\beta < \bar{\beta}$, where there is confinement, and are not for $\beta > \bar{\beta}$, where colour is deconfined.

We have made the test for the abelian projection defined by dyagonalizing the Polyakov line

$$P(\vec{x},t) = T \int \exp\left\{i\int_t^\infty A_0(\vec{x},\tau)d\tau\right\} \tag{50}$$

The behaviour of ρ_b across $\bar{\beta}$ is shown in Fig.3. A clear evidence exists of condensation of monopoles in the vacuum below $\bar{\beta}$, which disappears for $\beta > \bar{\beta}$.

Fig.3 shows the behaviour of ρ for a $12^3 \times 4$ lattice and a $16^3 \times 6$ lattice. The negative peaks occur at the appropriate decoupling temperatures. The area of the peak is roughly proportional to N_s^3 at fixed N_T. However it is roughly the same for the two lattices in Fig.3, which correspond to the same physical volume, showing that the effect is not an artefact of the lattice. The cluster property has also been tested. We therefore conclude that QCD vacuum is a dual superconductor.

Of course other abelian projections can produce the same effect: if the guess of 't Hooft[1] is true more or less all of them should work. It would be interesting to check the maximal abelian projection, but we have not yet done that for technical reasons. Indeed while for gauges in which the field φ is explicitely known the gauge invariant construction Eq.(13) can easily be translated on the lattice, and the modified action S_b (Eq.(37)) can be immediatly written, the maximal abelian gauge requires a maximization process to define S_b at each updating. Work is in progress to overcome this difficulty.

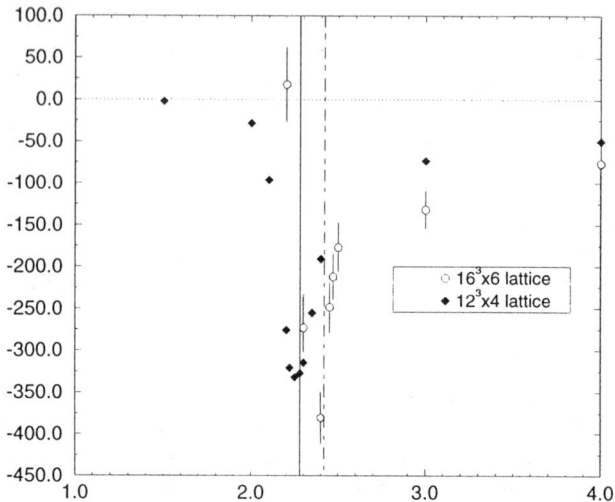

Fig. 3. $SU(2)$ gauge theory: $\rho_b(\beta)$ for the abelian projection diagonalizing Polyakov loop. The vertical lines indicate the deconfining transition.

6. Conclusions

We have constructed a disorder parameter for monopole condensation. We have checked our construction by comparison with the rigorous results for special forms of the action. We have found that in the abelian projection obtained by diagonalizing Polyakov loop, monopole do condense in the QCD vacuum, which is therefore a dual superconductor.

Further study on $SU(3)$ gauge theory and with alternative abelian projections is on the way, as well as on other models (x-y model in $3d$) in which there is a phase transition due to condensation of solitons.

1. G. 't Hooft, in *High Energy Physics*, EPS International Conference, Palermo 1975, ed. A. Zichichi; G. 't Hooft, *Nucl Phys.* **B 190** (1981) 455.
2. S. Mandelstam, *Phys. Rep.* **23C** (1976) 245.
3. A.B. Abrikosov *JETP* **5** (1957) 1174.
4. H.B. Nielsen, P. Olesen, *Nucl. Phys.* **B61** (1973) 45.
5. Y. Nambu *Phys. Rev.* **D 10** (1974) 4262.
6. G. Veneziano, *Nuovo Cimento* **57 A** (1968) 190.
7. A. Di Giacomo, M. Maggiore and Š. Olejník, *Phys. Lett.* **B236** (1990) 199; *Nucl. Phys.* **B347** (1990) 441.
8. R.W. Haymaker, J. Wosiek, *Phys. Rev.* **D 36** (1987) 3297.
9. F. Gliozzi. These proceedings.
10. L.Del Debbio, A.Di Giacomo, G.Paffuti: IFUP-TH 16/94,HEP-LAT 9403013, and in preparation.
11. L.Del Debbio, A.Di Giacomo, G.Paffuti and G. Pieri: IFUP-TH 30/94,HEP-LAT 9405025 and in preparation.
12. A. Di Giacomo, *Acta Phys. Polonica* **B25** (1994) 215.
13. S. Weinberg, *Progr. of Theor. Phys. Suppl.* No. 86 (1986) 43.
14. H. Georgi, S. Glashow *Phys. Rev. Lett.* **28** (1972) 1494.
15. G. 't Hooft *Nucl. Phys.* **B79** (1974) 276.
16. A.M. Polyakov *JEPT Lett.* **20** 894 (1974).
17. J. Arafune, P.G.O. Freund, G.J. Goebel *Journ. Math. Phys.* **16** (1974) 433.
18. L. Del Debbio, A. Di Giacomo, M. Maggiore, Š. Olejník, *Phys. Lett.* **B267** (1991) 254.
19. A.S. Kronfeld, G. Schierholz, U.J. Wiese, *Nucl. Phys.* **B293**(1987) 461.
20. A.S. Kronfeld, M.L. Laursen, G. Schierholz, U.J. Wiese, *Phys. Lett.* **B198** (1987) 516; T. Suzuki, *Nucl. Phys.* **B 30** (Proc. Suppl.) (1993) 176 and references therein.
21. T.L. Ivanenko, A.V. Pochinskii, M.I. Polikarpov, *Phys. Lett.* **B 252** (1990) 252.
22. D.A. Brown, R.W. Haymaker, V. Singh, *Phys. Rev.* **D 47** (1993) 1715; P. Cea, L. Cosmai, *Nuovo Cimento* **106 A** (1993) 1361.
23. L.P. Kadanoff, H.Ceva, *Phys. Rev.* **B3** (1971) 3918; E.C. Marino, B. Schroer, J.A. Swieca, *Nucl. Phys.* **B 200** (1982) 473; A. Liguori, M. Mintchev, M. Rossi, *Phys. Lett.* **B 305** (1993) 52.
24. J. Fröhlich and P.A. Marchetti, *Euro. Phys. Lett.* **2** (1986) 933; *Commun. Math. Phys.*112 (1987) 343.
25. L. Polley, U. Wiese, *Nucl. Phys.* **B356** (1991) 629.

QUANTUM BEHAVIOUR OF THE FLUX TUBE:

A comparison between QFT predictions
and lattice gauge theory simulations

F. GLIOZZI

Dipartimento di Fisica Teorica dell' Università di Torino
Via P.Giuria 1, I-10125 Torino, Italy

ABSTRACT

We review some universal features of the colour flux tube of gauge theories in the confining phase predicted by the infrared conformal limit of the underlying string theory. In particular we discuss shape effects in Wilson loops and rederive in a general way the logarithmic growth of the mean square width of the flux tube as a function of the interquark separation. Recent data on $3D$ Z_2 gauge theory, combined with high precision data on the interface physics of the $3D$ Ising model fit nicely to this behaviour over a range of more than two orders of magnitude.

1. Introduction

1.1. The String Picture of the Flux Tube

Experience on lattice gauge theory has shown that the relevant degrees of freedom of whatever gauge system in the confined phase are concentrated, in the infrared region, inside string-like flux tubes connecting among them the sources of the gauge field. This suggests, according to an old conjecture[1,2], that the vacuum expectation values of large Wilson loops $W(C)$, once the integration on the irrelevant degrees of freedom has been performed, are given by an effective string theory describing the dynamics of the flux tube as follows

$$\langle W(C) \rangle = \sum_{\{\Sigma : \partial \Sigma = C\}} e^{-S(\Sigma)} , \tag{1}$$

where Σ is a surface bounded by the loop C and S is a (largely unknown) string action. Within $SU(N)$ gauge theories there is an internal support of this idea: the $1/N$ expansion of weak coupling perturbation theory can be interpreted, according to 't Hooft analysis[2], as a topological expansion of a string theory in the number of handles of Σ in which $1/N$ acts as the string coupling g. For large N the vacuum expectation value $\langle W(C) \rangle$ is dominated by planar Feynman graphs. In the dual language of strings this implies that the sum in Eq.(1) is saturated by surfaces with no handles and the evaluation of $\langle W(C) \rangle$ is greatly simplified.

It has been recently shown [3] that in two-dimensional QCD on a Riemann surface the coefficients of the $1/N$ expansion have a simple, exact interpretation in terms of a string theory.

1.2. Condensation of Handles

For finite N, or for discrete gauge groups, the string enters a non–perturbative regime where the creation of handles becomes an important process. For this reason a stringy description of such gauge systems, which are the most interesting from the physical point of view, is often considered hopeless.

There is however an old argument [4], corroborated by recent numerical simulations [5,6], which suggests that this non–perturbative regime has actually a simple description. Indeed it has been shown long ago [4] that in critical string theory the sum of the divergent part of the handle insertions in any string amplitude can be reabsorbed in the renormalization of the string tension σ and of the string coupling g. An indirect check of this property comes from the simplest gauge system, namely the three-dimensional Z_2 gauge model. Here, using the duality relation with the Ising model, it is possible to describe the properties of the effective string in terms of the fluid interfaces between domains of opposite magnetization. It is also possible to study the topological properties of these surfaces and it turns out[5] that the number of handles is very large and proportional to the area of the surface. Such a condensation of handles seems an ubiquitous phenomenon of the physics of random surfaces with fluctuating topology[6,7]; its meaning is rather simple: the mean handle is microscopic and its size ℓ does not depend on that of the surface. As a consequence, at a scale larger than ℓ there are very few visible handles, so the effective string coupling g, being proportional to the amplitude of handle formation, is vanishing, like the perturbative coupling in the large N gauge systems.

The above argument strongly suggests that at large distances the functional integral given in Eq.(1) is dominated, like in the $1/N$ expansion, by surfaces with no handles, so the effective string description of the infrared region should greatly simplify.

1.3. α' Expansion and the Universality Class of the Flux Tube

Linear confinement implies that the action S is proportional, in the infrared region, to the area of the surface Σ swept by the flux tube, then the string action should be well approximated by the Nambu-Goto action, even if this simple string picture does not work at short distances [8].

A further simplification of the string action comes from the so called α' expansion, where $\alpha' = 1/2\pi\sigma$: for large rectangular Wilson loops of size $L_1 \times L_2$ one can expand the action S in the natural adimensional parameter $1/(\sigma L_1 L_2)$, obtaining

$$S = \sigma L_1 L_2 + \frac{1}{2} \int_0^{L_1} d\xi_1 \int_0^{L_2} d\xi_2 \sum_i^{D-2} (\nabla h_i(\xi_1, \xi_2))^2 + O\left(\frac{1}{\sigma L_1 L_2}\right) \quad , \qquad (2)$$

where h_i are the two-dimensional fields describing the transverse displacements of the flux tube. Thus the effective string theory reduces in the infrared limit to a massless gaussian model.

The above conclusion can be actually obtained also by more general considerations. Indeed strong coupling (β) expansions of $\langle W(C) \rangle$ in whatever gauge theory show that the flux tube undergoes a transition towards a rough phase for β larger than a given β_r (the roughening point). It is widely believed that this transition is described by the universality class of Kosterlitz and Thouless[9] (KT). Thus the effective string action, even if it is substantially unknown, should belong to the same universality class of the XY model. The renormalization group equations of the KT universality class can be simply expressed in terms of two couplings:

$$\dot{x}(t) = -y(t)^2 \quad , \tag{3}$$
$$\dot{y}(t) = -x(t)y(t) \quad , \tag{4}$$

where t is the RG scale parameter, $x = \pi T - 2$, (T is the temperature of the XY model) and y is the vortex fugacity. The inverse temperature can be identified in the infrared limit with the surface stiffness κ, which in the continuum limit coincides with the string tension. These RG equations describe the transition from a smooth phase (with free vortices) to a rough phase where the effective vortex fugacity flows to zero and then the the theory becomes at large scale a massless free field theory.

So, we have reached a conclusion similar, but more general than that obtained by the α' expansion of the Nambu-Goto action: for large enough Wilson loops it is not necessary to know the specific form of the string action because it flows towards a free massless theory. This limit action is of course critical in the whole rough phase, then it produces universal (i.e. gauge group independent) finite size effects. Moreover, because this theory is also infrared divergent, such finite size effects are expected to be rather strong. We shall discuss two of these universal effects: the shape dependence of the Wilson loops and the growth of the mean width of the flux tube.

2. Shape Effects in the Wilson Loops

2.1. The Functional Form of the Quantum Fluctuations

The contribution of the first few terms of the α' expansion to the vacuum expectation values of the Wilson loops can be calculated exactly[10]. Here we consider only the contribution of the gaussian term, which is the one associated to the universal effects.

For large enough Wilson loops we can parametrize $\langle W(L_1, L_2) \rangle$ as follows

$$-\log\langle W(L_1, L_2) \rangle = \sigma L_1 L_2 + p(L_1 + L_2) + c + q(L_1, L_2) + O(\alpha') \tag{5}$$

where the string tension σ and the two parameters c and p are the only free parameters of this equation; $q(L_1, L_2)$ is the universal contribution due to the gaussian functional integral and is a know function of the shape (L_1, L_2) which does not depend on the gauge system, but only on the choice of the boundary conditions. For instance, for fixed boundary conditions we have

$$q(x, y) = \frac{D-2}{2} \log\left(\frac{\eta(iy/x)}{\sqrt{x}}\right) \quad , \tag{6}$$

where the Dededekind eta function is given by

$$\eta(\tau) = q^{\frac{1}{24}} \prod_{n=1}^{\infty} (1 - q^n) \quad , q = e^{2\pi i \tau} \quad . \tag{7}$$

The older fits of Wilson loop data put arbitrarily $q(x, y) = 0$ with unacceptable χ^2 values. Insertion of the quantum correction given in Eq.(6) improves considerably the fits, even if discrepancies have been observed for various gauge systems both in $D = 3$ and $D = 4$ dimensions [11,12].

It has been suggested that these discrepancies might be due to the choice of fixed b.c.[13]. Actually the gaussian model is a member of a one-parameter family of conformal field theories (those with central charge $c = 1$) having the same local behaviour, but different topological modes (winding modes around a circle of radius r which is the parameter of this family). If the free theory describing the infrared limit of the effective string is allowed to have such topological modes, it is possible to introduce new kinds of boundary conditions which give a much better description of the Wilson loop. Roughly speaking, such a topological modification of the gaussian model allows one to treat in a different manner quarks line and antiquarks, while in the ordinary gaussian model there is no way to distinguish a flux tube connecting two quarks from that connecting a quark-antiquark pair, which from the point of view of the physics of the gauge system are of course completely different.

2.2. Fluid Interfaces

In order to test more accurately the above universal asymptotic behaviour it would be convenient to study the simplest gauge system, i.e. the $3D$ Z_2 gauge model. It is dual to the $3D$ Ising model where one could use non-local cluster algorithms for numerical simulations which yield a drastic reduction of the statistical error compared with the old Metropolis or Heat Bath methods. Direct measurements of expectation values of the Wilson loops with this new technique have not yet been performed, however there is a related observable that has been evaluated in this way. Let us see briefly the method.

Denote by $\sigma_l = \pm 1$ the link variables of the Z_2 gauge system and by $\tilde{\sigma} = \pm 1$ the site variables of the dual Ising model. One can relate the expectation value of the plaquette $P = \sigma_1 \sigma_2 \sigma_3 \sigma_4$ in terms of dual variables as follows

$$\langle P \rangle_{gauge} = \langle \exp(-2\tilde{\beta} \tilde{\sigma}_l) \rangle_{spin} \quad , \tag{8}$$

with $\tilde{\sigma}_l = \tilde{\sigma}_1 \tilde{\sigma}_2$, where l is the link of the dual lattice orthogonal to the plaquette P and punching the center of the plaquette; $\tilde{\beta}$ is related to β by

$$\tilde{\beta} = -\frac{1}{2} \log (\tanh \beta) \quad . \tag{9}$$

More generally, denoting by Σ an arbitrary surface with boundary $\partial\Sigma$, we have

$$\langle W(\partial\Sigma) \rangle_{gauge} = \langle \prod_{l \in \Sigma} \exp(-2\tilde{\beta} \tilde{\sigma}_l) \rangle_{spin} \quad , \tag{10}$$

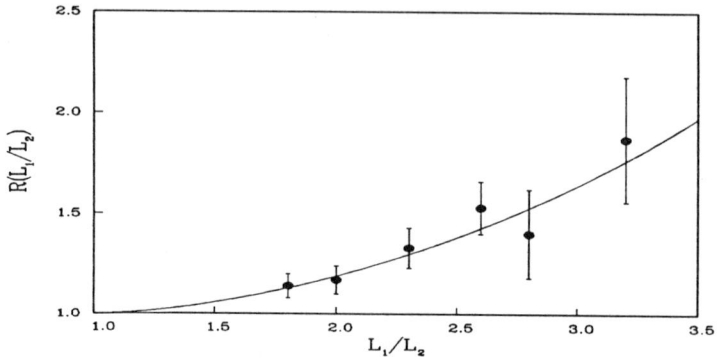

Fig. 1. Finite size effects in fluid interfaces.

where the product is over all the dual links cutting the surface Σ. In the dual Ising model with periodic b.c. in x and y directions one can also realize a sort of Wilson loop which has no analogue in the gauge system, by choosing as Σ a toroidal surface wrapped around the periodic directions (for instance a plane orthogonal to the z axis). Then $\langle \prod_{l \in \Sigma} \exp(-2\tilde{\beta}\tilde{\sigma}_l) \rangle$ denotes the interface partition function. This observable is particularly well suited to study the shape dependent effects, because there are no complications due to the boundary sources. The size of this periodic Wilson loop is four times larger than the maximal ordinary loop one can measure in the corresponding gauge system. The functional form of Eq.(5) is now modified because the perimeter term is absent, of course, and the gaussian term is slightly different with respect to Eq.(6) owing to the periodic boundary conditions. One gets

$$q(x,y) = \log\left(\sqrt{\Im m \tau}|\eta(\tau)|^2\right) \quad , \tau = i\frac{x}{y} \quad . \tag{11}$$

A set of numerical simulations on the Ising model[14] and on the three-state Potts model[15] gave a nice, accurate confirmation of this behaviour for a wide range of τ.

In particular, the ratio $R = Z(L_1, L_2)/Z(\tilde{L}, \tilde{L})$ of interfacial partition functions for systems with the same area $(L_1 L_2 = \tilde{L}^2)$, but different shape, is only a function of $L1/L2$ and is directly expressible in terms of Eq.(11) with no free parameters. In Fig.1 a comparison is made between such theoretical prediction and a set of numerical simulations in the Ising model at $\beta = .2275$ (black dots).

3. Width of the Colour Flux Tube

3.1. Thickness of the Flux Tube as an Effect of String Fluctuations

One of the simplest and general consequences of the existence of a thin flux tube connecting a quark pair in the confining phase is that this flux tube fluctuates. The effect of these quantum fluctuations is to produce an effective square width of the flux tube which grows logarithmically with the interquark distance.

Such a logarithmic behaviour has been predicted many years ago by Lüscher, Münster and Weisz [16] in the framework of the effective string picture of the gauge systems. In the next subsection we shall rederive and refine this universal law by directly using some exact results on two dimensional free field theory in a finite box.

Let us begin by defining the observable we want to discuss. The square width (or the square gyration radius) of the flux tube generated by a planar Wilson loop $W(C)$ is defined as the sum of the mean square deviation of the transverse coordinates of the underlying string, i. e.

$$w^2 = \frac{1}{A} \sum_{i=1}^{D-2} \int_{\mathcal{D}} d^2\xi \langle \left(h_i(\xi_1, \xi_2) - h_i^{CM}\right)^2 \rangle \quad , \tag{12}$$

where $\mathcal{D}$ is the planar domain bounded by C, A its area $A = \int_{\mathcal{D}} d^2\xi$ and h_i^{CM} is the transverse coordinate of the center of mass of the flux tube, given by

$$h_i^{CM} = \frac{1}{A} \int_{\mathcal{D}} d^2\xi \, h_i(\xi_1, \xi_2) \quad . \tag{13}$$

Combining Eq.(12) with Eq.(13) we get

$$
\begin{aligned}
w^2 &= \frac{1}{2(A)^2} \sum_{i=1}^{D-2} \int_{\mathcal{D}} d^2\xi \int_{\mathcal{D}} d^2\xi' \langle (h_i(\xi) - h_i(\xi'))^2 \rangle \\
&= \frac{1}{\sigma(A)^2} \sum_{i=1}^{D-2} \int_{\mathcal{D}} d^2\xi \int_{\mathcal{D}} d^2\xi' \, (G_i(\xi, \xi') - G_i(\xi, \xi)) \quad ,
\end{aligned} \tag{14}
$$

where $G_i(\xi, \xi')$ is the Green function of the field h_i defined as

$$G_i(\xi, \xi') = \frac{1}{\sigma} \langle h_i(\xi) \, h_i(\xi') \rangle \quad . \tag{15}$$

According to the discussion of §2, we expect that the 2D field theory describing the dynamics of the flux tube flows, for a domain $\mathcal{D}$ big enough, to a massless gaussian limit, where the Green function fulfills the free field equation

$$- \Delta G(\xi, \xi') = \delta^{(2)}(\xi - \xi') \quad . \tag{16}$$

For simplicity we omitted the transverse index i. It is convenient to adopt complex coordinates by defining $z = \xi_1 + i\xi_2$; then the solution $G_\infty(z, z')$ of the above equation in the infinite plane has the well known form

$$G_\infty(z, z') = -\frac{1}{2\pi} \log |z - z'| + c \quad , \tag{17}$$

where c is an arbitrary constant. Note that for $z \sim z'$ the free Green function is ultraviolet divergent , then Eq.(14) is ill-defined and needs regularization. Using for instance the point-splitting method one gets at once for a square Wilson loop $L \times L$

$$w^2 = \frac{1}{2\pi\sigma} \log\left(L/R_c\right) \quad , \tag{18}$$

where the UV cut-off has been absorbed in the definition of the scale R_c.

3.2. Green Functions in a Finite Box

Actually the above calculation has a flaw, because the theory is only defined in a finite box, while it is used the propagator of the infinite plane. The Green function on a box has a very different form which depends on the shape of the box and on the choice of the boundary conditions. This fact might in principle modify the functional form of the width. Actually we shall see that this is not the case.

The problem of finding the free Green function on an arbitrary, simply connected region $\mathcal{D}$ of the plane with fixed boundary conditions has been already solved by the mathematicians of the last century. Accordingly, it can be reformulated in a way where z and z' play an apparently asymmetric role: find a real function $f_{z'}(z, \mathcal{D})$ which is harmonic in the punctured set $\mathcal{D} \setminus \{z'\}$,vanishes at the boundary $\partial\mathcal{D}$ and diverges logarithmically as $-\log|z - z'|$ for $z \to z'$. Such a problem can be solved once it is found a conformal mapping $z \to \ell$ of $\mathcal{D}$ onto the unit circle $|\ell| = 1$ which maps z' into the origin $\ell = 0$. Denoting by $\ell_{z'}(z)$ the analytic function providing us with such a mapping, it is immediate to verify that

$$f_{z'}(z, \mathcal{D}) = \log|\ell_{z'}(z)| \quad , \tag{19}$$

and that the function

$$G_{\mathcal{D}}(z, z') = \frac{1}{2\pi} f_{z'}(z, \mathcal{D}) \tag{20}$$

satisfies Eq.(16). For, note that $G_{\mathcal{D}}$ is the real part of an analytic function, then $\Delta G_{\mathcal{D}} = 0$ for $z \neq z'$; since z' is mapped into the origin $\ell = 0$, then $G_{\mathcal{D}}$ diverges logarithmically for $z \to z'$ and is normalized as required by the delta function; finally, being the boundary $\partial\mathcal{D}$ mapped into the unit circumference, $G_{\mathcal{D}}$ vanishes at this boundary as it should.

For our purposes, the relevant property of the conformal mapping ℓ is that one can perform an arbitrary scale transformation $z \to \lambda z$ without destroying the conformal character of ℓ. More precisely we can write

$$f_{z'}(z, \mathcal{D}) = f_{\lambda z'}(\lambda z, \mathcal{D}_\lambda) \quad , \tag{21}$$

where $\mathcal{D}_\lambda$ denotes the scaled domain. A direct consequence is that it is always possible, for any finite $\mathcal{D}$, to fix the area of the scaled domain $\mathcal{D}_\lambda$ to an arbitrary value, say 1, without changing the Green function. It follows that the integration of the finite part $G(z, z')$ in Eq.(14) cannot depend on the size of the domain $\mathcal{D}$ but only on its shape. Surprisingly enough, the logarithmic growth of the square width w^2 comes from the UV divergent part. Let us see why. According to the point spitting method, we regularize $G(\xi,\xi)$ with the replacement

$$G(\xi,\xi) \to G(\xi,\xi + \varepsilon) \quad , \tag{22}$$

where ε is the ultraviolet cut-off. Combining the logarithmic divergence of G with its scaling property expressed in Eq.(21), we can write

$$G(\xi',\xi' + \varepsilon) \sim -\frac{1}{2\pi} \log\left(\varepsilon/L\right) + c \quad , \tag{23}$$

where L is a typical linear dimension of the domain $\mathcal{D}$. Inserting such an expression in Eq.(14) we get the logarithmic law of Eq.(18) where now R_c is a calculable function of the shape of the domain.

The physical meaning of such a behaviour is now clear: the dynamics of the flux tube is described by a truly free-field theory only at large distances; the cut-off ε sets up the ultraviolet scale R_c below which the free-field approximation breaks down. Obviously this scale cannot depend on the infrared scale L, then a variation of L cannot be balanced by a variation of ε. In other terms the ultraviolet divergence breaks the scale invariance of the problem and then a variation of L gives rise to an observable effect.

The volunterous reader may check the above arguments in a particularly simple example, given by the Green function on a disc of radius L

$$G_{disc}(z, z') = -\frac{1}{2\pi} \log\left| L \frac{z - z'}{L^2 - \bar{z}'z} \right| \quad . \tag{24}$$

For a rectangular box of size $L_1 \times L_2$, which is the most interesting case for the lattice gauge models, we have

$$G_{L_1,L_2}(z, z') = -\frac{1}{2\pi} \log\left| \frac{\sigma(z - z')\sigma(z + z')}{\sigma(z - \bar{z}')\sigma(z + \bar{z}')} \right| \quad , \tag{25}$$

where $\sigma(z)$ is the Weierstrass sigma function [17] for the rectangle of sides $2L_1$ and $2L_2$. Inserting this expression in Eq.(14) it turns out that the scale R_c does not depend very much on the modulus $\tau = iL_2/L_1$ for not too elongated rectangles.

3.3. Numerical Simulations

Though the logarithmic growth of the squared width of the flux tube is the most important and model-independent quantum effect predicted by the effective string description, it has not yet been observed until now. This is due to two main reasons. First, this effect is an infrared phenomenon that can be seen, as we have explained, only for very large Wilson loops, which are at the boundary of the sizes reached by the present numerical simulations. Secondly, a logarithmic growth can be checked numerically with reasonable confidence only if one can deal with a set of very precise data in a wide range of the loop sizes, which is again a condition which the current lattice simulations for $SU(2)$ and $SU(3)$ gauge theories have not yet attained.

It has been recently observed[18] that in the simplest gauge system, i.e. the Z_2 3D gauge model, it is nowadays possible to overcome these difficulties by exploiting the duality relation between this model and the ordinary 3D Ising model explained in §2.2 .

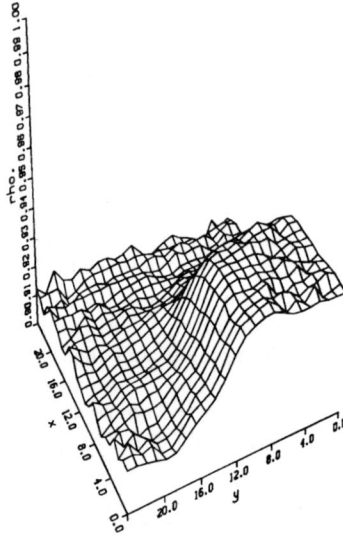

Fig. 2. Density of a Z_2 flux tube evaluated at a distance of two lattice spacings from a Wilson loop 24×24 sitting on the xy plane. Only the upper left quadrant is reported

Indeed, using the one-to-one mapping of physical observables of the gauge system into the corresponding spin observables it is possible to replace the ordinary Metropolis or heath-bath method with a non-local cluster algorithm (for instance the Swendsen and Wang method). This allows us to probe the structure of the flux tube with an unprecedented accuracy. The procedure is the following. A Wilson loop $W(C)$ is realized in the spin lattice by frustrating all the links cutting a given surface Σ bounded by C. These frustrated links modify the vacuum state so that the vacuum expectation value $\langle P \rangle_{W}$ of the plaquette, or better its spin analogue, is no longer translational invariant, but becomes a function of its relative position with respect to $W(L_1, L_2)$ and is related to the expectation value in the ordinary vacuum by

$$\langle P \rangle_{W} = \langle W(L_1, L_2) P \rangle / \langle W(L_1, L_2) \rangle \quad . \tag{26}$$

The difference between the expectation value of the plaquette in the vacuum modified by the presence of $W(L_1, L_2)$ and the expectation value in the ordinary vacuum can be considered as a measure of the density of the flux tube. Choosing for instance as a probe a plaquette $P_\parallel$ parallel to the plane of the Wilson loop we can write

$$\rho_\parallel(x, y, z) = \langle P_\parallel \rangle_W - \langle P \rangle \quad . \tag{27}$$

Other orientations of the plaquette give approximately the same distribution : The flux density $\rho_\parallel$ can be viewed, up to an obvious normalization, as the probability for

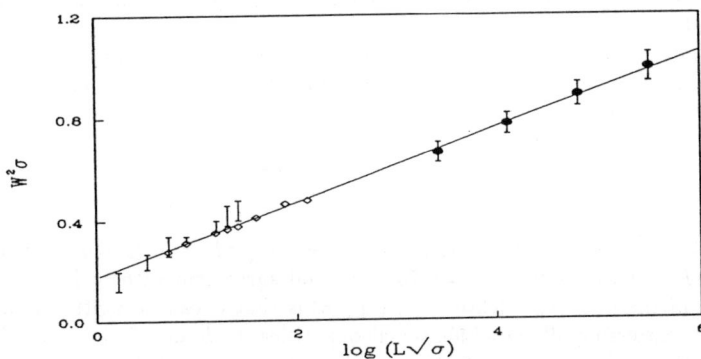

Fig. 3. Squared width of the flux tube in units of sigma.

unit of volume of finding the fluctuating string at the point of coordinates (x, y, z). The main difference between the flux density in Z_2 gauge model and that in theories with a continuous gauge group is that in the discrete case there is no self-energy peak associated to the quark lines ; as a consequence, the formation of the flux tube can be seen more neatly. Starting from a point on the perimeter of the Wilson loop an moving inwards, one finds that the flux density increases rapidly in a few lattice spacings and then it reaches a stable plateau. The same feature can be found in any plane parallel to the Wilson loop, with a value of the plateau which is only a function of the distance from the plane of the loop as Fig. 2 shows.

Then, assuming that the Wilson loop is located in the plane $z = 0$, the mean square width of the flux tube can be defined as

$$w^2 = \int d x^3 \, z^2 \rho_\parallel(x, y, z) \, / \int d x^3 \rho_\parallel(x, y, z) \quad , \tag{28}$$

where the integration volume is bounded by the plateau described above.

---

*This seems not to be the case in four dimensional systems[19,20]

We performed two sets of simulations. The first was at $\beta = 0.7516$ corresponding to[†]$\sigma a^2 = 0.0107(1)$. We considered six different squared Wilson loops of sides ranging from $L = 16$ to $L = 40$ and fitted the data to the two-parameter formula

$$w^2 = a \log(L) + b \quad . \tag{29}$$

We got $b = -17.4 \pm 4.0$ and $a = 14.4 \pm 1.2$. The theoretical value a_{th} of the parameter a, fixed by Eq.(18) to be $a_{th} = \frac{1}{2\pi\sigma}$, matches nicely with this value, indeed we have $a_{th} = 14.8$. The second set of simulations was made at $\beta = 0.7460$ corresponding to $\sigma a^2 = 0.0189(1)$ with squared Wilson loops of sides ranging from $L = 15$ to $L = 60$. Fits to Eq.(29) gave $b = -5.8 \pm 2.0$ and $a = 7.7 \pm 0.6$ while the theoretical value is $a_{th} = 8.42$.

An internal consistency check of these numerical data comes from a comparison between Eq.(18) and Eq.(29). We get that the physical adimensional quantity $\sqrt{\sigma}R_c$ is expressed in terms of a and b as

$$\sqrt{\sigma}R_c = \frac{\exp -b/a}{\sqrt{2\pi a}} \quad . \tag{30}$$

Using the fitted values of the parameters, we get $\sqrt{\sigma}R_c = 0.35 \pm 0.11$ at $\beta = 0.7516$ and $\sqrt{\sigma}R_c = 0.31 \pm 0.09$ at $\beta = 0.7560$ in good agreement with scaling.

One of the important features of Eq.(18) is that it can be written in a universal form by expressing all the dimensional quantities w, L and R_c in units of $\sqrt{\sigma}$. Accordingly we report in Fig. 3 data from different β's and and also recent very accurate data[21] for the mean square width of fluid interfaces in the dual Ising model (black dots). The straight line represents the logarithmic fit to our data at $\beta = 0.7460$ (rhombs). Within the statistical accuracy these data clearly support a logarithmic widening of the flux tube in a range of quark separation L of more than two orders of magnitude, starting about $L_{min}\sqrt{\sigma} \simeq 1.7$.

In order to compare these results with analogous data for other gauge groups, we can, with an abuse of language, express $\sqrt{\sigma}$ in the same physical units of the QCD string . Then the crossover to the infrared logarithmic behaviour observed in Fig. 3 corresponds to $L_{min} = .75$ fm while the maximal probed elongation corresponds to more than 100 fm.

The data on $4D$ $SU(2)$ flux tube[19,20] cover now a distance up to 2 fm, but are still affected by strong systematic errors and are compatible also with a constant width for separations larger than 1 fm. They are scattered mainly above the straight line of Fig. 3, suggesting a value of the UV scale a bit smaller than the value of the Z_2 system, while the crossover distance is approximately the same.

Such a similarity of the cross-over distance L_{min} in so different gauge systems has a plausible explanation. According to Eq.(2), L_{min} should be controlled by the first non gaussian term in the α' expansion of the effective string action. This term does not depend on the nature of the gauge group, so it should produce universal effects.

[†]The high precision of this value is a consequence of the use of the non-local cluster algorithm.

Actually it has been recently[22] shown that this term describes accurately some finite size effects in fluid interfaces which cannot explained in the free field limit.

4. References

1. H.B. Nielsen and P.Olesen, *Nucl. Phys.* **B61** (1973) 45 .
2. G.'t Hooft, *Nucl. Phys.* **B72** (1994) 461.
3. D.J. Gross and W. Taylor, *Nucl. Phys.* **B400** (1993) 181; *Nucl. Phys.* **B403** (1993) 395.
4. M.Ademollo, A.D'Adda, R.D'Auria, P.Di Vecchia, F. Gliozzi, E.Napolitano and S.Sciuto, *Nucl. Phys.* **B94** (1975) 221.
5. M.Caselle, F.Gliozzi and S.Vinti, DFTT 12-93, *Nucl. Phys.* (Proc. Suppl) **B34** (1994) 726.
6. V. Dotsenko, G. Harris, E.Marinari, E. Martinec, M. Picco and P. Windey, *Phys. Rev. Lett.* **71** (1993) 811.
7. C. Jeppesen and J.H.Ipsen, *Europhys. Lett.* **22** (1993) 713
8. A.M.Polyakov, *Nucl. Phys.* **B268**, 406 (1986)
9. J.M.Kostelitz and D.J. Thouless, *J. Phys.* **C6** (1973) 1181; J.M.Kostelitz, *J. Phys.* **C7** 1974 1046.
10. K.Diez and T.Filk, *Phys. Rev.* **D 27** (1983) 2944.
11. F.Gutbrod, *Z.Phys.* **C 37** (1987) 143; *Phys. Lett.* **B163** (1987) 389.
12. M.Caselle, R.Fiore, F.Gliozzi, P.Provero and S. Vinti, *Int. J. Mod. Phys.* **6** (1991) 4885 .
13. M.Caselle, R.Fiore, F.Gliozzi, P.Provero and S. Vinti, *Phys. Lett.* **B272** (1991) 272.
14. M.Caselle, F.Gliozzi and S.Vinti, *Phys. Lett.* **B302** (1993) 74.
15. P.Provero and S.Vinti, preprint DFTT 58/93 to be published on Physica **A**
16. M.Lüscher, G.Münster and P. Weisz, *Nucl. Phys.* **B180** [FS2] (1981) 1.
17. L.V. Ahlfors, *Complex Analysis* (McGraw-Hill, London, 1979) p. 273.
18. M.Caselle, F.Gliozzi, U.Magnea and S. Vinti, DFTT XX-94, to be published on *Nucl. Phys.* **B** (Proc. Suppl) (1995) .
19. L.Del Debbio, A.Di Giacomo and Yu. Simonov, preprint IFUP-TH 11/94 (1994).
20. G.S.Bali, C. Schlichter and K. Schilling, preprint CERN-TH.7413/94 (1994).
21. M.Hasenbusch and K.Pinn, *Physica A* **192** (1993) 342.
22. M.Caselle, R.Fiore, F.Gliozzi, M.Hasenbusch, K.Pinn and S.Vinti, preprint CERN-TH 7291/94 (1994), to be published on *Nucl. Phys.* **B**.

AN EFFECTIVE QUARK/ANTIQUARK POTENTIAL
FOR THE CONSTITUENT QUARK MODEL

F. ZACHARIASEN[*]

California Institute of Technology, Pasadena, CA 91125

ABSTRACT

We use Dual QCD to derive an effective potential to order (quark mass)$^{-2}$ for a constituent quark and anti-quark. This is done by expanding the Dual QCD Lagrangian to second order in the $q\bar{q}$ spins and velocities around the static central potential, in which the quarks are both spinless and stationary. The field equations are then used to eliminate the dual gluon fields and the Higgs fields of Dual QCD in favor of quark variables for an arbitrarily but slowly moving $q\bar{q}$ pair with a Dirac string of arbitrary shape connecting them. The result is a Lagrangian, and therefore, a potential, which depends only on the $q\bar{q}$ positions, velocities and spins.

Dual QCD contains only three parameters, which can be determined from the vacuum energy density, the string tension, and the strength of the Coulomb singularity of the central potential. The only free parameters in the spin and velocity dependent part of the effective potential are, therefore, the masses of the c and b quarks. When inserted into a Schrödinger equation these potentials provide a complete effective constituent quark theory which can be used to calculate $q\bar{q}$ energy levels in terms of the masses and the masses can thereby be fixed (agreement with experiment is excellent). The various potentials - spin-spin, spin-orbit, and spinless velocity dependent - can also in principle be compared to lattice calculations of the same quantities. For the spin-orbit case, for example, the agreement is good, although lattice results are not yet precise enough for a real comparison to be made. For the potentials proportional to velocity squared lattice results do not yet exist.

1. Introduction and Preliminaries

This paper is devoted to the calculation of the complete dynamical potential between an arbitrarily moving heavy quark anti-quark pair. The potential is given as an explicit function of the quark coordinates and velocities, includes all spin effects, and is accurate to order (mass)$^{-2}$, or quark velocity squared. (This is as far as one can go with a calculation that does not allow for the explicit radiation of gluons, and is therefore the extent to which the concept of a potential has a meaning.) If the potential we obtain is used in a Schrödinger equation (or, if desired, in any other wave equation such as, for example, a Salpeter equation to take into account the relativistic kinetic energy of the quarks) then we have a complete dynamical theory of constituent quark-antiquark pairs.

[*] Work supported in part by the U.S. Dept. of Energy under Grant No. DE-FG03-92-ER40701.

Our calculation of the potential is based on the dual QCD Lagrangian.[1,2] This Lagrangian can be viewed as an effective theory of long distance QCD, which incorporates in a very simple way most of what we believe to be true about real QCD. The dual field theory is relativistic, unitary, renormalizable, obeys non-Abelian dual gauge invariance and automatically gives rise to confinement of color via a non-Abelian dual Meissner effect. The theory is described in terms of a dual color octet vector potential C_μ^a (the quanta of which are dual gluons) as well as a set of three octet scalar "Higgs fields" $B_i^a(i = 1, 2, 3; a = 1...8)$. The Higgs mechanism, through the scalar fields B, gives mass to all eight dual gluons, as well as to the Higgs fields themselves, thus confining color flux into Z_N flux tubes and providing a dual superconductivity explanation of confinement. (At sufficiently high temperature, as in ordinary superconductivity, the spontaneous symmetry breaking disappears, so does the dual gluon mass and with it confinement).[3]

The introduction of quarks into dual QCD requires some delicacy, analogous to the introduction of magnetic monopoles in ordinary electrodynamics. Each quark must have attached to it a Dirac string, and these strings can cause technical problems in calculations. In Reference 4 Dirac has described how these difficulties can be dealt with in electrodynamics, and how the electrodynamic interaction of a pair of oppositely charged particles of arbitrary mass can be described using the dual vector potential, with an arbitrarily moving Dirac string connecting the arbitrarily moving charges. We can carry over these same techniques to dual QCD in order to describe a quark-antiquark pair, and this is in fact what makes it possible to derive the dynamical quark anti-quark potential (to order $(\text{mass})^{-2}$).

The paper is organized as follows. We first write down the dual QCD Lagrangian for a quark and an antiquark connected by a Dirac string of arbitrary shape. We then solve the dual QCD field equations for the various fields and for stationary quarks; this gives us our zero order solution and the static quark anti-quark central potential.[1] We then expand the dual QCD Lagrangian in inverse powers of quark mass to order $(\text{mass})^{-2}$, and calculate the corrected dual QCD fields in terms of the quark positions, velocities, and spins. These results are then used to eliminate the fields from the problem, and to calculate the energy of the system in terms only of the quark variables. This gives rise to a quark-antiquark potential containing (besides the static central potential obtained in zero order) a spin-spin potential V_{spin}, a spin-orbit potential $V_{spin-orbit}$, and a spin independent potential V_{v^2} of order quark velocity squared. All of these potentials are explicitly given in terms of the quark variables. The parameters they depend on are only the constituent quark masses m_1 and m_2; the other three parameters in dual QCD[1] are fixed from the vacuum energy density and the behavior of the central potential at zero and infinite distance (this last parameter is of course just the string tension).

The calculation we describe here has been partially carried out before, in that the various spin and velocity dependent potentials have been evaluated assuming a particular motion of the quark-antiquark pair, namely that they travel in circles.[5,6,7] Needless to say, our results here, which are valid for general quark motion, reduce to the old results when circular motion is imposed.

In dual QCD the dynamical field is C_μ^a, the vector potential dual to the ordinary vector potential A_μ^a. In the absence of quarks, the Lagrangian for C_μ^a is given by[1]

$$\mathcal{L} = 2trace \left[\frac{1}{2}(\vec{H}^2 - \vec{D}^2) + \frac{1}{2}(\mathcal{D}_\mu \vec{B})^2 \right] - W(B) , \qquad (1.1)$$

where the trace is over color indices, and where

$$\mathcal{D}_\mu \vec{B} = \partial_\mu \vec{B} - ig[C_\mu, \vec{B}] . \qquad (1.2)$$

$\vec{D}$ and $\vec{H}$ are the non-Abelian generalizations of the color electric displacement and magnetic field:

$$\vec{D} = -\vec{\nabla} \times \vec{C} - \frac{1}{2}ig[\vec{C}, \times \vec{C}] , \qquad (1.3)$$

and

$$\vec{H} = -\vec{\nabla}C_0 - \partial_0 \vec{C} - ig[\vec{C}, C_0] , \qquad (1.4)$$

where $C_\mu = \Sigma C_\mu^a \frac{1}{2}\lambda_a$ and $\frac{1}{2}\lambda_a$ are the generators of SU(3). In the infrared limit, the dual coupling constant $g = \frac{2\pi}{e}$ where e is the ordinary Yang-Mills coupling constant. The quantity $\vec{B}$ represents the three scalar octets necessary to give mass to all color components of C_μ, $\vec{B} = (B_1, B_2, B_3)$. The function $W(\vec{B})$ is the counter-term needed for renormalization and plays the role of a Higgs potential; its explicit form is given in Reference 1. Since they couple to the dual potentials the scalar fields $\vec{B}$ carry color magnetic charge.

Our first use of eq. (1.1) was to calculate the field configurations associated with quantized color electric flux tubes.[8] These, with one unit of quantized Z_3 flux, can be viewed as the fields between a static quark and antiquark at infinite separation. (Note that in the gauge chosen in reference 8, the Dirac strings - see below - attached to the quarks at $\pm\infty$ are taken to extend to $\pm\infty$ respectively, and do not join the two quarks).

For the calculation of flux tubes, we make the simplest color ansatz that produces a closed set of non-trivial field equations.[8] The fields $\vec{D}, \vec{H}$, and $\vec{C}$ are all proportional to the color matrix $Y = \lambda_8/\sqrt{3}$. The field C_0 is zero. Two of the three B fields can be chosen equal: $B_1 = B_2 \equiv B$. B_1, B_2, and B_3 are chosen to be in the color directions $(\lambda_7, -\lambda_5, \lambda_2)$ respectively. For this choice the function $W(B)$ becomes

$$
\begin{aligned}
W(B) = &-\frac{16}{9}\lambda(-\tilde{F}_0^2)(B_1^2 + B_2^2 + B_3^2) \\
&+ \frac{14}{3}\lambda(B_1^2 + B_2^2 + B_3^2)^2 \\
&+ \frac{22}{3}\lambda(B_1^4 + B_2^4 + B_3^4) .
\end{aligned}
$$

There are two parameters, which we call λ and $-\tilde{F}_0^2$,[1] in the Higgs potential, so, including g, we have three parameters overall. These three parameters can be roughly determined from the solutions for the dual QCD flux tube and the zero and infinite distance limits of the central potential.[8] First, from the form of $W(\vec{B})$, we can compute the vacuum energy density to be $\epsilon_{\text{vac}} = -\lambda(\tilde{F}_0^2)^2/9$. Through the use of the trace anomaly ϵ_{vac} can be related to the magnetic condensate: $\epsilon_{\text{vac}} = -\frac{11}{32}G_2$. Next, we can calculate the string tension σ in the flux tube which is also the coefficient of the linear (in separation) term in the central potential. Finally we can obtain g from the value of $\alpha_s = \pi/g^2$ given by the Coulomb singularity in the central potential. Putting all of this together, and using the values $G_2 = (330\text{MeV})^4, \sigma = (427\text{MeV})^2$, and $\alpha_s = .39$ we obtain $\lambda = 1.61, g^2/\lambda = 5$ and $\sqrt{-\tilde{F}_0^2} = 420$ MeV. With these parameters the flux tube comes out to have a radius of about half a Fermi. These rough values can serve as a guide to our more precise fits to the energy levels in the $c\bar{c}$ and $b\bar{b}$ systems, once we have calculated all the terms in the order $(mass)^{-2}$ heavy quark potential.

We next wish to extend this to a system consisting of a classical heavy quark of charge e and an antiquark of charge $-e$ at finite separation having masses m_1 and m_2, and spins $\vec{\sigma}_1$ and $\vec{\sigma}_2$. The quark charge density must also lie in the Y color direction in order to absorb the flux of $\vec{D}$. Because quarks in our dual theory are like magnetic monopoles in ordinary electrodynamics, we must modify (1.3) so that Gauss' law is satisfied. This is achieved by adding a string field $\vec{D}_s$ to (1.3)[1]

The string field is chosen to connect the two-quarks (a different gauge choice from that made in the flux tube case). It satisfies

$$\vec{\nabla} \cdot \vec{D}_s = \rho(\vec{x}, t) = e[\delta^3(\vec{x} - \vec{x}_1(t)) - \delta^3(\vec{x} - \vec{x}_2(t))]Y , \qquad (1.5)$$

where $\vec{x}_{1,2}(t)$ are the positions of the two-quarks at time t. Thus $\vec{D}_s$ must lie in the Y color direction as well. Explicitly, we have[4]

$$\vec{D}_s(\vec{x}, t) = -e \int_{\tau_2}^{\tau_1} d\tau \frac{d\vec{y}(\tau, t)}{d\tau} \delta^3(\vec{x} - \vec{y}(\tau, t))Y , \qquad (1.6)$$

where τ parametrizes position along the string $\vec{y}(\tau, t)$ and $\vec{y}(\tau_{1,2}, t) = \vec{x}_{1,2}(t)$. The moving string induces a magnetization $\vec{H}_s$ as well and we can write[4]

$$\vec{H}_s = e \int_{\tau_2}^{\tau_1} d\tau \frac{d\vec{y}(\tau, t)}{dt} \times \frac{d\vec{y}(\tau, t)}{d\tau} \delta^3(\vec{x} - \vec{y}(\tau, t)). \qquad (1.7)$$

These two fields satisfy

$$\vec{\nabla} \times \vec{H}_s - \partial_0 \vec{D}_s = \vec{j} \qquad (1.8)$$

where

$$\vec{j} = e[\vec{v}_1 \delta^3(\vec{x} - \vec{x}_1(t)) - \vec{v}_2 \delta^3(\vec{x} - \vec{x}_2(t))] \tag{1.9}$$

is the quark current. (Here $\vec{v}_{1,2} = \frac{d}{dt}\vec{x}_{1,2}(t)$). There is also, due to the quark spin, a magnetization

$$\vec{M} = \left[\frac{e\vec{\sigma}_1}{2m_1}\delta^3(\vec{x} - \vec{x}_1) - \frac{e\vec{\sigma}_2}{2m_2}\delta^3(\vec{x} - \vec{x}_2)\right] Y , \tag{1.10a}$$

and due to the quark motion an induced polarization $\vec{P}$:

$$\vec{P} = \left[\vec{v}_1 \times \frac{e\vec{\sigma}_1}{2m_1}\delta^3(\vec{x} - \vec{x}_1) - \vec{v}_2 \times \frac{e\vec{\sigma}_2}{2m_2}\delta^3(\vec{x} - \vec{x}_2)\right] Y . \tag{1.10b}$$

Both $\vec{P}$ and $\vec{M}$ are in the Y color direction.

The electric displacement $\vec{D}$ and the magnetic field $\vec{H}$ satisfy Maxwell's equations (since all of these fields are in the same color direction, this sector of the problem – that is, the sector excluding $\vec{B}$ – is Abelian):

$$\vec{\nabla} \cdot \vec{D} = \rho - \vec{\nabla} \cdot \vec{P} \tag{1.11}$$

$$\vec{\nabla} \times \vec{H} - \partial_0 \vec{D} = \vec{j} + \vec{\nabla} \times \vec{M} . \tag{1.12}$$

The solutions to these equations are expressed in terms of the dual potentials through

$$\vec{D} = -\vec{\nabla} \times \vec{C} + \vec{D}_s - \vec{P} \tag{1.13a}$$

$$\vec{H} = -\vec{\nabla}C_0 - \partial_0\vec{C} + \vec{H}_s + \vec{M} . \tag{1.13b}$$

These solutions can be inserted into eq. (1.1), and, as we shall see, we can retain the flux tube color structure of $\vec{B}$ and the fact that $B_1 = B_2 = B$.

The commutation relations $[Y, \lambda_5] = -i\lambda_4$, $[Y, \lambda_7] = -i\lambda_6$, and $[Y, \lambda_2] = 0$ yield

$$\mathcal{D}_\mu \vec{B} = ((\lambda_7 \partial_\mu - \lambda_6 g C_\mu)B, (-\lambda_5 \partial_\mu + \lambda_4 g C_\mu)B, \lambda_2 \partial_\mu B_3) . \tag{1.14}$$

The color traces are $2Tr\, Y^2 = 4/3$ and $2Tr\, \lambda_a \lambda_b = 4\delta_{ab}$. These equations give the final form of $\mathcal{L}$ in terms of the coefficients $C_\mu, \vec{H}$ and $\vec{D}$ of the color matrix Y and the functions B and B_3.

The resulting Lagrangian is, after evaluation of the color trace in eq. (1.1),

$$\mathcal{L} = \frac{2}{3}\vec{H}^2 - \frac{2}{3}\vec{D}^2 - 4g^2 B^2(\vec{C}^2 - C_0^2) + 4B\nabla^2 B - 4B\ddot{B} + 2B_3\nabla^2 B_3 - 2B_3\ddot{B}_3 - W(\vec{B}) , \tag{1.15}$$

where the double dot denotes the second derivative with respect to time. The non-quadratic terms in the fields reflect the underlying non-Abelian nature of the theory entering via the $(\mathcal{D}_\mu \vec{B})^2$ term in $\mathcal{L}$.

Note that it might appear from the commutation relations and (1.14) that we need to introduce components of $\vec{B}$ along the color directions λ_4 and λ_6. However, it is consistent to set these amplitudes equal to zero provided we choose $\vec{\nabla} \cdot \vec{C} = 0$, as we do.

2. Derivation of Equations for the Potential

Given the expression for the Lagrangian density, eq. (1.15), we may substitute the solutions (1.13) and (1.14) into it, to obtain the Lagrangian L. The first two terms of L are

$$\frac{2}{3} \int d^3\vec{x}(\vec{H}^2 - \vec{D}^2)$$

$$
\begin{aligned}
= \frac{2}{3} \int d^3\vec{x}\{ & (-\partial_0\vec{C} - \vec{\nabla}C_0 + \vec{H}_s)^2 \\
& + 2\vec{M} \cdot (-\partial_0\vec{C} - \vec{\nabla}C_0 + \vec{H}_s) \\
& + \vec{M}^2 - \vec{P}^{\,2} \\
& - (-\vec{\nabla} \times \vec{C} + \vec{D}_s)^2 \\
& + 2\vec{P} \cdot (-\vec{\nabla} \times \vec{C} + \vec{D}_s)\} \, .
\end{aligned}
\tag{2.1}
$$

Note that the cross terms

$$\int d^3\vec{x}(\vec{P} \cdot \vec{D}_s + \vec{M} \cdot \vec{H}_s)$$

$$= -e \int_{\tau_2}^{\tau_1} d\tau \frac{d\vec{y}(\tau,t)}{d\tau} \cdot (\vec{P}(\vec{y},t) - \frac{d\vec{y}(\tau,t)}{dt} \times \vec{M}(\vec{y},t))$$

$$= 0 \, ,$$

$$\tag{2.2}$$

because of the fact that $\vec{P}(\vec{y}) = \frac{d\vec{y}}{dt} \times \vec{M}(\vec{y})$. Furthermore, to second order in $(\text{mass})^{-1}$ we can omit the term $\vec{P}^{\,2}$. Thus altogether

$$
\begin{aligned}
L = \frac{2}{3} \int d^3\vec{x}\{ & -(-\vec{\nabla} \times \vec{C} + \vec{D}_s)^2 \\
& + 2C_0\vec{\nabla} \cdot \vec{M} + (-\partial_0\vec{C} - \vec{\nabla}C_0 + \vec{H}_s)^2 \\
& + 2\vec{P} \cdot (-\vec{\nabla} \times \vec{C}) - 2\vec{M} \cdot \partial_0\vec{C} + \vec{M}^2 \\
& + 6g^2 B^2(C_0^2 - \vec{C}^2) \\
& + 6(\partial_0 B)^2 - 6(\vec{\nabla}B)^2 + 3(\partial_0 B_3)^2 - 3(\vec{\nabla}B_3)^2 \\
& - \frac{3}{2}W(B)\} \, .
\end{aligned}
\tag{2.3}
$$

The classical field equations following from L are

$$\nabla^2 C_0 - \vec{\nabla} \cdot \vec{H}_s + \vec{\nabla} \cdot \dot{\vec{C}} - 6g^2 B^2 C_0 - \vec{\nabla} \cdot \vec{M} = 0 , \qquad (2.4a)$$

$$-\vec{\nabla} \times (\vec{\nabla} \times \vec{C}) - \ddot{\vec{C}} + \vec{\nabla} \times \vec{D}_s + \partial_0(-\vec{\nabla} C_0 + \vec{H}_s) ,$$

$$-6g^2 B^2 \vec{C} - \partial_0 \vec{M} = 0 , \qquad (2.4b)$$

$$\nabla^2 B - \ddot{B} - g^2(\vec{C}^2 - C_0^2)B = \frac{1}{2}\frac{\partial W}{\partial B} , \qquad (2.5a)$$

and

$$\nabla^2 B_3 - \ddot{B}_3 = \frac{\partial W}{\partial B_3} . \qquad (2.5b)$$

The value of L for $\vec{M} = \vec{v}_1 = \vec{v}_2 = 0$ is the negative of the central potential $V_0(R)$. We look for static solutions (i.e., $\dot{\vec{C}} = \dot{B} = \dot{B}_3 = 0$) with $C_0 = 0$ as well.[1,9] These solutions, which we denote $\vec{\bar{C}}, \bar{B}$ and $\bar{B}_3$, depend, of course, parametrically on the separation $R = |\vec{x}_1 - \vec{x}_2|$ of the quark sources. Thus the energy $V_0(R)$ is simply

$$V_0(R) = -\int d^3\vec{x} \mathcal{L}_0(\vec{\bar{C}}, \bar{B}, \bar{B}_3) , \qquad (2.6)$$

where

$$\mathcal{L}_0 = \frac{2}{3}(\vec{D}_s - \vec{\nabla} \times \vec{\bar{C}})^2 - 4g^2 \bar{B}^2 \vec{\bar{C}}^2$$

$$+4\bar{B}\nabla^2 \bar{B} + 2\bar{B}_3 \nabla^2 \bar{B}_3 - W(\vec{\bar{B}}) , \qquad (2.7)$$

is the static spinless Lagrangian density obtained from (2.3) by setting $\vec{v}, \vec{M}, C_0$, and the time derivatives to zero. The field equations obeyed by $\vec{\bar{C}}, \bar{B}$ and $\bar{B}_3$ are immediately derived from (2.7), or equivalently from (2.4) and (2.5). Thus

$$-\vec{\nabla} \times (\vec{\nabla} \times \vec{\bar{C}}) + \vec{\nabla} \times \vec{D}_s - 6g^2 \bar{B}^2 \vec{\bar{C}} = 0 , \qquad (2.8a)$$

$$\nabla^2 \bar{B} - g^2 \vec{\bar{C}}^2 \bar{B} = \frac{1}{2}\frac{\partial W}{\partial \bar{B}} , \qquad (2.8b)$$

$$\nabla^2 \bar{B}_3 - \bar{B}_3 = \frac{\partial W}{\partial \bar{B}_3} . \qquad (2.8c)$$

To solve these it is convenient to define[9]

$$\vec{\bar{C}} = \vec{c} + \vec{C}_D , \qquad (2.9)$$

where $\vec{C}_D$ is the Dirac monopole potential of the two-quarks satisfying[4]

$$\vec{\nabla} \times (\vec{\nabla} \times \vec{C}_D) + \vec{\nabla} \times \vec{D}_s = 0 \ . \tag{2.10a}$$

The solution to eq. (2.10a) is

$$C_D = -\frac{e}{4\pi} \int_{\vec{x}_2}^{\vec{x}_1} d\vec{y} \times \vec{\nabla} \frac{1}{|\vec{x} - \vec{y}|} \ . \tag{2.10b}$$

In cylindrical coordinates and for a straight line string joining the two quarks located on the z axis at $z = R_1$ and $z = -R_2$,

$$\vec{D}_s = e\{\theta(z - R_1) - \theta(z + R_2)\}\delta(x)\delta(y)\hat{e}_z \ , \tag{2.10c}$$

and

$$\vec{C}_D \equiv \frac{e}{4\pi\rho} \left\{ \frac{z - R_1}{\sqrt{\rho^2 + (z - R_1)^2}} - \frac{z + R_2}{\sqrt{\rho^2 + (z + R_2)^2}} \right\} \hat{e}_\varphi \ . \tag{2.10d}$$

The equation satisfied by $\vec{c}$ is easily seen to be

$$\nabla^2 \vec{c} - 6g^2 \bar{B}^2(\vec{c} + \vec{C}_D) = 0 \ , \tag{2.11}$$

where as noted earlier we have made the gauge choice $\vec{\nabla}\cdot\vec{C} = \vec{\nabla}\cdot\vec{c} = 0$. Eq. (2.11) has the virtue, important for numerical solution, of not containing the singular quantity $\vec{D}_s$. Using the decomposition (2.9), we can rewrite eq. (2.6) in the simpler form

$$V_0(R) = -\int d^3\vec{x} \left\{ \frac{2}{3}(\vec{D}_c - \vec{\nabla} \times \vec{c})^2 - 4g^2 \bar{B}^2(\vec{c} + \vec{C}_D)^2 \right.$$
$$\left. + 4\bar{B}\nabla^2\bar{B} + 2\bar{B}_3\nabla^2\bar{B}_3 - W(\vec{B}) \right\} \ , \tag{2.12}$$

where $\vec{D}_c$ is the Coulomb field of the two-quarks:

$$\vec{D}_c = \vec{D}_{c_1} + \vec{D}_{c_2} = \frac{e}{4\pi} \frac{(\vec{x} - \vec{x}_1)}{|\vec{x} - \vec{x}_1|^3} - \frac{e}{4\pi} \frac{(\vec{x} - \vec{x}_2)}{|\vec{x} - \vec{x}_2|^3} \ . \tag{2.13}$$

The static solutions have the following features.[9] For $R \to 0$, $\vec{C} \to \vec{C}_D$ and $\vec{D} = \vec{D}_s - \vec{\nabla} \times \vec{C}$ becomes a pure dipole field. As R increases, $\vec{D}$ evolves from a pure dipole to a squashed (in the axial direction) dipole and then, as R increases still further, to a flux tube for very large R. This flux tube coincides with our original flux tube,[8] except that this solution is in the gauge where the Dirac string joins the two-quarks while our original solution was in the gauge where the strings went from the charges to $\pm\infty$.

As $R \to 0$, the central potential $V_0(R)$ obtained from eq. (2.12) approaches a Coulomb potential while as $R \to \infty$ it becomes linear in R. An analytic fit to the central potential obtained from eq. (2.12) using the numerical solutions of the static field equations in eq. (2.8) and the value $g^2/\lambda = 5$ determined from the flux tube solution, is

$$V_0(R) = -\frac{4}{3}\frac{\alpha_s}{R}e^{-0.655x} + 1.0324(-\tilde{F}_0^2)R - 0.520\left(-\frac{\tilde{F}_0^2}{\lambda}\right)^{1/2}, \qquad (2.14)$$

where $\alpha_s = e^2/4\pi = \pi/g^2, x = \sqrt{-\lambda\tilde{F}_0^2}R$ and λ and $\tilde{F}_0^2$ are parameters characterizing the "Higgs potential" $W(\vec{B})$.[1] The exponential cutoff on the Coulomb term in eq. (2.14) reflects the inability of the color electric field to penetrate into the dual superconducting vacuum.

We next want to calculate the (mass)$^{-2}$ corrections to $V_0(R)$ produced by the spin and velocity of the quarks. It is easy to show that when the quark sources move, the static fields $\vec{C}, B$ and B_3 move rigidly with the sources through first order in $\vec{v}$.

To determine L to second order it now suffices to replace $\vec{C}, B$ and B_3 by $\vec{\bar{C}}, \bar{B}$ and $\bar{B}_3$. This is because L is stationary with respect to variations around $\vec{\bar{C}}, \bar{B}$ and $\bar{B}_3$ since these satisfy the static field equations (2.8). Hence, there are no terms linear in the differences $\vec{C} - \vec{\bar{C}}, B - \bar{B}$ and $B_3 - \bar{B}_3$, which are themselves already of second order.

The field equation for C_0 is (2.4a) with $\vec{\nabla} \cdot \vec{C} = 0$:

$$-\nabla^2 C_0 + 6g^2\bar{B}^2 C_0 = -\vec{\nabla} \cdot \vec{H}_s - \vec{\nabla} \cdot \vec{M} . \qquad (2.15)$$

Using this and a certain amount of algebra we can obtain from (2.3) the second order Lagrangian, which turns out to be

$$
\begin{aligned}
L_2 = \int d^3\vec{x} \Bigg\{ &\frac{2}{3}(-\partial_0\vec{\bar{C}} + \vec{H}_s)^2 \\
&+ \frac{4}{3}(-\vec{P} \cdot \vec{\nabla} \times \vec{\bar{C}} - \vec{M} \cdot \partial_0\vec{\bar{C}}) \\
&+ \frac{2}{3}M^2 \\
&+ \frac{2}{3}C_0(\vec{\nabla} \cdot \vec{M} + \vec{\nabla} \cdot \vec{H}_s) \\
&+ 4(\partial_0\bar{B})^2 \\
&+ 2(\partial_0\bar{B}_3)^2 \Bigg\} .
\end{aligned}
\qquad (2.16)
$$

From this we can extract the three second order potentials $V_{spin}, V_{spin-orbit}$ and V_{v^2}

(the term independent of spin but quadratic in the quark velocities). We note that the V's are by definition the negative of the appropriate parts of the Lagrangian. Setting $\vec{v}$ and the time derivatives equal to zero in eq. (2.16) gives

$$V_{spin} = -\frac{2}{3} \int d^3\vec{x} \{ C_0 \vec{\nabla} \cdot \vec{M} + \vec{M}^2 \} . \tag{2.17}$$

The $\vec{M}^2$ term here is evidently just part of the perturbation theory, or one gluon exchange, result; $C_0 \vec{\nabla} \cdot \vec{M}$ contains the non-perturbative effects of the theory as well as the rest of perturbation theory.

Similarly the linear terms in $\vec{v}$ give

$$V_{spin-orbit} = \int d^3\vec{x} \left\{ -\frac{4}{3}(\vec{P} \cdot \vec{\nabla} \times \vec{C} + \vec{M} \cdot \partial_0 \vec{C}) + \frac{4}{3} C_0 \vec{\nabla} \cdot \vec{H}_s \right\} . \tag{2.18}$$

To this we must add by hand the contribution of the Thomas precession,

$$V_{spin-orbit}^{Thomas} = -\left(\frac{\vec{\sigma}_1 \cdot \vec{L}}{4m_1^2} + \frac{\vec{\sigma}_2 \cdot \vec{L}}{4m_2^2} \right) \frac{1}{R} \frac{\partial V_0(R)}{\partial R} , \tag{2.19}$$

to account for the relativistic effect of spin.

In order to evaluate the nonperturbative contributions to these two potentials it is convenient to decompose C_0 into

$$C_0 \equiv c_0 + C_M , \tag{2.20}$$

where C_M satisfies

$$\nabla^2 C_M = -\vec{\nabla} \cdot \vec{M} . \tag{2.21}$$

Explicitly,

$$C_M = \frac{e}{4\pi} \left\{ \frac{\vec{\sigma}_1 \cdot (\vec{x} - \vec{x}_1)}{2m_1 |\vec{x} - \vec{x}_1|^3} - \frac{\vec{\sigma}_2 \cdot (\vec{x} - \vec{x}_2)}{2m_2 |\vec{x} - \vec{x}_2|^3} \right\} . \tag{2.22}$$

Eq. (2.15) then becomes an equation for c_0:

$$\nabla^2 c_0 - 6g^2 B^2 (c_0 + C_M) = 0 , \tag{2.23}$$

The calculation of V_{spin} and $V_{spin-orbit}$ thus requires the numerical solution of (2.23) for c_0, inserting the solution (via eq. (2.20)) into eqs. (2.17) and (2.18), and adding the Thomas term.

The contribution of C_M in eqs. (2.17) and (2.18) for V_{spin} and $V_{spin-orbit}$ is simply the Breit-Fermi expression for these two potentials. The function c_0 provides the non-perturbative contribution resulting from the non-Abelian nature of original Lagrangian, and thereby expresses the effects of confinement.

Finally we read off the expression for V_{v^2} by setting $\vec{M} = 0$ in the expression for L_2 and in the field equation (2.4a). This yields

$$
V_{v^2} = -\int d^3x \left\{ \frac{2}{3}(-\partial_0\vec{C} + \vec{H}_s)^2 \right.
$$
$$
+ \frac{2}{3}C_0\vec{\nabla}\cdot\vec{H}_s
$$
$$
+ 4(\partial_0 B)^2
$$
$$
\left. + 2(\partial_0 B_3)^2 \right\} ,
$$
(2.24)

where now, from (2.4a), and because we use the gauge $\vec{\nabla}\cdot\vec{C} = 0$,

$$
\nabla^2 C_0 - 6g^2\bar{B}^2 C_0 = \vec{\nabla}\cdot\vec{H}_s .
$$
(2.25)

We can define c_0' by

$$
C_0 = c_0' + C_{0D} ,
$$
(2.26)

where C_{0D} satisfies

$$
\nabla^2 C_{0D} = \vec{\nabla}\cdot\vec{H}_s .
$$
(2.27)

The solution of (2.27) is[4]

$$
C_{0D} = \frac{e}{4\pi}\int_{\tau_2}^{\tau_1} d\tau \frac{d\vec{y}(\tau,t)}{d\tau} \times \frac{d\vec{y}(\tau,t)}{dt} \cdot \frac{(\vec{x} - \vec{y}(\tau,t))}{|\vec{x} - \vec{y}|^3} ,
$$
(2.28)

and eq. (2.25) becomes

$$
\nabla^2 c_0' - 6g^2\bar{B}^2(c_0' + C_{0D}) = 0 .
$$
(2.29)

We recall that we have written

$$
\vec{C} = \vec{c} + \vec{C}_D ,
$$
(2.30)

in the course of solving the central potential problem. We furthermore note the identity

$$\vec{H}_{BS} = \vec{H}_s - \partial_0 \vec{C}_D - \vec{\nabla} C_{0D} \, , \tag{2.31}$$

where $\vec{H}_{BS}$ is the Biot-Savart field, which is

$$\vec{H}_{BS} = \frac{e}{4\pi} \left\{ \frac{\vec{v}_1 \times (\vec{x} - \vec{x}_1)}{|\vec{x} - \vec{x}_1|^3} - \frac{\vec{v}_2 \times (\vec{x} - \vec{x}_2)}{|\vec{x} - \vec{x}_2|^3} \right\} \, . \tag{2.32}$$

Using eq. (2.29) and eq. (2.31) and some algebra we can then show that (2.24) can be rewritten in the form

$$V_{v^2} = -\int d^3 \vec{x} \left\{ \frac{2}{3}(\vec{H}_{BS}^2 + (\partial_0 \vec{c})^2 - 2\vec{H}_{BS} \cdot \partial_0 \vec{c}) \right.$$

$$\left. - \frac{2}{3}\vec{H}_s \cdot \vec{\nabla} c_0' \right. \tag{2.33}$$

$$\left. + 4(\partial_0 B)^2 + 2(\partial_0 B_3)^2 \right\} \, .$$

Finally we use these forms, namely (2.17), (2.18) and (2.33) to explicitly compute all of the second order potentials.

3. Summary and Conclusion

The entire heavy quark potential to order $(mass)^{-2}$ can in general be written in the conventional form [12]

$$V = V_0(R) + \left(\frac{\vec{\sigma}_1 \cdot \vec{L}}{4m_1^2} + \frac{\vec{\sigma}_2 \cdot \vec{L}}{4m_2^2} \right) \frac{1}{R}[V_0'(R) + 2V_1'(R)]$$

$$+ \frac{(\vec{\sigma}_1 + \vec{\sigma}_2) \cdot \vec{L}}{4m_1 m_2} \frac{1}{R} V_2'(R)$$

$$+ \frac{1}{4m_1 m_2} \left(\frac{\vec{\sigma}_1 \cdot \vec{R}\vec{\sigma}_2 \cdot \vec{R}}{R^2} - \frac{\vec{\sigma}_1 \cdot \vec{\sigma}_2}{3} \right) V_3(R) \tag{3.1}$$

$$+ \frac{\vec{\sigma}_1 \cdot \vec{\sigma}_2}{12 m_1 m_2} V_4(R) + \frac{L^2}{4R^2} \left[\left(\frac{1}{m_1} + \frac{1}{m_2} \right)^2 V_+(R) + \left(\frac{1}{m_1} - \frac{1}{m_2} \right)^2 V_-(R) \right]$$

$$+ \frac{[\vec{R} \cdot (v_1 + v_2)]^2}{4R^2} V_\parallel(R) + \frac{[\vec{R} \cdot (v_1 - v_2)]^2}{4R^2} V_L(R) \, .$$

Within the classical approximation to dual QCD the eight potentials appearing in eq. (3.1) have been evaluated as functions of R, the quark-antiquark separation, by first solving numerically the differential equations (2.11), (2.23) and (2.29) for the functions $\vec{c} = c(\rho, z)\hat{e}_\varphi, c_0(\rho, z)$ and $c'_0(\rho, z)$ respectively and then inserting these solutions into the integrals (2.12), (2.17), (2.18) and (2.33). The potentials have an overall power of R (determined by dimensions) and are otherwise functions of $\alpha_s, g^2/\lambda$, which we fix to be 5, and the dimensionless distance $x = R\sqrt{-\lambda\tilde{F}_0^2}$. Analytic fits are made to the potentials computed in this way; we restate these here for convenience:

$$V_0(R) = -\frac{4}{3}\frac{\alpha_s}{R}e^{-.655x} + 1.0324(-\tilde{F}_0^2)R - 0.520\left(-\frac{\tilde{F}_0^2}{\lambda}\right)^{1/2}, \tag{3.2a}$$

$$V'_1(R) = \frac{4}{3}\frac{\alpha_s}{R}\sqrt{-\lambda\tilde{F}_0^2} \; 0.655(1 - e^{-0.655x}) + 1.0324\tilde{F}_0^2, \tag{3.2b}$$

$$V'_2(R) = \frac{4}{3}\frac{\alpha_s}{R}\sqrt{-\lambda\tilde{F}_0^2}\left(0.655 + \frac{e^{-0.655x}}{x}\right) \tag{3.2c}$$

$$V_3(R) = 12\pi\alpha_s(-\lambda\tilde{F}_0^2)^{3/2}(0.1018 + 0.0534x - 0.007x^2) \cdot \frac{e^{-0.360x}}{x^3} \tag{3.2d}$$

$$V_4(R) = \frac{32\pi\alpha_s}{3}\delta^3(\vec{R}) + 12\pi\alpha_s(-\lambda\tilde{F}_0^2)^{3/2}(0.0311 - 0.00361x)\frac{e^{-0.530x}}{x} \tag{3.2e}$$

$$V_+(R) = -\frac{\alpha_s}{R}(\frac{2}{3}e^{-1.458x} - 1.433x + .3419x^2) \tag{3.2f}$$

$$V_-(R) = -\frac{1}{2}V_0(R) \tag{3.2g}$$

$$V_\|(R) = V_-(R) + \frac{R}{2}\frac{\partial V_0(R)}{\partial R} \tag{3.2h}$$

$$V_L(R) = -\frac{4\alpha_s}{3R}e^{-.534x} + .091\tilde{F}_0^2 \tag{3.2i}$$

(The identities (3.2g) and (3.2h) are proved in Ref. 13.)

As to the comparison of these potentials with experiment, we first comment that $V_0(R)$ agrees extremely well with phenomenological central potentials, as shown in Fig. 1. Second we remark that the spin-orbit potentials $V'_1(R)$ and $V'_2(R)$ satisfy, to the limit of our numerical accuracy, the Gromes relation [10]

$$V'_1(R) = V'_2(R) - \frac{\partial}{\partial R}V_0(R)$$

which is derived using the general QCD expressions for V_0, V'_1 and V'_2 together with invariance under Lorentz boosts. Thirdly as seen in Fig. 1, our spin-orbit potentials are in good agreement with lattice calculations of these;[11] in particular it is worth noting that single gluon exchange predicts for $V'_1(R)$ zero while both our result and the lattice result show a nearly constant negative value.

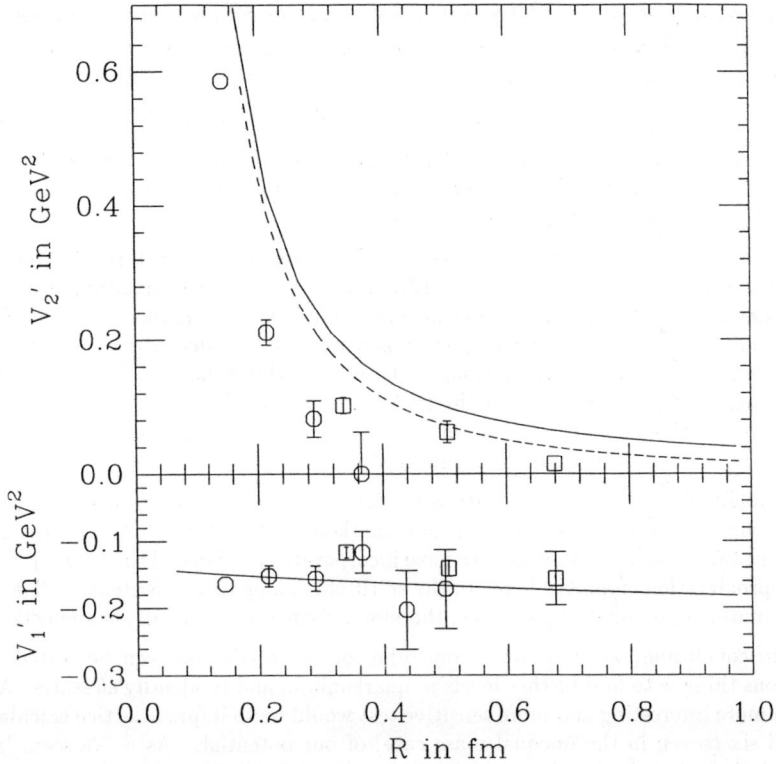

Fig. 1 Comparison of our computed spin-orbit potentials (solid lines) with lattice calculations[11] (indicated by the data points). The dotted line shows the single gluon exchange result for $V_2'(R)$; $V_1'(R)$ is zero for single gluon exchange.

To obtain the energy levels of the $c\bar{c}$ and $b\bar{b}$ systems, the analytic fits to six potentials (the V_- term vanishes for equal quark masses) are used perturbatively in the Schrödinger equation describing the central quark-antiquark interaction, and the energy levels are evaluated numerically. The results depend on the quark masses m_b and m_c, and the parameters λ, $-\tilde{F}_0^2$ and g, the dual coupling constant, related in the infrared limit to α_s by $\alpha_s = \pi/g^2$. Values for these last three parameters are roughly known a priori from our previous fits which lead us to choose $g^2/\lambda = 5$ (As remarked earlier, choosing $g^2/\lambda = 10$ or 2 changes the fits little). The other two parameters are varied together with m_b and m_c to provide a best fit to the known $c\bar{c}$ and $b\bar{b}$ levels. The resulting best fit parameters are $\lambda = 1.788, \sqrt{-\tilde{F}_0^2} = 455 MeV, m_c =$

$1.320 GeV, m_b = 4.750 GeV$. The values of λ and $\sqrt{-\tilde{F}_0^2}$ are quite compatible with the estimates obtained from vacuum energy density computed from the trace anomaly together with the value of the magnetic condensate, and the string tension. We found these parameters to be $\lambda \sim 1.61$ and $\sqrt{-\tilde{F}_0^2} \sim 420 MeV$, as mentioned in Section 1.

In the literature on phenomenological heavy quark potentials several "improvements" have sometimes been made. We next explain why we have chosen not to incorporate any of these in our potentials.

The first "improvement" is to include the so called Lüscher term.[14] This term, which consists of a contribution of $\pi^2/12R$ to $V_0(R)$, results from quantum excitations of a Lagrangian describing a string of flux joining the two quarks.[15] It is evidently valid only when the quark separation greatly exceeds the flux tube width, if then.[16] This is the case only at separations so large that the R part of $V_0(R)$ completely dominates any $1/R$ terms. What the Lüscher term turns into for smaller R, where, if it existed in the same form, it might influence the fits, is completely unknown. We, therefore, see no valid reason for including it in $V_0(R)$.

Finally, there are various relativistic corrections, configuration mixing and other phenomena sometimes included in heavy quarkonium level fitting. All of these effects (except the Thomas term, which we have incorporated) go beyond the $(mass)^{-2}$ order of approximation employed here. In the spirit of sticking to a consistent, well defined, systematic approximation, we have, therefore, chosen to ignore all such effects.

In conclusion, we may ask if any other of our predictions can be tested. One obvious thing is to find further levels in quarkonium, and to identify $c\bar{b}$ states. A perhaps more interesting and more sensitive test would be to improve lattice calculations of all six (seven in the unequal mass case) of our potentials. As we've seen, lattice calculations do exist for the two spin-orbit potentials $V_1'(R)$ and $V_2'(R)$, and these in fact agree with our predictions for these potentials rather well. We would like to see similar comparisons for the other potentials.

References

1. M. Baker, J. S. Ball, and F. Zachariasen, *Phys. Reports* **209** (1991) 73.

2. S. Maedan and T. Suzuki, *Prog. Theor. Phys.* **81** (1989) 229.

3. M. Baker, J.S. Ball and F. Zachariasen, *Phys. Rev. Lett.* **61** (1988) 521.

4. P.A.M. Dirac, *Phys. Rev.* **74** (1948) 817.

5. M. Baker, J.S. Ball and F. Zachariasen, *Phys. Rev.* **D44** (1991) 3949.

6. M. Baker, J.S. Ball and F. Zachariasen, *Phys. Lett.* **B283** (1992) 360.

7. M. Baker, J.S. Ball and F. Zachariasen, *Phys. Rev.* **D47** (1993) 3021.

8. M. Baker, J.S. Ball and F. Zachariasen, *Phys. Rev.* **D41** (1990) 2612.

9. M. Baker, J.S. Ball and F. Zachariasen, *Phys. Rev.* **D44** (1991) 3328; S. Maedan, Y. Matsubara and T. Suzuki, *Prog. Theor. Phys.* **84** (1990) 130.

10. D. Gromes, *Z. Phys.* **C26** (1984) 401.

11. A. Huntley and C. Michael, *Nucl. Phys.* **B286** (1987) 211.

12. W. Lucha, F. Schöbert and D. Gromes, *Phys. Rep.* **200** (1990) 127; E. Eichten and F. Feinberg, *Phys. Rev.* **23**(1981) 2724.

13. A. Barchielli, N. Brambilla and G. Prosperi, *Nuovo Cimento* **103** (1990) 59.

14. P. Fishbane, P. Kaus and S. Meshkov, *Phys. Rev.* **D33** (1986) 852.

15. M. Lüscher, K. Symanzik and P. Weisz, *Nucl. Phys.* **B173** (1980) 365.

16. It has never been shown that even the Abelian Higgs model turns into a string Lagrangian at large separation, though one intuitively would expect this to be so. See J. Polchinski and A. Strominger, Phys. Rev. Lett. **67** (1991) 1681.

MATRIX APPROACH TO SOLUTION OF
THE SPINLESS SALPETER EQUATION

LEWIS P. FULCHER

Physics and Astronomy, Bowling Green State University
Bowling Green, Ohio 43403 U.S.A.

ABSTRACT

A new procedure for applying the Rayleigh-Ritz Galerkin method to the solution of the spinless Salpeter equation for quark-antiquark systems is presented. Key ingredients in this new procedure are analytic expression for all of the basic matrix elements required for the calculation and computer algorithms to determine the matrix elements of the functional forms of the momentum and coordinate operators required for the potential and kinetic energy operators of the Salpeter Hamiltonian.

1. Introduction

Since the pioneering work of Stanley and Robson[1] and of Godfrey and Isgur[2], it has been clear that calculations of the properties of quark-antiquark systems must simultaneously address a number of relativistic corrections.[3] This problem is made more challenging by the fact that most of these relativistic corrections, involve nonlocal operators because of the square root functions that appear in the relativistic energy expressions. Thus, it is important to have at our disposal a variety of techniques for dealing with the nonlocal relativistic kinetic energy operator. Two of the methods most frequently used to attack this problem are to work in momentum space, where the square root becomes a straightforward algebraic operation, and to use the Rayleigh-Ritz-Galerkin method[4,5], where one expands the wave function in a complete set of basis functions. In order to work in momentum space, one must face the challenge presented by the Fourier transform of the confining potential.[6-8] A considerable amount of recent theoretical work has developed subtraction procedures for solving the singular integral equations that arise in this approach.

I want to report substantial progress in developing a new procedure to simplify the implementation of the Rayleigh-Ritz-Galerkin method for relativistic potential models[9,10]. The economies achieved with the new procedure are largely a consequence of the widespread use of analytic results for the matrix elements of the potential energy operators and the square of the momentum operator. One of the reasons for my interest in relativistic potential models is a result obtained by Gara, Durand, Durand and Nickish[11] in their reduced Bethe-Salpeter analysis of the meson spectrum. They reported that the trend of their 2S-1S separations for decreasing constituent mass was contrary to the observed results and encountered a problem with the slopes of their Regge trajectories. Their results were supported by the calculation of Lucha, Rupprecht and Schöberl[12].

My report follows that of Dr. Fred Zachariasen on an effective quark-antiquark potential based on dual QCD.[13] One of my applications will be the dual QCD potential discussed by Dr. Zachariasen. In the following talk Dr. Jim Ball will discuss calculations of the properties of heavy-quark systems with the dual QCD potential. He has also done some relativistic calculations with some of the techniques discussed here.

2. Relativistic Kinematics and the Spinless Salpeter Equation

The relativistic potential models that we discuss will be based on the spinless Salpeter equation[14]

$$H \psi_n = E_n \psi_n ; \quad H = \sqrt{m_1^2 + p^2} + \sqrt{m_2^2 + p^2} + V(r) , \tag{1}$$

where we will only consider the central potential, since all of our comparison with experiment will deal with spin averages. Equations (1) represent one of the three-dimensional reductions of the Bethe-Salpeter equation, where all of the momentum dependence that accompanies the reduction of the interaction kernels is neglected[15]. To solve Eqs. (1) we follow the method of Rayleigh, Ritz and Galerkin[4,5] and expand in terms of a complete set of basis functions. Thus

$$| \psi_n > = \sum_k C_k^n | U_k > ; \quad \sum_k H_{ik} C_k^n = E_n C_i^n . \tag{2}$$

As a prototype for potential model problems, we use the Cornell potential[16],

$$V(r) = Ar - \kappa/r . \tag{3}$$

The constants A and κ are flavor independent. Outside the heavy-quark sector additive constants will be allowed. All of the linear potential matrix elements and the Coulomb potential matrix elements required to solve the second of Eqs. (2) can be evaluated analytically with the basis chosen below. We will also show that the matrix elements of operator p^2 can be obtained analytically. Since we develop an algorithm to compute the matrix elements of the square root operators in the kinetic energy part of Eq. (1), *all of the matrix elements necessary to solve the spinless Salpeter equation for the Cornell potential can be done analytically.* Below we show that one can generalize the approach to include other static potentials.

We use the following basis functions to solve Eqs. (1),

$$U_{nl}(\mathbf{r}) = R_{nl}(r) Y_{lm}(\hat{\mathbf{r}}); \quad R_{nl}(r) = N_{nl} \beta^{3/2} (2\beta r)^l e^{-\beta r} L_n^{2l+2}(2\beta r), \quad (4)$$

where $N_{nl}^2 = 8 (n!) / \Gamma(n + 2l + 2)$, L_n^α denotes a Laguerre polynomial[17,18], and β is a factor that sets the scale of length. The basis functions of Eq. (4) satisfy an orthonormality relation. In order to evaluate the matrix elements of the potential and kinetic energy operators below, one makes frequent use of the recurrence relation and the sum relation,

$$(x - 1 - \alpha - 2n) L_n^\alpha + (n + 1) L_{n+1}^\alpha + (n + \alpha) L_{n-1}^\alpha = 0; \quad L_n^\alpha(x) = \sum_{m=0}^n L_m^{\alpha-1}(x). \quad (5)$$

How do we know that the set of functions listed in Eq. (4) is complete? One can show that the set of basis functions satisfies a second-order differential equation of the general Sturm-Liouville type[19]. It is important to appreciate that the basis functions of Eq. (4) are not those of the bound-state Coulomb problem, which do not form a complete set.

3. Matrix Elements

Using the recurrence relation of Eqs. (5) and the orthogonality relation, one can evaluate the matrix elements of the linear potential[10], that is,

$$\langle Ar \rangle_{nn'} = \frac{A}{2\beta} \left[(2l + 3 + 2n)\delta_{nn'} - \sqrt{n' (n' + 2l + 2)}\, \delta_{n', n+1} - \sqrt{n (n + 2l + 2)}\, \delta_{n', n-1} \right], \quad (6)$$

where the symmetry under the interchange of n and n' is manifest. Thus the linear potential is represented by a tridiagonal matrix.

The sum property of Eqs. (7) and orthogonality are the main ingredients required for the evaluation of the matrix elements of the Coulomb potential. One can obtain a general analytic expression[9] for the Coulomb potential matrix elements. For S states this analytic expression takes an especially simple form,

$$\langle \kappa / r \rangle_{nn'} = \kappa\beta \sqrt{\frac{(n + 1)(n + 2)}{(n' + 1)(n' + 2)}}, \quad (l = 0) \quad (7)$$

where $n \leq n'$.

Using techniques similar to those discussed above, one can derive a general analytic expression[10] for the matrix elements of the operator p^2. For S states, we have

$$(p^2)_{nn'} = \beta^2 \sqrt{\frac{(n+1)(n+2)}{(n'+1)(n'+2)}} \left[2 + \frac{4n}{3} - \delta_{nn'} \right], \quad (l = 0) \tag{8}$$

where $n \leq n'$. From Eq. (8) and its counterparts for higher values of l, it is clear that we can use analytic expressions to generate the matrix elements of the square of the kinetic energy operator $E_p^2 = m^2 + p^2$. Thus, we can obtain the matrix elements of the nonlocal kinetic energy operators of Eqs. (1), provided that we can find an algorithm to take the square roots of the squares of the single-particle kinetic energy operators there.

In order to calculate the matrix elements of the kinetic energy operator for a given constituent quark of mass m, one begins by calculating the matrix elements of the square of the kinetic energy operator. Then use a library subroutine (EVCSF in IMSL) to diagonalize the matrix representative of E_p^2. Its relationship[20] to the diagonal matrix of eigenvalues Λ is given by

$$(E_p^2)_l = U \Lambda U^{-1}, \tag{9}$$

where U is the matrix of normalized eigenfunctions. Form the square root of the matrix of eigenvalues by taking the positive square roots of the eigenvalues along the diagonal of Λ. Use the matrix of normalized eigenvalues to restore the square root operator to the original basis. Thus,

$$(E_p)_l = U \Lambda^{1/2} U^{-1}. \tag{10}$$

The algorithm that we have outlined brings to mind immediately two questions:

1. What matrix size is required to produce accurate results for the operator E_p?

2. Is it possible to understand in a simple way why the algorithm works?

The stability of the kinetic energy matrix elements for different matrix sizes is investigated in Table I, where calculations with 20 x 20, 30 x 30 and 40 x 40 matrices are compared. The differences are very slight, and the 40 x 40 result agrees with the numerical result from momentum space integrations[9]. In order to ensure a comfortable margin of error, we choose 40 x 40 matrices to carry out the kinetic energy diagonalizations for both of the constituent quarks in Eqs. (1). We give an answer to the second question below when discussing a square root algorithm for the potential energy term $r^{1/2}$.

TABLE I. Matrix elements for the kinetic energy operator for $\beta = 2.0$ GeV and m = 5.0 GeV.

Element	Matrix Size			Numerical Result
	20 x 20	30 x 30	40 x 40	
KE_{00}	5.3613	5.3613	5.3613	5.3613
KE_{10}	0.3861	0.3861	0.3861	0.3861
KE_{20}	0.2325	0.2325	0.2324	0.2324
KE_{30}	0.1540	0.1539	0.1539	0.1539
KE_{40}	0.1086	0.1085	0.1085	0.1085

4. Results for the Cornell Potential

After calculating the potential and kinetic energy matrix elements of the Hamiltonian operator in Eqs. (1), we must diagonalize the Hamiltonian to obtain the energy eigenvalues. Thus, our eigenvalue calculation requires 3 matrix diagonalizations in general, although 2 would suffice for those cases where the two constituents have equal masses. In Fig. 1 we examine the scale dependence of the lowest S state eigenvalues for the upsilon system potential. From a comparison of the full and dashed curves one can see that the acceptable range of β increases as the matrix size increases. We have also carefully investigated the appropriate size of the Hamiltonian matrix to ensure accurate eigenvalues. For most purposes a 20 x 20 Hamiltonian matrix should suffice[10].

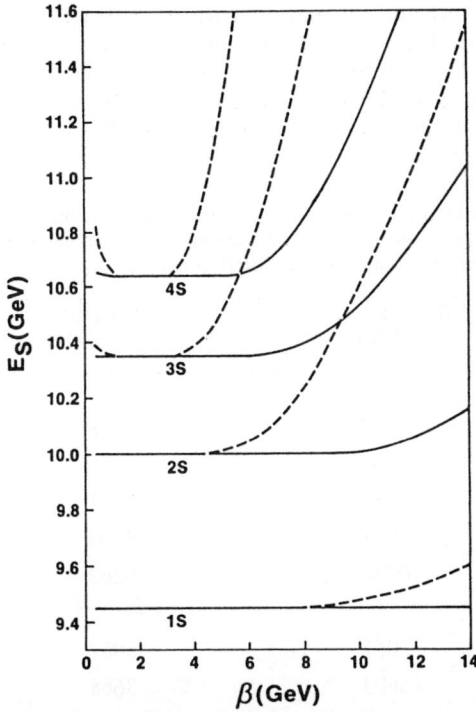

Figure 1.   Scale dependence of low-lying S eigenvalues for upsilon. The full curves were obtained with a 20x20 Hamiltonian matrix, and the dashed curves with a 10x10.

Our results for the spin-averaged energies of charmonium and the upsilon system are shown in Table 2. Since the goal of the calculation was a comparison of the Schrödinger results with the Salpeter results[10,14], in each case the parameters were adjusted to fit the measured values for the 1S, 1P and 2P states of the upsilon system and the 1S state of charmonium. The two sets of results for the upsilon system presented in Table 2 are very similar. The two sets of results are also very similar for the 4S, 5S and 6S states, which of course supports the usual claim that the upsilon system is a good example of a nonrelativistic system. Differences between the Schrödinger and Salpeter results become apparent in the charmonium system, which is reflected in the χ^2 values listed in Table 2. The differences of the χ^2 values are even more pronounced in the heavy flavor systems and the light quark systems[10]. For the Schrödinger equation, $\chi^2_{hvlt} = 16.3$ and $\chi^2_{lt} = 44.7$. For the Salpeter equation, $\chi^2_{hvlt} = 4.1$ and $\chi^2_{lt} = 1.1$. Thus, in each of the three sectors, heavy, heavy-light and light, the Salpeter results agree with experiments better than the Schrödinger, and our calculation with the Cornell potential suggests that it is possible to find a signature of the nonlocal, relativistic kinetic energy operator of Eqs. (1). The importance of documenting successes of relativistic potential models is underlined by the difficulties Lucha, Rupprecht and Schöberl[12] encountered while trying to examine the relativistic consistency of several potential models.

TABLE II.   Spin averaged-energies for heavy quarkonium systems. The results for upsilon were obtained with β = 2.0 GeV and those for charmonium with β = 1.5 GeV. Values for the parameters are listed in Ref. 10.

States	Schrödinger (MeV)	Salpeter (MeV)	Experiment (MeV)
b$\bar{\text{b}}$ states			
1S	9448	9448	9448 $\pm$ 5
2S	10007	9999	10017 $\pm$ 5
3S	10356	10351	10351 $\pm$ 5
1P	9901	9900	9900 $\pm$ 1
2P	10261	10262	10260 $\pm$ 1
c$\bar{\text{c}}$ states			
1S	3067	3067	3067 $\pm$ 2
2S	3693	3668	3663 $\pm$ 5
1P	3497	3504	3525 $\pm$ 1
χ^2	827	459	

5. Generalization to Other Potentials

Now we address the question of whether the procedure we have developed for the Cornell potential can be generalized to other static potentials. It is clear that this should be the case for powers of r since the matrix elements for these operators can be constructed from those of r in Eq. (6) by matrix multiplication. Other interesting phenomenological potentials are those of Song and Lin[21] and Lichtenberg[22], namely,

$$V_{SL}(r) = a\, r^{1/2} - b\, r^{-1/2}; \qquad V_L = a\, r^{3/4} - b\, r^{-3/4}. \tag{11}$$

We illustrate the versatility of the procedure developed for the Cornell potential by describing an algorithm developed to obtain matrix elements for $r^{1/2}$.

First generate values for the matrix elements of r from the analytic expressions of Eq. (6). Then use a library subroutine to diagonalize the matrix representative of r. Form the matrix of eigenvalues Λ', which is connected to the original representation of r by the relation,

$$r = U' \Lambda' U'^{-1} , \tag{12}$$

where U' is the matrix of normalized eigenvectors for the operator r. Take the square roots of the eigenvalues listed in Λ'. Then the matrix elements of the potential $r^{1/2}$ are given by

$$r^{1/2} = U' \Lambda'^{1/2} U'^{-1} . \tag{13}$$

The stability of the matrix elements generated with the algorithm of Eqs. (12) and (13) is examined in Table 3. Since the numbers of Table 3 show no discernible difference between the 40 x 40 and the 60 x 60 results, we conclude that the matrix elements obtained with the 40 x 40 calculation are sufficiently accurate for our purposes. This conclusion is supported by a comparison of the Salpeter eigenvalues for the $r^{1/2}$ potential obtained with the 40 x 40 matrix of Table III with those obtained with the 60 x 60 matrix there. The 1S - 4S eigenvalues and the 1P - 4P eigenvalues agree with each other to 4 decimal places, far beyond the accuracy we need for comparison with experiments.

TABLE III. Matrix elements of the potential $r^{1/2}$. The scale $\beta = 2.0$ GeV.

	40 x 40	60 x 60
$< r^{1/2} >_{00}$	0.83084	0.83084
$< r^{1/2} >_{01}$	-0.23984	-0.23984
$< r^{1/2} >_{02}$	-0.04240	-0.04240
$< r^{1/2} >_{03}$	-0.01642	-0.01642

To conclude this section, we make some comments on why the square root algorithm works, and what kinds of success we can expect in generalizing the Cornell potential procedure. Using the recurrence relation of Eqs. (5), we can see that the result of multiplying the operator r by one of the basis functions of Eq. (4) is

$$r_{op} \mid R_{nl} > = \frac{(2n + 2l + 3)}{\beta} \mid R_{nl} > - \frac{(n + 1)}{\beta} \mid R_{n+1, l} > - \frac{(n + 2l + 2)}{\beta} \mid R_{n-1, l} > , \qquad (14)$$

that is, both a rotation and a dilation of the original vector. The eigenvectors of the operator r can be expanded in terms of the original basis vectors as follows,

$$\mid k > = \sum_n b_n \mid R_{nl} > , \qquad (15)$$

The result of the operation of r on one of the eigenvectors is only a dilation (or contraction) but no rotation, and since the eigenvectors are also eigenvectors for any function of r_{op}, we have that

$$r_{op} \mid k > = r_k \mid k > ; \qquad f(r_{op}) \mid k > = f(r_k) \mid k > . \qquad (16)$$

The second of Eqs. (16) states that we now know the result of the operator $f(r)$ acting on each member of a complete set of vectors that span the Hilbert space. This amount of information should suffice to determine the action of a physically reasonable operator in the entire Hilbert space. One then uses the unitary transformation of Eq. (13) to restore the matrix elements to the original basis. Thus, we expect an algorithm like that of Eqs. (12) and (13) to be valid for a large number of different static potentials. In each case one should check the stability of the matrix elements by comparing results obtained with different sizes.

6. Dual QCD Potential

The effective quark-antiquark potential of the dual QCD model of Baker, Ball and Zachariasen[13,23] is an ambitious attempt to describe the interaction of an arbitrary quark-antiquark pair with only two potential parameters. Although these authors have succeeded in deriving the potentials to order v^2/c^2, here we will focus on the central potential, that is,

$$V_D(r) = Ar - \kappa e^{-\eta r}/r - C, \qquad (17)$$

where $\eta = 0.655 (A / 1.2323\kappa)^{1/2}$ and $C = 0.5591 (\kappa A)^{1/2}$. The Yukawa factor in Eq. (17) presents a new challenge to our matrix methods. Although a power series expansion of $e^{-\eta r}$ is a viable possibility, a more elegant procedure begins with adding and subtracting a Coulomb potential to Eq. (17). Then we need an algorithm to calculate the difference function $f(r) = (1 - e^{-\eta r}) / r$. The algorithm proceeds in a similar manner to that described in Sect. 5 for $r^{1/2}$, except that of course the function f of Eqs. (16) is the difference function f. The stability of the matrix elements for the difference function f is examined in Table IV. The results there show no discernible difference between the 40 × 40 calculation and the 60 × 60 calculation, as well as excellent agreement with analytic results for the matrix elements. It is also straightforward to obtain reliable results for the matrix elements of the potential of Eq. (17) with numerical integration[24].

TABLE IV. Matrix elements for the difference between the Yukawa function and the Coulomb potential. The potential parameters were A = 0.1885 GeV^2 and κ = 0.4913.

Matrix Elements	40 x 40	60 x 60	Analytic
< f(r) >$_{00}$	0.32086	0.32086	0.32086
< f(r) >$_{01}$	0.02293	0.02293	0.02293
< f(r) >$_{02}$	0.00180	0.00180	---------
< f(r) >$_{03}$	0.00014	0.00014	---------

Using the dual potential of Eq. (17) in the Salpeter equation, it is possible to obtain a fit to low-lying S and P state energies of the upsilon system and charmonium that is comparable to that of Table II for the Cornell potential. The parameters required to achieve this are A = 0.219 GeV^2, κ = 0.445, m_b = 4.7405 Gev and m_c = 1.3307 GeV. After the parameters of the potential have been determined in heavy quark systems, the real challenge that faces the dual potential is its test in heavy-light systems and light systems, where the relativistic kinematics of Eqs. (1) are essential. It is worth remembering that the dual QCD potential does not contain any further adjustable parameters. Of course, relativistic corrections to the potential will play an important role in these systems also. Our first attempts to address this challenge are shown in Figures 2 and 3, where the dependence of the heavy B-flavor and charmed 1S masses are shown as a function of the light quark mass. In these figures, it is apparent that the Salpeter eigenvalues are 300 or 400 MeV closer than the Schrödinger eigenvalues, however a considerable gap between experiment and theory remains. Work is in progress to see if including the leading relativistic potential corrections will close this gap.

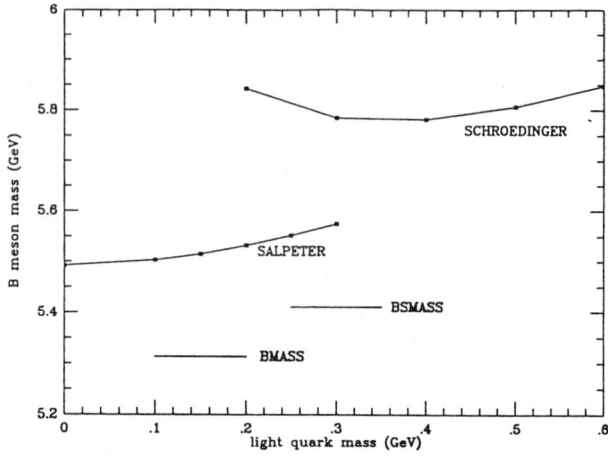

Figure 2.   The B meson mass as a function of the light quark constituent mass. The scale factor
β = 1.0 GeV.

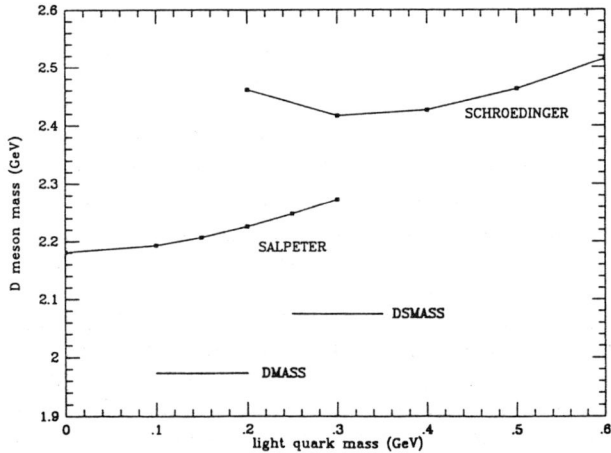

Figure 3.   The D meson mass as a function of the light quark constituent mass. The scale factor
β = 1.0 GeV.

7. Summary and Outlook

We have presented the essentials of a promising new procedure for solving the spinless Salpeter with the method of Rayleigh, Ritz and Galerkin. Since the analytic expression for all of the basic matrix elements are straightforward and computer algorithms required to compute nonlocal kinetic energy operators operate efficiently, the new procedure should be of interest to a number of researchers who work with various three-dimensional reductions of the Bethe-Salpeter equation. The next logical step in the development of the new procedure would be to address the momentum-dependent potential energy operators of equations such as Blankenbecler-Sugar[25], Kadyshevsky[26] or the equation encountered by Gara et al.[11]. In this extension a top priority would be to verify that the matrix representatives of each of the momentum-dependent operators or position-dependent operators are stable, so that their combination could be addressed as an exercise in matrix multiplication.

REFERENCES

1. D. P. Stanley and D. Robson, *Phys. Rev.* **D21** (1980) 3180.
2. S. Godfrey and N. Isgur, *Phys. Rev.* **D32** (1985) 189.
3. D. B. Lichtenberg, *Int. J. Mod. Phys.* **A2** (1987) 1669.
4. I. Stakgold, *Green's Functions and Boundary Value Problems* (Wiley, New York, 1979).
5. P. M. Morse and H. Feshbach, *Methods of Theoretical Physics* (McGraw-Hill, New York, 1953).
6. K. Maung, D. Kahana and J. Norbury, *Phys. Rev.* **D47** (1993) 1182.
7. A. J. Sommerer, J. R. Spence and J. P. Vary, *Phys. Rev.* **C49** (1994) 513.
8. F. Gross, *Phys. Rev.* **186** (1969) 1448; F. Gross and J. Milana, *ibid* **D43** (1991) 2401.
9. L. P. Fulcher, Z. Chen and K. C. Yeong, *Phys. Rev.* **D47** (1993) 4122.
10. L. P. Fulcher, *Phys. Rev.* **D50** (1994) 447.
11. A. Gara, B. Durand, L. Durand and L. J. Nickisch, *Phys. Rev.* **D40** (1989) 843; A. Gara, B. Durand and L. Durand, *ibid* **42** (1990) 1651.
12. W. Lucha, H. Rupprecht and F. Schöberl, *Phys. Rev.* **D46** (1992) 1088.
13. M. Baker, J. S. Ball and F. Zachariasen, *Phys. Rep.* **209** (1991) 73; *Phys. Rev.* **D44** (1991) 3328.
14. S. Jacobs, M. G. Olsson and C. Suchyta, *Phys. Rev.* **D33** (1986) 3338.
15. D. A. Owen, *Foundations of Physics* **24** (1994) 273; M. Halpert and D. A. Owen, *J. Phys.* **G20** (1994) 51.
16. E. Eichten *et. al.*, *Phys. Rev.* **D17** (1978) 3090; **D21** (1980) 203.
17. E. Rainville, *Special Functions* (Macmillan, New York, 1960).
18. N. Lebedev, *Special Functions and Their Applications* (Prentice-Hall, Englewood Cliffs, NJ, 1965).
19. J. Mathews and R. L. Walker, *Mathematical Methods of Physics* (Benjamin, New York, 1970).
20. L. J. Nickisch, L. Durand and B. Durand, *Phys. Rev.* **D30** (1984) 660.
21. X. Song and H. Lin, *Z. Phys.* **C34** (1987) 223.
22. D. B. Lichtenberg et al., *Z. Phys.* **C41** (1989) 615.
23. M. Baker, J. S. Ball and F. Zachariasen, CALT-68-1887 preprint (1994).
24. J. S. Ball, private communication.
25. R. Blankenbecler and R. Sugar, *Phys. Rev.* **142** (1966) 1051.
26. V. G. Kadyshevsky, *Nucl. Phys.* **B6** (1968) 125.

QUARKONIUM MASSES DETERMINED BY
THE DUAL QCD EFFECTIVE QUARK-ANTIQUARK POTENTIAL

JAMES S. BALL

Physics Department, University of Utah
Salt Lake City, Utah 84112, USA

ABSTRACT

An effective potential to order (quark mass)$^{-2}$ for a constituent quark and anti-quark has been derived from Dual QCD. The resulting potentials - central, spin-spin, spin-orbit and spinless velocity dependent are completely determined in terms of two parameters. Two additional parameters, the two quark masses, are required to calculate $q\bar{q}$ energy levels. We have made a "best" fit to all of the observed $c\bar{c}$ and $b\bar{b}$ bound states using the Schrödinger equation and the Salpeter equation which incorporates some relativistic effects. Once the four parameters are determined we can then predict the masses of a number of states that have not yet been observed. The quality of the fit obtained with the Schrödinger equation is very good, fitting the 17 observed states with an average error of .36%. Finally, we find that the Schrödinger fit is improved by using a running coupling constant at small R (one new parameter). Using the Salpeter equation to treat the quark kinematics relativistically requires some modification of the methods used to calculate the s-wave mass splitings. This is because the wave functions are singular at $R = 0$ and δ-function that appears in our hyperfine (spin-spin) potential requires a more sophisticated treatment. Examining the origin of the δ-function, which arises when one takes the nonrelativistic limit of the Dirac equation, provides a prescription for the hyperfine calculation in relativistic wave equations. When the Salpeter equation is used to fit the spectrum the results are significantly poorer than the Schrödinger fit. Finally, using our central potential in the Dirac equation for the heavy-light quark systems of $D - D^*$ and $B - B^*$ we obtain encouraging results.

1. Introduction

This talk will be organized as follows: The next section will discuss the potential obtained from dual QCD as described in a previous talk at this conference (F. Zachariasen) and it will present empirical formulas suitable for use in the Schrödinger equation or its relativistic generalizations. We will also consider a simple form which incorporates a coupling constant that runs at small R (dual QCD should not be applicable to this region).

The next section will describe our fit to the energy levels of the known $c\bar{c}$ and $b\bar{b}$ states and predictions for the as yet unobserved states. We also predict the masses of the $^1S_0$ and $^3S_1$ states in $c\bar{s}$ and $b\bar{s}$ systems, though, because the s quark is very light, we do not expect our predictions to be very reliable.

Section 4 will discuss the method of solution of the Salpeter equation and problems associated with the behavior of the wave functions at the origin. Our proposed resolution of these problems is then used to fit the existing data.

The final section will describe our preliminary investigations of heavy-light systems employing our potential in the Dirac equation.

2. Dual QCD Potentials

The entire heavy quark potential to order $(mass)^{-2}$ can in general be written in the form[1]

$$
\begin{aligned}
V = \ & V_0(R) + \left(\frac{\vec{\sigma}_1 \cdot \vec{L}}{4m_1^2} + \frac{\vec{\sigma}_2 \cdot \vec{L}}{4m_2^2}\right) \frac{1}{R}[V_0'(R) + 2V_1'(R)] + \frac{(\vec{\sigma}_1 + \vec{\sigma}_2) \cdot \vec{L}}{4m_1 m_2} \frac{1}{R} V_2'(R) \\
& + \frac{1}{4m_1 m_2}\left(\frac{\vec{\sigma}_1 \cdot \vec{R}\vec{\sigma}_2 \cdot \vec{R}}{R^2} - \frac{\vec{\sigma}_1 \cdot \vec{\sigma}_2}{3}\right) V_3(R) + \frac{\vec{\sigma}_1 \cdot \vec{\sigma}_2}{12m_1 m_2} V_4(R) \\
& + \frac{L^2}{4R^2}\left[\left(\frac{1}{m_1} + \frac{1}{m_2}\right)^2 V_+(R) + \left(\frac{1}{m_1} - \frac{1}{m_2}\right)^2 V_-(R)\right] \\
& + \frac{[\vec{R} \cdot (\vec{v}_1 + \vec{v}_2)]^2}{4R^2} V_{\parallel}(R) + \frac{[\vec{R} \cdot (\vec{v}_1 - \vec{v}_2)]^2}{4R^2}\left(-\frac{4\alpha_s}{3R}\right) \ .
\end{aligned}
\tag{1}
$$

Within the classical approximation to dual QCD, we[2] have evaluated the seven potentials appearing in eq. (1) as functions of R, the quark-antiquark separation, by first evaluating them numerically from dual QCD differential equations and then by constructing analytic fits to these data. The potentials have an overall power of R (determined by dimensions) and are otherwise functions of λ and $\tilde{F}_0^2$. We have fixed $g^2/\lambda = 5$ and as a result $\alpha_s = \frac{\pi}{5\lambda}$. Analytic fits are

$$
V_0(R) = -\frac{4}{3}\frac{\alpha_s}{R}e^{-.655x} + 1.0324(-\tilde{F}_0^2)R - 0.520\left(-\frac{\tilde{F}_0^2}{\lambda}\right)^{1/2},
\tag{2}
$$

$$
V_1'(R) = \frac{4}{3}\frac{\alpha_s}{R}\sqrt{-\lambda\tilde{F}_0^2}\ 0.655(1 - e^{-0.655x}) + 1.0324\tilde{F}_0^2,
\tag{3}
$$

$$
V_2'(R) = \frac{4}{3}\frac{\alpha_s}{R}\sqrt{-\lambda\tilde{F}_0^2}\left(0.655 + \frac{e^{-0.655x}}{x}\right),
\tag{4}
$$

$$
V_3(R) = 12\pi\alpha_s(-\lambda\tilde{F}_0^2)^{3/2}(0.1018x + 0.0534x - 0.007x^3) \cdot \frac{e^{-0.360x}}{x^3},
\tag{5}
$$

$$
V_4(R) = \frac{32\pi\alpha_s}{3}\delta^3(\vec{R}) + 12\pi\alpha_s(-\lambda\tilde{F}_0^2)^{3/2}(0.0311 - 0.00361x)\frac{e^{-0.530x}}{x},
\tag{6}
$$

$$
V_+(R) = -\frac{\alpha_s}{R}(\frac{2}{3}e^{-1.458x} - 1.433x + 0.342x^2),
\tag{7}
$$

$$
V_-(R) = -\frac{1}{2}V_0(R),
\tag{8}
$$

$$
V_{\parallel}(R) = V_-(R) + \frac{R}{2}\frac{\partial V_0(R)}{\partial R},
\tag{9}
$$

where the dimensionless distance $x = R\sqrt{-\lambda\tilde{F}_0^2}$.

The relativistic invariance of the Lagrangian density imposes several exact relations between the various terms in Eq.(1). The first is the Gromes' relation[3] for the spin-orbit potentials. This requires that

$$V_1'(R) = V_2'(R) - \frac{\partial V_0(R)}{\partial R} .$$ (10)

Our numerical results agree with this to the accuracy of our numerical calculation, and the analytic fits to V_1', V_2' and V_0 satisfy it exactly. Two additional relations for V_- and $V_\parallel$ have been derived by Barchielli, Brambilla and Prosperi[4]. These are

$$V_- = -\frac{1}{2}V_0$$ (11)

and

$$V_\parallel = V_- + \frac{R}{2}\frac{\partial V_0}{\partial R} .$$ (12)

These too are satisfied by our numerical results although $V_\parallel$ is rather small in magnitude and is therefore not as accurately determined as our other potentials. It should also be noted that neither of these potentials are necessary for our quarkonium fits with equal mass quarks.

Another quark-antiquark potential that we have calculated via dual QCD is the potential associated with the creation of a pair within a flux tube and the subsequent separation of the pair to form closed ends on each of the divided flux tubes. No fields remain in the original region of pair creation. In this calculation one considers a fixed length L of an infinitely long flux tube. Initially, the energy is simply σL. When the quarks have been created but are close together, they interact with a Coulomb-like potential with all of the quark flux absorbed by the antiquark. The flux-tube remains nearly unchanged. As the pair separates further, an increasing amount of quark flux is absorbed by the flux tube in which it resides and less flux passes through the region between the pair. This is the result of the screening properties of the QCD vacuum state. At large separations the only change in the energy is due to the decreasing amount of flux tubes remaining in the length L. If we define the potential $V_b(R)$ as the change in energy in going from the initial state to a configuration in which the quarks are separated a distance R, we find that it is very well fit by the following expression:

$$V_b(R) = -\frac{4}{3}\frac{\alpha_s}{R}e^{-.655x} - 1.0324(-\tilde{F}_0^2)R.$$ (13)

This is essentially $V_0(R)$ with the opposite sign for the string tension because the flux tube length decreases with increasing R. From this potential and knowing the distribution of color fields in a flux-tube, it should be possible to calculate the probability of pair production per unit length in a long flux tube.

3. Fits and Predictions for $c\bar{c}$ and $b\bar{b}$ Energy Levels

The procedure for obtaining a best fit to the energy levels of the known $c\bar{c}$ and $b\bar{b}$

states is as follows. We define an effective χ^2 to be:

$$\chi^2 = \sum \left(\frac{(experiment - theory)}{(.01 \times experiment)} \right)^2 . \tag{14}$$

This would be the actual χ^2 if the experimental statistical error was in fact 1% or equivalently what might be expected to be equal to the number of degrees of freedom if the theory was good to 1%. Our four parameters, λ, $\sqrt{-\tilde{F}_0^2}$, and the two-quark masses m_c and m_b are then varied to minimize the effective χ^2. Our procedure is the following: Using our central potential we solve the Schrödinger equation to determine the eigenvalues and the wave functions for the necessary orbital angular momentum states. The spin and angular momentum dependent potentials are then used perturbatively to calculate the energies of the individual states and the χ^2 is evaluated. The four parameters are then varied to minimize χ^2. It should be emphasized that these are our only parameters and the dependence of the potentials on these parameters is completely determined by dual QCD.

Once the best fit parameters are determined we can predict the unobserved energy levels. Our best fit to the seventeen observed states is given in Table 1. The resulting χ^2 is 1.70, corresponding to an average least-square error of .36%.

Table 1. Schrödinger equation fit masses of all observed $c\bar{c}$ and $b\bar{b}$ states below threshold[5]. Parameters are $\lambda = 1.770$, $\sqrt{-\tilde{F}_0^2} = 455$MeV, $m_c = 1.320$GeV and $m_b = 4.750$GeV.

State	Pred. Mass (GeV)	Exper. Mass (GeV)
$\eta_c(1S)$	2.969	2.980
$\psi(1S)$	3.114	3.097
$\psi(2S)$	3.687	3.686
$\chi_{c_0}(1P)$	3.440	3.415
$\chi_{c_1}(1P)$	3.502	3.511
$\chi_{c_2}(1P)$	3.542	3.556
$h_c(1P)$	3.514	3.526
$\Upsilon(1S)$	9.464	9.460
$\Upsilon(2S)$	9.994	10.023
$\Upsilon(3S)$	10.344	10.355
$\Upsilon(4S)$	10.635	10.580
$\chi_{b_0}(1P)$	9.861	9.860
$\chi_{b_1}(1P)$	9.891	9.892
$\chi_{b_2}(1P)$	9.914	9.913
$\chi_{b_0}(2P)$	10.223	10.232
$\chi_{b_1}(2P)$	10.248	10.255
$\chi_{b_2}(2P)$	10.268	10.268

The predicted energy levels for the as yet unobserved (nearly) stable states of these systems are shown in Table 2.

Table 2. Predicted masses of the unobserved $c\bar{c}$ and $b\bar{b}$ states

State	Pred. Mass (GeV)
$\eta_c(2S)$	3.581
$(^1D_2)c\bar{c}\ (n = 1)$	3.840
$(^3D_1)c\bar{c}\ (n = 1)$	3.830
$\eta_b(1S)$	9.330
$\eta_b(2S)$	9.931
$\eta_b(3S)$	10.294
$\eta_b(4S)$	10.591
$(^1P_1)b\bar{b}\ (n = 1)$	9.900
$(^1P_1)b\bar{b}\ (n = 2)$	10.256
$(^1D_2)b\bar{b}\ (n = 1)$	10.153
$(^1D_2)b\bar{b}\ (n = 2)$	10.456
$\chi_{b_0}(3P)$	10.519
$\chi_{b_1}(3P)$	10.542
$\chi_{b_2}(3P)$	10.561
$(^3D_1)b\bar{b}\ (n = 1)$	10.144
$(^3D_2)b\bar{b}\ (n = 1)$	10.152
$(^3D_3)b\bar{b}\ (n = 1)$	10.159
$(^3D_1)b\bar{b}\ (n = 2)$	10.447
$(^3D_2)b\bar{b}\ (n = 2)$	10.455
$(^3D_3)b\bar{b}\ (n = 2)$	10.462

Finally, we have also computed the $^1S_0$ and $^3S_1$ states of the $c\bar{s}$ and $b\bar{s}$ systems. (These are known as D_s, D_s^*, B_s and B_s^* respectively). Our results for these fits are shown in Table 3, and use a value $m_s = 430\text{MeV}$.

Table 3. Predicted masses of observed $c\bar{s}$ and $b\bar{s}$ states[5]

	$c\bar{s}$		$b\bar{s}$	
	Theory	Expt.	Theory	Expt.
$^1S_0$	2.385	1.969	5.820	$5.359 \to 5.409$
$^3S_1$	2.542	2.536	5.877	$5.406 \to 5.456$

This mass is very low, and it is not surprising that the fit in this case is much poorer than for the $c\bar{c}$ and $b\bar{b}$ systems.

The dual QCD region of applicability is in the IR or long range behavior and therefore provides no information about short range effects. From what is known about QCD, one would certainly expect the coupling constant to run at very short distances reflecting the fact that the theory is assymptotically free. We have investigated what such a modification might do to our fits to the data by using a modified formula for the coupling constant for the central potential.

$$\alpha_s(R) = \begin{cases} \alpha_s & R > R_0 \\ \frac{\alpha_s}{1+\frac{R-R_0}{R_0}\beta\ln(R/R_0)} & R < R_0 \end{cases} \tag{15}$$

β was calculated from the β-function for QCD for no flavors and R_0 was arbitrarily fixed at $1/\text{GeV}$. The resulting χ^2, obtained from fitting the data with this modification, was 1.2, certainly a noticeable improvement over our fixed α_s results. This shows that the data certainly favors a running coupling constant.

4. Fits to the $c\bar{c}$ and $b\bar{b}$ Energy Levels Using the Salpeter Equation

There are several important reasons for seeking a relativistic treatment of quark-antiquark boundstate. First of all, for quarks moving in a coulomb potential $v/c \simeq \alpha$ where α is the coefficient of the $1/R$ singularity, this number is about .5 for our potential, making even the $b\bar{b}$ states quite relativistic. This certainly causes some doubt as to the reliability of calculations that employ the Schrödinger equation, and quantities that depend on short range behavior are particularly suspect. A second reason is the need to treat lower mass quarks where relativistic effects will be even more important.

The Salpeter equation

$$(2\sqrt{p^2 + m^2} + V)\Psi = E\Psi \tag{16}$$

represents a simple generalization to treat the quark kinematics relativistically. We have used the method proposed by Fulcher[6] (described in the previous talk), employing a finite number of basis functions to obtain a matrix representation of the operator in the left of Eq.(16). The next step in the calculation of quarkonium energy levels is the perturbative calculation of the splitting produced by the potentials given in Eqs.(3-9). Because of the δ-functions that appear in V_4 and in the last term in Eq.(1) this is not a trivial step.

It has been shown by Durand et al.[7] that the Salpeter wave function behaves like $1/R^{1-\gamma}$ for a potential that behaves like α/R near the origin.

$$\gamma = \frac{\alpha}{2}\tan(\frac{1}{2}\pi\gamma) \tag{17}$$

Note that as $\alpha \to 4/\pi$, γ goes to zero. For larger values of α, the solution of Eq.(17) shifts to another branch of the tangent and no normalizable wave function exists. Fortunately, our value of $\frac{4}{3}\alpha_s$ is about .5 so that a solution exists but $\gamma \simeq .8$. If one uses a running coupling constant to weaken the singularity at the origin, the power behavior disappears, but Durand[8] has shown that a logarithmic singularity remains and $\Psi(0)$ is still infinite.

We have developed a method to regularized the δ-function terms. Recall that the δ-function in V_4 is the result of taking the nonrelativistic limit of the Dirac equation in which the vector potential of the nucleus is treated as a perturbation. The hyperfine matrix element is

$$\Delta E \sim \int d^3 r \bar{\Psi} \alpha \cdot \vec{A} \Psi \tag{18}$$

where

$$\vec{A} = \frac{\vec{\mu} \times \vec{r}}{r^3} \tag{19}$$

where $\vec{\mu}$ is the nuclear magnetic moment. Writing this matrix element in terms of the large g and small f components of the spin we obtain:

$$\Delta E \sim \int dr f(r) g(r). \tag{20}$$

Note that this is finite even though, like the Salpeter wave function, the Dirac wave functions are singular at the origin. If we eliminate the small component, we find

$$\Delta E \sim -\int dr \frac{2m}{(E + m - V)} g(r) \frac{dg}{dr}, \tag{21}$$

where V is the Coulomb potential. In this form it is clear that V in the denominator is regulating the singularity at the origin. Taking the nonrelativistic limit we obtain

$$\Delta E \sim -\int dr g(r) \frac{dg}{dr} = \frac{1}{2} g(0)^2. \tag{22}$$

This would be a problem if it were not for the fact that g now is the Schrödinger wave function which is well behaved at the origin. This is exactly the same term that the δ-function potential produces. Evidently the correct replacement for the δ-function for a relativistic wave function is

$$\delta^3(\vec{r}) \rightarrow -\frac{4m}{(E + m - V)} \Psi(r) \frac{d\Psi}{dr}. \tag{23}$$

This prescription has the further advantage that no new parameters have been introduced.

The fit to the data using the Salpeter eq. and the potential with a running coupling is shown in Table 4.

Table 4. Salpeter equation fit masses of all observed $c\bar{c}$ and $b\bar{b}$ states below threshold[5]. Parameters are $\lambda = 1.642$, $\sqrt{-\tilde{F}_0^2} = 493\text{MeV}$, $m_c = 1.255\text{GeV}$, and $m_b = 4.662\text{GeV}$. $R_0 = 1.0$ fixed

State	Pred. Mass (GeV)	Exper. Mass (GeV)
$\eta_c(1S)$	2.972	2.980
$\psi(1S)$	3.106	3.097
$\psi(2S)$	3.661	3.686
$\chi_{c_0}(1P)$	3.445	3.415
$\chi_{c_1}(1P)$	3.511	3.511
$\chi_{c_2}(1P)$	3.552	3.556
$h_c(1P)$	3.522	3.526
$\Upsilon(1S)$	9.518	9.460
$\Upsilon(2S)$	10.005	10.023
$\Upsilon(3S)$	10.361	10.355
$\Upsilon(4S)$	10.665	10.580
$\chi_{b_0}(1P)$	9.843	9.860
$\chi_{b_1}(1P)$	9.864	9.892
$\chi_{b_2}(1P)$	9.880	9.913
$\chi_{b_0}(2P)$	10.216	10.232
$\chi_{b_1}(2P)$	10.237	10.255
$\chi_{b_2}(2P)$	10.252	10.268

The value of χ^2 was 2.7, significantly worse that our earlier results ($\chi^2 = 1.2$). Qualitatively, the Salpeter fit was somewhat better for the $c-\bar{c}$ states and considerably worse for the $b - \bar{b}$ states.

Aside from the fact that the Salpeter eq. doesn't help the fitting of the data, it has several other defects. First of all, it only applies to scalar particles and provides no guidance about how spin should be introduced. Secondly, we have simply assumed that our potential is the fourth component of a four-vector. If part of our V_0 was a scalar rather than a vector (indistinguishable for the Schrödinger eq. and in our derivation of the potential) it should have been added to m in Eq. (16) with only the vector term in V. Finally, since the Salpeter eq. doesn't come from any correct relativistic equation it provides no guidance as to recoil corrections to the potential such as factors of m/E.

5. Dirac Equation

In contrast to the Salpeter eq., the Dirac eq. does provide a correct relativistic treatment for atomic systems. Although the mass ratios are not as large for heavy-light systems, they probably are large enough to keep the heavy quark in the non-relativistic regime. In a recent paper, Mur et al[9] give theoretical arguments for the use of the Dirac equation to describe heavy-light quark-antiquark systems in QCD. We begin with the investigation states of a light quark bound to a c or b quark, and determine the eigenvalues and eigenvectors of the Dirac eq. in the limit that the heavy mass is infinite. The hyperfine splitting will then be calculated perturbatively, using the heavy quark mass obtained in our Schrödinger fit. The radial Dirac eq. in matrix form for the small and large components

$$
\begin{pmatrix} V_v + m + V_s & -\frac{d}{dr} - \frac{l+1}{r} \\ \frac{d}{dr} - \frac{l}{r} & V_v - m - V_s \end{pmatrix} \begin{pmatrix} g \\ f \end{pmatrix} = E \begin{pmatrix} g \\ f \end{pmatrix}
\tag{24}
$$

where V_s is the scalar potential and V_v is the 4th component of a 4-vector. The conventional view is that the confining potential is a scalar and that the Coulomb term is a vector. For our potentials the fact that the string tension appears in V_1' tends to support this division, although in this application we really have no choice because there is no solution to the Dirac eq. for a linear potential in V_v.

Each of the four matrices on the left of Eq.(24) is expressed as an NxN matrix using the same basis function as in the solution of the SP equation. The resulting eigenvalues and eigenvectors are then used to calculate the hyperfine splitting. For a pure Coulomb potential ($\alpha = .5$ and m=1) using a 20x20 representation for the individual matrices, we obtained .866028 for the groundstate energy to be compared to .866025 for the exact result. For the linear potential we compared our results with those of Mur et al[9] and found excellent agreement.

In the calculation of the masses of the heavy-quark light-quark systems, we first adjusted the light-quark mass (the only free parameter in this calculation) to give the correct center of mass of the s-wave states and then calculated the splitting using this mass and the heavy-quark mass.

The our results are given in Table 5.

Table 5. Predicted masses of B, B^*, D and D^* mesons[5]. m_u is the light quark mass.

state	m_u=84MeV Theory	Expt.	state	m_u=251MeV Theory	Expt.
B	5.277	5.279	D	1.814	1.869
B^*	5.325	5.325	D^*	2.029	2.010

The results are very good except for the fact that one would like the same light quark mass for both the $D - D^*$ system and the $B - B^*$ system. Including a nonrelativistic kenetic energy term for the heavy quark is a simple process and might both improve the fit to the D and D^* masses and change the required light-quark mass.

6. References

1. W. Lucha, F. Schöberl and D. Gromes,*Phys. Rep.* **200** (1990) 127; E. Eichten and F. Feinberg,*Phys. Rev.*23 (1981) 2724.
2. M. Baker, J.S. Ball and F. Zachariasen,*Phys. Rev.* **D44** (1991) 3949; M. Baker, J.S. Ball and F. Zachariasen,*Phys. Lett.* **B283** (1992) 360; M. Baker, J.S. Ball and F. Zachariasen,*Phys. Rev.* **D47** (1993) 3021.

3. D. Gromes, *Z. Phys.* **C26** (1984) 401.
4. A. Barchielli,N. Brambilla and G. M. Prosperi, *Nuovo Cimento***103** (1990) 59
5. Particle Data Group, *Phys. Rev.***D50** (1994) 1173
6. L. P. Fulcher,Z. Chen and K. C. Yeong, *Phys. Rev.* **D47** (1993) 4122.
7. L. J. Nickisch, L. Durand and B. Durand, *Phys. Rev.* **D30** (1984) 660.
8. L. Durand, *Phys. Rev.* **D32** (1985) 1257.
9. V. D. Mur, V. S. Popov, Yu. A. Simonov and V. P. Yurov,*J. Expt. Theor. Phys.* **78** (1994) 1. (HEP-PH-9401203)

FLUX TUBES AND HEAVY QUARK SYMMETRY

M. G. OLSSON

Physics Department, University of Wisconsin, 1150 University Ave.
Madison, WI 53706, USA

ABSTRACT

The relativistic flux tube model has achieved considerable success in providing a physical picture of the dynamic confinement of meson states. We discuss here the extension of the model to unequal quark masses, and some evidence for the running of heavy quark masses. Predictions for the Isgur-Wise function agree very well with recent experimental results from semi-leptonic B decay.

1. Introduction

Progress in understanding the $q\bar{q}$ color interaction has been rapid in recent years. Both numerical lattice simulations and explicit dynamical pictures, such as monopole condensation models and Wilson loop formulations, clearly indicate the presence of a flux tube field configuration for large quark separations. The relativistic flux tube (RFT) model begins with this picture and seeks to establish detailed predictions for meson dynamics. We first establish some basic facts concerning confinement models. Next some recent technical progress is described concerning mesons with quarks of finite but unequal mass. Finally, we explore predictions for the Isgur-Wise function describing semi-leptonic heavy quark decay. Recent high quality experimental data allows a detailed test of the RFT predictions for this form factor.

The general shape of the static potential is now well established. In Fig. 1 we show a recent numerical lattice simulation by the Hamburg group[1]. A fit to the potential of the form

$$V(r) = -\frac{\kappa}{r} + ar + C \qquad (1)$$

represents the "data" quite well, where a is the tube tension. The fit implies that C is large and negative and that $\kappa \simeq 0.25$. We will return later to the significance of C. The value of κ is consistent with the Lüscher term[2] of $\kappa = \pi/12 \simeq 0.26$ and is typical of lattice results in the quenched approximation. The phenomenological result for κ, required to account for heavy onia spectroscopy, is about a factor of two larger. It may be that the discrepancy can be accounted for by light quark loops in the confining interaction[3].

Over ten years ago a beautiful qualitative picture of dynamic confinement was proposed by Buchmüller[4] to explain the spin dependence of the long range interaction. He observed that with a chromoelectric flux tube there can be no coupling of the color magnetic moments and hence the only possible spin dependence arises from the kinematic "Thomas" spin-orbit term. At the time this picture was widely considered

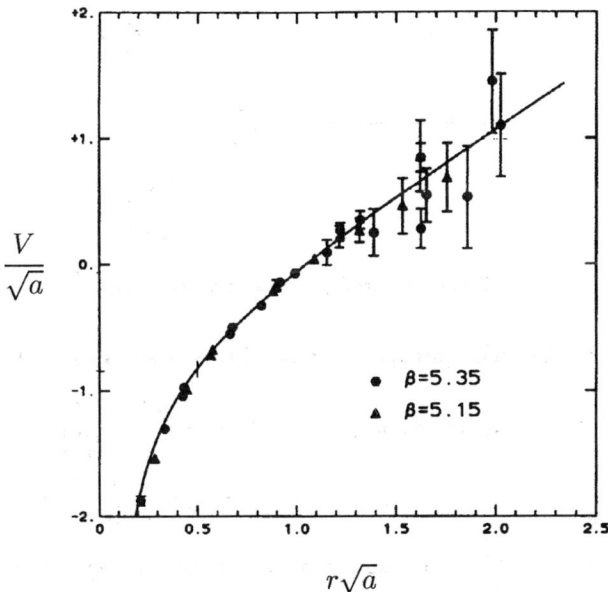

Fig. 1. Lattice QCD calculation of the static interquark potential. The energy and quark separation are scaled to dimensionless quantities by the tension a.

as justification of the concept of "scalar confinement" but as we will soon establish, the tube model itself should have been taken more seriously.

The technique of the Wilson loop expansion for slowly moving quarks has been valuable in understanding the implications of QCD for hadron states. This program originally addressed only spin-dependence[5] but more recently has been used to extract rigorous spin-independent relativistic corrections[6]. The results were surprising in that the spin-independent corrections were nothing like those expected from scalar confinement[7,8].

The RFT model on the other hand accurately accounts for all of the relativistic corrections[8-10]. The spin-independent corrections for large angular momentum have their origin in the angular momentum of the flux tube about the center of momentum of the heavy quarks at its ends[8]. The RFT model is thus intrinsically differs from the potential model in that the momentum of the interaction (tube) is critical.

Most scalar potential models of confinement also suffer from a catastrophic behavior first pointed out by Gara et al.[11]. When the quark motion becomes relativistic a cancellation occurs destroying linear Regge behavior—a universal observation for light quark states. To see in essence how this arises consider a fermion interacting with a scalar field as shown in Fig. 2.

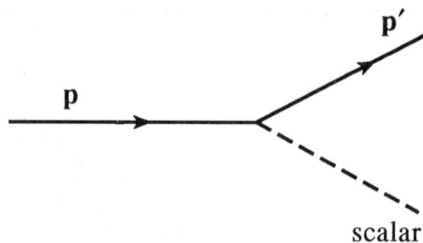

Fig. 2. Fermion interacting with an exchanged scalar field

The vertex is described (dropping numerical factors and spinors) by

$$
\bar{u}(\mathbf{p}',\tau')\,\mathbf{1}\,u(\mathbf{p},\tau)\,K_x(\mathbf{p}',\mathbf{p}) \;\simeq\; \left(1,\; \frac{-\boldsymbol{\sigma}\cdot\mathbf{p}'}{E'+m}\right)\left(\frac{1}{\frac{-\boldsymbol{\sigma}\cdot\mathbf{p}}{E+m}}\right)K_s
$$
$$
= \left(1 - \frac{\boldsymbol{\sigma}\cdot\mathbf{p}'\,\boldsymbol{\sigma}\cdot\mathbf{p}}{(E'+m)(e+m)}\right)K_s \tag{2}
$$

where $K_s = (\mathbf{p}' - \mathbf{p})^{-4}$ is the usual confinement kernel. Spin averaging and for relativistic quarks we obtain

$$
\bar{u}\,\mathbf{1}\,u K_s \rightarrow \frac{(1 - \hat{\mathbf{p}}'\cdot\hat{\mathbf{p}})}{|\mathbf{p}' - \mathbf{p}|^4}\;. \tag{3}
$$

For collinear quarks the confining kernel is singular but the numerator vanishes in the collinear case which spoils the linear Regge behavior.

The RFT model avoids this problem neatly[9]. When the quarks become relativistic the tube begins to dominate the meson rotational dynamics and the Nambu-Goto string emerges. Normal Regge trajectories are thus assured.

2. Asymmetric RFT's

The RFT model can be formulated in a covariant and gauge independent manner in terms of the Wilson action[12]. However, the most straightforward way to obtain the governing equations with spinless quarks[8,9] is to read off the relevant kinematical quantities from Fig. 3. The angular momentum and energy equations are[8,13]

$$
J = W_{r1}\gamma_{\perp 1}v_{\perp 1}r_1 + 2ar_1^2 f(v_{\perp 1}) + (1 \rightarrow 2) \tag{4}
$$
$$
H = W_{r1}\gamma_{\perp 1} + ar_1\frac{\arcsin(v_{\perp 1})}{v_{\perp 1}} + (1 \rightarrow 2) \tag{5}
$$

where $W_{r1} = \sqrt{p_{r1}^2 + m_1^2}$, $\gamma_\perp = (1 - v_\perp^2)^{-1/2}$, and $4v_\perp f(v_\perp) = \left(\frac{\arcsin v_\perp}{v_\perp} - \gamma_\perp^{-1}\right)$. For an asymmetric meson we must be careful to keep the center of momentum at rest.

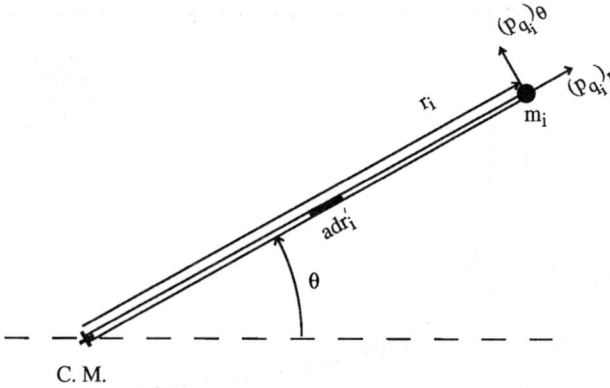

Fig. 3. One segment of the quark/flux tube, extending from the center of momentum to one quark.

This is done by the conditions

$$p_{r1} = p_{r2} = p_r , \qquad (6)$$

$$P_\perp = 0 = W_{r1}\gamma_{\perp 1}v_{\perp 1} + \frac{ar_1}{v_{\perp 1}}\left(1 - \gamma_{\perp 1}^{-1}\right) - (1 \rightarrow 2) . \qquad (7)$$

Finally we impose the straight flux tube condition

$$\frac{v_{\perp 1}}{r_1} = \frac{v_{\perp 2}}{r_2} . \qquad (8)$$

2.1. The Classical Solution:

An analytic classical solution for circular motion can be found[13] for the equations (4–8) including the condition that the CM remain fixed. For circular motion $p_r = 0$ and hence $W_r = m$. The $P_\perp = 0$ condition then follows from the condition of radial force balance between the quarks[13]. The circular classical solution minimizes the energy for a given angular momentum. An example of the classical solution is shown by the solid curve of Fig. 4 when $m_1 = 0$ and $m_2 = 1.5$ Gev. The figure illustrates the transition of the Regge slope, with energy of the Light Degrees of Freedom (LDF), from the heavy-light regime at small J to the relativistic light-light case at large J. The slope decreases from two to unity in Nambu slope units of $(2\pi a)^{-1}$. The classical solution is a valuable check on our quantum solution.

2.2. The Quantized Solution:

The RFT equations (5–8) can also be quantized. To explain the method we first consider for simplicity the equal mass case. For $m_1 = m_2 = m$ the RFT equations

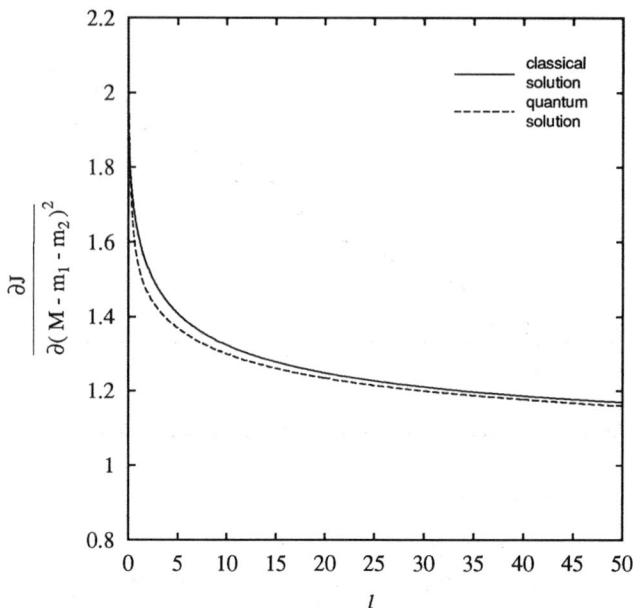

Fig. 4. Classical and quantized solutions for the LDF Regge slope as a function of angular momentum. Here $m_1 = 0$ and $m_2 = 1.5$ GeV and the transition from $h\ell$ to $\ell\ell$ dynamics is evident.

schematically become

$$J(W_r, r_1, v_\perp)\psi = \sqrt{\ell(\ell+1)}\,\psi\,, \tag{9}$$

$$H(W_r, r_1, v_\perp)\psi = M\psi\,. \tag{10}$$

In the above we have expressed the angular momentum and energy equations as eigenvalue equations and in coordinate representation the radial momentum square is taken as

$$p_r^2\psi = -\frac{1}{r}\frac{d^2}{dr^2}(r\psi)\,. \tag{11}$$

If it were possible to eliminate the operator $v_\perp$ between the J and H equations the solution would be straightforward. Unfortunately this is not possible and so we are forced to develop methods to numerically achieve the same end. Our method[9], very briefly, begins by expanding the wavefunction in a complete set of basis states and truncating the infinite series at N basis states. The J and H equations then become finite matrix equations. We solve the J equation for the $v_\perp$ matrix which is subsequently used to eliminate $v_\perp$ dependence in the H equation. We then diagonalize H to obtain the meson mass eigenstates and the corresponding wavefunctions. We confine ourselves to Hermitian (i.e., symmetric) representations of the operator $v_\perp$

and symmetrizing the J and H equations. Technically this is done by iteration from the non-symmetrical equations.

The asymmetrical calculation (where $m_1 \neq m_2$) is more difficult but proceeds along similar lines. In this case we do not *a priori* know the location of the center of momentum and we then have two unknown operators $v_{\perp 1}$ and $v_{\perp 2}$. There is however an additional relation, $P_\perp = 0$, which together with the J equation are (again schematically)

$$\left[J(W_{ri}, r, v_{\perp i}) - \sqrt{\ell(\ell+1)} \right] \psi = 0 , \tag{12}$$
$$P_\perp(W_{ri}, r, v_{\perp i}) \psi = 0 .$$

Starting with known equal mass solutions we solve the above two equations iteratively for the two operators (now matrices) $v_{\perp 1}$ and $v_{\perp 2}$, and substitute into and then diagonalize the Hamiltonian to obtain the desired solutions. Although the calculations are more involved than solving (say) the Schrödinger equation they are in principle very similar. Our numerical algorithms are reliable and run quickly on a modern work station.

3. Running Heavy Quark Parameters

In order to compare our model with actual meson states we must augment the confining tube by adding a short range "Coulombic" term to the Hamiltonian. In addition we can also add a constant term C giving

$$H = H_{\text{quarks}} + H_{\text{tube}} - \frac{\kappa}{r} + C . \tag{13}$$

From previous experience we expect κ to be about 0.5 and it will turn out that the fitted physical states do not depend on the value of C. We fix the constant term C and the light quark mass m_{ud} at convenient values and vary the parameters a and κ (different for each spectroscopy) and the quark masses m_s, m_c and m_b. The result is an essentially perfect fit for any value of C between 0 and -1 GeV. The results of the fit are shown in Figs. 5 to 7. In Fig. 5 we observe that κ decreases from heavy-light mesons to $b\bar{b}$ states. If we make the reasonable assumption that the QCD scale increases as the reduced mass rises then we can conclude that κ falls as the QCD scale increases, as expected. Quantitative calculations are difficult due to the non-perturbative range of scale variation. In Fig. 6 and Fig. 7 we show the corresponding variation of m_c and m_b for the different spectroscopies.

Again we see that the heavy quark masses depend on QCD scale but this dependence changes with C. In both Fig. 6 and Fig. 7 the heavy quark mass falls with increasing scale if C is sufficiently negative. Finally we note that in the fits the strange quark exceeds the light quark mass by about 140 MeV independent of the values of C or the light quark mass.

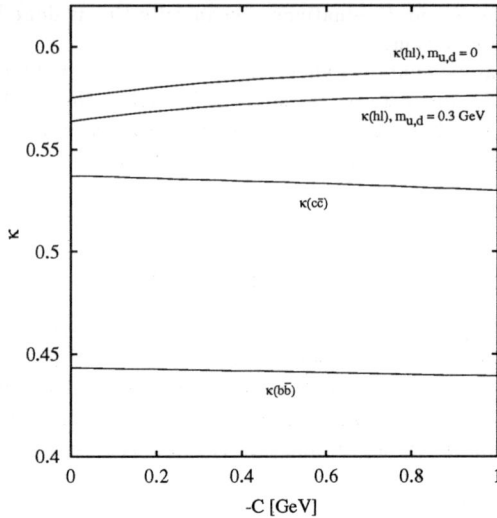

Fig. 5. Strong coupling constant κ as a function of C in the heavy-light (for $m_{ud} = 0$ and 0.3 GeV) $c\bar{c}$ and $b\bar{b}$ mesons. The values of κ found are roughly independent of C, but systematically fall with running QCD scale (reduced mass).

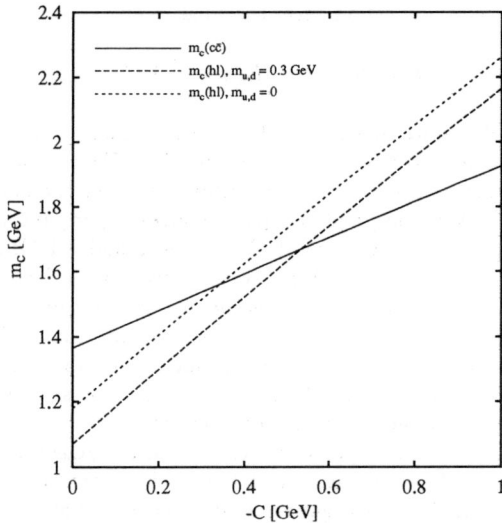

Fig. 6. Mass of the c quark as a function of C in the heavy-light (for $m_{ud} = 0$ and 0.3 GeV, dashed lines) and $c\bar{c}$ mesons. For C less than about -0.5 GeV, m_c falls with increasing QCD scale (reduced mass).

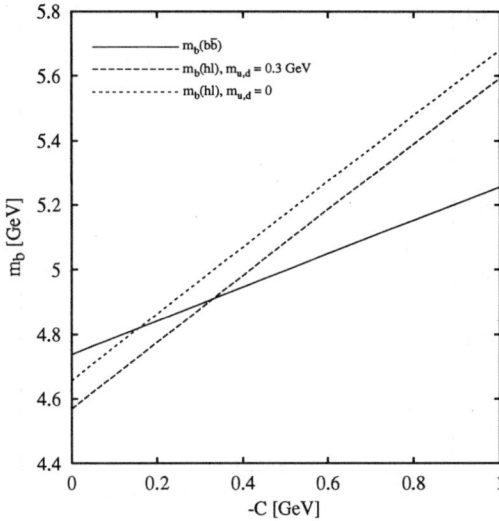

Fig. 7. Mass of the b quark as a function of C. See Fig. 6.

4. Exclusive Semi-Leptonic Decay the the Isgur-Wise Function

As is well known[14] the semi-leptonic decays of B mesons into D or D^* mesons can be expressed (in the heavy quark limit) in terms of a single form factor, the Isgur-Wise (IW) function $\xi(\omega)$. Once the heavy-light ground state wavefunction and light degrees of freedom energy E_q are known an overlap integral can be done to compute[15,16] the IW function as a function of the four vector velocities of the two mesons $\omega = v \cdot v'$

$$\xi(\omega) = \frac{2}{\omega+1} \left\langle j_0 \left(2E_q \sqrt{\frac{\omega-1}{\omega+1}} r \right) \right\rangle_{\text{ground state}} . \qquad (14)$$

Using our heavy-light fit to the observed spin averaged states we evaluate[16] the IW function which is plotted in Fig. 8. As can be seen the prediction is in excellent agreement with the experimental data. The data shown is from CLEO[17] and ARGUS[18]. The error bars shown are only for the more recent CLEO data.

5. Conclusions and Outlook

The absence of spin correlations at large quark separations was the original reason for choosing scalar confinement but one must then hope that the spin-independent relativistic corrections are correct. As it turns out, they disagree with rigorous QCD predictions and hence the concept of scalar confinement must be discarded. The

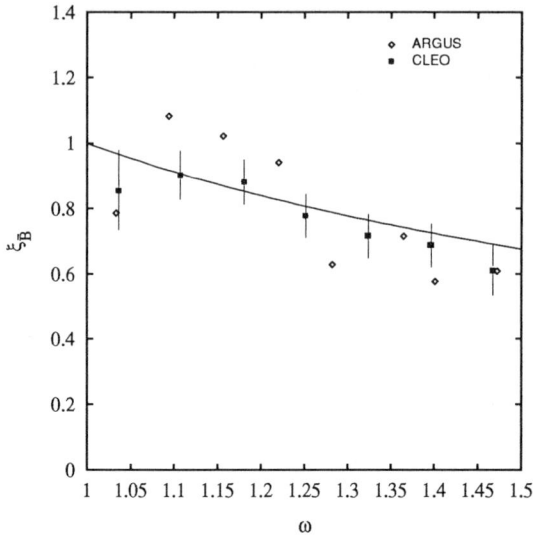

Fig. 8. The isgur-Wise function prediction using the spinless RFT model.

relativistic flux tube on the other hand reproduces all spin and spin-independent relativistic corrections expected by QCD.

We have emphasized here one technical step in the development of the RFT model. Mesons constructed of unequal mass quarks pose a new problem of how to fix the CM of the meson. These asymmetric mesons are of great interest. Even mesons containing a charm or bottom quark must eventually be treated with both quarks moving if one is to understand spin-dependence or the $1/M$ corrections to the Isgur-Wise function. The difficulties imposed in fixing the CM have been completely solved for spinless quarks.

The major future direction for the RFT model is to implement the program outlined in Ref. 10 for relativistic fermions confined by the flux tube. Fortunately, all of the spinless quark formalism remains valid without change. As discussed in Ref. 10, the tube can be introduced into the kinetic part of the Salpeter equation leaving the interaction kernel to account only for short-range perturbative interactions.

Acknowledgments

This research was supported in part by the U.S. Department of Energy under Contract No. DE-AC02-76ER00881 and in part by the University of Wisconsin Research Committee with funds granted by the Wisconsin Alumni Research Foundation.

1. K.D. Born et al., *The Interquark Potential: A QCD Lattice Analysis*, *Phys. Lett.* **B329** (1994) 325.
2. M. Lüscher, K. Symanzi, and P. Weisz, *Nucl. Phys.* **B173** (1980) 365; M. Lüscher, *ibid.* **B180** (1981) 317; J.D. Stack and M. Stone, *Phys. Lett.* **100B** (1981) 476; J. Kogut, *Phys. Rev. Lett* **47** (1981) 1089.
3. M.G. Olsson and C.J. Suchyta III, *Phys. Rev. Lett.* **57** (1986) 37.
4. W. Buchmüller, *Phys. Lett.* **112B** (1982) 479.
5. E. Eichten and F. Feinberg, *Phys. Rev.* **D 23** (1981) 2724; D. Gromes, *Z. Phys.* **C22** (1984) 265; **26**(1984) 401.
6. A. Barchielli, E. Montaldi and G.M. Prosperi, *Nucl. Phys.* **B296**(1988) 625; **B303** (1988) 752(E); A. Barchielli, N. Brambilla and G.M. Prosperi, *Nuovo Cim.* **103A** (1989) 59.
7. N. Brambilla and G.M. Prosperi, *Phys. Lett.* **B236**, 69 (1990).
8. C. Olson, M. G. Olsson and K. Williams, *Phys. Rev.* **D45** (1992) 4307.
9. D. LaCourse and M.G. Olsson, *Phys. Rev.* **D39** (1989) 2751.
10. M. G. Olsson and K. Williams, *Phys. Rev.* **D48** (1993) 417.
11. A. Gara, B. Durand and L. Durand, *Phys. Rev.* **D40**, 843 (1980).
12. A. Yu. Dubin, A.B. Kaidalov and Yu. A. Simonov, *Journal of Nuclear Physics* **56/12** (1993) 213; *Phys. Lett.* **323B** (1994) 41.
13. M.G. Olsson and S. Veseli, *The Asymmetric Flux Tube*, Univ. of Wisconsin-Madison preprint MAD/PH/816 (1994), to appear in *Phys. Rev.* **D**
14. N. Isgur and M. B. Wise, *Phys. Lett.* **232B** (1989) 113; **237B** (1990) 527.
15. M. Sadzikowski and K. Zalewski, *Z. Phys.* **C59** (1993) 677.
16. M.G. Olsson and S. Veseli, *Relativistic Flux Tube Model Calculation of the Isgur-Wise Function*, Univ. of Wisconsin-Madison preprint MAD/PH/851 (1994), to appear in *Phys. Rev.* **D**.
17. B. Barish et al., CLEO Collaboration, Cornell Nuclear Studies Wilson Lab preprint HEPEX 9406005 (1994).
18. H. Albrecht et al., ARGUS Collaboration, *Z. Phys.* **C57** (1993) 533.

Non–perturbative determination of the strong coupling constant

ROBERTO PETRONZIO

Dipartimento di Fisica, Università di Roma Tor Vergata
Via della Ricerca Scientifica 00133 Roma

Abstract

I review the calculation of the running coupling constant in quenched SU(2) based on a finite volume recursive technique. The results show that a precision higher than the one attainable with high energy experiments is reachable.

Perturbative QCD allows to extract from high energy experiments good estimates of the strong coupling constant. In particular, the determination coming from the hadronic decay width of the Z_0 is theoretically well understood and reaches a precision of the order of 6–7 percent for α_s renormalized at the energy of the Z_0 mass. Lower energy measurements, for example from scaling violations in deep inelastic scattering, are affected by systematic errors coming from the ignorance of higher order terms in the perturbative expansion and from the correlation between the value extracted for Λ — the QCD scale — and the form of parton structure functions.

A higher precision for α_s at LeP could resolve the slight discrepancy with the value extracted from multijets events and allow tests of the standard model based on the correlation between the value of the top mass and the value of α_s implied by LeP data. Indeed from global fits to the data, the best value of α_s increases by about 1–2 percent with the top mass ranging from 140 to 180 GeV. Significant tests of the standard model seem to require a one percent precision. An improuvement by a factor 30 in Lep statistics and a corresponding reduction of systematic errors are unrealistic: in this talk I will present a prototype estimate of α_s based on lattice QCD simulation for the pure gauge SU(2) case where the desired precision can be reached.

To make the discussion accessible to non experts, I include the following essential dictionary:

LATTICE : space time discretization, suitable for a numerical estimate of functional integrals.

LATTICE SPACING "a" : minimal distance between lattice points which

1. sets the units of length

2. acts as an ultraviolet cutoff

BETA : parameter proportional to the inverse of the bare coupling constant, $\beta = 1/g_0^2$ which acts as the "inverse temperature" of the system.

SCALING : independence of dimensionless physical quantities from the cutoff. For small values of the bare coupling scaling implies a crucial relation between g_0 and the lattice spacing based on the asymptotic freedom property of the theory.

$$g_0(a)^2 = \frac{1}{b_0 \ln(1/a\Lambda)} \tag{1}$$

CONTINUUM LIMIT : $a \to 0 \iff \beta \to \infty$

PHYSICAL UNITS : are obtained by calculating a dimensionful physical quantity on the lattice

$$M_{lattice}^2 = M_{physical}^2 a^2. \tag{2}$$

In the perturbative region for the bare coupling, one can obtain the coupling renormalized at the lattice spacing scale.

$$\alpha_{\overline{MS}}(\pi/a) = g_0^2/4\pi(1 + Cg_0^2/4\pi) \tag{3}$$

where C is a calculable and calculated constant.

This procedure is not viable because it needs to determine the lattice spacing a from a low energy physical quantity in the region where g_0 is very small. Indeed, the momentum $q = \pi/a$ should be large enough to keep under control the higher order terms in the perturbative connection ($q > 60 GeV \Rightarrow a < 0.01 \ Fermi$) and the physical lattice size La should be large enough to accomodate the determination of the low energy physical quantity without finite size effects ($La > 1 \ Fermi$). The two conditions together lead to values of L of order one hundred, far beyond present and foreseeable computer capabilities. One can cover the region from non perturbative energies ($\simeq 500$ MeV) to the Z_0 mass with a recursive finite size scaling technique. The goal of the method is to reconstruct the running coupling constant from low energies (where it is normalized) to high energies (where it is used to check the standard model). The method consists of three steps:

1. the definition of a coupling constant running with the lattice size L,

2. a perturbative calculation connecting α calculated on the lattice with $\alpha_{\overline{MS}}$,

3. a finite size iteration scheme with an extrapolation to zero lattice spacing.

1. The definition

There are two definitions of finite size renormalized coupling which have been used in the literature:

1. with "sources"[1],

2. with correlations[2].

The first is based on the Schroedinger functional (SF) and consists in calculating the response of the free energy to variations of an external chromoelectric field fixed at the initial and the final time. The results discussed in this talk refer to the second definition, which is based on the correlation of Polyakov loops at a time distance equal to half the lattice size with twisted periodic boundary conditions. These conditions fix the values of the gauge field at positions which differ in a given coordinate by L, the lattice size, to be not identical as in ordinary periodic boundary conditions, but related by a global gauge transformation (the twist matrix) which in general depends upon the direction. This choice of boundary conditions allows to perform perturbative calculations on a finite volume without the problem of toron degeneracy, i.e. of gauge field configurations degenerate on a finite lattice with the perturbative vacuum (zero gauge field) and to compute the relation between the "twisted Polyakov" definition (α_{TP}) and the more familiar $\alpha_{\overline{MS}}$. The coupling is defined through a ratio of correlations of Polyakov loops. The latter are defined as:

$$P(t) = \sum_{x,y} \prod_{i=1,L/a} U_z(z_i,\ldots)\Omega_z \qquad (4)$$

where Ω_z is a unitary matrix different from the unit matrix if the direction is a twisted one. The observable which reduces to α_s up to a constant at tree level is given by:

$$\mathcal{O} = \langle P_T(0)P_T(t)\rangle/\langle P(0)P(t)\rangle \qquad (5)$$

where the suffix T indicates that the corresponding Polyakov is taken along a twisted direction. A ratio of correlations is needed to cancel spurious linear divergences occurring in Polyakov loops: the quantity in the above expression behaves properly in the continuum limit and is affected by the divergences associated only with the renormalization of α_s.

2. The perturbative calculation

An explicit one loop calculation on a finite volume relates the renormalized finite size coupling to the bare coupling through a perturbative expansion with corrections due to lattice artifacts which die as the square (in the TP case) of the inverse lattice size:

$$g_{TP}^2(L) = g_0^2 + g_0^4(2b_0\ln(L) + r_0 + O(\ln(L)/L^2 + \ldots)) = g_{lattice}^2(L) + g_0^4 r_0 \quad (6)$$

where

$$g_{lattice}^2 \equiv -\frac{1}{2b_0\ln(L\Lambda_{lattice})} \qquad (7)$$

The coefficient of the logarithmic term is known from continuum theory and is equal to the one loop coefficient of the beta function. The last equality gives the definition of $\Lambda_{lattice}$ and the order g_0^4 term fixes the relations between $\Lambda_{lattice}$ and Λ_{TP}. The latter is defined through the equation:

$$g_{TP}^2 \equiv -\frac{1}{2b_0 \ln(L\Lambda_{TP})} = -\frac{1}{2b_0[\ln(L\Lambda_{lattice}) + \ln(\Lambda_{TP}/\Lambda_{lattice})]} \qquad (8)$$

By expanding the denominator and taking into account that $g_{lattice}^2(1) \equiv g_0^2$ one gets:

$$\frac{r_0}{2b_0} = \ln(\Lambda_{TP}/\Lambda_{lattice}) \qquad (9)$$

By using the known relation between $\Lambda_{lattice}$ and $\Lambda_{\overline{MS}}$ one gets[2]:

$$\Lambda_{TP}/\Lambda_{\overline{MS}} = 1.6136(2) \qquad (10)$$

Figure 1 shows the behaviour of the order g_0^4 term in expression (6) after the subtraction of the continuum value of the logarithmic term. The asymptotic large L value can be determined with a very high precision.

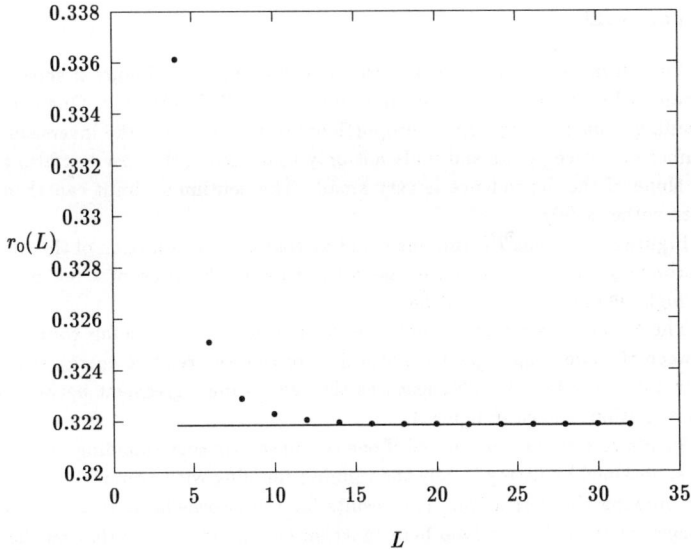

Figure 1: Fit of the coefficient r_0

3. The iterative method

The reconstruction of the running coupling constant as a function of the energy scale is performed through a recursive finite size scaling technique. One starts from an input value for the renormalized coupling on a lattice of size L. By doubling the number of lattice points at a fixed value of β one obtains the change of the renormalized coupling by a change of the scale by a factor two. This change is repeated for several values of L readjusting each time the value of β in order to keep constant the value of the input renormalized coupling, i.e. of the total volume in physical units. The extrapolation to zero lattice spacing at fixed physical volume is equivalent to the limit of an infinite number of lattice points and can be performed rather safely if the dependence is smooth. Such an extrapolation provides the value of an output renormalized coupling. By tuning again the value of β, one can iterate the procedure by choosing as the input value of the renormalized coupling the output value of the previous step. This procedure allows to reconstruct the running coupling constant down to an energy scale which should be normalized by the calculation of an independent physical quantity.

4. The results

The results are summarized in the following figures[3]. Figure 2 shows the extrapolation curves to the continuum limit of the TP definition. One expects a leading behaviour for large L proportional to the square of the inverse of the number of lattice points and finds not only that such is the case but also that the slope of the dependence is very small. The continuum limit can then be taken rather safely.

Figure 3 shows the TP running coupling constant as a function of the energy in arbitrary units. One can see that a precision of the order of a few percent at "high energies" can indeed be reached.

The SF definition gives a similar curve. In order to obtain a fair comparison between the two couplings, it is natural to rescale the relative energy scale by the ratio of the two Λ. This leads to the remarkable agreement between the two definitions shown in figure 4.

Such a rescaling is not needed if one compares the corresponding beta functions, obtained by fitting locally the running coupling with a smooth curve and then making the derivative. The results for the two definitions are in figure 5, together with the two loop beta function computed in perturbation theory. At the highest values of the couplings one sees a departure from perturbation theory.

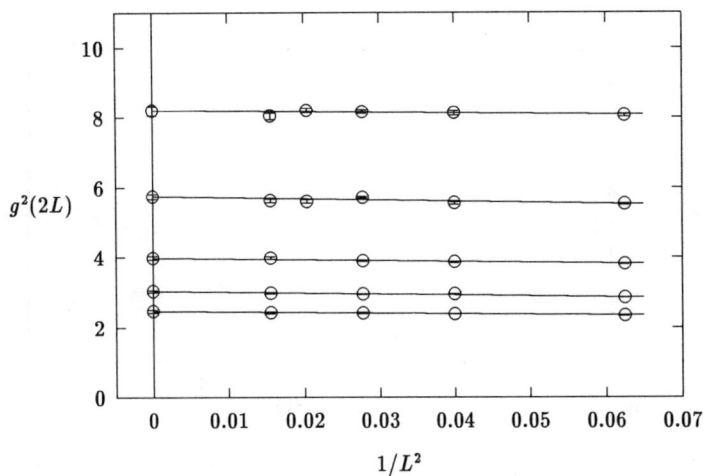

Figure 2: Extrapolation to the continuum limit

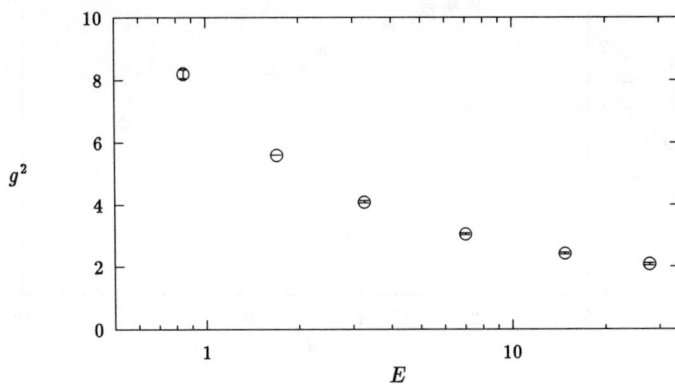

Figure 3: α_{TP} as a function of an arbitrary energy scale

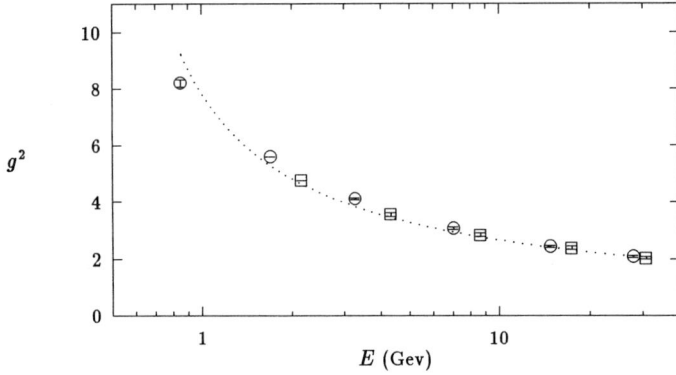

Figure 4: α_{TP} (o) and α_{SF} ($\square$) after energy rescaling

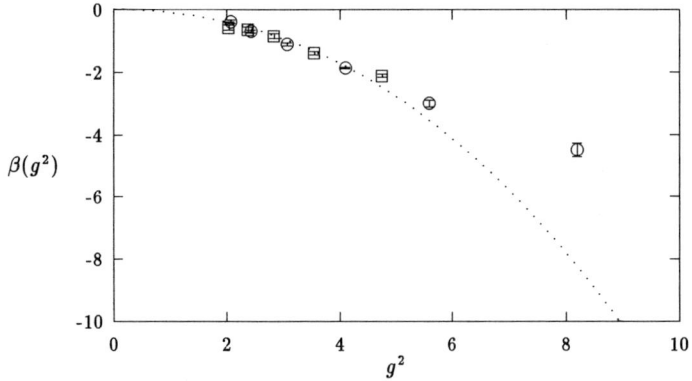

Figure 5: Non perturbative beta function for α_{TP} (o) and α_{SF} ($\square$).

5. Conclusions

The determination of the strong coupling constant in lattice gauge theories can reach a precision not attainable with high energy experiments. The error on Λ comes essentially from the calibration of the energy scale and can be kept within ten percent which means a value for α, at LeP energies at one percent. The calculations described in this talk are a feasibility study for the simple quenched $SU(2)$ case: the two definitions agree remarkably well on the final result and show the universality of the continuum limit[4]. The extension to fermions appear more difficult for the TP definition which however shows a nicer approach to the continuum limit with respect to the SF definition without improuvements. The overall computer time involved in the TP calculation has been of approximately four months at an average speed of 1.5–2 Gigaflops. The ape100 or ape1000 series of machines make the extension to the unquenched $SU(3)$ case realistic.

6. References

1. M. Lüscher, R. Narayanan, R. Sommer, P. Weisz and U. Wolff, Nucl. Phys. B (Proc. Suppl.) 30 (1993) 139
2. G.M. de Divitiis, R. Frezzotti, M. Guagnelli and R. Petronzio, Nucl. Phys. B422 (1994) 382
3. G.M. de Divitiis, R. Frezzotti, M. Guagnelli and R. Petronzio, Rome preprint (1994), hep–lat/9407028, accepted for publication on Nucl. Phys. B
4. G.M. de Divitiis, R. Frezzotti, M. Guagnelli, M. Lüscher, R. Petronzio, R. Sommer, P. Weisz and U. Wolff, preprint: DESY 94-196, CERN-TH 7447/92, HUB-IEP 94/16 , ROM2F 94/46, submitted for publication to Nucl. Phys. B

CHIRAL SYMMETRY BREAKING

IN THE CONFINING QCD VACUUM

YU. A. SIMONOV

Institute of Theoretical and Experimental Physics
117259, Moscow, B. Cheremushkinskaya 25, Russia

ABSTRACT

Chiral symmery breaking due to topological charges in the confining QCD vacuum is discussed.

1. Introduction

The phenomenon of chiral symmetry breaking (CSB) is of nonperturbative origin, and moreover most probably is associated with topological properties of the QCD vacuum. The latter statement is supported by the Banks-Casher relation[1]

$$< \bar{q}q > = -\frac{\pi\nu(0)}{V_4}$$

where $\nu(\Lambda)$ is the density of quark modes with eigenvalue Λ: $i\hat{D}u_n(\Lambda) = \Lambda_n u_n(\Lambda)$. Hence it is zero modes (or rather quasizero modes) which ensure CSB. By the Atiya–Singer theorem zero quark modes are connected with topological charges, and therefore a very natural physical picture of CSB emerges, where an ensemble of topological charges in the QCD vacuum provides local quark zero modes, which mix and shift due to interaction between top charges and finally a global density $\nu(\Lambda)$ emerges with $\nu(0) \neq 0$. This picture was suggested and realized in the framework of the instanton model[2,3]. Recently magnetic monopoles were considered and shown to produce CSB due to the similar mechanism[4], since magnetic monopoles also produce (infinitely many) quark zero modes.

But we know that instantons do not confine[5], and there should be additional confining configurations in the QCD vacuum. How would those modify the CSB picture described above? It appears that this general treatment of vacuum fields, comprising both topological charges and confining fields drastically improves the original model[6]:

i) One obtains a gauge–covariant description of CSB and in particular gauge–invariant expressions for effective chiral Lagrangians (ECL) in contrast to that obtained earlier in[3,7].

ii) Confining background slightly squeezes and stabilizes the instanton size at the value of the gluonic correlation length $Tg \approx 0.2fm$. Stability of instanton-antiinstanton gas is model independent and does not depend now on $I\bar{I}$ interaction, in contrast to what happened without confinement[3].

From physical point of view the important new element appears: confinement radius (hadron size) R_c is much larger than "chiral radius" $R_{ch} \approx \rho$, $R_c \sim 1 fm$, $R_{ch} \approx 0.2 fm$.

Chiral phenomena (creation of chiral mass m_{ch} and chiral condensate $< \bar{q}q >$) can be seen to occur at distances R_{ch}, while confinement contributes at larger distances, $r \sim R_c$ and only starts to set in at $r \sim T_g$. Hence many chiral and confinement phenomena can be considered separately, e.g. numerically confinement does not change much m_{ch} and $< \bar{q}q >$, while chiral physics (instanton induced interaction) does not change much ρ– meson mass.

However there are important issues where both chiral and confinement configurations work together.

First of all, this is the fact that confinement is strongly correlated with CSB: in the temperature deconfinement also chiral symmetry is restored. In case of instanton-produced CSB a natural explanation was given in[8] based on our generalized picture, where confinement is needed to stabilize molecular $(I\bar{I})$ quasizero modes, which otherwise produce a full delocalization of quarks. Recently an importance of $I\bar{I}$ molecules in the chiral phase transition for a model vacuum has been demonstrated[9].

Secondly, our generalized approach allows to solve an old standing problem of a two–faced nature of pions. In particular, as it was discussed in[10], chiral and confinement configurations produce two types of poles for a pseudoscalar, the first – a Goldstone pole at $m = 0$, the second – a $q\bar{q}$ pole at $m_a \approx \sqrt{\sigma}$, while in nature one observes only one pole.

What happens in our generalized approach with these poles? We show below, that only the Goldstone pole survives, while the second cancels out in the general expression containing both types of poles.

We give below a schematic exposition of our method[11], based on the effective action deduced there, and exemplify some results for chiral mass, ECL and double pion problem.

2. Effective action for quarks in the field of instantons and confining background

We neglect in what follows perturbative gluonic fields and represent A_μ as

$$A_\mu(x) = B_\mu(x) + \sum_{i=i}^{N} A_\mu^{(i)}(x) \tag{1}$$

where $A_\mu^{(i)}$ is the (anti)instanton field in the singular gauge and $B_\mu(x)$ is due to confining configurations. The superposition ansatz (1) is supposed to be reasonable for a dilute gas of instantons, which can be proved a posteriori.

Another important approximation which we use following[3,7,11] is that we omit in the spectral representation of the quark Green's function $S^{(i)}$ on the i-th instanton in the background B_μ all terms except (quasi) zero mode $u_0^{(i)}(x)$ and the tail of high excited states is given by free Green's function:

$$S^{(i)}(x,y) = \sum_n \frac{u_n^{(i)}(x)u_n^{(i)+}(y)}{\lambda_n^{(i)} - im} \simeq S_0(x,y) + \frac{u_0^{(i)}(x)u_0^{(i)+}(y)}{\lambda_0^{(i)} - im} \tag{2}$$

From (1) and (2) one can rigorously deduce the quark partition function for any number of flavours N_f (see details in Appendix C of[11])

$$Z_{QCD} = \int D\Psi D\Psi^+ DB_\mu e^{-S(B)+\int d^4x \Psi_f^+ i\hat{D}(B)\Psi_f} \times \tag{3}$$

$$< \prod_{f=1}^{N_f} \prod_{i=1}^{N} (2im_f - \int d^4x \Psi_f^+ i\hat{D}u_0^{(i)}(x) \int d^4y u_0^{(i)+}(y) i\hat{D}\Psi_f(y) >_{R_i,\Omega_i}$$

where $< ... >_{R_i,\Omega_i}$ denotes averaging over instanton positions R_i and color orientations Ω_i. To keep all formalism gauge–covariant, one can write explicitly for the quark zero mode

$$u_0^{(i)}(x) = \Phi(x,R_i)\Omega_i\varphi_0(x-R_i) \tag{4}$$

where $\Phi(x,y) = Pexp \int_y^x B_\mu dz_\mu$ is parallel transporter, and $\varphi_0(x)$ is the known 'tHooft solution[12].

1) Consider first one-flavour case, $N_f = 1$.

In this case the partition function becomes[11]

$$Z = const \int DB_\mu e^{-S(B)} D\Psi D\Psi^+ exp\{\int \Psi^+(x)[i\hat{D}\delta(x-y)+iM(x,y)]\Psi(y)dxdy\} \tag{5}$$

where

$$M(x,y) = \frac{\varepsilon N}{2V N_c} \int dR i\hat{D}\varphi_0(x-R)\Phi(x,R,y)\varphi_0^+(y-R)i\hat{D} \tag{6}$$

and ε is to be found from the "gap equation"

$$\frac{N}{2} = Tr(\frac{M^2}{-\hat{D}^2 + M^2}). \tag{7}$$

Here Tr implies summation over color, Lorentz indices and coordinates (or momenta).

In the limit of no background, $B_\mu \equiv 0$, Eq.(7) reduces to the old gap equation of[3,7].

$$\int \frac{d^4p}{(2\pi)^4} \frac{M^2(p)}{p^2 + M^2(p)} \frac{4V N_c}{N} = 1. \tag{8}$$

The mass creation is what characterizes CSB for $N_f = 1$; the mass (6) is however nonlocal operator with nonlocality of the order of $\rho \equiv R_{ch}$, which is the chiral radius, and as we have discussed in Introduction, $R_{ch} \sim T_g \sim 0.2fm$. This value is much smaller than typical hadronic radius, of the order of the confinement radius, $R_c \sim 1fm$.

Therefore chiral phenomena and confinement can be separated at least for some observables. In particular, in the expressions (6) and (7) the integrals are saturated at distances $x \sim \rho \ll R_c$ and confinement forces are inessential there (remember that e.g. Wilson loops $< W(C) > \approx 1$ for small contours C). Also quark condensate is defined by small distances $x \leq \rho$

$$< \bar{q}q >= iTr < (-i\hat{D} - i\hat{M})^{-1} >_B \to - \int \frac{d^4 p}{(2\pi)^4} \frac{M(p)}{p^2 + M^2(p)} \qquad (9)$$

where the last equality in (9) coincides with the corresponding in[3].

One can show that the gap equation (7) and our arguments about range separation hold true also for $N_f = 2$.

2) The case of two flavours , $N_f = 2$.

Physically one expects in this case a new phenomenon - that of Nambu–Goldstone bosons. Indeed, using the steepest descent method of[7], exact in the thermodynamic limit, $N \to \infty, V \to \infty, N/V = const$, one introduces eight bosonic fields $\pi_i, \sigma_i, \sigma, \eta, i = 1, 2, 3(\pi_i, \eta-$ are pseudoscalars, σ_i, σ – scalars) as auxiliary fields in the Hubbard – Stratonovich transformation and obtains the effective action

$$S_{eff} = \int dx dy \Psi_f^+(x) W_{fg}(x, y) \Psi_y(y) + \frac{N}{2V} \int dx (\sigma^2(x) + \eta^2(x)) \qquad (10)$$

where we have used notations

$$W_{fg}(x, y) = [i\hat{D} + i(1 + \sigma + \eta)UV \frac{1 - \gamma_5}{2} \hat{M} + \qquad (11)$$

$$+(1 + \sigma - \eta)VU^+ \frac{1 + \gamma_5}{2} \hat{M}]_{fg}$$

and $U = exp(i\pi_i \tau_i)$, $V = exp(i\sigma_i \tau_i)$, while $\hat{M}(x, y)$ is defined in (6).

To check the Nambu–Goldstone boson masses one can integrate over Ψ, Ψ^+ in $\int D\Psi D\Psi^+ exp W_{eff}$ and obtain effective action for bosons only:

$$S_{bos} = tr \ln W + \frac{N}{2V} \int dx (\sigma^2(x) + \eta^2(x)) \qquad (12)$$

Expansion of S_{bos} (12) in powers of bosonic field yields the masses of boson quanta as coefficients of corresponding fields squared.

It is remarkable and gratifying to find that only pions are massless (we always consider chiral limit, $m_f \equiv 0$), while η, σ, σ_i have masses of the order of $\rho^{-1} \approx 1 Gev$. This is in agreement with analogous procedure made in[7] for the case $B_\mu = 0$.

It is instructive to find the η–mass which amounts to the solution of the $U(1)$ problem in the case of two flavours. Similarly to[7] one obtains

$$m_{\eta'}^2 = \frac{2N_f N}{F_\pi^2 V}, \quad N_f = 2 \qquad (13)$$

where F_π can be expressed in terms of the trace over $\hat{M}$ and $(\hat{D} + M)^{-1}$ (see[11] for details). Nothing that[7]

$$N =< (N_+ - N_-)^2 >=< Q^2 > \qquad (14)$$

one reproduces in (13) the well-known Witten-Veneziano formula.

3. ECL, the $q\bar{q}$ correlators and the double nature of the pion

As discussed above, only π_i have zero masses, while η, σ, σ_i are not Goldstone particles. For them our approximation (2) may be not good and therefore we should exclude them from our treatment, keeping only pions. As a result the ECL (12) becomes

$$W(\pi) = -\operatorname{Tr}\ln(i\hat{D}(B) + i\hat{M}e^{i\pi_i\tau_i\gamma_5}). \tag{15}$$

In the limit $B_\mu \to 0$ it coincides with ECL obtained in[3] and proposed also elsewhere (see refs. in[11]).

Expanding (15) in powers of the pion field one obtains the standard form of chiral Lagrangian[13], where coefficients are now expressed through the traces of operators $(\hat{D} + \hat{M})^{-1}$ and $\hat{M}$. In this way all chiral dynamics at low energies is fixed in our formalism.

Let us now turn to the $q\bar{q}$ correlators. Calculating those with the effective action S_{eff} one obtains for $\prod(x,y) = <\Psi^+(x)\Gamma\Psi(x)\Psi^+(y)\Gamma\Psi(y)>$

$$\prod(x,y) = \frac{1}{Z}\int D\pi_i DB_\mu e^{-S(B)-W(\pi)}\{Tr(S(x,y:\pi)\Gamma S(y,x,\pi)\Gamma) \tag{16}$$

$$-TrS(x,x,\pi)TrS(y,y,\pi)\}$$

where

$$S(x,y,\pi) = (i\hat{D} + iMe^{i\pi_i\tau_i\gamma_5})_{xy}^{-1} \tag{17}$$

Two terms on the r.h.s. of (16) represent one-quark-loop and two-quark-loop contributions to $\prod$. Inside each loop there are B_μ exchanges, leading to confinement (area law for the loop) and pion exchanges. From $W(\pi)$ it is easy to deduce that pion coupling $g_{\pi q\bar{q}} = \frac{M}{F_\pi} = 0(N_c^{-1/2})$.

The one–loop contribution is thus the usual quark model correlator corrected by including pion exchanges (relatively weak). For PS channel ($\Gamma = \gamma_5$ or $\gamma_5\gamma_\mu$) it yields a pole at around $200 \div 350$ MeV. The two –loop term in (16) contains a one-pion exchange between two loops with a zero –mass pole – a Goldstone pole. To resolve this doubling of pion poles, one can look more carefully into the structure of the pion exchange propagator (obtained from expansion of $W(\pi)$ (15)) and finds that $\Pi(x,y) \to \pi_5(k)$ can be written as

$$\pi_5(k) = \pi_5^{(0)}(k) - \pi_5^{(0)}(k)\frac{M^2(k)}{N(k)}\pi_5^{(0)}(k) \tag{18}$$

The denominator $N(k)$ can be seen to be proportional to k^2 at $k \to 0$ and moreover represented as

$$N(k) = C + M^2(0)\pi_5^{(0)}(k) \tag{19}$$

The "quark–model pole" appears in the one–loop correlator $\pi_5^{(0)}(k) = \frac{\lambda^2}{k^2+m_a^2}$, and one can see that this pole exactly cancels in the total correlator $\pi_5(k)$, while the pole at $k = 0$ due to $N(k)$ survives, maintaining in this way the Goldstone theorem.

4. Discussion

We have derived above and in[11] the gauge invariant expressions for chiral Lagrangians and correlators which contain contribution of both confinement and topological charges (instantons), leading to CSB. Both $U_A(1)$ and SU(2) breaking is maintained and Nambu-Goldstone bosons are explicitly described. Several types of ECL are deduced containing or not containing quark degrees of freedom. The coefficients of ECL introduced phenomenologically in[13] can now be computed explicitly.

We have shown how the long–standing problem of two natures of pion, and of two corresponding poles can be resolved in the formalism combining contributions both from confinement (quark model) and CSB (Goldstone degrees of freedom).

We have not touched here the important question of connection between CSB and confinement, seen e.g. in the temperature deconfinement. For the instanton mechanism, dealt with here, a possible explanation as due to $I\bar{I}$ molecules is given in[8] and supported by recent calculations in[9]. In case of magnetic monopoles (τ –monopoles [4]) the connection is simple since monopoles are carriers of both confinement and top charges.

5. References

1. T.Banks and A.Casher,*Nucl. Phys.* **B169** (1980) 102.
2. D.G.Caldi, *Pys. Rev. Lett.* **39**(1977) 121.
   E.V.Shuryak, *Nucl. Phys.* **203** (1982) 93,116, 140.
3. D.I.Dyakonov and V.Yu. Petrov, *Nucl. Phys.***B272** (1986) 457.
4. A.Gonzalez-Arroyo and Yu. A.Simonov (in preparation).
5. C.G.Callan, R.Dashen and D.J.Gross, *PL* **B66** (1977) 375.
6. N.O.Agasyan, *JETP* (subm.to)
   N.O.Agasyan and Yu.A.Simonov, *Phys. Lett. B* (to be subm. to).
7. D.I.Dyakonov and V.Yu. Petrov, *LINP* 1153 (1986)
   also in: Hadron matter under extreme conditions, Kiev, 1986 (in Russian).
8. Yu.A.Simonov, *Yad. Fiz.* **54** (1991) 224,
   JETP Letters **53** (1991) 10.
9. E.-M.Ilgenfritz and E.V.Shuryak, *Phys. Lett.* **B325** (1994) 263.
10. A.Manohar and H.Georgi, *Nucl. Phys.* **B234** (1984) 189.
11. Yu.A.Simonov, *HEP-PH 9401320, Yad. Fiz.* **58** (1994) N8.
12. G.t'Hooft, *Phys. Rev.* **D14** (1976) 3432.
13. J.Gasser and H.Leutwyler, *Ann. Phys.* **158** (1984) 142.

ALL AROUND THE SPINLESS SALPETER EQUATION

WOLFGANG LUCHA

Institut für Hochenergiephysik, Österreichische Akademie der Wissenschaften,
Nikolsdorfergasse 18, A-1050 Wien, Austria

and

FRANZ F. SCHÖBERL

Institut für Theoretische Physik, Universität Wien,
Boltzmanngasse 5, A-1090 Wien, Austria

ABSTRACT

We review some important topics related to the semirelativistic description of bound states by the spinless Salpeter equation: the special case of the Coulomb interaction, numerical approximation methods, and a way to avoid the problematic square-root operator of the relativistic kinetic energy.

1. Introduction

The "spinless Salpeter equation" represents a well-defined standard approximation to the Bethe–Salpeter formalism for the description of bound states within relativistic quantum field theories. It may be derived from the Bethe–Salpeter equation[1]

1. by eliminating, in full accordance with the spirit of an instantaneous interaction, any dependence on timelike variables, which leads to the Salpeter equation,[2] and

2. by neglecting any reference to the spin degrees of freedom of the involved bound-state constituents and restricting exclusively to positive-energy solutions.

The Hamiltonian governing the dynamics of the quantum system under consideration

- incorporates relativistic kinematics by involving the square-root operator of the relativistic kinetic energy, $\sqrt{\mathbf{p}^2 + m^2}$ for particles of mass m and momentum $\mathbf{p}$, but

- describes the forces acting between the bound-state constituents by an arbitrary coordinate-dependent static interaction potential $V(\mathbf{x})$.

For the case of bound states consisting of two particles of equal mass m, the generic Hamiltonian H in the center-of-momentum frame of these constituents, expressed in terms of the relative momentum $\mathbf{p}$ and the relative coordinate $\mathbf{x}$, reads

$$H = 2\sqrt{\mathbf{p}^2 + m^2} + V(\mathbf{x}) . \tag{1}$$

The equation of motion involving this type of Hamiltonian, $H|\psi\rangle = E|\psi\rangle$, with energy eigenvalues E and corresponding Hilbert-space eigenstates $|\psi\rangle$, has been widely used, e. g., for the (semi-)relativistic description of hadrons as bound states of quarks within the framework of potential models.[3,4,5]

2. The Spinless Relativistic Coulomb Problem

Without doubt, a central rôle in physics is played by the Coulomb potential $V_C(r)$. This Coulomb potential is a spherically symmetric potential, i. e., one which depends only on the radial coordinate $r \equiv |\mathbf{x}|$; its interaction strength is parametrized by some coupling constant κ:

$$V(\mathbf{x}) = V_C(r) = -\frac{\kappa}{r}, \quad \kappa > 0. \tag{2}$$

The bound-state problem defined by the semirelativistic Hamiltonian of Eq. (1) with the Coulomb potential (2) is what we call the "spinless relativistic Coulomb problem."

2.1. (Mostly) Analytic Results

Over the past years, the spinless relativistic Coulomb problem has been the subject of intense study. We summarize, in chronological order, the knowledge gained so far:

- Herbst,[6] in a rigorous mathematical discussion, developed the complete spectral theory of the one-particle counterpart of the operator (1), (2)—from which one may directly deduce for the two-particle semirelativistic Coulombic Hamiltonian under consideration its essential self-adjointness for $\kappa \leq 1$ and the existence of its Friedrichs extension up to some critical value, viz., $\kappa_{cr} = 4/\pi$, of the coupling constant—and derived some strict lower bound on the energy E_0 of the ground state which translates in the two-particle case to

$$E_0 \geq 2m\sqrt{1 - \left(\frac{\pi\kappa}{4}\right)^2} \quad \text{for } \kappa < \frac{4}{\pi}.$$

- Durand and Durand[7] presented in an involved analysis of the spinless relativistic Coulomb problem for the particular case of vanishing orbital angular momentum of the bound-state constituents the explicit construction of the analytic solution for the corresponding wave function.

- Castorina et al.[8] generalized the behaviour of this wave function near the origin, that is, for small relative distances of the bound-state constituents, to arbitrary values of the orbital angular momentum.

- Hardekopf and Sucher[9] carried out a comprehensive numerical investigation of one- and two-particle relativistic wave equations for both spin-0 as well as spin-$\frac{1}{2}$ particles. Their numerical results for the ground-state energy E_0 of the spinless relativistic Coulomb problem are consistent with the lower bound of Herbst and provide no evidence that this ground-state energy indeed approaches zero as the coupling constant κ rises to its critical value $\kappa_{cr} = 4/\pi$, that is, $E_0(\kappa = \kappa_{cr}) \neq 0$.

- Martin and Roy[10] improved the lower bound of Herbst somewhat to

$$E_0 \geq 2m\sqrt{\frac{1+\sqrt{1-\kappa^2}}{2}} \quad \text{for } \kappa < 1 .$$

- Raynal et al.[11] made use of a generalized version of the "local-energy" theorem:[12] Assume (i) that the Fourier transform $\tilde{V}(\mathbf{p})$ of the interaction potential $V(\mathbf{x})$ is strictly negative, except at infinity, as is the case for the Coulomb potential (2), (ii) that the spectrum of the Hamiltonian under consideration, H, is discrete, and (iii) that the ground state of the Hamiltonian exists. Define the local energy $\mathcal{E}(\mathbf{p}) \equiv T(\mathbf{p}) + \int d^3q\, \tilde{V}(\mathbf{p}-\mathbf{q})\, \phi(\mathbf{q})/\phi(\mathbf{p})$; here $T(\mathbf{p})$ represents the kinetic energy, which in our case is, of course, given by $T(\mathbf{p}) = 2\sqrt{\mathbf{p}^2 + m^2}$, and $\phi(\mathbf{p})$ denotes some suitably chosen, positive trial function, $\phi(\mathbf{p}) > 0$. Then the lowest-lying eigenvalue of the Hamiltonian H, E_0, is bounded by

$$\inf_{\mathbf{p}} \mathcal{E}(\mathbf{p}) \leq E_0 < \sup_{\mathbf{p}} \mathcal{E}(\mathbf{p}) .$$

  With the help of this theorem, Raynal et al. succeeded in restricting numerically the ground-state energy eigenvalue of the semirelativistic Hamiltonian (1), (2), considered as a function of the coupling strength κ, to some remarkably narrow band. In particular, at the critical value $\kappa_{\mathrm{cr}} = 4/\pi$ of this coupling constant κ they found $0.4825 < E_0(\kappa = \kappa_{\mathrm{cr}})/2m < 0.4843$, which proved the nonvanishing of the ground-state energy eigenvalue E_0 at the critical coupling constant.

- Le Yaouanc et al.[13] gave the systematic series expansion of the eigenvalues of the Hamiltonian (1), (2) in powers of the coupling constant κ up to and including the order $O(\kappa^7)$ for arbitrary states with vanishing orbital angular momentum. For the ground-state energy E_0 this expansion reads

$$\frac{E_0}{2m} = 1 - \frac{\kappa^2}{8} - \frac{5\,\kappa^4}{128} + \frac{\kappa^5}{12\,\pi} + \frac{\kappa^6}{64} \ln \kappa + O(\kappa^6) .$$

  Higher orders of this expansion in the form given by Le Yaouanc et al., however, involve the logarithmic derivative of the gamma function, the digamma function, as well as its derivatives, the polygamma functions. Up to order $O(\kappa^4)$ the above result coincides with the improved lower bound of Martin and Roy.

The presence of non-analytic $\ln \kappa$ terms in the expansion quoted by Le Yaouanc et al. gives a clear hint at the non-analytic nature to be expected for the energy eigenvalues of the spinless relativistic Coulomb problem as functions of the coupling strength κ.

2.2. Analytic Upper Bounds

In view of the fact that exact solutions to the spinless relativistic Coulomb problem are still lacking, we seek for an analytic upper bound on the ground-state energy level of the semirelativistic Coulombic Hamiltonian (1), (2). To this end, we make use of a rather standard variational technique, which proceeds along the steps of the following, extremely simple recipe:[4]

1. Choose a suitable set of trial states $\{|\lambda\rangle\}$. The different members of this set are distinguished from each other by some sort of variational parameter λ.

2. Compute the set of expectation values of the Hamiltonian under consideration, H, with respect to these trial states $|\lambda\rangle$ in order to obtain $E(\lambda) \equiv \langle\lambda|H|\lambda\rangle$.

3. Determine, from the first derivative, that value $\lambda_{\min}$ of the variational parameter λ which minimizes the resulting, λ-dependent expression $E(\lambda)$.

4. Compute $E(\lambda)$ at the point of the minimum $\lambda_{\min}$ to find in this way the minimal expectation value $E(\lambda_{\min})$ of the Hamiltonian H in the Hilbert-space subsector of the chosen trial states $|\lambda\rangle$.

This minimum $E(\lambda_{\min})$ provides, of course, only an upper bound to the proper energy eigenvalue E of the Hamiltonian H: $E \leq E(\lambda_{\min})$.

Let us apply the above simple recipe to the case of the Coulomb potential, Eq. (2). For the Coulomb potential, the most reasonable choice of trial states is obviously the one for which the coordinate-space representation $\psi(\mathbf{x})$ of the states $|\lambda\rangle$ for vanishing radial and orbital angular momentum quantum numbers is given by the hydrogen-like trial functions

$$\psi(\mathbf{x}) = \sqrt{\frac{\lambda^3}{\pi}} \, \exp(-\lambda r) \, , \quad \lambda > 0 \, .$$

For this particular set of trial functions we obtain for the expectation values we shall be interested in, namely, the ones of the square of the momentum $\mathbf{p}$ and of the inverse of the radial coordinate r, respectively, evaluated with respect to the trial states $|\lambda\rangle$: $\langle\lambda|\mathbf{p}^2|\lambda\rangle = \lambda^2$ and $\langle\lambda|1/r|\lambda\rangle = \lambda$.

Now, as an immediate consequence of the fundamental postulates of any quantum theory, the expectation value of some given Hamiltonian H taken with respect to any normalized Hilbert-space state and, therefore, in particular, taken with respect to any of the above trial states must necessarily be larger than or equal to that eigenvalue E_0 of the Hamiltonian H which corresponds to its ground state: $E_0 \leq E(\lambda) \equiv \langle\lambda|H|\lambda\rangle$. The application to the semirelativistic Hamiltonian of Eq. (1) yields for the right-hand side of this inequality

$$E(\lambda) = 2 \left\langle \lambda \left| \sqrt{\mathbf{p}^2 + m^2} \right| \lambda \right\rangle + \langle\lambda|V(\mathbf{x})|\lambda\rangle \, . \tag{3}$$

In order to obtain a first crude estimate, we may take advantage of some trivial but nevertheless fundamental inequality. This inequality relates the expectation values of both the first and second powers of any self-adjoint but otherwise arbitrary operator $\mathcal{O} = \mathcal{O}^\dagger$ taken with respect to arbitrary Hilbert-space vectors $|\rangle$ (in the domain of $\mathcal{O}$) normalized to unity; it reads

$$|\langle\mathcal{O}\rangle| \leq \sqrt{\langle\mathcal{O}^2\rangle} \, .$$

Applying this inequality to the kinetic-energy part of the above expression for $E(\lambda)$, we may replace, in turn, $E(\lambda)$ by an upper bound which can be evaluated much easier than $E(\lambda)$ itself:

$$E(\lambda) \leq 2\sqrt{\langle\lambda|\mathbf{p}^2|\lambda\rangle + m^2} + \langle\lambda|V(\mathbf{x})|\lambda\rangle .$$

Identifying in this—as far as its evaluation is concerned, simplified—upper bound the until-now general potential $V(\mathbf{x})$ with the Coulomb potential (2) and inserting both of the λ-dependent expectation values given above implies $E(\lambda) \leq 2\sqrt{\lambda^2 + m^2} - \kappa\,\lambda$.

From this intermediate result, upon inspection of the limit $\lambda \to \infty$, we may state already at this rather early stage that, for the semirelativistic Hamiltonian (1), (2) to be bounded from below at all, the Coulombic coupling strength κ has to stay below a certain critical value: $\kappa \leq 2$.

The value of the variational parameter λ which minimizes the latter upper bound may be determined from the first derivative of this expression with respect to λ:

$$\lambda_{\min} = \frac{m\,\kappa}{2\sqrt{1 - \dfrac{\kappa^2}{4}}} .$$

For this particular value of λ, by shuffling together all our previous inequalities, we find that the energy eigenvalue corresponding to the ground state of the semirelativistic Hamiltonian (1) with the Coulomb potential (2), E_0, is bounded from above by[14]

$$E_0 \leq 2\,m\sqrt{1 - \frac{\kappa^2}{4}} .$$

The reality of this upper bound requires again $\kappa \leq 2$.

It is straightforward to improve this crude estimate by direct evaluation of Eq. (3). For our choice of trial functions, the resulting upper bound on the ground-state energy E_0 may be expressed in terms of the hypergeometric function F:[15]

$$E_0 \leq \left[\frac{128}{15\,\pi}\,F\left(-\frac{1}{2}, 2; \frac{7}{2}; 1 - \frac{m^2}{\lambda^2}\right) - \kappa\right]\lambda .$$

Quite obviously, in this case the minimizing value of the variational parameter λ must be determined numerically. As before, the limit $\lambda \to \infty$ tells us that now $\kappa \leq 16/3\,\pi$.

3. Numerical Approximation Methods

For an arbitrary interaction potential, it is, in general, not possible to find a closed solution to the spinless Salpeter equation. Therefore, several numerical approximation methods for the (numerical) solution of this equation have been introduced.[16,17,18,19,20] All of these numerical schemes aim at the conversion of the spinless Salpeter equation into an equivalent matrix eigenvalue problem, and confine themselves to the case of a spherically symmetric potential $V(r)$. Here, we briefly sketch the maybe most efficient among them, namely, the "semianalytical matrix method" developed by Lucha et al.[19]

For states of definite orbital angular momentum ℓ described by some wave function $\psi(\mathbf{x})$, we define the reduced radial wave function $u(r)$ by

$$\psi(\mathbf{x}) = \frac{u(r)}{r}\, \mathcal{Y}_{\ell m}(\theta, \phi)\ ;$$

here $\mathcal{Y}_{\ell m}(\theta, \phi)$ are the spherical harmonics of angular momentum ℓ and projection m. For this reduced radial wave function $u(r)$, an integral representation[16] of the spinless Salpeter equation may be found. Furthermore, since we consider only the case of equal (nonvanishing) masses m of the bound-state constituents, we may eliminate from the kinetic-energy part of this integral equation any dependence on the mass m by scaling the radial variable r like $x := m\, r$. Introducing the scaled reduced radial wave function $\tilde{u}(x) := u(x/m) = u(r)$ as well as the dimensionless energy eigenvalue $\tilde{E} := E/m$ and the dimensionless interaction potential $\tilde{V}(x) := V(x/m)/m = V(r)/m$, this particular integral representation of the spinless Salpeter equation becomes[16]

$$\left[\tilde{E} - \tilde{V}(x)\right]\tilde{u}(x) = \frac{2}{\pi}\int_0^\infty dy\, G_\ell(x,y)\left[-\frac{d^2}{dy^2} + \frac{\ell(\ell+1)}{y^2} + 1\right]\tilde{u}(y)\ , \qquad (4)$$

where the kernel G_ℓ is defined by

$$G_\ell(x,y) = 2^\ell\, z^{\ell+1}\left(\frac{1}{z}\frac{\partial}{\partial z}\right)^\ell \frac{1}{z}\left[(s-z)^{\ell/2} K_\ell\left(\sqrt{s-z}\right) - (s+z)^{\ell/2} K_\ell\left(\sqrt{s+z}\right)\right]\ ,$$

with $s \equiv x^2 + y^2$, $z \equiv 2\, x\, y$. Here K_ℓ is the modified Bessel function[15] of the second kind of order ℓ.

In the free case, i. e., for $V(r) \equiv 0$, as well as for potentials which are less singular than the Coulomb potential, i. e., $V(r) \propto 1/r^\eta$ with $\eta < 1$, the reduced radial wave function $u(r)$ behaves for small r, $r \to 0$, asymptotically like $u(r) \propto r^{\ell+1}$. Accordingly, we make the ansatz $\tilde{u}(x) = x^{\ell+1}\, w(x)$. With this substitution, the integro-differential equation (4) is equivalent to

$$\left[\tilde{E} - \tilde{V}(x)\right] x^{\ell+1}\, w(x) = \frac{2}{\pi}\int_0^\infty dy\, G_\ell(x,y)\, y^{\ell+1}\left[1 - \frac{d^2}{dy^2} - \frac{2(\ell+1)}{y}\frac{d}{dy}\right] w(y)\ .$$

In order to solve this scaled form of the spinless Salpeter equation, we expand the solution $w(x)$ of this integro-differential equation into a complete orthonormal system $\{f_n(x),\ n = 0, 1, 2, \ldots\}$ of basis functions for $L_2(R^3)$:

$$w(x) = \sum_{n=0}^N \lambda_n\, f_n(x)\ ,$$

with some set of real coefficients λ_n. For finite N, this expansion represents, of course, only an approximation to the exact solution $w(x)$. The crucial point[19] of the present approach is our rather sophisticated choice of the basis functions f_n which, at least in principle, allows for a thorough analytical treatment of the spinless Salpeter equation: $f_n(x) := \sqrt{2}\,\exp(-x)\,L_n(2x)$, where $L_n(x)$ are the Laguerre polynomials.[15]

After some straightforward algebra, the scaled spinless Salpeter equation may be cast into the form of a matrix eigenvalue equation for the coefficient vector $\lambda \equiv \{\lambda_n\}$ in the above expansion of the solution $w(x)$. In self-explanatory matrix notation, this eigenvalue equation is given by[19]

$$\tilde{E}\,\lambda = \left(P^{(\ell)}\right)^{-1}\left[\left(T^{(\ell)}\right)^T + V^{(\ell)}\right]\lambda\,, \tag{5}$$

where the "power matrix" $P_{nm}^{(\ell)}$ and the "potential matrix" $V_{nm}^{(\ell)}$ are defined by

$$P_{nm}^{(\ell)} := \int_0^\infty dx\, x^{\ell+1}\, f_n(x)\, f_m(x) = P_{mn}^{(\ell)}\,,\quad V_{nm}^{(\ell)} := \int_0^\infty dx\, x^{\ell+1}\, \tilde{V}(x)\, f_n(x)\, f_m(x) = V_{mn}^{(\ell)}\,,$$

and the "kinetic matrix" $T_{nm}^{(\ell)}$ represents the action of the kinetic term in the spinless Salpeter equation on the vector of basis functions f_n. In this way, the solution of the spinless Salpeter equation can be reduced to a simple matrix eigenvalue problem. The eigenvalues of this equation are the energies E of the bound state under consideration. The corresponding eigenvectors $\{\lambda_n\}$ give the radial wave functions $u(r)$ according to

$$u(r) = (mr)^{\ell+1}\sum_{n=0}^N \lambda_n\, f_n(mr)\,.$$

For $N = \infty$ this treatment would be exact. For $N < \infty$ it provides an approximation to the exact solution of increasing accuracy with increasing N, that is, with increasing size of the involved matrices. As is evident from the above construction, this procedure works analytically, at least for all potentials of the type "power times exponential," i. e., for all those potentials which involve only terms of the form $r^n \exp(-b_n\, r)$, with (maybe vanishing) constants b_n.

Let us illustrate this simple prescription for the case of bound states with vanishing orbital angular momentum ℓ of their constituents, that is, $\ell = 0$. All we need are the explicit expressions of the kinetic matrix $T^{(0)}$ and of the inverse of the power matrix $P^{(0)}$ for $\ell = 0$. The first few entries $T_{nm}^{(0)}$ in the kinetic matrix $T^{(0)}$ read[19]

$$T_{nm}^{(0)} = \frac{2}{\pi}\,\frac{8}{(2\,n + 2\,m + 3)!!}\,S_{nm}\,,$$

where the matrix S is given by

$$S = \begin{pmatrix} 1 & -3 & -5 & -21 & \cdots \\ -1 & 81 & -375 & -1029 & \cdots \\ -1 & -225 & 13125 & -77175 & \cdots \\ -3 & -441 & -55125 & 3565485 & \cdots \\ \vdots & \vdots & \vdots & \vdots & \ddots \end{pmatrix}\,.$$

(Note the range of the indices n and m: $n, m = 0, 1, \ldots, N$.) The elements $P_{nm}^{(0)}$ of the power matrix $P^{(0)}$ read explicitly[19]

$$P_{nm}^{(0)} \equiv \frac{1}{2} \int_0^\infty dx\, x\, \exp(-x)\, L_n(x)\, L_m(x) = \frac{1}{2} \begin{cases} 2n+1 & \text{for } m = n \\ -m & \text{for } m = n+1 \\ -n & \text{for } m = n-1 \\ 0 & \text{else} \end{cases}$$

$$= \frac{1}{2} \begin{pmatrix} 1 & -1 & 0 & 0 & \cdots \\ -1 & 3 & -2 & 0 & \cdots \\ 0 & -2 & 5 & -3 & \cdots \\ 0 & 0 & -3 & 7 & \cdots \\ \vdots & \vdots & \vdots & \vdots & \ddots \end{pmatrix}.$$

The inverse of this matrix, required for the matrix eigenvalue equation (5), depends explicitly on the size N of the involved matrices and is given by[19]

$$\left(P^{(0)}\right)_{nn}^{-1} = \sum_{k=n}^{N} \frac{2}{k+1} \qquad \text{for } n = 0, 1, \ldots, N,$$
$$\left(P^{(0)}\right)_{nm}^{-1} = \left(P^{(0)}\right)_{mn}^{-1} = \left(P^{(0)}\right)_{nn}^{-1} \qquad \text{for } n = 0, 1, \ldots, N,\ m = 0, 1, \ldots, n-1.$$

For instance, for $N = 3$ this inverse reads

$$\left(P^{(0)}\right)_{4\times4}^{-1} = \frac{1}{6} \begin{pmatrix} 25 & 13 & 7 & 3 \\ 13 & 13 & 7 & 3 \\ 7 & 7 & 7 & 3 \\ 3 & 3 & 3 & 3 \end{pmatrix}.$$

Consider, for example, the so-called funnel potential $V_F(r)$, which depends on just two parameters, namely, on the Coulomb coupling constant κ and on the slope a of the linear term: $V(r) = V_F(r) \equiv -\kappa/r + a\,r$. This funnel-shaped potential represents the prototype[3] of all "realistic," that is, phenomenologically acceptable, "QCD-inspired" static interquark potentials proposed for the description of hadrons as bound states of (constituent) quarks in the framework of potential models, the quarks inside a hadron being bound by the strong interactions arising from quantum chromodynamics. Now, according to the above definition, the scaled form of the funnel potential, $\tilde{V}_F(x)$, reads $\tilde{V}_F(x) = -\kappa/x + (a/m^2)\,x$, which entails for the corresponding potential matrix

$$V_F^{(\ell)} = -\kappa\, P^{(\ell-1)} + \frac{a}{m^2}\, P^{(\ell+1)}.$$

In particular, specializing again to the case $\ell = 0$, the matrix elements of the funnel potential, taken with respect to $\ell = 0$ states, are given by

$$V_F^{(0)} = -\kappa\, P^{(-1)} + \frac{a}{m^2}\, P^{(1)}.$$

Since, according to the orthonormalization condition for the basis functions $\{f_n(x)\}$, the power matrix $P^{(-1)}$ is identical to the unit matrix, $P^{(-1)} = 1_{N \times N}$, we have

$$V_{\mathrm{F}}^{(0)} = -\kappa + \frac{a}{m^2} P^{(1)} .$$

Working out in detail the $\ell = 1$ power matrix $P_{nm}^{(1)}$, we obtain[19]

$$
\begin{aligned}
P_{nn}^{(1)} &= \frac{3\,n\,(n+1)+1}{2} & \text{for } n = 0, 1, \ldots, N , \\
P_{n,n+1}^{(1)} &= -n\,(n+2) - 1 & \text{for } n = 0, 1, \ldots, N-1 , \\
P_{n,n+2}^{(1)} &= \frac{n\,(n+3)+2}{4} & \text{for } n = 0, 1, \ldots, N-2 , \\
P_{nm}^{(1)} &= 0 & \text{else .}
\end{aligned}
$$

Explicitly, the first entries of $P^{(1)}$ read

$$
P^{(1)} = \frac{1}{2}
\begin{pmatrix}
1 & -2 & 1 & 0 & \cdots \\
-2 & 7 & -8 & 3 & \cdots \\
1 & -8 & 19 & -18 & \cdots \\
0 & 3 & -18 & 37 & \cdots \\
\vdots & \vdots & \vdots & \vdots & \ddots
\end{pmatrix} .
$$

In summary, shuffling together the above results for the kinetic matrix $T^{(0)}$, for the inverse of the power matrix $P^{(0)}$, and for the potential matrix $V^{(0)}$ of the potential under consideration one ends up with a well-defined matrix eigenvalue equation, which represents the $\ell = 0$ special case of the general matrix form (5) of the spinless Salpeter equation.

4. Effectively Semirelativistic Hamiltonians of Nonrelativistic Form

Almost all of the troubles which one encounters when trying to apply the spinless Salpeter equation are obviously brought about by the nonlocality of the "square-root" operator of the relativistic kinetic energy, $\sqrt{\mathbf{p}^2 + m^2}$, in the Hamiltonian H, Eq. (1). In contrast to the nonrelativistic limit, obtained from the expansion of the square root up to the lowest $\mathbf{p}^2$-dependent order, $\sqrt{\mathbf{p}^2 + m^2} = m + \mathbf{p}^2/(2\,m) + \ldots$, the presence of this relativistic kinetic-energy operator prevents, in general, a thoroughly analytic discussion; one is forced to rely on numerical solutions of the problem. This (intrinsic) difficulty of any (semi-)relativistic formalism may be circumvented by approximating the Hamiltonian (1) by the corresponding "effectively semirelativistic" Hamiltonian,[21] which retains apparently the easier-to-handle nonrelativistic kinematics but resembles its relativistic counterpart to the utmost possible extent by replacing some of its basic parameters by effective ones which depend, in a well-defined manner, on the square of the relevant momentum $\mathbf{p}$. The main idea[3,4] of the construction[21] of these "effectively semirelativistic" Hamiltonians is as follows.

Applying again the above-mentioned fundamental inequality

$$|\langle \mathcal{O} \rangle| \le \sqrt{\langle \mathcal{O}^2 \rangle} \; ,$$

which holds for any self-adjoint operator $\mathcal{O} = \mathcal{O}^\dagger$ and arbitrary Hilbert-space vectors $|\rangle$ (in the domain of $\mathcal{O}$) normalized to unity, to the relativistic kinetic-energy operator $\sqrt{\mathbf{p}^2 + m^2}$ yields

$$\left\langle \sqrt{\mathbf{p}^2 + m^2} \right\rangle \le \sqrt{\langle \mathbf{p}^2 \rangle + m^2} \; .$$

By employing this inequality, we obtain for an arbitrary expectation value $\langle H \rangle$ of the semirelativistic Hamiltonian H, Eq. (1),

$$
\begin{aligned}
\langle H \rangle &= 2 \left\langle \sqrt{\mathbf{p}^2 + m^2} \right\rangle + \langle V \rangle \le 2 \sqrt{\langle \mathbf{p}^2 \rangle + m^2} + \langle V \rangle \\
&= 2 \frac{\langle \mathbf{p}^2 \rangle + m^2}{\sqrt{\langle \mathbf{p}^2 \rangle + m^2}} + \langle V \rangle = \left\langle 2 \frac{\mathbf{p}^2 + m^2}{\sqrt{\langle \mathbf{p}^2 \rangle + m^2}} + V \right\rangle \; .
\end{aligned}
\tag{6}
$$

From now on we specify the generic Hilbert-space vectors in all expectation values to be the eigenstates of our Hamiltonian H. In this case the expectation value of H, $\langle H \rangle$, as appearing, e. g., in Eq. (6), becomes the corresponding semirelativistic energy eigenvalue E, i. e., $E \equiv \langle H \rangle$, and the inequality (6) tells us that this energy eigenvalue is bounded from above by[3,4]

$$E \le \left\langle 2 \frac{\mathbf{p}^2 + m^2}{\sqrt{\langle \mathbf{p}^2 \rangle + m^2}} + V \right\rangle \; .$$

The operator within brackets on the right-hand side of this inequality may be regarded as some "effectively semirelativistic" Hamiltonian H_{eff} which possesses, quite formally, the structure of a nonrelativistic Hamiltonian,[3,4]

$$H_{\mathrm{eff}} \equiv 2 \frac{\mathbf{p}^2 + m^2}{\sqrt{\langle \mathbf{p}^2 \rangle + m^2}} + V = 2\,\hat{m} + \frac{\mathbf{p}^2}{\hat{m}} + V_{\mathrm{eff}} \; ,
\tag{7}$$

but involves, however, the effective mass[3,4]

$$\hat{m} = \frac{1}{2} \sqrt{\langle \mathbf{p}^2 \rangle + m^2}
\tag{8}$$

as well as the effective nonrelativistic potential[3,4]

$$V_{\mathrm{eff}} = \frac{2\,m^2}{\sqrt{\langle \mathbf{p}^2 \rangle + m^2}} - \sqrt{\langle \mathbf{p}^2 \rangle + m^2} + V = 2\,\hat{m} - \frac{\langle \mathbf{p}^2 \rangle}{\hat{m}} + V \; .
\tag{9}$$

The effective mass $\hat{m}$ given by Eq. (8) and the constant, i. e., coordinate-independent, term in the effective potential V_{eff} of Eq. (9), $2\,\hat{m} - \langle \mathbf{p}^2 \rangle / \hat{m}$, obviously depend on the expectation value of the square of the momentum $\mathbf{p}$, that is, on $\langle \mathbf{p}^2 \rangle$, and will therefore differ for different energy eigenstates.

Motivated by our above considerations, we propose to approximate the true energy eigenvalues E of the semirelativistic Hamiltonian H of Eq. (1) by the corresponding "effective" energy eigenvalues E_{eff}, defined as the expectation values of some effective Hamiltonian $\tilde{H}_{\text{eff}}$ taken with respect to the eigenstates $|\rangle_{\text{eff}}$ of its own, $E_{\text{eff}} = \langle \tilde{H}_{\text{eff}} \rangle_{\text{eff}}$, where the effective Hamiltonian $\tilde{H}_{\text{eff}}$, as far as its structure is concerned, is given by Eqs. (7) through (9) but is implicitly understood to involve the expectation values of $\mathbf{p}^2$ with respect to the effective eigenstates $|\rangle_{\text{eff}}$ (that is, $\langle \mathbf{p}^2 \rangle_{\text{eff}}$ in place of $\langle \mathbf{p}^2 \rangle$):[21]

$$\tilde{H}_{\text{eff}} = 4\,\tilde{m} + \frac{\mathbf{p}^2 - \langle \mathbf{p}^2 \rangle_{\text{eff}}}{\tilde{m}} + V , \quad \text{with } \tilde{m} = \frac{1}{2}\sqrt{\langle \mathbf{p}^2 \rangle_{\text{eff}} + m^2} .$$

Accordingly, the effective energy eigenvalues E_{eff} are given by a rather simple formal expression, viz., by[21]

$$E_{\text{eff}} = 4\,\tilde{m} + \langle V \rangle_{\text{eff}} . \tag{10}$$

We intend to elaborate our general prescription for the construction of effectively semirelativistic Hamiltonians $\tilde{H}_{\text{eff}}$ in more detail for the particular case of power-law potentials depending only on the radial coordinate $r \equiv |\mathbf{x}|$, i. e., for potentials of the form $V(r) = a\,r^n$ with some constant a. The reason for this restriction is twofold:

1. On the one hand, for power-law potentials the general virial theorem[22,23] in its nonrelativistic form[3,4] appropriate for the present case,

$$\left\langle \frac{\mathbf{p}^2}{\tilde{m}} \right\rangle_{\text{eff}} = \frac{1}{2} \left\langle r \frac{dV(r)}{dr} \right\rangle_{\text{eff}} ,$$

enables us to replace the expectation value of the potential in (10) immediately by a well-defined function of the expectation value of the squared momentum:

$$a \langle r^n \rangle_{\text{eff}} = \frac{2}{n} \frac{\langle \mathbf{p}^2 \rangle_{\text{eff}}}{\tilde{m}} .$$

This implies for the effective energy eigenvalues

$$E_{\text{eff}} = 4\,\tilde{m} + \frac{2}{n} \frac{\langle \mathbf{p}^2 \rangle_{\text{eff}}}{\tilde{m}} . \tag{11}$$

2. On the other hand, we take advantage of the fact that for power-law potentials it is possible to pass, without change of the fundamental commutation relations between coordinate variables and their canonically conjugated momenta, from the dimensional phase-space variables adopted at present to new, dimensionless phase-space variables and to rewrite the Hamiltonian in form of a Hamiltonian which involves only these dimensionless phase-space variables.[3] The eigenvalues ϵ of this dimensionless Hamiltonian are, of course, also dimensionless.[3] Applying this procedure, we find for the effective energy eigenvalues

$$E_{\text{eff}} - 4\,\tilde{m} + \frac{\langle \mathbf{p}^2 \rangle_{\text{eff}}}{\tilde{m}} = \left\langle \frac{\mathbf{p}^2}{\tilde{m}} + a\,r^n \right\rangle_{\text{eff}} = \left(\frac{a^2}{\tilde{m}^n} \right)^{\frac{1}{2+n}} \epsilon .$$

Combining both of the above expressions for E_{eff}, we obtain a relation which allows us to determine $\langle \mathbf{p}^2 \rangle_{\text{eff}}$ unambiguously in terms of the dimensionless energy eigenvalues ϵ:[21]

$$\langle \mathbf{p}^2 \rangle_{\text{eff}}^{2+n} = \frac{1}{4} \left(\frac{n}{2+n} \right)^{2+n} a^2 \, \epsilon^{2+n} \left(\langle \mathbf{p}^2 \rangle_{\text{eff}} + m^2 \right) . \tag{12}$$

For a given power n this equation may be solved for $\langle \mathbf{p}^2 \rangle_{\text{eff}}$. Insertion of the resulting expression into Eq. (11) then yields the corresponding eigenvalue E_{eff} of the effectively semirelativistic Hamiltonian $\tilde{H}_{\text{eff}}$.

Bound-state solutions are usually characterized by some radial quantum number n_r and orbital angular-momentum quantum number ℓ. For instance, for the harmonic oscillator, i. e., for $n = 2$, the dimensionless energy eigenvalues ϵ are given by[3] $\epsilon = 2\,N$ with $N = 2\,n_r + \ell + \frac{3}{2}$.

For the Coulomb potential $V(r) = -\kappa/r$, that is, for $n = -1$, Eq. (12) reduces to a linear equation for the expectation value $\langle \mathbf{p}^2 \rangle_{\text{eff}}$. Inserting the well-known expression[3] for the dimensionless energy eigenvalues ϵ of the (nonrelativistic) Coulomb problem, $\epsilon = -(2\,N)^{-2}$ with $N = n_r + \ell + 1$, we obtain from this linear equation for $\langle \mathbf{p}^2 \rangle_{\text{eff}}$

$$\langle \mathbf{p}^2 \rangle_{\text{eff}} = \frac{\kappa^2 \, m^2}{16\,N^2 - \kappa^2} \, ,$$

and, after inserting this expression into Eq. (11), for the effective energy eigenvalue[21]

$$E_{\text{eff}} = \frac{m}{N} \frac{8\,N^2 - \kappa^2}{\sqrt{16\,N^2 - \kappa^2}} \, .$$

The capability of this effective formalism to imitate the semirelativistic treatment becomes apparent by inspecting, for instance, the behaviour of the energy eigenvalues E_{eff} for large orbital angular momenta ℓ. For the linear potential $V(r) = a\,r$, that is, for $n = 1$, the effective formalism yields (in the ultrarelativistic limit, i. e., for $m = 0$) linear Regge trajectories:[21] $E_{\text{eff}}^2 = 9\,a\,\ell$. This fits very well to the exact semirelativistic result,[24,25] $E_{\text{SR}}^2 = 8\,a\,\ell$, but is in clear contrast to the nonrelativistic approach, which gives[3,4] $E_{\text{NR}}^2 = 9\,(a^2/4\,m)^{2/3}\,\ell^{4/3}$.

5. Summary and Conclusion

In spite of the regrettable fact that, at present, not even for the Coulomb potential the exact solution for the ground-state energy is known, the spinless Salpeter equation is certainly a useful tool for the semirelativistic description of bound states consisting of scalar bosons only as well as of the spin-averaged spectra of bound states consisting of fermionic constituents.

Acknowledgements

We would like to thank J. Ball, L. P. Fulcher, and I. W. Herbst for some interesting discussions.

References

1. E. E. Salpeter and H. A. Bethe, *Phys. Rev.* **84** (1951) 1232.
2. E. E. Salpeter, *Phys. Rev.* **87** (1952) 328.
3. W. Lucha, F. F. Schöberl, and D. Gromes, *Phys. Rep.* **200** (1991) 127.
4. W. Lucha and F. F. Schöberl, *Int. J. Mod. Phys.* **A7** (1992) 6431.
5. W. Lucha, H. Rupprecht, and F. F. Schöberl, *Phys. Rev.* **D46** (1992) 1088.
6. I. W. Herbst, *Commun. Math. Phys.* **53** (1977) 285; **55** (1977) 316 (addendum).
7. B. Durand and L. Durand, *Phys. Rev.* **D28** (1983) 396; and erratum (to be published).
8. P. Castorina, P. Cea, G. Nardulli, and G. Paiano, *Phys. Rev.* **D29** (1984) 2660.
9. G. Hardekopf and J. Sucher, *Phys. Rev.* **A30** (1984) 703;
   G. Hardekopf and J. Sucher, *Phys. Rev.* **A31** (1985) 2020.
10. A. Martin and S. M. Roy, *Phys. Lett.* **B233** (1989) 407.
11. J. C. Raynal, S. M. Roy, V. Singh, A. Martin, and J. Stubbe, *Phys. Lett.* **B320** (1994) 105.
12. M. F. Barnsley, *J. Phys.* **A11** (1978) 55.
13. A. Le Yaouanc, L. Oliver, and J.-C. Raynal, Orsay preprint LPTHE Orsay 93/43 (1993); *Ann. Phys. (N. Y.)* (to be published).
14. W. Lucha and F. F. Schöberl, Vienna preprint HEPHY-PUB 596/94 (1994) (unpublished);
    W. Lucha and F. F. Schöberl, Vienna preprint HEPHY-PUB 606/94 (1994); *Phys. Rev. D* (in print).
15. *Handbook of Mathematical Functions*, eds. M. Abramowitz and I. A. Stegun (Dover, New York, 1964).
16. L. J. Nickisch, L. Durand, and B. Durand, *Phys. Rev.* **D30** (1984) 660; **D30** (1984) 1995 (erratum).
17. S. Jacobs, M. G. Olsson, and C. Suchyta III, *Phys. Rev.* **D33** (1986) 3338; **D34** (1986) 3536 (erratum).
18. L. Durand and A. Gara, *J. Math. Phys.* **31** (1990) 2237.
19. W. Lucha, H. Rupprecht, and F. F. Schöberl, *Phys. Rev.* **D45** (1992) 1233.
20. L. P. Fulcher, Z. Chen, and K. C. Yeong, *Phys. Rev.* **D47** (1993) 4122;
    L. P. Fulcher, *Phys. Rev.* **D50** (1994) 447.
21. W. Lucha, F. F. Schöberl, and M. Moser, Vienna preprint HEPHY-PUB 594/93 (1993) (unpublished);
    W. Lucha and F. F. Schöberl, Vienna preprint HEPHY-PUB 611/94 (1994).
22. W. Lucha and F. F. Schöberl, *Phys. Rev. Lett.* **64** (1990) 2733.
23. W. Lucha and F. F. Schöberl, *Mod. Phys. Lett.* **A5** (1990) 2473.
24. J. S. Kang and H. J. Schnitzer, *Phys. Rev.* **D12** (1975) 841;
    A. Martin, *Z. Phys.* **C32** (1986) 359;
    Yu. A. Simonov, *Phys. Lett.* **B226** (1989) 151.
25. W. Lucha, H. Rupprecht, and F. F. Schöberl, *Phys. Lett.* **B261** (1991) 504.

WILSON LOOPS, $q\bar{q}$ and $3q$ POTENTIALS,

BETHE–SALPETER EQUATION

N. BRAMBILLA and G.M. PROSPERI *

Dipartimento di Fisica dell'Università di Milano and I.N.F.N.

Via Celoria 16, 20133 Milano

ABSTRACT

The derivation of the $q\bar{q}$ and the $3q$ potential for two dynamical quarks in a Wilson–loop context is reviewed. Some improvements are introduced. Only the usual assumptions in the evaluation of the Wilson loop integrals and expansions in the quark velocities are required for the result. It is shown that under the same assumptions it is possible to obtain the relativistic flux–tube lagrangian and a $q\bar{q}$ Bethe–Salpeter equation with a confining kernel for spinless quarks.

1. Introduction

In this paper first we review the derivation of the $q\bar{q}$ and the $3q$ semirelativistic potentials for dynamical quarks as has been given in preceding papers[1] (for a general review on the subject see[2]) and introduce some significant improvements. Then we show that, under the same assumptions and in the case of spinless quarks, a Bethe–Salpeter equation with a confining kernel can be obtained.

The basic objects considered in the derivation are the appropriate Wilson loop integrals $W_{q\bar{q}}$ and W_{3q} and the basic assumptions are:

i) the quantities $i \ln W$ can be expressed as the sum of a short range contribution $i \ln W^{SR}$ and a long range one $i \ln W^{LR}$;

ii) the SR–term can be obtained simply from a perturbative expansion and the LR–term from a strong coupling expansion (in practice by the area law).

The improvement consists in the fact that an ad–hoc explicit instantaneous approximation is no longer required and only expansions in the quark velocities are used. Furthermore, the $O(\alpha_s^2)$ contribution is explicitly taken into account in the static part of the potential and it is shown that a covariant Lorentz gauge as well as the Coulomb gauge can be used.

As it is well known, the arguments in favour of the two assumptions are asymptotic freedom and the observation that the SR–part of the potential vanishes for $r \to \infty$, while the LR–part vanishes for $r \to 0$. Obviously, with the simple additivity assumption i), the resulting potential or kernel is expected to be inaccurate at intermediate distances; interferences of the two mechanisms should be important there. However, no attempt is made in this paper to use a more sophisticated approximation scheme of the type proposed e.g. in Refs.[3] (see also[4]).

---

*Presenting author

In Sec. 2 we discuss the evaluation of the Wilson loop integrals, in Secs. 3 and 4 we derive the $q\bar{q}$ and the $3q$ potentials respectively, in Secs. 5 and 6 we sketch the derivation of the flux–tube lagrangian and of the Bethe–Salpeter equation.

2. Wilson loop integrals

For the $q\bar{q}$ case the basic object is

$$W_{q\bar{q}} = \frac{1}{3} \left\langle \text{Tr P} \exp \left(ig \oint_\Gamma dx^\mu \, A_\mu(x) \right) \right\rangle . \tag{1}$$

Here the integration loop Γ is assumed to be made by an arbitrary world line Γ_1 between an initial position y_1 at the time t_i and a final one x_1 at the time t_f for the quark $(t_i < t_f)$, a similar world line Γ_2 described in the reverse direction from x_2 at the time t_f to y_2 at the time t_i for the antiquark and two straight lines at fixed times which connect x_1 to x_2, y_2 to y_1 and close the contour. As usual $A_\mu(x) = \frac{1}{2}\lambda_a A^a_\mu(x)$, P prescribes the ordering of the color matrices (from right to left) according to the direction fixed on the loop and the angular brackets denote the functional integration.

Integrating explicitly the fermion fields, for any functional of the gauge field alone one obtains

$$\langle f[A] \rangle = \frac{\int \mathcal{D}[A] M_f(A) f[A] e^{iS[A]}}{\int \mathcal{D}[A] M_f(A) e^{iS[A]}} , \tag{2}$$

where $S[A]$ denotes the pure gauge action plus the gauge–fixing terms and $M_f[A]$ is the fermionic determinant

$$
\begin{aligned}
M_f[A] &= \text{Det} \prod_j [1 + igA(i\partial - m_j)^{-1}] = 1 + \sum_j [g \int d^4x \text{Tr}(iA(x)S_F^{(m_j)}(0)) \\
&- \frac{1}{2}g^2 \int d^4x \int d^4y \text{Tr}(A(x)iS_F^{(m_j)}(x-y)A(y)iS_F^{(m_j)}(y-x)) + ...].
\end{aligned} \tag{3}
$$

Using the above equations and writing the gauge field lagrangian as the sum of the free and the interaction parts, $\mathcal{L}(A) = \mathcal{L}_0 + \mathcal{L}_{\text{int}}$, we have the perturbative expansion

$$
\begin{aligned}
W_{q\bar{q}}^{\text{pert}} &= \frac{1}{3} \sum_{n=0}^\infty \sum_{p=0}^\infty \frac{1}{n!p!} \langle \text{TrP}(ig \oint dz^\mu A_\mu)^n (i \int d^4x \mathcal{L}_{\text{int}}(x))^p \rangle_0 = \\
&= \frac{1}{3} \sum_{n=0}^\infty \sum_{p=0}^\infty \frac{(ig)^n i^p}{p!} \int d^4x_1 ... \int d^4x_p \oint ... \oint_{z_1 > z_2 > ... > z_n} dz_1^{\mu_1} ... dz_n^{\mu_n} \\
&\langle \text{Tr}[A_{\mu_1}(z_1) ... A_{\mu_n}(z_n)] \mathcal{L}_{\text{int}}(x_1) ... \mathcal{L}_{\text{int}}(x_p) \rangle_0,
\end{aligned} \tag{4}
$$

where, due to (3), the single terms must be understood as expansions in g in turn. Then, identifying W^{SR} with W^{pert} according to assumption ii), we obtain in graphical terms (we omit graphs that are obtained by permutation of other ones or completely cancelled by renormalization)

$$i \ln W_{q\bar{q}}^{SR} = $$

$$(5)$$

where the external circuit stands for the Wilson loop Γ, and the inserted lines for ordinary free propagators. Notice the term which includes a quark-antiquark loop, which obviously comes from (3).

The various quantities occurring in (5) have been extensively studied from the point of view of renormalization[5]. To our knowledge however no explicit evaluation in closed form has been given other than in very special cases[6]. For the purpose of the derivation of a semirelativistic potential, an evaluation in terms of an expansion in the quark velocities shall be sufficient.

Let $(z_j^0 = t_j,\ z_j = z_j(t))$ be the equation for the world lines of the quark and the antiquark and set $\dot{z}_j^\mu = dz_j^\mu/dt = (1, \dot{z}_j)$. The first-order term in $\alpha_s = g^2/4\pi$ can be written explicitly as

$$(i \ln W^{SR})_{q\bar{q}}^{(1)} = \frac{4}{3} g^2 \int_{t_i}^{t_f} dt_1 \int_{t_i}^{t_f} dt_2 \dot{z}_1^\mu(t_1) \dot{z}_2^\nu(t_2) D_{\mu\nu}(z_1(t_1) - z_2(t_2)), \qquad (6)$$

where the limit for large $t_f - t_i$ has been understood and the contribution from the equal-time lines are neglected. Performing the change of variables $t = \frac{t_1 + t_2}{2}$, $\tau = t_1 - t_2$, expanding z_1 and z_2 around t,

$$z_1(t_1) = z_1(t) + \frac{1}{2}\tau \dot{z}_1(t) + \frac{1}{8}\tau^2 \ddot{z}_1(t) + \ldots, \qquad z_2(t_2) = z_2(t) - \frac{1}{2}\dot{z}_2(t) + \frac{1}{8}\tau^2 \ddot{z}_2(t) - \ldots$$

$$(7)$$

and integrating over τ (between $-\infty$ and $+\infty$), in the Coulomb gauge we obtain immediately

$$(i \ln W^{SR})^{(1)} = \int_{t_i}^{t_f} dt \{ -\frac{4}{3} \frac{\alpha_s}{r} [1 - \frac{1}{2}(\delta^{hk} + \hat{r}^h \hat{r}^k)) \dot{z}_1^h \dot{z}_2^k + \ldots] \}, \qquad (8)$$

with $r = z_1 - z_2$ and $\hat{r} = r/r$. If we had worked, e.g., in the Feynman gauge, we would have obtained

$$(i \ln W^{SR})_{q\bar{q}}^{(1)} = \int_{t_i}^{t_f} dt \{ -\frac{4}{3} \frac{\alpha_s}{r} [1 - \dot{z}_1 \cdot \dot{z}_2 + \frac{1}{8}((\dot{z}_1 + \dot{z}_2)^2 + r \cdot \ddot{r}) - \frac{1}{8 r^2}((r \cdot (\dot{z}_1 + \dot{z}_2))^2 + \ldots] \}$$

$$(9)$$

from which (8) can be recovered by eliminating the acceleration term by partial integration. This is a consequence of the gauge invariance of the Wilson integral.

In a similar way, after renormalization, we can obtain for the α_s^2 term in the static limit

$$(i \ln W^{\mathrm{SR}})^{(2)}_{q\bar{q}} = \int_{t_i}^{t_f} dt \left\{ -\frac{4}{3} \frac{\alpha_s^2}{4\pi} \frac{1}{r} [(\frac{66 - 4N_f}{3})(\ln \mu r + \gamma) + A] + \dots \right\}. \tag{10}$$

In Eq.(10) μ is the renormalization scale and A is a constant that depends on the renormalization convention. In the $\overline{\mathrm{MS}}$ scheme $A = \frac{5}{6}(\frac{66 - 4N_f}{3}) - 8$.

Let us come to the LR part of the Wilson integral. We shall make the assumption

$$i \ln W^{\mathrm{LR}}_{q\bar{q}} = \sigma S_{\min} + \frac{1}{2} C P, \tag{11}$$

where $S_{\min}$ denotes the minimal surface enclosed by the loop Γ and P its lenght. Eq. (11) is suggested by the pure lattice gauge theory and it is believed to be true in the so–called quenched approximation, i.e. when we replace $M_f(A)$ by 1 in (2). Corrections to the pure potential theory (pair creation effects) should be introduced for this fact but they shall not be considered here.

In more explicit terms (11) can be written as

$$i \ln W^{\mathrm{LR}}_{q\bar{q}} = \sigma \min \int_{t_i}^{t_f} dt \int_0^1 ds [-(\frac{\partial x}{\partial t})^2 (\frac{\partial x}{\partial s})^2 + (\frac{\partial x^\mu}{\partial t} \frac{\partial x_\mu}{\partial s})^2]^{\frac{1}{2}} + \frac{1}{2} C \sum_{j=1,2} \int_{t_i}^{t_f} dt [\dot{z}_j^\mu \dot{z}_{j\mu}]^{\frac{1}{2}}, \tag{12}$$

where the minimum is taken over all surfaces of equation $x^\rho = x^\rho(t, s)$ having Γ as contour. Obviously $x^0 = t$, $\mathbf{x}(t, 1) = \mathbf{z}_1(t)$ and $\mathbf{x}(t, 0) = \mathbf{z}_2(t)$.

By solving the appropriate Euler equations and expanding in the velocities, we obtain

$$\mathbf{x}_{\min}(t, s) = s \mathbf{z}_1(t) + (1 - s) \mathbf{z}_2(t) - \frac{1}{2} s(1 - s)[\eta + \frac{1}{3}(1 + s)\zeta] + \dots \tag{13}$$

with $\eta = (\dot{\mathbf{r}} \cdot \dot{\mathbf{z}}_2 - \mathbf{r} \cdot \ddot{\mathbf{z}}_2)\mathbf{r} + (\mathbf{r} \cdot \dot{\mathbf{r}})\dot{\mathbf{z}}_2 - 2(\mathbf{r} \cdot \dot{\mathbf{z}}_2)\dot{\mathbf{r}} + r^2\ddot{\mathbf{z}}_2$ and $\zeta = -(\mathbf{r} \cdot \dot{\mathbf{r}})\dot{\mathbf{r}} + r^2\ddot{\mathbf{r}} + (\dot{r}^2 - \mathbf{r} \cdot \ddot{\mathbf{r}})\mathbf{r}$.

Actually it can be checked that the $O(v^2)$ term in (13) does not contribute to $S_{\min}$ at order v^2 (such a term is however important in principle for the evaluation of the functional derivatives). Replacing (13) in (12), finally we have

$$i \ln W^{\mathrm{LR}}_{q\bar{q}} = \int_{t_i}^{t_f} dt \, \sigma r \int_0^1 ds \, [1 - (s\dot{\mathbf{z}}_{1\mathrm{T}} + (1 - s)\dot{\mathbf{z}}_{2\mathrm{T}})^2]^{\frac{1}{2}} + \frac{1}{2} C \sum_{j=1}^2 \int_{t_i}^{t_f} (1 - \dot{z}_j^h \dot{z}_j^h)^{\frac{1}{2}} =$$

$$= \int_{t_i}^{t_f} dt \, \sigma r \, [1 - \frac{1}{6}(\dot{\mathbf{z}}_{1\mathrm{T}}^2 + \dot{\mathbf{z}}_{2\mathrm{T}}^2 + \dot{\mathbf{z}}_{1\mathrm{T}} \cdot \dot{\mathbf{z}}_{2\mathrm{T}})] + \dots + \frac{1}{2} C \sum_{j=1}^2 \int_{t_i}^{t_f} (1 - \frac{1}{2} \dot{z}_j^h \dot{z}_j^h + \dots), \tag{14}$$

where $\dot{\mathbf{z}}_{j\mathrm{T}}$ denotes the *transversal part* of $\dot{\mathbf{z}}_j$, $\dot{z}_{j\mathrm{T}}^h = (\delta^{hk} - \hat{r}^h \hat{r}^k)\dot{z}_j^h$.

In conclusion, we can write

$$i \ln W_{q\bar{q}} = (i \ln W^{\mathrm{SR}}_{q\bar{q}})^{(1)} + (i \ln W^{\mathrm{SR}}_{q\bar{q}})^{(2)} + \dots + i \ln W^{\mathrm{LR}}_{q\bar{q}}, \tag{15}$$

with the various terms as given by (8), (10), (14).

Let us turn to the three–quark system. In this case the basic quantity is

$$W_{3q} = \frac{1}{3!} \left\langle \varepsilon_{a_1 a_2 a_3} \varepsilon_{b_1 b_2 b_3} \left[\mathrm{P} \exp \left(ig \int_{\overline{\Gamma}_1} dx^{\mu_1} A_{\mu_1}(x) \right) \right]^{a_1 b_1} \right.$$

$$\left. \left[\mathrm{P} \exp \left(ig \int_{\overline{\Gamma}_2} dx^{\mu_2} A_{\mu_2}(x) \right) \right]^{a_2 b_2} \left[\mathrm{P} \exp \left(ig \int_{\overline{\Gamma}_3} dx^{\mu_3} A_{\mu_3}(x) \right) \right]^{a_3 b_3} \right\rangle . \qquad (16)$$

Here a_j, b_j are colour indices, $j = 1, 2, 3$ and $\overline{\Gamma}_j$ denote the curve made by: the world lines Γ_j for the quark j between the times t_i and t_f ($t_i < t_f$), a straight line on the surface $t = t_i$ merging from an arbitrary fixed point I (which we also denote by y_M) and connected to the world line, another straight line on the surface $t = t_f$ connecting the world line to a second fixed point F (also denoted as x_M).

Under the assumptions i) and ii) we can write in place of (6) and (11)

$$i \ln W_{3q} = \frac{2}{3} g^2 \sum_{i<j} \int_{\Gamma_i} dx_i^\mu \int_{\Gamma_j} dx_j^\nu \, i D_{\mu\nu}(x_i - x_j) + \sigma S_{\min} + \frac{1}{3} C P . \qquad (17)$$

Here the perturbative term is taken at the lowest order in α_s and $S_{\min}$ denotes the minimum among all the surfaces made by three sheets having the curves $\overline{\Gamma}_1, \overline{\Gamma}_2$ and $\overline{\Gamma}_3$ as contours and joining on a line Γ_M connecting I with F (the minimum is understood at fixed $\overline{\Gamma}_j$ as the surfaces and Γ_M change). Obviously, P denotes the total length of $\overline{\Gamma}_1, \overline{\Gamma}_2$ and $\overline{\Gamma}_3$. Notice that a priori the constants σ and C occurring in (17) could be different from those occurring in (11); however, the fact that when two quarks coincide the potential derived from (17) must coincide with that derived from (11) (in a colour singlet state two quarks are equivalent to an antiquark) grants that they must be actually equal.

The right–hand side of (17) can be evaluated as an expansion in $\dot{z}_j$ on the same foot used for Eqs.(8), (10) and (14). In particular up to the second order in the velocities, $S_{\min}$ coincides with the surface described by the equations

$$\mathbf{x}_j^{\min}(t, s) = s \mathbf{z}_j(t) + (1 - s) \mathbf{z}_M(t), \qquad j = 1, 2, 3 . \qquad (18)$$

Here $\mathbf{z}_M(t)$ is constructed from the positions $\mathbf{z}_1(t)$, $\mathbf{z}_2(t)$ and $\mathbf{z}_3(t)$ of the three quarks according to the following rule: if no angle in the triangle made by $\mathbf{z}_1(t)$, $\mathbf{z}_2(t)$ and $\mathbf{z}_3(t)$ exceeds 120^0 (configuration I), $\mathbf{z}_M(t)$ coincides with the point inside the triangle which sees the three sides under the same angle 120^0; if one of the three angles in the triangle is $\geq 120^0$ (configuration II), $\mathbf{z}_M(t)$ coincides with the corresponding vertex, let us say $\mathbf{z}_{\bar{j}}(t)$.

In conclusion, the result is

$$i \ln W_{3q} = \int_{t_i}^{t_f} dt \left\{ \sum_{j<l} \left[-\frac{2}{3} \frac{\alpha_s}{r_{jl}} + \frac{1}{2} \frac{2}{3} \frac{\alpha_s}{r_{jl}} (\delta^{hk} + \hat{r}_{jl}^h \hat{r}_{jl}^k) \dot{z}_j^h \dot{z}_l^k \right] + \right.$$

$$\left. + \sigma \sum_{j=1}^{3} r_j \left[1 - \frac{1}{6} (\dot{z}_{jT_j}^2 + \dot{z}_{MT_j}^2 + \dot{z}_{jT_j} \cdot \dot{z}_{MT_j}) \right] + \frac{C}{3} \sum_{j=1}^{3} \int_{t_i}^{t_f} dt (1 - \frac{1}{2} \dot{z}_j^h \dot{z}_j^h) \right\}, \qquad (19)$$

where $\mathbf{r}_{jl} = \mathbf{r}_j - \mathbf{r}_l \equiv \mathbf{z}_j - \mathbf{z}_l$, $\mathbf{r}_j = \mathbf{z}_j - \mathbf{z}_M$ and the transversal prescription T_j is now referred to $\mathbf{r}_j$. Furthermore we can notice that the quantity $\dot{z}_M$ can be obtained by deriving the equation $\sum_{j=1}^3 (\mathbf{r}_j/r_j) = 0$. We have indeed $\sum_{j=1}^3 \frac{1}{r_j}(\delta^{hk} - \hat{r}_j^h \hat{r}_j^k)\dot{z}_j^k = \sum_{j=1}^3 \frac{1}{r_j}(\delta^{hk} - \hat{r}_j^h \hat{r}_j^k)\dot{z}_M^k$. Obviously in configuration II we have $\dot{z}_M = \dot{z}_j$.

3. Quark–antiquark potential

The starting point is the gauge invariant quark-antiquark $(q_1, \bar{q}_2)$ Green function (for definiteness let us assume the two particles to have different flavours)

$$G(x_1, x_2; y_1, y_2) = \frac{1}{3}\langle 0|\mathrm{T}\psi_2^c(x_2)U(x_2, x_1)\psi_1(x_1)\overline{\psi}_1(y_1)U(y_1, y_2)\overline{\psi}_2^c(y_2)|0\rangle =$$

$$= \frac{1}{3}\mathrm{Tr}\langle U(x_2, x_1)S_1^F(x_1, y_1|A)U(y_1, y_2)C^{-1}S_2^F(y_2, x_2|A)C\rangle, \qquad (20)$$

where c denotes the charge-conjugate fields, C is the charge-conjugation matrix, U the path-ordered gauge string $U(b, a) = \mathrm{P} \exp\left(ig \int_a^b dx^\mu A_\mu(x)\right)$ (the integration path being the straight line joining a to b), S_1^F and S_2^F the quark propagators in an external gauge field A^μ.

We assume $x_1^0 = x_2^0 = t_f$, $y_1^0 = y_2^0 = t_i$ (with $t_f - t_i > 0$ and large) and note that S_j^F are 4×4 Dirac type matrices. Then, performing a Foldy–Wouthuysen transformation on G, we can replace S_j^F with a Pauli propagator K_j (a 2×2 matrix in the spin indices) and obtain a two-particle Pauli-type Green function K. Solving the Schrödinger-like equation for K_j by the path–integral technique and replacing it in the expression of K, we obtain even this quantity in the form of a path integral on the world lines of the two quarks (see Ref.[1] for details):

$$K(\mathbf{x}_1, \mathbf{x}_2, \mathbf{y}_1, \mathbf{y}_2; t_f - t_i) = \int_{\mathbf{z}_1(t_i)=\mathbf{y}_1}^{\mathbf{z}_1(t_f)=\mathbf{x}_1} \mathcal{D}[\mathbf{z}_1, \mathbf{p}_1] \int_{\mathbf{z}_2(t_i)=\mathbf{y}_2}^{\mathbf{z}_2(t_f)=\mathbf{x}_2} \mathcal{D}[\mathbf{z}_2, \mathbf{p}_2]$$

$$\exp\{i \int_{t_i}^{t_f} dt \sum_{j=1}^2 [\mathbf{p}_j \cdot \dot{\mathbf{z}}_j - m_j - \frac{\mathbf{p}_j^2}{2m_j} + \frac{\mathbf{p}_j^4}{8m_j^3}]\}\langle \frac{1}{3}\mathrm{Tr}\,\mathrm{T}_s\,\mathrm{P}\exp\{ig \oint_\Gamma dx^\mu A_\mu(x)$$

$$+ \sum_{j=1}^2 \frac{ig}{m_j}\int_{\Gamma_j} dx^\mu(S_j^l\hat{F}_{l\mu}(x) - \frac{1}{2m_j}S_j^l\varepsilon^{lkr}p_j^k F_{\mu r}(x) - \frac{1}{8m_j}D^\nu F_{\nu\mu}(x))\}\rangle. \qquad (21)$$

Here T_s is the time-ordering prescription for the spin matrices; P, Tr, Γ, Γ_1 and Γ_2 are defined as in Eq.(1). Furthermore, as usual $F^{\mu\nu} = \partial^\mu A^\nu - \partial^\nu A^\mu + ig[A^\mu, A^\nu]$, $\hat{F}^{\mu\nu} = \frac{1}{2}\varepsilon^{\mu\nu\rho\sigma}F_{\rho\sigma}$ and $D^\nu F_{\nu\mu} = \partial^\nu F_{\nu\mu} + ig[A^\nu, F_{\nu\mu}]$, $\varepsilon^{\mu\nu\rho\sigma}$ being the four-dimensional Ricci symbol.

In order to show that the interaction between q_1 and $\bar{q}_2$ can be described in terms of a semirelativistic potential we must check that at the order $\frac{1}{m^2}$ we can write

$$\left\langle \frac{1}{3}\mathrm{Tr}\,\mathrm{T}_s\,\mathrm{P}\exp\ldots\right\rangle = \mathrm{T}_s \exp\left[-i\int_{t_i}^{t_f} dt\,V^{q\bar{q}}(\mathbf{z}_1(t), \mathbf{z}_2(t), \mathbf{p}_1(t), \mathbf{p}_2(t), \mathbf{S}_1, \mathbf{S}_2)\right], \qquad (22)$$

for some $V^{q\bar{q}}$. Expanding the logarithm on the left-hand side of (22), this is equivalent to state

$$i \ln W_{q\bar{q}} + i \sum_{j=1}^{2} \frac{ig}{m_j} \int_{\Gamma_j} dx^\mu \left(S_j^l \langle\langle \hat{F}_{l\mu}(x) \rangle\rangle - \frac{1}{2m_j} S_j^l \epsilon^{lkr} p_j^k \langle\langle F_{\mu r}(x) \rangle\rangle - \right.$$

$$\left. - \frac{1}{8m_j} \langle\langle D^\nu F_{\nu\mu}(x) \rangle\rangle \right) - \frac{1}{2} \sum_{j,j'} \frac{ig^2}{m_j m_{j'}} T_s \int_{\Gamma_j} dx^\mu \int_{\Gamma_{j'}} dx'^\sigma \, S_j^l \, S_{j'}^k$$

$$\left(\langle\langle \hat{F}_{l\mu}(x) \hat{F}_{k\sigma}(x') \rangle\rangle - \langle\langle \hat{F}_{l\mu}(x) \rangle\rangle \langle\langle \hat{F}_{k\sigma}(x') \rangle\rangle \right) + \ldots = \left[\int_{t_i}^{t_f} dt \, V^{q\bar{q}} \right], \quad (23)$$

with the notation

$$\langle\langle f[A] \rangle\rangle = \frac{\frac{1}{3} \langle \mathrm{Tr} \, \mathrm{P} \, [\exp(ig \oint_\Gamma dx^\mu \, A_\mu(x))] \, f[A] \rangle}{\frac{1}{3} \langle \mathrm{Tr} \, \mathrm{P} \exp(ig \oint_\Gamma dx^\mu \, A_\mu(x)) \rangle} \quad (24)$$

and $W_{q\bar{q}}$ as given by (15).

Notice that after replacing $\dot{z}_j$ by $\frac{p_j}{m_j}$ in Eqs.(8), (10) and (14), the expression resulting for $i \ln W_{q\bar{q}}$ is already of the desired form. Concerning the spin–dependent part we observe that the occurring field expectation values can be expressed in terms of $i \ln W_{q\bar{q}}$ by the functional derivatives

$$g \langle\langle F_{\mu\nu}(z_1) \rangle\rangle = \frac{\delta(i \ln W_{q\bar{q}})}{\delta S^{\mu\nu}(z_1)}, \quad (25)$$

$$g^2 \left(\langle\langle F_{\mu\nu}(z_1) F_{\rho\sigma}(z_2) \rangle\rangle - \langle\langle F_{\mu\nu}(z_1) \rangle\rangle \langle\langle F_{\rho\sigma}(z_2) \rangle\rangle \right) =$$

$$= \frac{\delta^2 \ln W_{q\bar{q}}}{\delta S^{\mu\nu}(z_1) \delta S^{\rho\sigma}(z_2)} = -ig \frac{\delta}{\delta S^{\rho\sigma}(z_2)} \langle\langle F_{\mu\nu}(z_1) \rangle\rangle, \quad (26)$$

where $\delta S^{\mu\nu}(z_j) = \frac{1}{2}(dz_j^\mu \delta z_j^\nu - dz_j^\nu \delta z_j^\mu)$ is the element of the surface spanned by the path $z_j(t)$ as a consequence of the variation $z_j(t) \to z_j(t) + \delta z_j(t)$.

The evaluation of the right hand side of (25) and (26) requires some care, since the functional derivatives may lower the order of magnitude in the velocities. However, it can be done without any additional assumptions and the results are

$$g \langle\langle F_{0k}(z_1) \rangle\rangle = (\frac{4}{3} \alpha_s \frac{1}{r^3} + \frac{\sigma}{r}) r^k + O(v^2), \quad (27)$$

$$g \langle\langle F_{hk}(z_1) \rangle\rangle = (\frac{4}{3} \frac{\alpha_s}{m_2} \frac{1}{r^3} + \frac{\sigma}{m_2} \frac{1}{r})(r^h p_2^k - r^k p_2^h) + O(v^3), \quad (28)$$

$$g^2 \left(\langle\langle F_{hk}(z_1) F_{lm}(z_2) \rangle\rangle - \langle\langle F_{hk}(z_1) \rangle\rangle \langle\langle F_{lm}(z_2) \rangle\rangle \right) =$$

$$= -\frac{4}{3} \frac{ig^2}{8\pi} \delta(t_1 - t_2) \left\{ \partial_l \partial_k \left[\frac{1}{r} \left(\delta^{hm} + \hat{r}^h \hat{r}^m \right) \right] - \partial_l \partial_h \left[\frac{1}{r} \left(\delta^{km} + \hat{r}^k \hat{r}^m \right) \right] - \right.$$

$$\left. - \partial_m \partial_k \left[\frac{1}{r} \left(\delta^{hl} + \hat{r}^h \hat{r}^l \right) \right] + \partial_m \partial_h \left[\frac{1}{r} \left(\delta^{kl} + \hat{r}^k \hat{r}^l \right) \right] \right\} + O(v^2), \quad (29)$$

$$\left(\langle\langle F_{\mu\nu}(z_1) F_{\rho\sigma}(z_1') \rangle\rangle - \langle\langle F_{\mu\nu}(z_1) \rangle\rangle \langle\langle F_{\rho\sigma}(z_1') \rangle\rangle \right) = 0 \tag{30}$$

and similar ones.

In the end one obtains the potential in the form of a static part, a spin–dependent part and a velocity–dependent one, $V^{q\bar{q}} = V_{\text{stat}}^{q\bar{q}} + V_{\text{sd}}^{q\bar{q}} + V_{\text{vd}}^{q\bar{q}}$, with

$$V_{\text{stat}}^{q\bar{q}} = -\frac{4}{3}\frac{\alpha_s}{r} + \sigma r - \frac{4}{3}\frac{\alpha_s^2}{4\pi}\frac{1}{r}[\frac{66-4N_f}{3}(\ln \mu r + \gamma) + A] \tag{31}$$

$$V_{\text{sd}}^{q\bar{q}} = \frac{1}{8}\sum_{j=1,2}\frac{1}{m_j^2}\nabla^2\left(-\frac{4}{3}\frac{\alpha_s}{r}+\sigma r\right) + \left(\frac{4}{6}\frac{\alpha_s}{r^3}-\frac{\sigma}{2r}\right)\sum_{j=1,2}\frac{1}{m_j^2}\mathbf{S}_j\cdot\mathbf{L}_j + \frac{1}{m_1 m_2}\frac{4}{3}\frac{\alpha_s}{r^3}$$

$$(\mathbf{S}_2\cdot\mathbf{L}_1+\mathbf{S}_1\cdot\mathbf{L}_2) + \frac{4\alpha_s}{3m_1 m_2}\left[(\frac{3}{r^5}(\mathbf{S}_1\cdot\mathbf{r})(\mathbf{S}_2\cdot\mathbf{r})-\frac{\mathbf{S}_1\cdot\mathbf{S}_2}{r^3}) + \frac{8\pi}{3}\delta^3(\mathbf{r})\mathbf{S}_1\cdot\mathbf{S}_2\right] \tag{32}$$

$$V_{\text{vd}}^{q\bar{q}} = \frac{1}{m_1 m_2}\{\frac{4}{3}\frac{\alpha_s}{r}(\delta^{hk}+\hat{r}^h\hat{r}^k)p_1^h p_2^k\}_{\text{Weyl}} - C\sum_j\frac{p_j^2}{4m_j^2}$$

$$- \sum_{j=1}^2\frac{1}{6m_j^2}\{\sigma r \mathbf{p}_{j\text{T}}\}_{\text{Weyl}} - \frac{1}{6m_1 m_2}\{\sigma r \mathbf{p}_{1\text{T}}\cdot \mathbf{p}_{2\text{T}}\}_{\text{Weyl}}. \tag{33}$$

At the order α_s the above potential coincides globally with that given in Ref.[1]. In particular V_{sd} was originally given by Eichten and Feinberg and corrected by Gromes[2] (see also Ref.[3]), while V^{vd} has been obtained for the first time in[1]. Notice that Eq.(33) differs from the corresponding one proposed under the ad hoc assumption of scalar confinement and does not present the phenomenological difficulties of this [7]. Notice also that the terms in C can be reabsorbed in a redefinition of the masses $m_j \rightarrow m_j' = m_j + \frac{C}{2}$. The $O(\alpha_s^2)$ term in (31) has been obtained for the first time in [6].

The $O(\alpha_s^2)$ contributions to V_{sd} and V_{vd} have been evaluated by Gupta et al.[8] in an S matrix context but they have not been included here. In fact such contributions are found to be important for an understanding of the fine and hyperfine structure of the meson spectrum. However, due to the ambiguities inherent in the derivation method, a consistent evaluation in the Wilson loop approach should be desirable. Calculations are in progress in this line.

Finally, let us come to the ordering in (33). Obviously, ordering is related to the discretization prescription in the definition of the path integral. If in the definition of the gauge field functional integration we identify the element $U_{n'n}$ of the colour group associated to the link between the contiguous sites n and n' with $\exp[ig(x_{n'}-x_n)^\mu A_\mu(\frac{x_{n'}+x_n}{2})]$, we obtain the Weyl ordering

$$\{X^{hk}(r) p_j^h p_{jl}^k\}_{\text{Weyl}} = \frac{1}{4}\{p_j^h, \{X^{hk}(r), p_{jl}^k\}\} =$$

$$= \frac{1}{4}(X^{hk}(r)p_j^h p_{jl}^k + p_j^h X^{hk}(r)p_{jl}^k + p_{jl}^k X^{hk}(r)p_j^h + X^{hk}(r)p_{jl}^k p_j^h). \tag{34}$$

4. Three-quark potential

The three-quark gauge invariant Green function can be written as (again we assume the quarks to have differents flavours)

$$G(x_1, x_2, x_3, y_1, y_2, y_3) = \frac{1}{3!} \varepsilon_{a_1 a_2 a_3} \varepsilon_{b_1 b_2 b_3}$$

$$\langle 0 | T\, U^{a_3 c_3}(x_M, x_3) U^{a_2 c_2}(x_M, x_2) U^{a_1 c_1}(x_M, x_1) \psi_{3c_3}(x_3) \psi_{2c_2}(x_2) \psi_{1c_1}(x_1)$$

$$\overline{\psi}_{1d_1}(y_1)\overline{\psi}_{2d_2}(y_2)\overline{\psi}_{3d_3}(y_3) U^{d_1 b_1}(y_1, y_M) U^{d_2 b_2}(y_2, y_M) U^{d_3 b_3}(y_3, y_M) | 0 \rangle \qquad (35)$$

and we assume $x_1^0 = x_2^0 = x_3^0 = x_M^0 = t_f$, $y_1^0 = y_2^0 = y_3^0 = y_M^0 = t_i$, $t_f - t_i$ large.
The integration over the fermionic variables is again trivial and one can write

$$G(\mathbf{x}_1, \mathbf{x}_2, \mathbf{x}_3, \mathbf{y}_1, \mathbf{y}_2, \mathbf{y}_3; \tau) = \frac{1}{3!} \varepsilon_{a_1 a_2 a_3} \varepsilon_{b_1 b_2 b_3} \left\langle \left(U(x_M, x_1) S_1^{\mathrm{F}}(x_1, y_1 | A) U(y_1, y_M) \right)^{a_1 b_1} \right.$$

$$\left(U(x_M, x_2) S_2^{\mathrm{F}}(x_2, y_2 | A) U(y_2, y_M) \right)^{a_2 b_2} \left. \left(U(x_M, x_3) S_3^{\mathrm{F}}(x_3, y_3 | A) U(y_3, y_M) \right)^{a_3 b_3} \right\rangle . (36)$$

From (36), we can proceed strictly as in Sec.3 and in conclusion we have to show that

$$i \ln W_{3q} + i \sum_{j=1}^{3} \frac{ig}{m_j} \int_{\Gamma_j} dx^\mu \left(S_j^l \langle\langle \hat{F}_{l\mu}(x) \rangle\rangle - \frac{1}{2m_j} S_j^l \varepsilon^{lkr} p_j^k \langle\langle F_{\mu r}(x) \rangle\rangle - \right.$$

$$\left. - \frac{1}{8m_j} \langle\langle D^\nu F_{\nu\mu}(x) \rangle\rangle \right) - \frac{1}{2} \sum_{j,j'} \frac{ig^2}{m_j m_{j'}} \mathrm{T}_s \int_{\Gamma_j} dx^\mu \int_{\Gamma_{j'}} dx^{l\sigma}\, S_j^l\, S_{j'}^k \cdot$$

$$\cdot \left(\langle\langle \hat{F}_{l\mu}(x) \hat{F}_{k\sigma}(x') \rangle\rangle - \langle\langle \hat{F}_{l\mu}(x) \rangle\rangle \langle\langle \hat{F}_{k\sigma}(x') \rangle\rangle \right) = \left[\int_{t_i}^{t_f} dt\, V^{3q}(z_j, \mathbf{p}_j, \mathbf{S}_j) \right] \qquad (37)$$

with W_{3q} given by (19) and

$$\langle\langle f[A] \rangle\rangle = \frac{\frac{1}{3!} \langle \varepsilon \varepsilon \{ \prod_j \mathrm{P}[\exp(ig \int_{\overline{\Gamma}_j} dx^\mu A_\mu(x))] \} f[A] \rangle}{\frac{1}{3!} \langle \varepsilon \varepsilon \{ \prod_j \mathrm{P} \exp(ig \int_{\overline{\Gamma}_j} dx^\mu A_\mu(x)) \} \rangle} . \qquad (38)$$

Again, after the replacement $\dot{z}_j \to p_j/m_j$, the quantity $i \ln W_{3q}$ is already of the desired form, while the field expectation values can be evaluated according to equations analogues to (25) and (26) and lead to similar expressions. The final result is again of the form $V^{3q} = V_{\mathrm{stat}}^{3q} + V_{\mathrm{sd}}^{3q} + V_{\mathrm{vd}}^{3q}$ with

$$V_{\mathrm{stat}}^{3q} = \sum_{j<l} \left(-\frac{2}{3} \frac{\alpha_s}{r_{jl}} \right) + \sigma(r_1 + r_2 + r_3) + C, \qquad (39)$$

$$V_{\mathrm{sd}}^{3q} = \frac{1}{8m_1^2} \nabla_{(1)}^2 \left(-\frac{2}{3} \frac{\alpha_s}{r_{12}} - \frac{2}{3} \frac{\alpha_s}{r_{31}} + \sigma r_1 \right) +$$

$$+ \left\{ \frac{1}{2m_1^2} \mathbf{S}_1 \cdot \left[(\mathbf{r}_{12} \times \mathbf{p}_1) \left(\frac{2}{3} \frac{\alpha_s}{r_{12}^3} \right) + (\mathbf{r}_{31} \times \mathbf{p}_1) \left(-\frac{2}{3} \frac{\alpha_s}{r_{31}^3} \right) - \frac{\sigma}{r_1} (\mathbf{r}_1 \times \mathbf{p}_1) \right] + \right.$$

$$+ \ \frac{1}{m_1 m_2} \mathbf{S}_1 \cdot (\mathbf{r}_{12} \times \mathbf{p}_2) \left(-\frac{2}{3} \frac{\alpha_s}{r_{12}^3} \right) + \frac{1}{m_1 m_3} \mathbf{S}_1 \cdot (\mathbf{r}_{31} \times \mathbf{p}_3) \left(\frac{2}{3} \frac{\alpha_s}{r_{31}^3} \right) \Bigg\} +$$

$$+ \ \frac{1}{m_1 m_2} \frac{2}{3} \alpha_s \left\{ \frac{1}{r_{12}^3} \left[\frac{3}{r_{12}^2} (\mathbf{S}_1 \cdot \mathbf{r}_{12})(\mathbf{S}_2 \cdot \mathbf{r}_{12}) - \mathbf{S}_1 \cdot \mathbf{S}_2 \right] + \frac{8\pi}{3} \delta^3(\mathbf{r}_{12}) \mathbf{S}_1 \cdot \mathbf{S}_2 \right\} +$$

$$+ \ \text{cyclic permutations} , \tag{40}$$

$$V_{\text{vd}}^{3q} = \sum_{j<l} \frac{1}{2 m_j m_l} \left\{ \frac{2}{3} \frac{\alpha_s}{r_{jl}} (\delta^{hk} + \hat{r}_{jl}^h \hat{r}_{jl}^k) p_j^h p_l^k \right\}_{\text{Weyl}} - \sum_{j=1}^{3} \frac{1}{6 m_j^2} \{ \sigma \, r_j \, \mathbf{p}_{jT_j}^2 \}_{\text{Weyl}} -$$

$$- \ \sum_{j=1}^{3} \frac{1}{6} \{ \sigma \, r_j \, \dot{\mathbf{z}}_{MT_j}^2 \}_{\text{Weyl}} - \sum_{j=1}^{3} \frac{1}{6 m_j} \{ \sigma \, r_j \, \mathbf{p}_{jT_j} \cdot \dot{\mathbf{z}}_{MT_j} \}_{\text{Weyl}} - \sum_{j} \frac{C}{6 m_j^2} p_j^2, \tag{41}$$

where the notations are the same as used in (19) and the ordering is as in (34). Notice, in particular, that the quantity $\dot{\mathbf{z}}_M$ in (41) is given by

$$\dot{\mathbf{z}}_M = \begin{cases} R^{-1} \sum_{j=1}^{3} \left(\mathbf{p}_{jT_j} / m_j r_j \right) & \text{type I configuration} \\ \mathbf{p}_j / m_j & \text{type II configuration} : (\mathbf{z}_M \equiv \mathbf{z}_j) , \end{cases} \tag{42}$$

R being the matrix with elements $R^{hk} = \sum_{j=1}^{3} \frac{1}{r_j} (\delta^{hk} - \hat{r}_j^h \hat{r}_j^k)$.

Notice that Eq.(40) properly refers to the configuration I case. In general one should write $V_{\text{sd}}^{\text{LR}} = -\sum_{j=1}^{3} \frac{1}{2 m_j^2} \mathbf{S}_j \cdot \nabla_j V_{\text{stat}}^{\text{LR}} \times \mathbf{p}_j$ (In comparing this with (40) one should keep in mind that the partial derivatives in $\mathbf{z}_M$ of $V_{\text{stat}}^{\text{LR}}$ vanish due to the definition of M).

We observe that the short range part in Eqs.(39)–(41) is of a pure two body type: in fact it is identical to the electromagnetic potential among three equally charged particles but for the colour group factor 2/3 and it is well known. Even the static confining potential in Eq.(39) is well known (for a review see e.g.[9]). Furthermore the long range part in Eq.(40) coincides with the expression obtained by Ford[10] starting from the assumption of a purely scalar Salpeter potential of the form

$$\sigma \left(r_1 + r_2 + r_3 \right) \beta_1 \beta_2 \beta_3 , \tag{43}$$

but to our knowledge it was not obtained consistently in a Wilson loop context before Ref.[1]. Eq.(41) has been given for the first time in Ref.[1]. Eq. (41) differs from the corresponding equation obtained from (43). The situation for the three quarks is so similar to that occurring for the quark–antiquark system. In Eqs.(39) and (41) the terms in C can be again eliminated by the redefinition of the masses $m_j \to m_j'' = m_j + \frac{C}{3}$. Notice however that m_j'' differs from m_j'.

5. Relativistic flux tube model

Let us now neglect in Eq.(21) the spin–dependent terms and replace the $\frac{1}{m^2}$ expansion by its exact relativistic expression

$$K(\mathbf{x}_1, \mathbf{x}_2; \mathbf{y}_1, \mathbf{y}_2; t_f - t_i) =$$

$$\int \mathcal{D}[\mathbf{z}_1, \mathbf{p}_1] \int \mathcal{D}[\mathbf{z}_2, \mathbf{p}_2] \exp\left\{i\left[\int_{t_i}^{t_f} dt \sum_{j=1}^{2}(\mathbf{p}_j \cdot \dot{\mathbf{z}}_j - \sqrt{m_j^2 + \mathbf{p}_j^2})\right] + \ln W_{q\bar{q}}\right\} \quad (44)$$

Let us further evaluate $i \ln W_{q\bar{q}}^{\text{SR}}$ by the original Eq.(6) and assume that a sensible approximation is obtained even in the relativistic case postulating the first line of Eq.(14) in the center-of-mass system of the two particles. Then, if we expand again the exponent in (44) around the stationary values $\mathbf{p}_j = \frac{m\mathbf{z}_j}{\sqrt{1-\dot{\mathbf{z}}_j^2}}$, in the gaussian approximation we obtain the ordinary lagrangian

$$L = -\sum_{j=1}^{2} m_j \sqrt{1 - \dot{\mathbf{z}}_j^2} + \frac{4}{3}\frac{\alpha_s}{r}\left[1 - \frac{1}{2}(\delta^{hk} + \hat{r}^h\hat{r}^k)\dot{z}_1^h\dot{z}_2^k\right] +$$
$$- \sigma r \int_0^1 ds[1 - (s\dot{\mathbf{z}}_{1\text{T}} + (1-s)\dot{\mathbf{z}}_{2\text{T}})^2]^{1/2}. \quad (45)$$

This coincides with the relativistic flux-tube lagrangian[11].

From (45) is not possible to obtain even a classical hamiltonian in a closed form, due to the complicate velocity dependence. However, in terms of an expansion in $\frac{\sigma}{m^2}$ we have (we assume $m_1 = m_2 = m$ for simplicity and have already eliminated the terms in C)

$$\mathcal{H}(\mathbf{r}, \mathbf{q}) = 2\sqrt{m^2 + q^2} + \frac{\sigma r}{2}\left[\frac{\sqrt{m^2 + q^2}}{q_\text{T}}\arcsin\frac{q_\text{T}}{\sqrt{m^2 + q^2}} + \sqrt{\frac{m^2 + q_r^2}{m^2 + q^2}}\right] +$$
$$+ \frac{\sigma^2 r^2}{16 q_\text{T}^2}\frac{m^2 + q_r^2}{\sqrt{m^2 + q^2}}\left[\frac{\sqrt{m^2 + q^2}}{q_\text{T}}\arcsin\frac{q_\text{T}}{\sqrt{m^2 + q^2}} - \sqrt{\frac{m^2 + q_r^2}{m^2 + q^2}}\right]^2 + \ldots(46)$$

with $\mathbf{r} = \mathbf{z}_{1\text{CM}} - \mathbf{z}_{2\text{CM}}$, $\mathbf{q} = \mathbf{p}_{1\text{CM}} = -\mathbf{p}_{2\text{CM}}$, $q_r = (\hat{\mathbf{r}} \cdot \mathbf{q})/\hat{\mathbf{r}}$ and $q_\text{T}^h = (\delta^{hk} - \hat{r}^h\hat{r}^k)q^k$. From this a quantum hamiltonian can be immediately obtained by setting

$$\langle \mathbf{k}'|H_{\text{FT}}|\mathbf{k}\rangle = \int \frac{d\mathbf{r}}{(2\pi)^3} e^{i(\mathbf{k}-\mathbf{k}')\cdot\mathbf{r}} \, \mathcal{H}(\mathbf{r}, \frac{\mathbf{k}' + \mathbf{k}}{2}), \quad (47)$$

in which the ordering prescription is again Weyl prescription. By an expansion in $\frac{1}{m^2}$ a semirelativistic hamiltonian can be obviously reobtained with a potential given by (31)–(33).

6. Bethe–Salpeter equation

Let us go back to the equation analogous to (20) for spinless quarks and in it use the covariant representation for the quark propagator in an external gauge field

$$\Delta^{\text{F}}(x, y|A_\mu) = \frac{i}{2}P_{xy}\int_0^\infty d\tau \int_{z(0)=y}^{z(\tau)=x} \mathcal{D}[z]\exp i \int_0^\tau d\tau'\{-\frac{1}{2}[\dot{z}^{2'} + m^2] + g\dot{z}^{\mu'}A_\mu(z')\} \quad (48)$$

In place of (21) we find

$$G_4(x_1, x_2; y_1, y_2) = (\frac{i}{2})^2 \int_0^\infty d\tau_1 \int_0^\infty d\tau_2 \int_{z_1(0)=y}^{z_1(t_1)=x_1} \mathcal{D}[z_1] \int_{z_2(0)=y_2}^{z_2(\tau_2)=x_2} \mathcal{D}[z_2]$$

$$\exp\frac{-i}{2}\{\int_0^{\tau_1} d\tau_1'[(\frac{dz_1}{d\tau_1'})^2 + m_1^2] + \int_0^{\tau_2} d\tau_2'[(\frac{dz_2}{d\tau_2'})^2 + m_2^2]\} \cdot \frac{1}{3}\langle \mathrm{Tr\,P} \exp ig \oint dz^\mu A_\mu(z)\rangle$$

(49)

where the equation of a path connecting y with x is written as $z^\mu = z^\mu(\tau)$, in terms of an arbitrary parameter τ (rather than the time t) and z' stands for $z(\tau')$. Notice the occurrence again in (49) of the Wilson loop integral $W_{q\bar{q}}$. Even in this case we can use the evaluation of $i\ln W_{q\bar{q}}^{\mathrm{SR}}$ as given by Eq. (5) or (6) and assume for $i\ln W_{q\bar{q}}^{\mathrm{LR}}$ the first line of (14) in the center–of–mass system.

Then, by appropriate manipulations of the resulting expression, one can obtain the Bethe–Salpeter equation (for $x_1^0 - x_2^0$ and $y_1^0 - y_2^0$ large)

$$G_4(x_1, x_2; y_1, y_2) = G_2(x_1 - y_1)\, G_2(x_2 - y_2) - i \int d^4\xi_1 d^4\xi_2 d^4\eta_1 d^4\eta_2$$

$$G_2(x_1 - \xi)\, G_2(x_2 - \xi_2)\, I(\xi_1, \xi_2; \eta_1, \eta_2)\, G_4(\eta_1, \eta_2; y_1, y_2),$$

(50)

with a kernel of the form $I(\xi_1, \xi_2; \eta_1, \eta_2) = I^{\mathrm{SR}}(\xi_1, \xi_2; \eta_1, \eta_2) + I^{\mathrm{LR}}(\xi_1, \xi_2; \eta_1, \eta_2)$. Here I^{SR} coincides with the ordinary perturbative kernel, while in the momentum representation I^{LR} can be written as (for simplicity we have neglected the perimeter term)

$$\tilde{I}^{\mathrm{LR}}(p_1', p_2'; p_1, p_2) = \int d^3r e^{i(k'-k)\cdot r} J(r, \frac{p_1' + p_1}{2}, \frac{p_2' + p_2}{2})$$

(51)

$(p_1' + p_2' = p_1 + p_2, \mathbf{p}_1 = -\mathbf{p}_2 = \mathbf{k}, \mathbf{p}_1' = -\mathbf{p}_2' = \mathbf{k}')$ with

$$J(\mathbf{r}, q_1, q_2) = 4\frac{\sigma r}{2}\frac{1}{q_{10} + q_{20}}[q_{20}^2\sqrt{q_{10}^2 - \mathbf{q}_T^2} + q_{10}^2\sqrt{q_{20}^2 - \mathbf{q}_T^2} +$$

$$+ \frac{q_{10}^2 q_{20}^2}{|\mathbf{q}_T|}(\arcsin\frac{|\mathbf{q}_T|}{|q_{10}|} + \arcsin\frac{|\mathbf{q}_T|}{|q_{20}|})] + O(\frac{\sigma^2}{m^4})$$

(52)

(having set $\mathbf{q}_1 = -\mathbf{q}_2 = \mathbf{q}$, $q_T^h = (\delta^{hk} - \hat{r}^h\hat{r}^k)q^k$).

Notice that, according to a standard procedure, the BS kernel $\tilde{I}$ can be associated with a relativistic potential (to be used in the Salpeter equation) given by

$$\langle \mathbf{k}'|V|\mathbf{k}\rangle = \frac{1}{(2\pi)^3}\frac{1}{4\sqrt{w_1(\mathbf{k})w_2(\mathbf{k})w_1(\mathbf{k}')w_2(\mathbf{k}')}}\tilde{I}_{\mathrm{inst}}(\mathbf{k}', \mathbf{k})$$

(53)

where $w_j(\mathbf{k}) = \sqrt{m_j^2 + \mathbf{k}^2}$ and the instantaneous kernel $\tilde{I}_{\mathrm{inst}}$ is obtained from $\tilde{I}$ by replacing p_{j0} and p_{j0}' by their mass shell values $\sqrt{m_j^2 + \mathbf{k}^2}$, $\sqrt{m_j^2 + \mathbf{k}^2}'$. The corresponding hamiltonian coincides with (46) at the classical level and differs from (47) only for the ordering prescription.

Going back to the expression of the kernel $\tilde{I}^{LR}$ as given by (52)–(51), one has to notice that this is highly singular for $\mathbf{k}' = \mathbf{k}$, due to the occurrence of the factor r in (52), and it must be appropriately regularized before being used in equation (50) (e.g., one can make the substitution $r \to re^{-\varepsilon r}$). This circumstance is related to the fact that, being confining, $\tilde{I}^{LR}$ should admit only bound states, while the inhomogeneous BS equation provides also a continuous two–particles spectrum. Therefore one should solve (50) for the regularized kernel and only at the end take the limit for $\varepsilon \to 0$ (we admit in this way that resonances would evolve in bound states).

7. Acknowledgements

We acknowledge warmly Prof. E.Montaldi and Dr. P.Consoli for having contributed to some of the results reported in this paper.

8. References

1. A. Barchielli, E. Montaldi and G.M. Prosperi, *Nucl. Phys.* **B296** (1988) 625; A. Barchielli, N. Brambilla and G.M. Prosperi, *Nuovo Cimento* **103A** (1990) 59; N. Brambilla, P. Consoli and G. M. Prosperi, *Phys. Rev.* **D 50** (1994).
2. W. Lucha, F.F. Schöberl and D. Gromes, *Phys. Rep.* **200** (1991) 127.
3. Y. Simonov, , HD-THEP-93-16; Y. Simonov, *Nucl. Phys.* **B324** (1989) 67; H. G. Dosch, this Conference
4. J. Ball, F. Zachariasen , this conference and quoted references.
5. A.M. Polyakov, *Nucl. Phys.* **B164** (1979) 171; V.S. Dotsenko and S.N. Vergeles, *Nucl. Phys.* **B169** (1980) 527; R. Brandt et al. *Phys. Rev.* **D24** (1981) 879; see also A. Bassetto et al., *Nucl. Phys.* **B408** (1993) 62 and references therein.
6. W. Fischler, *Nucl. Phys.* **B129**(1977) 157; T. Appelquist et al. *Phys. Rev.* **D17** (1978) 2074; A. Billoire, *Phys. Lett.* **92B** (1980) 343; F. J. Yndurain and S. Titard, *Phys. Rev.* **D 49** (1994) 6007.
7. A. Gara et al. *Phys. Rev.* **D42** (1990) 1651; **40** (1989) 843; J.F. Lagae, *Phys. Rev.* **D45** (1992) 305; **45** (1992) 317; N. Brambilla and G.M. Prosperi, *Phys. Lett.* **B236** (1990) 69.
8. S.N. Gupta and S. Radford, *Phys. Rev.* **D24** (1981) 2309; S.N. Gupta et al. *Phys. Rev.* **D34** (1986) 201, F. Halzen et al. MAD/PH/706.
9. J.M. Richard, Phys. Rep. **212** (1992) 1.
10. J.I. Ford, *Journ. Phys.* **G 15** (1989) 1641.
11. C. Olson, M.G. Olsson and K. Williams, *Phys. Rev.* **D45** (1992) 4307; N. Brambilla and G.M. Prosperi, *Phys. Rev.* **D47** (1993) 2107.

RIGOROUS QCD ANALYSIS OF HEAVY QUARKONIUM STATES

F.J. YNDURÁIN

Departamento de Física Teórica, C-XI
Universidad Autónoma de Madrid
Canto Blanco, E-28049 Madrid, Spain

ABSTRACT

No detailed report is presented because the talk summarized the papers of S. Titard and the speaker, *Phys. Rev.* **49**, 6007 (1994) and FTUAM 94-6, to appear in *Phys. Rev.* In them it is presented a calculation of the $n = 1$ $c\bar{c}$ and $n = 1, 2$ $b\bar{b}$ systems with inclusion of radiative and leading nonperturbative corrections.

THE NEW MESONS, ARE THEY DEUSONS OR GLUEBALLS?

NILS A. TÖRNQVIST

Research Institute for High Energy Physics, SEFT, University of Helsinki
PB 9, FIN-00014 Helsinki, Finland

ABSTRACT

The new mesons which cannot find a place within the $q\bar{q}$ model, the $f_0(1370)$, $f_0(1525)$, $f_0(1590)$, $f_2(1520)$, $\eta(1440)$ and $f_1(1420)$, are discussed and it is suggested that some of these are deuteronlike two-meson bound states, *deusons*. The contribution from one pion exchange to the binding energy between two ground state mesons is calculated. The calculations are most reliable in the heavy meson sector where one-pion exchange alone is strong enough to form six deuteron-like $B\bar{B}^*$ and $B^*\bar{B}^*$ composites, with different spin parities, bound by approximately 50 MeV. Similarly six composites of $D\bar{D}^*$ and $D^*\bar{D}^*$ states bound by pion exchange alone can be expected near the thresholds.

In the light quark sector pion exchange alone is not strong enough to form bound meson-meson states. But, if supplemented with shorter range attraction, the quantum numbers and masses of the candidates listed above are generally the same as those expected for light deusons (or threshold effects due to virtually bound meson-meson states) composed of either $(\rho\rho \pm \omega\omega)/\sqrt{2}$, $K\bar{K}^*$ or $K^*\bar{K}^*$. Possibily the X(3100) could be the first heliumlike four-meson state composed of one K^* and three ω's or ρ's, while X(3250) could be composed of $K^*\bar{K}^*\rho\omega$.

A crucial test of the scheme is to find one of the predicted $D\bar{D}^*$ or $D^*\bar{D}^*$ deusons. Today the Fermilab $p\bar{p}$ accumulator is probably the best place to look for them.

1. Introduction

In meson spectroscopy it is by now a well known experimental fact (See Ref.[1]) that there exist several mesonic states, which cannot find a place within the conventional $q\bar{q}$ model. In recent papers [2] I suggested that they may be "deuteronlike meson-meson states", Fig.1, for which I use the acronym *deusons* (for reasons explained in more detail in Ref.[3]). These are rather loosely bound states of two mesons, where pion exchange plays a dominant rôle.

In Fig.2 the natural parity I=0 ω-like mesons are plotted in a Chew-Frautchi plot of spin vs. squared mass. The expected $q\bar{q}$ mesons with the orbital excitations are the empty squares on the linear Regge trajectories. As can be seen there are clearly much too many f_0 and f_2 mesons in the 1370-1710 MeV region. A similar plot is shown in Fig. 3 for the unnatural parity I=0 $(u\bar{u} \pm d\bar{d})/\sqrt{2}$-like (Fig. 3a) and $s\bar{s}$-like (Fig. 3b) mesons. One sees that the $f_1(1420)$ and the $\eta(1440)$ (which may be 2 resonances) do not have a natural place in the $q\bar{q}$ model.

Are these states gluonium or "molecular multiquark" states or what? Many predictions for 12 new heavy deusons composed of charmed or beauty mesons are given in the ZPC paper of Ref.[2] and below. The predictions of heavy deusons are less

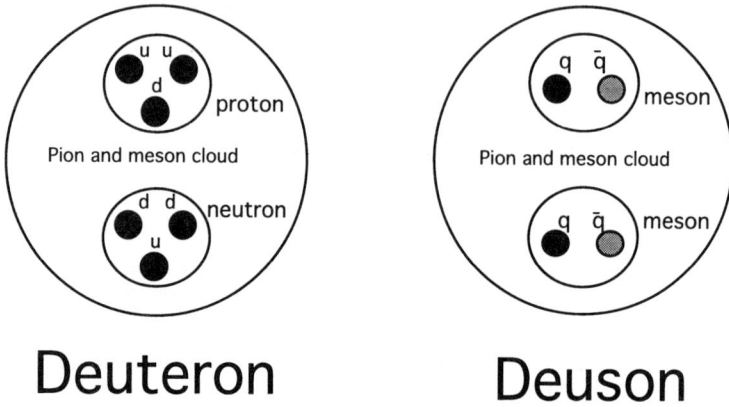

Fig. 1. A simple picture of the deuteron and a deuson.

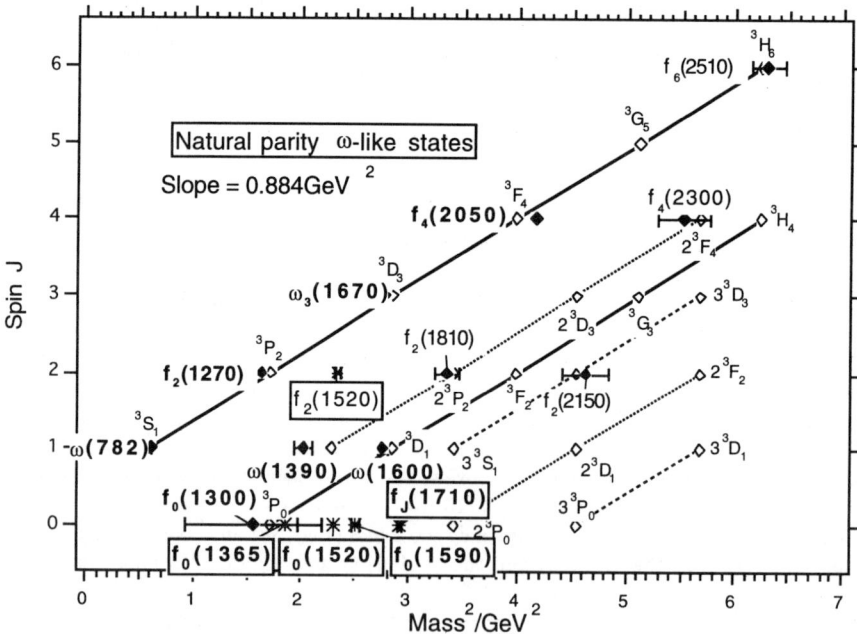

Fig. 2. The natural parity, I=0, ω-like mesons plotted in a Chew-Frautchi plot. The linear trajectories with open squares indicate where $(u\bar{u} \pm d\bar{d})/\sqrt{2}$ mesons are expected. The extra non-$q\bar{q}$ candidates are in boxes. See text.

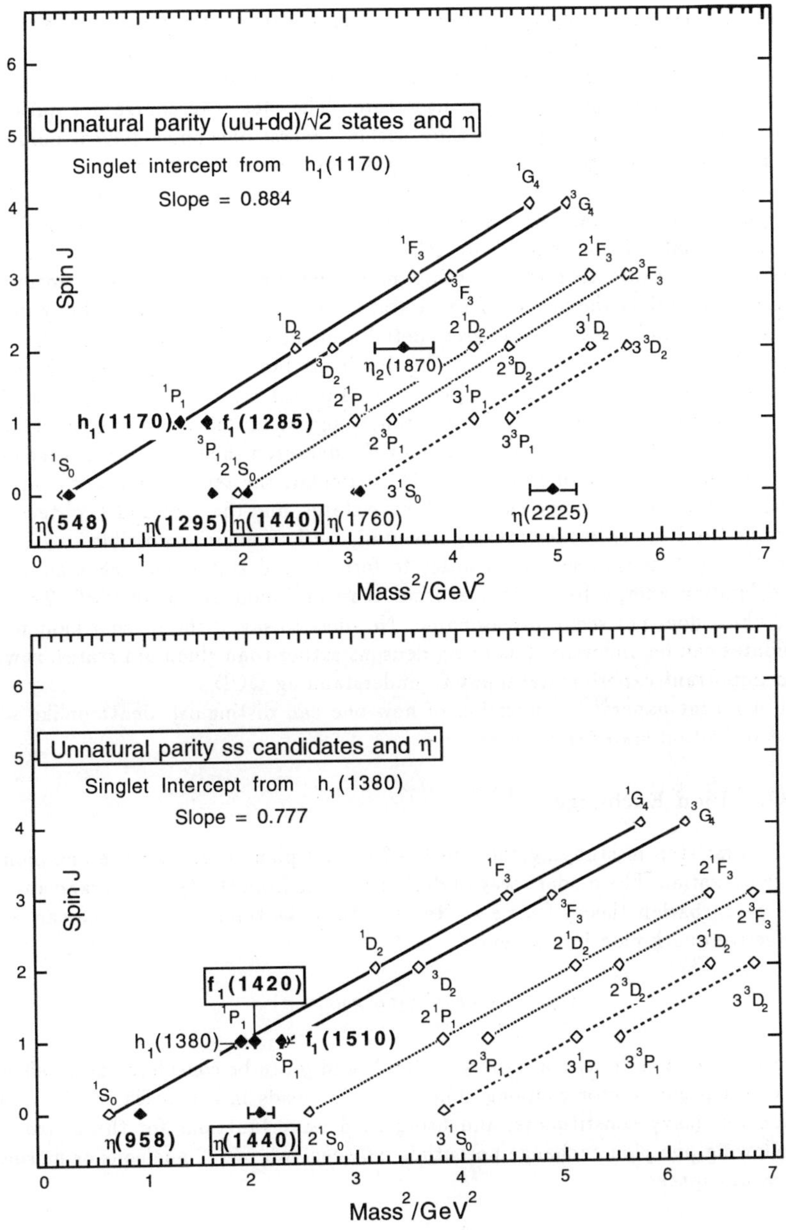

Fig. 3. The unnatural parity, I=0 mesons $(u\bar{u} \pm d\bar{d})/\sqrt{2}$ (a) and $s\bar{s}$ (b) mesons and non-$q\bar{q}$ candidates plotted in a Chew-Frautchi plot as in fig 2.

ambiguous than for the light sector, since a nonrelativistic approximation should be sufficient, and the potential term becomes relatively stronger (compared to the kinetic term). The question whether such bound states actually can exist has been studied surprisingly sparsely in the literature. The idea is of course not quite new, but has been discussed only in passing within general phenomenological models for meson-meson bound states (See Refs.[4]–[14]), where pion exchange is not given special attention. Recently, after my first letter[2] Ericson and Karl[15] has also studied the strength of pion exchange with similar conclusions, and Manohar and Weise[16] have studied especially flavour exotic two B^*-meson bound states.

When one wants to study whether pion exchange can bind two hadrons the deuteron is certainly the prime reference state, the existence of which nobody doubts. The deuteron has been studied in great detail over the years (See Ref.[17] and the recent reviews[18,19]). There one knows that the dominant binding energy comes from pion exchange between two colourless qqq clusters - a proton and a neutron. Thus in short, I adopt a similar formalism which has been so successful for describing the deuteron to the meson-meson case. As is the case for the deuteron the tensor force turns out to be important also in the mesonic case. In particular, the tensor force can contribute directly attractively to single channel P-wave states, and the $1/r^2$ and $1/r^3$ terms can overcome the P-wave barrier making pseudoscalar I=0 deusons possible.

In the light deuson sector, in order to form bound states one needs about half of the binding energy from other sources than one pion exchange ("σ", 2π, ρ, ω, etc.), which does not seem unreasonable. Needless to say, if the present light non-$q\bar{q}$ candidates can be understood as being deusons rather than gluonium states, it would be an important experimental input to understanding QCD.

In a recent paper[20] the question of how one can distinguish deuteronlike states from two-hadron states is discussed.

2. One-Pion Exchange

The first step in studying the effects of virtual pions is to find the one-pion exchange potential. The modern way of deriving this is from the QCD Lagrangian using chiral perturbation theory (See e.g. Ref.[16]). To make things simple, one can derive an effective quark pion interaction

$$\mathcal{L}_{int} = \frac{g}{f}\bar{q}(x)\gamma^\mu\gamma_5\vec{\tau}q(x)\partial_\mu\pi(x) , \qquad (1)$$

where f is the pion decay constant ≈ 132MeV and g can be considered as an effective pion quark pseudovector coupling constant. This leads in a nonrelativistic approximation for heavy constituents, and using $SU6$ wave functions for the hadrons, to the following one-pion exchange potentials between two I=1/2 nucleons or mesons in momentum space

$$V_\pi(NN) = \frac{25}{9}\frac{g^2}{f^2}(\vec{\tau}_1\cdot\vec{\tau}_2)(\vec{\sigma}_1\cdot\vec{q})(\vec{\sigma}_2\cdot\vec{q})\frac{1}{\vec{q}^{\,2}+m_\pi^2} , \qquad (2)$$

$$V_\pi(VV) = -V_\pi(V\bar{V}) = \frac{g^2}{f^2}(\vec{\tau}_1 \cdot \vec{\tau}_2)(\vec{\Sigma}_1 \cdot \vec{q})(\vec{\Sigma}_2 \cdot \vec{q})\frac{1}{\vec{q}^2 + m_\pi^2} , \qquad (3)$$

$$V_\pi(PV \to VP) = \frac{g^2}{f^2}(\vec{\tau}_1 \cdot \vec{\tau}_2)(\vec{\epsilon}_1 \cdot \vec{q})(\vec{\epsilon}_2^* \cdot \vec{q})\frac{1}{\vec{q}^2 + \mu^2} , \qquad (4)$$

where the constituents are assumed to be isodoublets, and the normalization $\tau^2 = \sigma^2 = 3$, $\Sigma^2 = 2$ is used. For $I = 1,0$ constituents (ρ, ω) only the isospin factor, $\vec{\tau}_1 \cdot \vec{\tau}_2$, needs to be replaced by an $SU3_f$ weight (See Ref.[2]). In the PV transition potential Eq. (4) m_π^2 is replaced by $\mu^2 = m_\pi^2 - (m_V - m_P)^2$, whereby one can take into account, at least approximately, the recoil effect due to the fact that the P and V constituents can have unequal masses.

These relations are not expected to be exact, but are generally believed to be right to within about 30%, which is sufficient for our present exploratory purpose.

Let us introduce the constant $V_0 = m_\pi^3 g^2/(12\pi f^2 9) \approx 1.3$ MeV, which is fixed by the πN coupling constant ($f_{\pi N}^2/(4\pi) = 0.08 = \frac{25}{9}\frac{g^2}{f^2}m_\pi^2$ from which $g \approx 0.6$). This constant V_0 measures the effective potential between two quarks, when ($S_{qq} = I_{qq} = 1$). It is independent of the mass of an I=0 spectator quark.

One can combine the formulas (2-4) to a single one by collecting all the constants into an overall number γ, the "relative coupling number". This number γ measures the relative strength of the potential compared to the contribution from one pair of quarks in a spin triplet and isospin triplet state for which $\gamma_{SI}^{qq} = -1$. For example for the deuteron channel $\gamma_{10}^{NN} = 25/3$ and for $D^*\bar{D}^*$ in I=0, S=0, $\gamma_{00}^{VV} = 6$. The larger this number γ is, the stronger is the attraction, and if it is negative there is repulsion. The universal one-pion exchange potential in momentum space can then be written compactly:

$$V_\pi(q) = -\gamma V_0 \frac{4\pi}{m_\pi} [D + S_{12}(q)]\frac{1}{\vec{q}^2 + m_\pi^2} , \qquad (5)$$

where D is a diagonal matrix whose first element is normalized to unity. This is done by convenience in order to fix an overall γ in the case when there are several channels. Often D is simply the unit matrix.

Taking the Fourier transform one gets the potential in configuration space (omitting the delta function piece in the central potential, which from the phenomenological point of view will be included in the short range potential and regularization scheme discussed later). Thus the general one-pion exchange potential as a function of r is:

$$V_\pi(r) = -\gamma V_0 [D \cdot C(r) + S_{12}(\hat{r}) \cdot T(r)] , \qquad (6)$$

where D is as above, $\hat{r}$ is the unit vector, and the r dependence is given by the functions

$$C(r) = \frac{\mu^2}{m_\pi^2}\frac{e^{-\mu r}}{m_\pi r} , \qquad (7)$$

$$T(r) = C(r)[1 + \frac{3}{\mu r} + \frac{3}{(\mu r)^2}] . \qquad (8)$$

Note that the r dependence in the tensor $T(r)$ contains singular r^{-2} and r^{-3} terms. These make the tensor potential dominate at small r like an axial dipole-dipole interaction. Because of this singular behaviour of the tensor potential it must be regularized at small distances. (See e.g. the discussion of Ericson and Rosa-Clot[18] and Ref.[2].) The perhaps most natural method is to introduce a form factor at each πN vertex, such as $(\Lambda^2 - \mu^2)/(\Lambda^2 + t)$, which in r-space can be looked upon as a spherical pion source with rms radius $R = \sqrt{10}/\Lambda$.

The matrix elements of the tensor force, S_{12} connect different spin-orbit configurations and are real numbers. E.g. for the deuteron which is our prime reference state has two spin-orbit states ($^3S_1$, $^3D_1$), these are

$$< {}^3S_1|S_{12}|^3S_1 >= 0, \; < {}^3S_1|S_{12}|^3D_1 >= \sqrt{8} \;,< {}^3D_1|S_{12}|^3D_1 >= -2 \;.$$

The deuteron potential $V_{10}(r)$ can thus be written in matrix form as:

$$V_{10}(r) = -\frac{25}{3}V_0 \left[\begin{pmatrix} 1 & 0 \\ 0 & 1 \end{pmatrix} C(r) + \begin{pmatrix} 0 & \sqrt{8} \\ \sqrt{8} & -2 \end{pmatrix} T(r) \right] \;, \qquad (9)$$

The overall strength of the potential is given by $\gamma_{10}^{NN} V_0$, which expressed in terms of the conventional pion nucleon coupling constant is $\gamma_{10}^{NN} V_0 = \frac{f_{\pi N}^2}{4\pi} \cdot f \approx 11.2$ MeV.

For the other three spin (S) and isospin (I) NN channels the potentials $V_{SI}(r)$ are one dimensional:

$$V_{00}(r) \;\; = \;\; -\frac{25}{3}V_0 C(r) \;, \qquad (10)$$

$$V_{01}(r) \;\; = \;\; +25 V_0 C(r) \;, \qquad (11)$$

$$V_{11}(r) \;\; = \;\; +\frac{25}{9}V_0 [C(r) + 2T(r)] \;. \qquad (12)$$

For the different meson-meson configurations similar forms of the potential can be derived some of which are given below.

Note that in (12) (the $^3P_0$ NN channel) the large tensor piece simply adds diagonally to the central potential giving a very strong potential, but being repulsive it produces of course no $^3P_0$ NN bound state. In the meson-meson analogue, there is a similar situation in the pseudoscalar channels, but with the essential difference that the potential is attractive.

From the deuteron we also learn that the central part of the one pion potential $\gamma_{10}^{NN} V_0 C(r)$ is insufficient to bind the deuteron. By scaling arguments it is easy to show[21] that a simple Yukawa potential can form a bound state only if the well depth parameter satisfies the bound

$$s = 0.5953 \cdot (\gamma_{SI}^{NN} V_0) M/\mu^2 > 1 \;. \qquad (13)$$

For the deuteron using the conventional pion nucleon coupling constant this well depth parameter is $s_d = 0.33$, i.e., it falls well below the bound by a large factor of 3. The nonexistence of a bound state in $|{}^1S_0 >$ is a manifestation of this fact, since there the central one pion potential (10) is present with the same strength as for the deuteron, but the tensor potential is absent. However, the ${}^1S_0$ is almost bound, only a few keV are missing, which implies that the remainder from short range forces $(2\pi, "\sigma"$ or quark-gluon exchange) gives additional attraction, strong enough to compensate for the hard core repulsion (from Fermi statistics of the 6 quark system, or ω etc. exchange).

Thus the tensor pion exchange potential is important in providing the deuteron binding[19]. Shorter range contributions than one pion exchange of course also contribute, although they are more difficult to estimate. The same applies to mesonic deusons as well: If one pion exchange is large and attractive one has a good reason to expect bound states to exist.

3. Numerical Matrix Method

When I started this work, I first got hold of old programs for solving the deuteron. These were typically 20 pages of fortran listings. Fortunately, I soon became aware of a matrix method (See Refs.[22,23] where one discretizes the r dependence to a finite dimensional vector, wherby the Hamiltonian becomes a finite matrix. Then the Schrödinger equation can be solved as a matrix equation, for which one can use efficient standard matrix routines for finding the eigenvalues and eigenvectors, and which requires only one single line in the program. Thus I scrapped the old deuteron programs and wrote a short single A4 page program, which solved the problem in no time without too much effort.

This method is particularly accurate for finding the ground state, which is precisely the problem of this paper. It is also particularly well suited when one is dealing with coupled channels, as is here the case. Namely, for a n-channel problem the matrix dimension of the Hamiltonian simply becomes n times larger, in a rather trivial way. E.g. for two channels as for the deuteron, 16 points on the r-axis are sufficient to obtain a very good approximation for the binding energy and wave functions, by diagonalizing a 32 dimensional matrix form of the Hamiltonian. The method is comprehensively discussed for the one-channel problem in Ref.[23].

The deuteron S and D wave functions obtained using this numerical method agree of course very well with standard solutions found in the literature, and with solutions obtained with much more detailed NN potentials (such as that of Ref.[24]) apart from a small region near $r = 0$. Here no other shorter range contributions than one-pion exchange are included, and the cut-off Λ is fixed to get the right binding energy. The deuteron quadrupole moment, D wave probability, rms radius and other static properties are correctly predicted close to their experimental values, since these depend very little on the short range part of the potential, once the binding energy is right.

4. DEUSONS

Parity forbids two pseudoscalars to be bound by one-pion exchange; therefore the lightest deusons are pseudoscalar-vector (PV) states. Again in such PV systems the pion is too light to be a constituent, because the small reduced mass of a πV system would give a too large kinetic term, which cannot be overcome by the potential. In other words, the well depth parameter Eq. (13) becomes very small. Thus the lightest deusons (or threshold effects due to virtually bound deusons) where pion exchange can play a dominant rôle are $K\bar{K}^*$ systems, i.e. the deuson spectrum can start at ≈ 1400 MeV.

Because of time limitations, I discuss here only vector-vector (VV) systems, but quote the results for both PV and VV deusons. See the original paper Ref.[3] for more details. For composites of two vector mesons $V\bar{V}$ the relative coupling number, γ_{SI}^{VV}, which measures the overall relative strength of one pion exchange in the different spin isospin states, are given by the formula $\gamma_{SI}^{VV} = (\vec{\Sigma}_1 \cdot \vec{\Sigma}_2)(\vec{\tau}_1 \cdot \vec{\tau}_2)$ which gives the strongest attraction in the spin and isospin singlet channels ($\gamma_{00}^{VV} = +6$), followed by the spin triplet, isosinglet channels ($\gamma_{10}^{VV} = +3$). For I=1 one has either repulsion or very weak attraction.

Table 1. The lowest spin parity VV states J^{PC}, the contributing spin orbit states, the relative magnitude of the tensor term, and the relative coupling numbers for I=0 and 1.

J^{PC}	States	$< |S_{12}| >$	γ_{S0}^{VV}	γ_{S1}^{VV}
0^{++}	$^1S_0\ ^5D_0$	See Eq. (14)	+6	−2
0^{-+}	$^3P_0$	+2	+3	−1
1^{++}	$^5D_1$	−1	−3	+1
CP-exotic: 1^{-+}	$^3P_1$	−1	+3	−1
1^{+-}	$^3S_1,\ ^3D_1$	See Eq. (15)	+3	−1
1^{--}	$^1P_1,\ ^5P_1,\ ^5F_1$	See Eq. (17)	+6	−2
2^{++}	$^1D_2,\ ^5S_2,\ ^5D_2,\ ^5G_2$		+6	−2
CP-exotic: 2^{+-}	$^3D_2$	−1	+3	−1

Table 1 lists the lowest spin parity VV states, J^{PC}, and the contributing partial waves. There are also listed the relative coupling number for isodoublet constituents $(K^*\bar{K}^*,\ D^*\bar{D}^*,\ B^*\bar{B}^*)$ which is given by $\gamma_{SI}^{VV} = [(S+1)S - 4] \cdot [(I+1)I - 3/2]$. From this table one sees that for S waves, the largest γ appears in the isoscalar 0^{++} channel, for which $\gamma_{00}^{VV} = +6$, and the next largest is in the isoscalar, 1^{+-} channel, where $\gamma_{10}^{VV} = +3$. In both of these channels the S wave can mix with a D wave: $^5D_0$, respectively $^3D_1$. Thus these are two-channel problems with the potentials:

$$V_{0^{++}} = -\gamma_{0I}^{VV} V_0 \left[\begin{pmatrix} 1 & 0 \\ 0 & -\frac{1}{2} \end{pmatrix} C(r) + \begin{pmatrix} 0 & \sqrt{\frac{1}{2}} \\ \sqrt{\frac{1}{2}} & 1 \end{pmatrix} T(r) \right], \qquad (14)$$

$$V_{1^{+-}} = -\gamma_{1I}^{VV} V_0 \left[\begin{pmatrix} 1 & 0 \\ 0 & 1 \end{pmatrix} C(r) + \begin{pmatrix} 0 & -\sqrt{2} \\ -\sqrt{2} & 1 \end{pmatrix} T(r) \right]. \qquad (15)$$

Here as in the following multichannel cases the overall γ_{SI}^{VV} is normalized by the first state which has the strongest central potential. Note that these potentials are similar to that for the deuteron in form, although the numbers have changed somewhat. For the single channel cases the most interesting is the pseudoscalar channel $^3P_0$:

$$V_{0-+} = -\gamma_{1I}^{VV} V_0 [C(r) + 2T(r)] = -\gamma_{1I}^{VV} V_0 \frac{e^{-m_\pi r}}{fr} [3 + \frac{6}{m_\pi r} + \frac{6}{(m_\pi r)^2}] . \quad (16)$$

This is a remarkably strong potential, since the large tensor part adds directly to the central. In addition the centrifugal P wave barrier can be compensated by the r^{-2} and r^{-3} terms in the tensor force. Thus one can expect pseudoscalar P wave bound states at least for $B^*\bar{B}^*$ and probably also for $D^*\bar{D}^*$.

For the three other single channel cases 1^{++}, 1^{-+} and 2^{+-} (the last two of which which are CP-exotic) the matrix element of S_{12} is -1. Therefore, one gets a much weaker potential and no bound states are expected. Finally there remains two spin parities which are of interest: 1^{--} and 2^{++}. For the 1^{--}, which is interesting since these states can be produced directly in e^+e^- annihilation, there are 3 channels, $(^1P_1, {}^5P_1, {}^5F_1)$ and the potential is:

$$V_{1--} = -\gamma_{1I}^{VV} V_0 \left[\begin{pmatrix} 1 & 0 & 0 \\ 0 & -\frac{1}{2} & 0 \\ 0 & 0 & -\frac{1}{2} \end{pmatrix} C(r) + \begin{pmatrix} 0 & +\sqrt{\frac{4}{5}} & -\sqrt{\frac{6}{5}} \\ +\sqrt{\frac{4}{5}} & -\frac{7}{5} & \sqrt{\frac{6}{25}} \\ -\sqrt{\frac{6}{5}} & \sqrt{\frac{6}{25}} & -\frac{8}{5} \end{pmatrix} T(r) \right] . \quad (17)$$

The overall $\gamma_{10}^{VV} = 6$ is large for $I = 0$, but the presence of at least a P wave centrifugal barrier in all partial waves makes bound states unlikely.

Finally for the 2^{++}, there are 4 channels $(^1D_2, {}^5S_2, {}^5D_2, {}^5G_2)$, i.e. there is even an S wave present, although it is repulsive for $I = 0$. The potential is:

$$V_{2++} = -\gamma_{2I}^{VV} V_0 \left[\begin{pmatrix} 1 & 0 & 0 & 0 \\ 0 & -\frac{1}{2} & 0 & 0 \\ 0 & 0 & -\frac{1}{2} & 0 \\ 0 & 0 & 0 & -\frac{1}{2} \end{pmatrix} C(r) + \begin{pmatrix} 0 & +\sqrt{\frac{1}{10}} & -\sqrt{\frac{1}{7}} & +\sqrt{\frac{9}{35}} \\ +\sqrt{\frac{1}{10}} & 0 & -\sqrt{\frac{7}{10}} & 0 \\ -\sqrt{\frac{1}{7}} & -\sqrt{\frac{7}{10}} & -\frac{3}{14} & -\frac{6\sqrt{5}}{35} \\ +\sqrt{\frac{9}{35}} & 0 & -\frac{6\sqrt{5}}{35} & +\frac{5}{7} \end{pmatrix} T(r) \right] (18)$$

Numerically by solving the Schrödinger equation one finds that in the pseudoscalar channel with the potential (16) and $\Lambda=1.2$ GeV there is a $B^*\bar{B}^*$ bound state with binding energy 59 MeV, i.e. an $\eta_b(10590)$. In $D^*\bar{D}^*$ there is almost a bound state at theshold (which becomes bound if Λ is increased to 1.5 GeV, or if one adds a very weak potential such as $-0.05[\text{GeV}] \cdot exp[-(r/0.9[\text{fm}])^2]$. In the light sector $K^*\bar{K}^*$ becomes bound if one adds a potential $-0.42[\text{GeV}] \cdot exp[-(r/0.9[\text{fm}])^2]$ giving roughly half of the binding.

In the scalar sector the potential (14) with $\Lambda=1.2$ GeV binds a $B^*\bar{B}^*$ $\chi_{b0}(10582)$ (see figure of wave function in Ref.[3]) with a 67 MeV binding energy, and a $D^*\bar{D}^*$ $\chi_{c0}(\approx$

4015) near threshold (with 1 MeV binding energy). Again for the $K^*\bar{K}^*$ system one must add a weak potential to the central piece such as $-0.1[\text{GeV}] \cdot exp[-(r/0.9[\text{fm}])^2]$ in order to have a bound state.

In the axial sector, 1^{+-}, the potential (15) supports a bound state in $B^*\bar{B}^*$, i.e. a $h_b(10608)$ (see figure of wave function in Ref.[3]) with a 41 MeV binding energy, and almost a $D^*\bar{D}^*$ bound state, a $h_c(\approx 4015)$, at threshold (which becomes bound if Λ is increased to 1.3 GeV). For $K^*\bar{K}^*$ one would have to add to the central piece some short range attraction, such as $-0.2[\text{GeV}] \cdot exp[-(r/0.9[\text{fm}])^2]$ to have a bound state.

Finally in the 4-dimensional case of $J^{PC} = 2^{++}$, which is difficult to analyse analytically, and where the S wave central potential is repulsive for I=0, one finds perhaps surprisingly that one also gets a bound state in $B^*\bar{B}^*$ a $\chi_{b2}(10602)$ with 47 MeV binding (see figure of wave function in Ref.[3]).

5. RESULTS

In summary I find that in particular in the heavy meson sector deusons must exist by pion exchange alone. Table 2 lists the quantum numbers of altogether 12 expected heavy deusons. Especially in the beauty sector the existence of these states seems impossible to avoid, since they are already bound by about 50 MeV including only one-pion exchange. In the charm sector pion exchange alone predicts states near the thresholds, and with some small contribution of shorter range these states should also exist.

Table 2. The predicted heavy deuson states (all with I=0) close to the $D\bar{D}^*$ and the $D^*\bar{D}^*$ thresholds and about 50 MeV below the $B\bar{B}^*$ and $B^*\bar{B}^*$ thresholds using one pion exchange only. With additional attraction of shorter range, the masses decrease.

Composite	J^{PC}	Deuson	Composite	J^{PC}	Deuson
$D\bar{D}^*$	0^{-+}	$\eta_c(\approx 3870)$	$B\bar{B}^*$	0^{-+}	$\eta_b(\approx 10545)$
$D\bar{D}^*$	1^{++}	$\chi_{c1}(\approx 3870)$	$B\bar{B}^*$	1^{++}	$\chi_{b1}(\approx 10562)$
$D^*\bar{D}^*$	0^{++}	$\chi_{c0}(\approx 4015)$	$B^*\bar{B}^*$	0^{++}	$\chi_{b0}(\approx 10582)$
$D^*\bar{D}^*$	0^{-+}	$\eta_c(\approx 4015)$	$B^*\bar{B}^*$	0^{-+}	$\eta_b(\approx 10590)$
$D^*\bar{D}^*$	1^{+-}	$h_{c0}(\approx 4015)$	$B^*\bar{B}^*$	1^{+-}	$h_b(\approx 10608)$
$D^*\bar{D}^*$	2^{++}	$\chi_{c2}(\approx 4015)$	$B^*\bar{B}^*$	2^{++}	$\chi_{b2}(\approx 10602)$

The widths are expected to be quite narrow. This is especially the case for the four $D\bar{D}^*$ and $B\bar{B}^*$ states, which because of parity cannot decay to $D\bar{D}$, respectively $B\bar{B}$. Of course, through annihilation of the heavy quarks, decays to light mesons are possible. However, this should be suppressed by form factors, when these states are much larger in size than normal $Q\bar{Q}$ states, provided they are weakly bound. The $D^*\bar{D}^*$ and $B^*\bar{B}^*$ deusons can generally decay into $D\bar{D}$ and $B\bar{B}$, which should be their main decay mode giving widths of a few tens of MeV.

The lighter the constituents are, the larger will the kinetic term of the Hamiltonian be compared to the potential term. Therefore, deusons with light mesons as constituents are much harder to bind, and one-pion exchange alone cannot support such bound states. With the pion itself as a constituent this is definitely a crucial obstacle, but also for $K\bar{K}^*$ systems one has a potential term roughly only half as strong as needed. In Table 3 the quantum numbers are listed, where one finds the largest attraction, together with the experimental non-$q\bar{q}$ candidates. The fact that the best non-$q\bar{q}$ candidates are included in this list, can be taken as support of the idea that some of these states actually are deusons, where pion exchange contribute about half of their binding energy. One should also remember that, even if the attraction is not strong enough to form a bound state, it can give rise to a threshold effect (a virtual bound state, like the $^1S_0$ NN channel) giving a phase shift first rising close to 90° and then falling. Since the $\eta(1440)$ and $f_1(1420)$ peaks are seen a little above the $\bar{K}K^*$ threshold they may, in fact, be such "virtually bound deusons".

Table 3. The meson-meson channels in the light meson sector where one-pion exchange is attractive. With additional attraction of shorter range, giving roughly half of the binding, bound states should exist. The experimental non-$q\bar{q}$ candidates which could be such deusons are listed in the last column. The iota peak (listed under $\eta(1440)$ in Ref.[1]) probably contains two states $\eta(1390)$ and $\eta(1490)$ and the theta peak, $f_J(1710)$, may contain both spin 0 and 2).

Composite	I J^{PC}	Threshold/MeV	Experimental non-$q\bar{q}$ state
$K\bar{K}^*$	0 0^{-+}	1390	$\eta(1390)$? in the "iota" peak
$K\bar{K}^*$	0 1^{++}	1390	$f_1(1420)$
$K^*\bar{K}^*$	0 0^{++}	1790	$f_0(1710)$ "theta"
$K^*\bar{K}^*$	0 0^{-+}	1790	$\eta(1760)$
$K^*\bar{K}^*$	0 1^{+-}	1790	
$K^*\bar{K}^*$	0 2^{++}	1790	$f_2(1710)$ "theta"
$(\rho\rho+\omega\omega)/\sqrt{2}$	0 0^{-+}	1540-1566	$\eta(1490)$? in the "iota" peak
$(\rho\rho-\omega\omega)/\sqrt{2}$	0 0^{++}	1540-1566	$f_0(1370)$, $f_0(1525)$, or $f_0(1590)$
$(\rho\rho+\omega\omega)/\sqrt{2}$	0 2^{++}	1540-1566	$f_2(1520)$ ("Ax")
$(K^*\rho-K^*\omega)/\sqrt{2}$	$\frac{1}{2}$ 0^{++}	1665-1678	

In addition there are two good non-$q\bar{q}$ candidates the X(3100), and X(3250) which are believed to be multiquark states. These could be the first heliumlike four-meson states composed primarily of one K^* and three ω's or ρ's, and $K^*\bar{K}^*\rho\omega$ respectively. Note that the binding energy increases with number of constituents[25], since the number of potential terms grow faster than the kinetic terms.

In two-meson channels with exotic flavour or CP quantum numbers pion exchange is generally repulsive or quite weak. Therefore, one does not expect that such exotic deusons exist, although B^*B^* may be an exception. Neither does the deuson model predict new non-$q\bar{q}$ states which should have been seen. E.g., for I=1 channels one pion exchange is generally a factor 3 weaker than for I=0.

Where could the predicted heavy deusons of Table 2 be produced and seen experimentally? Unfortunately, this will not be easy, but two good places seems to be: $p\bar{p}$ in flight at the Fermilab antiproton accumulator, and Υ decay looking at final states including e.g. J/ψ ω (for the $D\bar{D}^*$ deusons) or $D\bar{D}$, $D\bar{D}^*$ (for the $D^*\bar{D}^*$ deusons).

To find these states would be important, not only because they would confirm the expectations from pion exchange and constrain the parameters of the model presented here. More importantly, if these deusons are found, they at the same time would give strong support for the interpretation that many of the present light non-$q\bar{q}$ candidates really are deuteronlike states. This would then imply that experimental evidence for baglike multiquark states and glueballs would have to be looked for at higher energies.

1. K. Hikasa et al., (The Particle Data Group), *Phys. Rev.* **D45** (1992) 1, and 1994 edition to appear in Phys. Rev. D.
2. N.A. Törnqvist, *Phys. Rev. Lett.* **67** 556 (1992); Zeit. Physik C **61** 525 (1994); *Proc. of Hadron 91*, College Park, Maryland USA, 1991, Eds. S. Oneda, D.C. Peaslee, World Scientific (1992), 795.
3. N.A. Törnqvist, *Proc. Int. Conf. High Energy Physics*, Dallas 1992, Ed. J.R. Sanford, AIP (Conf. Proc.) No **272**, 784.
4. M.B. Voloshin and L.B. Okun, **JETP Lett. 23** (1976) 333.
5. M. Chaichian, R. Kögerler, and M. Roos, *Nucl. Phys.* **B141** 110 (1978).
6. F. Gutbrod, G. Kramer and Ch. Rumpf, *Z. Physik* **C1** 391 (1979).
7. K. Dooley, E.S. Swanson and T. Barnes, *Phys. Lett.* bf B275 478 (1992).
8. T. Barnes and E.S. Swanson,*Phys. Rev.* **D46** 131 (1992).
9. E.S. Swanson, *Ann. Phys.* (New York) **220** 73 (1992).
10. R. Longacre, *Phys. Rev.* **D42** 874 (1990).
11. J. Weinstein and N. Isgur, *Phys. Rev. Lett.* **48** 656 (1982); Phys. Rev. **D27** 588 (1983).
12. J. Weinstein and N. Isgur,*Phys. Rev.* **D41** (1990) 2236 (1991).
13. J. Weinstein and N. Isgur,*Phys. Rev.* **D43** 95.
14. K. Maltman and N. Isgur,*Phys. Rev. Lett.* **50** 1827 (1983); Phys. Rev. **D29** 952 (1984); N. Isgur, *Acta Phys. Austriaca Suppl.* **XXVII** 177-266 (1985) .
15. T.E.O. Ericson and G. Karl, *Physics Letters* B **309** 426 (1993).
16. A.V. Manohar and M.B. Wise, *Nucl. Phys.* B **399** 17 (1993).
17. N.K. Glendenning and G. Kramer, *Phys. Rev.* **126** (1962) 2159.
18. T.E.O. Ericson and M Rosa-Clot, *Ann.Rev.Nucl.Part.Sci.* **35** (1985) 271, and *Nucl. Phys.* **A405** (1983) 497.
19. T.E.O. Ericson and M. Weise,*Pions and Nuclei*, Clarendron Press (1988).
20. N.A. Törnqvist, "How to Parametrize an S-wave Resonance and How to Identify Two-Hadron Composites", Helsinki preprint HU-SEFT-R-1994-03.
21. J.M. Blatt and V.F. Weisskopf,*"Theoretical Nuclear Physics"*, John Wiley & Sons 1952, (page 56).
22. S. Boukraa and J.-L.Basdevant, *J. Math. Phys.* **30** (1989) 1060.
23. J.M Richard, *Phys. Rep.* **212** (1992) 1.
24. M. Lacombe, B. Loiseau, J.M. Richard, R. Vinh Mau, J. Côté, P. Pirès and R. de Tourreil, *Phys. Rev.* **C21** (1990) 861.
25. J.L. Basdevant, A. Martin, J.M. Richard, *Nucl. Phys.* **B340**, 60 (1990).

THRESHOLD PRODUCTION OF CHARMED AND B MESONS

IN e^+e^- ANNIHILATION

NINA BYERS

Physics Department, UCLA, Los Angeles, CA 90024

ABSTRACT

After an historical introduction reviewing the successes and failures for heavy quarkonium spectroscopy of the nonrelativistic quark model including $(v/c)^2$ corrections, the discussion is widened to include light quark pair creation; i.e., dynamical quarks. We find that the simple extension of the QCD inspired potential model which includes dynamical quarks first proposed by Eichten et al.[2] is remarkably successful in accounting both for the observed narrow state heavy quarkonium spectroscopy *and* experimental data on threshold production of charmed and B mesons in e^+e^- annihilation. We have studied various different models of light quark pair creation. In addition to the original Cornell model, we find that a variant of it can also account for the data. The cross section data are mainly inclusive cross section measurements. The two models give distinguishably different predictions for the energy dependence of production cross sections for individual channels, and measurement of the energy dependence of production cross sections for individual channels may distinguish between these models.

1. Introduction

In this talk I would like to discuss three sorts of QCD inspired phenomenological models which describe $Q\overline{Q}$ physics at low energies.[1] The simplest is the nonrelativistic naive quark model (NR NQM). By naive quark model (NQM), I refer to treatments in which the problem is treated strictly as a two body problem. As I remind you below, the NR NQM initially had many successes. However, there were also some failures which were put right when $(v/c)^2$ corrections were taken into account. I will discuss separately two types of $(v/c)^2$ corrections; (i) those applied to the NQM and (ii) extension of the NQM to take into account light quark pair creation. This is one of the two main points of my talk, namely that light quark pair creation effects are $(v/c)^2$ corrections to the NQM and are of the same order as effects due to the spin dependence of the $Q\overline{Q}$ force and kinematic $(v/c)^2$ corrections. All these effects are important in obtaining good agreement of quark model predictions with experimental data. There are two types of light quark pair creation effects to be discussed here; namely, (1) those that occur in heavy quarkonium states below flavor threshold (vacuum polarization effects) and (2) those responsible for production of charmed and B mesons at low energies. The effects of virtual light quark pairs on the narrow $Q\overline{Q}$ states below flavor threshold can be included through modification of the two-body $Q\overline{Q}$ potential. If one confines one's attention only to these states, then the phenomenological $Q\overline{Q}$ potential obtained by fits to data includes both the two-body potential and the effect on the spectrum of virtual $q\overline{q}$ pairs. Such a treatment, however, does not allow for relating the effects of virtual $q\overline{q}$ pair creation with actual $q\overline{q}$ pair creation as seen in the productions of charmed and B mesons. It is also somewhat inadequate in that it is difficult to obtain the configuration mixing necessary to account for certain spectroscopic results without explicit inclusion of four-body $Q\overline{q}q\overline{Q}$ virtual states.

In section II is a brief summary of the successes, and failures, of the NR NQM; and, in section III, the improvement when $(v/c)^2$ corrections and the spin-dependence of the $Q\overline{Q}$ interaction are taken into account. Section IV presents some QCD inspired potential models of light quark pair creation. Section V outlines the method used to include these dynamical quarks in calculating both the spectroscopy of the narrow $Q\overline{Q}$ states below flavor threshold and the production cross sections for charmed and B mesons above flavor threshold. Of the various possible models we studied in detail, two seem reasonably successful in accounting for both the observed spectroscopy of the narrow states and the observed threshold production cross sections in e^+e^- annihilation. These models are the Cornell model [2] and an extension of it studied by Zambetakis [5]. Section VI compares predictions of these models with available production cross section data. These are mainly includive cross section data. Both models give remarkably good accounts of these data but neither are precise fits to the data. This section also presents exclusive production cross sections; for these the models predict different energy dependences. Measurement of exclusive production cross sections may indicate which of these models is the more correct one.

2. Successes and Failures of the Nonrelativistic Naive Quark Model (NR NQM).

This section is something of an historical account which is incomplete and included here to give background to the assertion that corrections to the NR NQM owing to light quark pair creation, and the coupled channel mixing it induces, are of the same order as those due to spin-dependent forces and other $(v/c)^2$ corrections.

2.1. Successes

Here are outlined the early outstanding successes of the NR NQM. Using a simple QCD inspired potential which behaves at short distances like a one gluon exchange potential and is linearly rising at large distances, the model has as parameters the strength of the Coulombic part and the slope of the linear part of the potential; in addition, the masses of the c and b quarks. With these few parameters, the model accounts quite well for a large number of data:

Masses of narrow heavy quarkonium states (spin averaged).

Leptonic decay rate ratios $\Gamma_{ee}(\psi')/\Gamma_{ee}(\psi)$, etc..

Allowed radiative transition rates (given by $< r >_{fi}$); e.g., $\psi' \to \chi_J$, $\chi_J \to J/\psi$, etc..

Allowed M1 radiative transition rates (given by $< 1 >_{fi}$) ; e.g., $J/\psi \to \eta_c$.

2.2. Failures

Among the successes above were some failures. These were:

$\Gamma_{ee}(J/\psi)$ too big.

E1 radiative transition rate for $\psi' \to \chi_0$ too big by a factor 2.

Forbidden M1 transitions such as $\psi' \to \eta_c$ observed.

$\psi(3770)$ observed in e^+e^- annihilation with mass too close to the ψ' to be understandable as a $^3S_1$ $c\bar{c}$ state.

In addition this simple model failed to account for the observed fine and hyperfine splittings in the spectra. The items above, and the fine structure of the mass spectra, are corrected by taking relativisitc corrections into account.

3. Relativistic Corrections.

There are two types of relativistic corrections: (i) $(v/c)^2$ corrections to the NQM; i.e., $(v/c)^2$ corrections to the nonrelativistic potential model in which nevertheless the model remains a two-body treatment of the problem, and (ii) extension of the model to take into account dynamical quarks; i.e., light quark pair production. Type (i) relativistic corrections have been discussed in this Conference explicitly by G. M. Prosperi, Yu-Bing Dong, L. Fulcher, F. Schoeberl, H. Sazdjian, Kuan-Ta Chao, and extensively in the literature; see,e.g., [7,8] and references cited therein. Spin-dependent forces are also regarded as type (i) $(v/c)^2$ corrections because they arise in this order in the Breit-Fermi reduction of a particle exchange diagram. Estimates of the spin-dependence of the $Q\overline{Q}$ force have been obtained other ways as well; see, e.g., [3,4] and the contributions of J. M. Ball and F. Zachariasen to this

Conference. The effects that I would classify as type (ii) relativistic corrections are those that arise owing to virtual light quark pairs. These are of the same magnitude as those of type (i) and arise in relativistic quantum field theory in order $(v/c)^2$. Below is a list of relativistic corrections which have been found to correct the 'failures' mentioned above. In most cases several effects contribute.

- a. spin-orbit $Q\overline{Q}$ interaction

- b. spin-spin $Q\overline{Q}$ interaction

- c. tensor forces

- d. spin-independent $(v/c)^2$ corrections to the $Q\overline{Q}$ Hamiltonian

- e. direct $Q\overline{Q}$ $^3D_1$ - photon coupling

- f. light quark pair creation

- g. coupled channel mixing.

In Table I. are indicated which of the above items tend to correct the so-called failures.

put right	by
Fine structure splitting	a
El rate $\psi' \to \chi_0$	a and g
$\Gamma_{ee}(J/\psi)$	g
Forbibben M1 rates	d and g
$\psi(3770)$ explained	c, e, and g
$\sigma(e^+e^- \to$ charmed mesons$)$	f

Table I.

I have not included hyperfine splittings in the above discussion. This is because I believe that remains an unsolved problem.

The bulk of this paper is about light quark pair creation. The foregoing was included to indicate that inclusion of light quark pair creation improves the the agreement of NQM results with experimental data in two ways. Not only does it extend the model to describe quarkonium decays to charmed and B mesons, it also improves the agreement with the narrow state spectroscopic data.

Before addressing explicitly light quark pair creation effects, it is appropriate here to remind you that relativistic effects such as spin-orbit couplings, as well as pair creation, reflect the Lorentz transformation properties of the interaction. For example, in a Breit-Fermi Hamiltonian the spin-orbit interaction coming from vector and scalar exchange have opposite sign and differ by a factor 3. Correspondingly pairs created from vector and scalar exchange are produced in 1^{--} and 0^{++} states, respectively. At present it is not clear that the spin-orbit interaction is correctly viewed as an effect found in a Breit-Fermi reduction of vector and scalar exchange interactions. However, when this assumption is made and the result compared with data the observed fine structure splittings indicate that the long range (linearly rising) part of the potential comes from scalar exchange. We included light quark pair creation in such a model and found this coupling to decay channels incapacitated the model. (See below.) On the other hand, models in which the quarkonium potential as a whole is assumed vector exchange can account well for the (spin-averaged) masses [5]; 'vector' models also account reasonably well for observed threshold charmed and B meson production cross section data. Therefore, it may be reasonable to assume there is some truth ion these 'vector' models. They are not inconsistent with the analyses of Eichten and Feinberg [3] and Gromes [4] who obtain in a $1/M^2$ expansion of QCD spin-dependent $Q\overline{Q}$ interactions similar to those obtained in the Breit-Fermi reduction of a vector short range and scalar long range potential.

4. Including Dynamical Quarks

Eichten et al. proposed a quarkonium potential model that includes dynamical quarks nearly 20 years ago. [2] Their framework, with various different dynamical assumptions, has been used by most authors writing on this subject since then. Straightforward inclusion of light quark pair creation in a $Q\overline{Q}$ model requires quantum field theory and immediately confronts us with a many body problem to solve. Eichten et al., hereafter referred to as the Cornell group, simplified the problem to manageable proportions. Before going into details about model predictions for charmed and B meson productions, I would like to briefly outline their framework. Its structure incorporates fundamental properties of QCD, namely confinement and asymptotic freedom. Consider the QCD vacuum polarization diagram shown below.

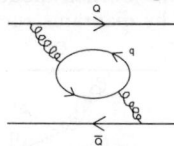

For loop momenta large compared to constitutent quark masses, this diagram can be treated perturbatively and is taken into account using the QCD running coupling constant. However, diagrams with small loop momenta should be treated nonperturbatively. For low momenta, higher order diagrams with multiple gluon exchanges should be taken into account. These were included by the Cornell group invoking duality. They used a phenomenological field theoretic model for $q\overline{q}$ creation and then took the strong interactions into account by replacing the $Q\bar{q}q\bar{Q}$ intermediate states by $Q\bar{q}$ and $q\bar{Q}$ mesons (bound by the same potential as binds $Q\bar{Q}$). This reduces the many body problem to a tractable pair of two-body problems.

First one solves a potential model for bound $Q\bar{Q}$, $Q\bar{q}$ and $q\bar{Q}$ states. The Cornell group used

$$V(r) = -\frac{\kappa}{r} + C + \frac{r}{a^2} . \tag{1}$$

Then, as in the Wigner-Weisskopf treatment of coupled channel problems in nuclear physics, one solves a coupled channel two-body problem that can be represented by the Hamiltonian

$$H = \begin{pmatrix} H_{Q\bar{Q}} & h^\dagger \\ h & H_{Q\bar{q}q\bar{Q}} \end{pmatrix} \tag{2}$$

which operates in a space of bound $Q\bar{Q}$ states and bound $Q\bar{q}$ and $q\bar{Q}$ states. In the Cornell model the $Q\bar{Q}$ states are eigenstates of

$$H_{Q\bar{Q}} = \frac{p^2}{M_Q} + V + 2M_Q . \tag{3}$$

and the $Q\bar{q}$ and $q\bar{Q}$ mesons eigenstates of the corresponding Hamiltonian (with relativistic correction for the light quark). In the calculations, however, measured (or to be measured) values of the $Q\bar{q}$ and $q\bar{Q}$ meson masses are used. The piece denoted by $H_{Q\bar{q}q\bar{Q}}$ in (2) is the kinetic energy operator for the two-meson states. The off-diagonal piece h couples the $Q\bar{Q}$ states to their two-body decay channels. (Final state interactions are neglected; i.e., interactions between the bound $Q\bar{q}$ and $q\bar{Q}$ mesons are neglected.)

To complete the framework described above, we need to put in a mechanism for pair creation. In the Cornell model, pairs are created by the same interaction that binds; i.e., by diagrams such as

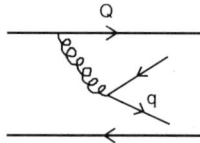

where the corkscrew line represents the propagator of an (instantaneous) interaction whose Fourier transform is the quarkonium potential V. The quark vertices, correspondingly, were taken as nonrelativistic reduction of a vector exchange. One may vary these assumptions and consider other dynamical models of quark pair creation. For example, one could assume scalar rather than vector exchange. As mentioned above, the fine structure in the mass spectra indicate that the long distance part of the potential may be due to scalar exchange. These two assumptions give quite different results; a vector interaction creates a pair with quantum numbers 1^{--} while a scalar creates a pair with the quantum numbers of the vacuum. We can report here calculations based upon various different assumptions about the dynamics of $q\bar{q}$ pair production. They are tabulated in Table II.

Model	h	⟨ℓℓℓℓℓ⟩	ex
Cornell		r/a^2	V
Zambetakis		$V(r)$	V
Grotch, Zamb, Byers		$-\kappa/a + fC$ $+(1-f)C + r/a^2$ $\{0 \leqslant f \leqslant 1\}$	V S
Grotch & Byers		order α_s^2 QCD	V
Le Yaouanc et al., ...		$\dfrac{m_q \gamma_{QPC}}{\Gamma^2}$	S

Table II.

The dynamical assumptions represented in Table II. are as follows. First, the Cornell model: the original Cornell model neglected pair creation from the Coulombic part of the potential and took pair creation only from the long distance part. This is indicated in the third column. In the fourth column is indicated whether the exchange is vector or scalar. The second row represents Zambetakis' extension of the original Cornell model to include the Coulombic part of the potential in the pair creation matrix elements. [5] The third row represents our calculations with Howard Grotch in which we took the Coulombic part of the potential to be vector exchange and the long range part to be scalar. There is an ambiguity in implementing this idea;

namely where does short become long? Quantitatively this translates into the question of how to treat the constant term C in (1). For vector coupling, C plays no role; the pair creation matrix elements only depend on the gradient of the potential. However with scalar coupling the potential behaves like a mass term and the pair creation matrix elements depend on V rather than $\vec{\nabla} V$. Thus the constant C can be significant. Fits to the data indicate C is large and negative; of the order of -800 MeV.[5] To take this ambiguiuty into account, we introduced another parameter f which specifies which fraction of C couples vectorially, and allowed it to vary between 0 and 1. The fourth row refers to calculations with Grotch of the contribution of the triple gluon coupling graph. We evaluated this graph in Coulomb gauge, and found its contributions were not large; they are indeed of order α_s^2. We also studied the other QCD graphs of this order and found none seems to be singular. We therefore conclude that such effects are not at this stage significant. In the last row is represented the QPC model of Le Yaouanc et al..[12] It is a model sometimes referred to as a flux tube or string breaking model. In it the $q\bar{q}$ are produced with the quantum numbers of the vacuum, and the pair creation matrix element is a constant which can be included in calculations like the above by taking it to be given by a constant potential $m_q \gamma_{QPC}/\sqrt{2}$. This model gives interesting results which have been discussed in a number of papers; see Ref.[12].

In order to get the matrix elements of h from the diagrams in Table II., one must sandwich them between $Q\overline{Q}$ wave functions and $Q\overline{q}$ and $q\overline{Q}$. (Generally the $Q\overline{Q}$ wave functions are taken to be eigenstates of (3), and for simplicity the $Q\overline{q}$ and $q\overline{Q}$ wave functions are taken to be Gaussian fits to bound state wave functions bound by the r/a^2 part of the potential.)

We report here the results of our studies of first four of these models. Mainly what we have to report are our results for the Cornell and Zambetakis models because (1) we found the order α_s^2 graphs were not likely to be large enough to be important; (2) the hybrid model in which the long range part of the potential was considered scalar exchange was unsuccessfull (explained below); and (3) both the Cornell and Zambetakis models were able to account for available data - a large amount of data on the narrow states below threshold and on production cross sections for charmed and B mesons above threshold. This is will be discussed in detail in the last section of this paper.

We end this section with a brief explanation of why we think that the hybrid model of short distance vector and long distance scalar exchange is not a viable model. Allowing f to vary between 0 and 1, we could not find values of the parameters in the potential (1) that allowed a fit to the narrow state spectroscopic data. The reason for this is that in the region of r where the quarkonium wave functions are large V changes sign, going from Coulombic at short distances to linearly rising at large distances, and consequently, when it is coupled as a scalar, its contributions to matrix elements of h are relatively small differences of two large numbers. Owing to this the matrix elements of h vary erratically as one goes from one of the low-lying quarkonium states to another; consequently their contributions to level shifts are unstable against small variations of the parameters.

5. Modification of NQM

Before discussing our numerical results, it may be of interest to outline here how the inclusion of dynamical quarks changes the NQM. This can best be seen in a Hamiltonian framework in which the light quarks have been integrated out and one has remaining an effective Hamiltonian in the $Q\overline{Q}$ space. This is a peculiar Hamiltonian because (i) it is energy-dependent; and (ii) for energies above threshold for meson production, it develops an anti-hermitian part. It is, however, quite interesting because it gives mass shifts and configuration mixings and also gives production cross sections for charmed and B mesons in e^+e^- annihilation.

In the space spanned by eigenstates of (3), hereafter called NQM states, the effective Hamiltonian is (see, e.g., Refs. [5,9])

$$H_{eff}(W) = \mathcal{M}^{bare} + \Omega(W) . \tag{4}$$

where $\mathcal{M}^{bare}$ is a diagonal matrix whose elements are the eigenvalues of (3). The Ω matrix is second order in h; the explicit expression for it is given below. It is perhaps useful to state some more of the properties of H_{eff} here.

(1). The physical masses M_N of the narrow quarkonium states are given by the solutions to

$$H_{eff}(M_N) a^N = M_N a^N \tag{5}$$

where a^N denotes a state vector in the space spanned by NQM states. A physical state Ψ_N may be expanded as

$$\Psi_N = \sum_i a_i^N \psi_i + \sum_n b_n^N \phi_n . \tag{6}$$

where ψ_i are NQM statesand ϕ_n are continuum two-meson states. The a_i^N are the components of the 'eigenvectors' a^N in (5). The problem (5) is not exactly an eigenvalue problem. The dependence on M_N is highly nonlinear. Nevertheless solutions can be found.[5]

(2). $H_{eff}(W)$ is not hermitian when W is above the threshold of the lowest mass two-meson continuum. The quarkonium states whose masses are above threshold are poles in the complex W plane. They are poles of the effective propagator $\mathcal{G}(W)$, where

$$\mathcal{G}(W) = (W - H_{eff}(W))^{-1} . \tag{7}$$

For W above threshold, Ω has the structure

$$\Omega(W) = R(W) - i\Gamma(W)/2 \tag{8}$$

with R and Γ real symmetric matrices. Unitarity requires the diagonal elements of Γ to be positive.

The matrix elements of $\Omega(W)$ are given by are given by

$$\Omega_{ji}(W) = \sum_n \frac{(\psi_j, h^\dagger \phi_n)(\phi_n, h\psi_i)}{(W - E_n + i\epsilon)} . \tag{9}$$

The sum over n denotes sum over channels and integration over channel phase space; E_n is the channel energy. The channels are labeled by f which specifies channel spins, flavor, masses, etc.. Because of this sum over channels, Ω is a sum of matrices

$$\Omega(W) = \sum_f \Omega^{(f)}(W) . \tag{10}$$

Each matrix element of $\Omega^{(f)}$ is an integral which becomes singular when $W \geq m_1 + m_2$ where m_1 and m_2 are the masses of the mesons in channel f. These singular integrals are evaluated in the usual way taking W in the upper half plane and then the boundary value on the real axis from above.

For W above flavor threshold, charmed and B meson production cross sections in e^+e^- annihilation may be calculated as follows. The cross section ratio

$$\Delta R^f = \frac{\sigma(e^+e^- \to f)}{\sigma(e^+e^- \to \mu^+\mu^-)} , \tag{11}$$

in one photon approximation, is given by the dispersive part of $\mathcal{G}(W)$. With a high speed computer and H_{eff} in matrix form (see below), $\mathcal{G}$ is easily calculated by matrix inversion. There is no coupling between the $c\bar{c}$ and $b\bar{b}$ subspaces, so one deals separately with these two subspaces and the calculations for the production cross sections of charmed and B mesons are done independently. The charm contribution to the cross section ratio ΔR_c can be expressed as a trace in the $c\bar{c}$ subspace; viz.,

$$\Delta R_c = -\frac{72\pi}{W^2} e_c^2 \, \text{Tr}(\, \mathcal{W} \, \frac{(\mathcal{G} - \mathcal{G}^\dagger)}{2i} \,)_{c\bar{c}} \qquad (12)$$

where $e_c = 2/3$, and the matrix $e_c^2 \, \mathcal{W}$ is bilinear in the $c\bar{c}$ -photon coupling. Nonrelativistically, matrix elements of $\mathcal{W}$ are given by $c\bar{c}$ NQM wave functions; viz.,

$$\mathcal{W}_{ij} = \psi_i(0)\psi_j(0) \, . \qquad (13)$$

(Relativistic corrections are important here, particularly for charm; see Ref. [10].) In their original papers, the Cornell group showed that

$$\mathcal{G} - \mathcal{G}^\dagger = \mathcal{G}^\dagger(\Omega - \Omega^\dagger)\mathcal{G} \, . \qquad (14)$$

Thus, from (12), (14), and (10), one sees that the exclusive production cross section for channel f is

$$\Delta R_c^{(f)} = -\frac{72\pi}{W^2} \, e_c^2 \, \text{Tr}(\, \mathcal{W} \, \mathcal{G}^\dagger \, Im\Omega^{(f)} \, \mathcal{G} \,)_{c\bar{c}} \, . \qquad (15)$$

Similarly, one calculates production cross sections for B mesons. From the above it is clear that this is a nonperturbative treatment of dynamical quarks.

With high speed computers it is straightforward to solve, as outlined above, for both the narrow states below threshold and the (resonant) cross sections above threshold. One first solves (3) for the NQM wave functions and masses. The matrix elements of the Ω matrix and (4) can then be explicitly evaluated. These matrix elements are discrete because, owing to confinement, the entire $Q\overline{Q}$ state space is spanned by a discrete set of eigenfunctions. With these matrices one can calculate all of the above quantities. Most computers are now able to work with matrices of (almost) arbitrarily high dimension. However, since we are interested only in the low lying quarkonium states, we do not need to calculate matrices of high dimension. The highly excited NQM states make negligible contributions.

On the other hand, another approximation involved in a numerical evaluation of $\Omega(W)$ may cause significant error in predicted masses of quarkonium resonances in the continuum. In calculating $\Omega(W)$ one generally approximates the infinite sum over channels f by a finite sum neglecting all channels with thresholds greater than some minimum energy E_{min} which is greater than the maximum value of W for which one is calculating the Ω matrix. Though the contribution from any one of these neglected channels is small, their cumulative effect can be appreciable because all these channels contribute coherently (negatively) to the diagonal elements of Re Ω; c.f., Eq. (9). (Cancellations are likely in contributions to the off-diagonal elements owing to random phases.) These cumulative mass shifts are small for W substantially below E_{min}; however, they may become appreciable when W comes close to E_{min}. *

6. Production Cross Section Calculations and Comparison with Experimental Data

In this section we report our results on threshold production cross sections, inclusive and exclusive, for charmed and B mesons in e^+e^- annihilation and compare them with available data. There are two sets of results; one obtained from the original Cornell model and the other from Zambetakis' extension in which pair creation occurs at short distances also. The available data are mainly inclusive cross section measurements. The agreement with data is about the same for both although it differs in detail in the two cases. The two models give significantly different results for exclusive channel cross sections. Measurement of these may distinguish between them. The degree of agreement of these model predictions with data is, in our view, significant because the calculated cross sections are obtained without adjustment of parameters or introduction of additional parameters. The parameters are the parameters of the potential κ, C, and $1/a^2$. We calculated the masses of the narrow charmonium states using (5) and determined these three parameters by fitting the observed masses of J/ψ, χ_{cog}, and ψ'. [†]Parameter values which fit these data are[‡]

model	$1/a^2$ in GeV^2	C in GeV	$\kappa(c\bar{c})$
Zambetakis	0.31	-0.97	0.49
Cornell	0.22	-0.85	0.52

Using these parameters, we calculated (spin averaged) Υ masses, charmonium and Υ leptonic widths, etc. and found good agreement with measured values [5].[§] Then we calculated charmed and B meson production cross sec-

---

*In our calculations we included only the channels with ground state pseudoscalar and vector mesons. This is a good approximation for the narrow states below flavor threshold; however, for values of W in the continuum we found it necessary, as did the Cornell group, to compensate for the neglect of excited meson decay channels by putting in a phenomenological mass shift to fit the observed resonance energies [2,11].

[†]After a detailed and extensive analysis, the Cornell group used the following constitutent quark masses $M_c = 1.84$ GeV, $M_b = 5.17$ GeV, and for light quarks, $m_u = m_d = 335$ MeV, and $m_s = 450$ MeV. We used the same values in our calculations.

[‡]The slope of the linear potential $1/a^2$ is significantly greater in Zambetakis' model because the short distance pair creation increases the coupling of $c\bar{c}$ to inelastic channels This stronger coupling to inelastic channels increases the separation of charmonium energy levels, and requires increased strength of the confining potential to bring the mass differences back to their observed values.

[§]For the Υ masses and widths, and for B meson production cross sections, the QCD

tions above flavor threshold evaluating (15) with the same parameters. We calculated charmed meson productionfrom threshold to 4.5 GeV, and for B from threshold to 11.1 GeV. In the numerical calculations we truncated the $c\bar{c}$ state space after the 4S and 2D NQM states, as in the original Cornell calculations. We found that including more states did not significantly alter the results. For B meson production, the $\Upsilon(4S)$ is the first $b\bar{b}$ resonance above threshold; we truncated after the 7S and 4D NQM states. Results were stable against increaseing the number of such states. However, we also found that the results changed little when we omitted the D-states and only changed noticeably near 11 GeV when we omitted the 7S state. The results presented here were obtained truncating $b\bar{b}$ states at the 6S. The reason it was not necessary to include D-states is that owing to the heavier b quark mass, the S-D mixing and photon-D state coupling are smaller than in the charm case.

6.1. *Charmed meson production.*

To understand the complicated energy dependence of the inclusive charm production cross section above threshold (3.73 GeV), it is useful to consider the singularities of the 1^{--} $c\bar{c}$ propagator $\mathcal{G}_{c\bar{c}}(W)$ in the complex energy plane; $\mathcal{G}(W)$ is given by (7). The singularities are cuts along the real axis and poles in the lower half plane. There is a cut for each open channel with a branch point at its threshold. The poles correspond to resonances in the cross section. The branch points and poles are shown in Fig. 1 where the positions of the singularities are determined by the experimental values of the masses and widths of the indicated states [13]; the distance below the real axis of each pole is the measured total half-width. The separations of the branch points and positions of the poles are to scale.

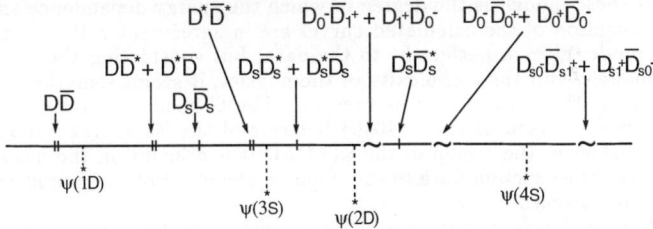

Figure 1.

The corresponding cross section measurements and theoretical curves for R are shown in Fig. 2. The first resonance above threshold is understood, from

---

logarithmic decrease of α_s was taken into account by using for $\kappa(b\bar{b})$ the values 0.46 and 0.48 for the Zambetakis and Cornell models, respectively.

the point of view of the Cornell model, as the $\psi(1D)$ though it is not a pure state but has admixtures of other nearby 1^{--} S and D states. Though this is a relatively narrow resonance as indicated by its proximity to the real axis in Fig. 1, it is not as prominent as the S-state peaks, $\psi(4040)$ and $\psi(4415)$, owing to the fact that nonrelativistically D states do not couple directly to the photon. As it is primarily a D-state, the $\psi(3770)$ owes its presence in an e^+e^- annihilation cross section to the fact that it has some configuration mixing with nearby S states (particularly the 2S) and a direct coupling to the photon in order a $(v/c)^2$. Similarly the $\psi(2D)$ appears in the cross section as a shoulder on the $\psi(3S)$ or $\psi(4040)$ peak owing to its admixture of 3S-state and direct coupling to photon.

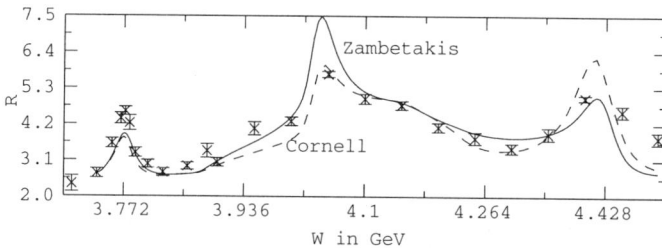

Figure 2.

The data points in Fig. 2 are from the *Review of Particle Properties.* [13] The curves are calculated using the Zambetakis and Cornell models with the contribution to R from other than charmed hadrons taken from the data to be 2.65; i.e., the curves in Fig. 2 are $\Delta R_c(W) + 2.65$.¶ What is remarkable about these models is the degree to which the energy dependence and overall normalization of the calculated curves are in agreement with the data. Neither model gives a perfect fit to the data; but considering the simplicity of the models, and the complexity of the physics, it seems remarkable that the models fit the data as well as they do. The Cornell model seems to fit the data in the region of the $\psi(4040)$ better and the Zambetakis model seems to do better in the region of the $\psi(4415)$. It is difficult on the basis of these inclusive cross section data to say if one or the other of these models is closer to being correct.

The models are more easily distinguished in their predictions for the exclusive cross sections. In Figs. 3 and 4 are shown the individual channel contributions to ΔR_c.

¶A phenomenological mass shift matrix was used as compensation for the neglect of excited meson decay channels; we only included the ground state psuedoscalar and vector meson decay channels in our calculations, c.f., section 5.

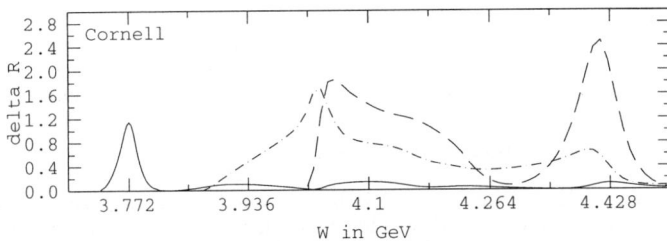

Figure 3.

The full curve is the contribution to R_c from $D\overline{D}$ channels, both charged and neutral; the dot-dashed curve from $D\overline{D}^* + D^*\overline{D}$; and the dashed curve from $D^*\overline{D}^*$.

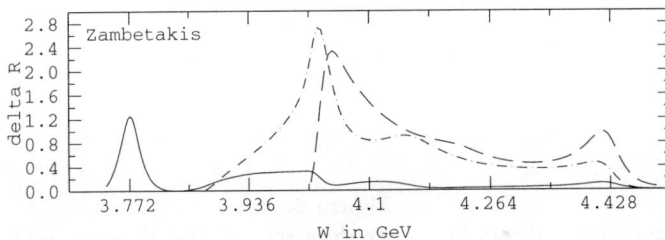

Figure 4.

The cross sections for D_s and D_s^* production are roughly 1/3 of those for non-strange D mesons. These are shown in Fig. 5.

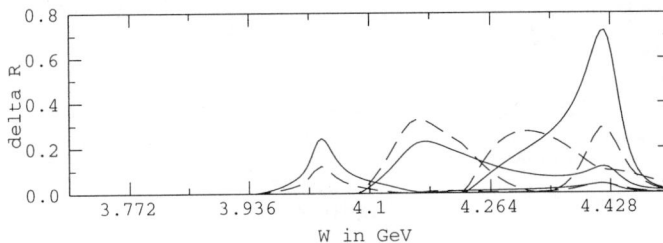

Figure 5.

In Fig. 5 the full curves are Cornell model predictions for $D_s\overline{D}_s$, $D_s\overline{D}_s^* +$ $D_s^*\overline{D}_s$, $D_s^*\overline{D}_s^*$., and the dashed curves Zambetakis model predictions. The curves for the different channels can be distinguished by the fact that they start at different thresholds. Note the scale change. This suppression of the strange mesons is natural to these models. It comes from the m_s being greater than m_u. The suppression is due, essentially, to two effects: (i) the pair creation amplitude is inversely proportional to the light quark mass, and (ii) the strange quark meson states are smaller (more tightly bound).

6.2. B meson production cross sections.

The singularity structure of $\mathcal{G}(W)_{b\bar{b}}$ relevant for our purpose is shown in Fig. 6. It appears simpler than Fig. 1 because the $B^* - B$ mass difference is smaller than $D^* - D$ and because we have indicated only the S-state resonances. In the Υ system both S-D mixing and direct photon-D-state coupling are small. These are effects of order $(v/c)^2$ and therefore smaller than in charmonium. Consequently D-state contributions are not significant in e^+e^- annihilation cross sections.

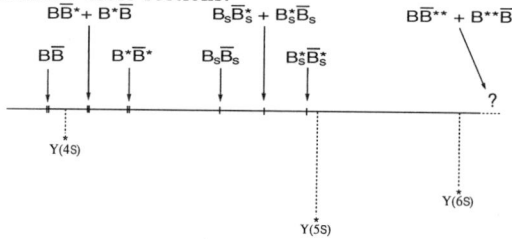

Figure 6.

Corresponding to this in Fig. 7, we show the inclusive B meson production cross section data[14] along with the calculated curves.

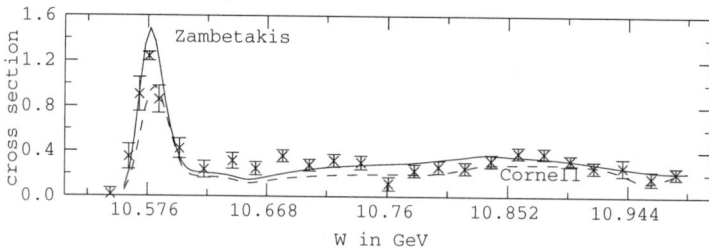

Figure 7.

Radiative and beam spreading corrections have been applied to the theoretical curves in Fig. 7 so that they can be compared with the data which are uncorrected for radiative corrections and beam spread.[||]

Owing to the fact that the $\Upsilon(4S)$ can only decay to $B\bar{B}$, it is a relatively narrow resonance and prominent in the data. The $\Upsilon(5S)$, on the other hand, can decay to all nine ground state psuedoscalar and vector mesons and is relatively broad. In Fig. 6 its position is far from the real axis and relative to its width, its mass is near many sharp thresholds; the mass and width are 10.865 GeV and 110 MeV, respectively. Because of these complications, the $\Upsilon(5S)$ does not have a normal Breit-Wigner shape and is hard to see in the cross sections. The $\Upsilon(6S)$ mass is quoted in Ref. [13] as 11.019 GeV. We have included in Fig. 7 data only out to 10.99 GeV because we think that our calculated values in the region of the 6S are unreliable owing to the fact that we have not included channels with B mesons in excited states - either with $\ell \neq 0$ or excited $\ell = 0$ mesons; and these are likely to have thresholds in the vicinity of the $\Upsilon(6S)$.

The curves in Fig. 7 have been calculated without any adjustment of the parameters of the model. It is, therefore, in our opinion remarkable that they agree with the data as well as they do. As indicated at the end of section 5, a mass shift matrix has been used to fix the peak values of the cross section at 10.580 GeV and 10.865 GeV. Aside from this, the energy dependence and the overall normalization of the calculated curves are determined by the models. The Zambetakis model (full curve) appears to fit the data around the $\Upsilon(4S)$ better than the Cornell model (dashed curve). However, in the region of the $\Upsilon(5S)$ the Cornell model may be a better fit.

As in the charm case, the energy dependence of the exclusive cross sections are quite different for the two models. The contributions to the cross section ratio R from the various channels are shown in Figs. 8 and 9; the full curves are for the $B\bar{B}$ channels, neutral plus charged; the dot-dashed curves are for $B\bar{B}^* + \bar{B}B^*$; and the dashed curves are for $B^*\bar{B}^*$. Notice that at the higher energies where these curves tend to flatten out they go over toward the proportion 1:4:7 which are the spin ratios, discussed in Ref. [2], for $B\bar{B}$, $B\bar{B}^* + \bar{B}B^*$, and $B^*\bar{B}^*$, respectively. In Figs. 8 and 9 we omitted the $\Upsilon(4S)$ resonance region because the cross section there is so much bigger than it is in the rest of the range.

[||]We thank Dave Besson, Persis Drell and Elliot Chu for applying the radiative and beam spread corrections to the calculated cross sections.

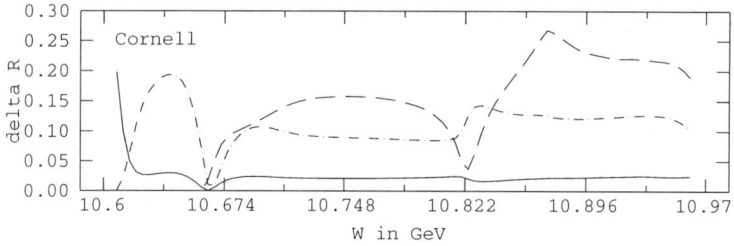

Figure 8.

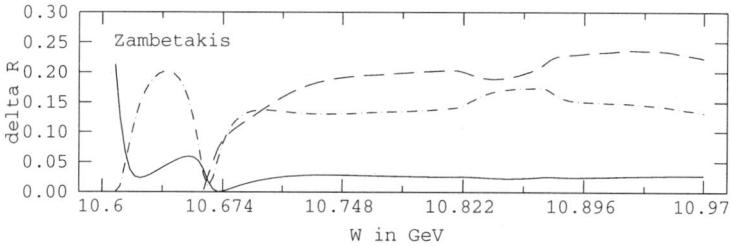

Figure 9.

Finally in Fig. 10 we show the contributions to R from B_s channels; the full curves are Cornell model predictions and dashed curves Zambetakis model. The curves for the various channels can be distinguished by their thresholds.

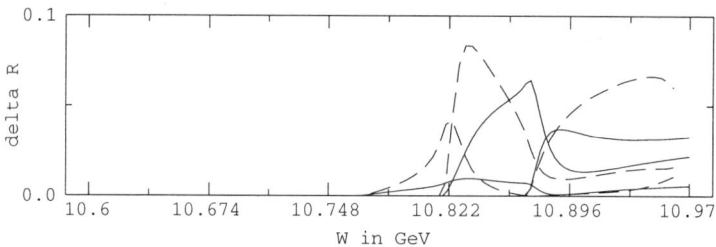

Figure 10.

One interesting feature of the curves in Fig. 10 is that both models predict relatively copious $B_s \overline{B}_s^* + \overline{B}_s B_s^*$ production in the $\Upsilon(5S)$ resonance region.

(These predictions are sensitive to the B_s mass and the $B_s^* - B_s$ mass difference.) As in the charm case, however, these models predict that the strange mesons are produced at a level of about $1/3$ that of the nonstrange mesons. Note the change of scale in Fig. 10.

7. Conclusions

The fact that both the the Cornell and Zambetakis models, as simplistic as they are, can account very well for charmonium and Υ narrow state spectroscopy and, without adjustment of parameters, give production cross sections which agree as well as they do with the data leads us to conclude that they merit further study. There are two directions for this. Experimentally, it would be interesting to compare calculated values for exclusive cross sections with measured values. From theoretical point of view, it may be worthwhile to evaluate taking predicted configuration mixing into account the allowed and forbidden M1 transition rates in charmonium and Υ and compare those with measured values.

8. acknowledgements

It is a pleasure to acknowledge contributions to the computational side of this work from Stanley Cohen and staff of Speakeasy Corporation, and from the UCLA Office of Academic Computing. This work was supported in part by the Department of Energy grant FG03-91ER40662.

9. references

1. In this paper we will use Q for c or b and q for u, d, or s.
2. E. Eichten, K. Gottfried, N. Kinoshita, K.D. Lane, and H.- M.Yan, Phys. Rev. D17, 3090 (1978); 21,313(E)(1980);and Phys. Rev. D21, 203 (1980).
3. Estia Eichten and Frank Feinberg, Phys. Rev. D23, 2734 (1981).
4. Dieter Gromes, Z.Phys. C26, 401 (1984). See also Phys. Lett. 202B, 262 (1988).
5. V. Zambetakis, UCLA Ph.D. thesis (1985); available as UCLA research report UCLA/85/TEP/2.
6. H. Grotch, V.Zambetakis and N. Byers. (unpublished)
7. Richard Lee McClary, UCLA thesis (1982); N. Byers and R. L. McClary, Phys. Rev. D28, 1692(1983)
8. N. Brambilla and G. M. Prosperi, Phys. Lett. B236,69 (1990).
9. T. Appelquist, R. M. Barnett, K. Lane in e^+e^- *Annihilation: New Quarks and Leptons*, Benjamin/Cummings pub., R. N. Cahn, ed..

10. V. A. Novikov et al., Phys. Rev. Lett. 38, 626 (1977) These authors pointed out that there is a a $(v/c)^2$ correction which gives direct coupling of the photon to $^3D_1$ states. There is in addition a one gluon exchange correction to the quark-photon vertex which the Cornell group includes by multiplying the $\ell = 0$ subspace of $\mathcal{W}$ by $(1 - 4\kappa/\pi)$.

11. N. Byers and E. Eichten *Heavy Flavour Production near Threshold in e^+e^- Annihilation* (unpublished). N. Byers and E. Eichten, Nucl. Phys. B(Proc. Suppl.) 16, 281 (1990).

12. Le Yaouanc, L. Olivier, O. Pine, and J. C. Raynal, Phys, Rev. D $\underline{8}$, 2223 (1973); See also S. Ono, A. I. Sanda, and N. A. Tornqvist,Phys. Rev. D $\underline{35}$, 907 (1987) and references cited therein.Studies of B production in this model have also been made by Martin and Ng; cf., A. D. Martin and C.-K. Ng,preprint DTP/88/6. Fits to the production cross sections are made with parameters in addition to those in the $Q\bar{Q}$ potential and the quark pair creation strength γ_{QPC}.

13. Particle Data Group, Phys. Rev. D50, 1177/(1994). The data points in Fig. 2 are those in that energy range in the graph on p.1334. We thank Michael Barnett and Tom Trippe for these selected data. The caption to this graph reads in part:"Systematic normalization errors are not included; they range from $\sim$5-20%, depending on experiment. We caution that especially the older experiments tend to have large normalization uncertainties." Presumably BEPC will in future remeasure these cross sections.

14. D. Z. Besson et al., Phys. Rev. Lett. 381 (1985). The constant non-$b\bar{b}$ background in R has been subtracted from these published data using the mean value 4.56 of the data points below beauty threshold, and the variance of these points has been added in quadrature with the published error to obtain the error bars in our graph.

THE "CHROMO"-GRAVITY ALGORITHM IN QCD

YUVAL NE'EMAN [*†]

Sackler Faculty of Exact Sciences, Tel-Aviv University
69978 Tel-Aviv, Israel

and

DJORDJE ŠIJAČKI [‡§]

Institute of Physics, P.O.B. 57, Belgrade, Yugoslavia

ABSTRACT

We identify an IR limit, in which the sum of all possible zero-colour gluon combinations exchanged between two hadrons is equivalent to a spin 2 zero-mass pole in the exchange channel, similar to a graviton. This component may explain many of the features relating to the soft hadron-hadron interactions and to confinement, for which Salam and others had suggested the action of a "strong gravity", extraneous to QCD. This also fits smoothly with the QCD treatment of high-energy scattering and the Harari-Freund conjecture in dual models, in which no quark lines passed in the exchange channel for the Pomeron, and thus overlaps with the studies of elastic scattering based on gluon "ladders" in QCD. Mixing with $J = 2$ quark-antiquark and massive gluonium poles could produce the observed pattern of intercepts.

1. Problematics of QCD in the "Soft" Regime and Confinement

The adoption of QCD and its incorporation in the Standard Model were the outcome of the success of asymptotic freedom (AF) in fitting the scaling results of deep inelastic electron-nucleon scattering. $SU(3)_{colour}$, on the other hand, was providing an explanation of some (paradoxical) key features of the hadron spectrum – otherwise nicely fitted by $SU(3)_{flavour}$ and the Non-Relativistic Quark Model (NRQM): the "wrong" spin-statistics correlation of the baryon (56 in the $SU(6)$) ground state and the zero-triality of the $SU(3)_{flavour}$ physical spectrum in its entirety. QCD, combining the AF of a Yang-Mills gauge theory with the algebraics of $SU(3)_{colour}$, promised the advantages of a perturbative system in the UV ("hard") region with an algebraic structure adapted to the spectrum. It was hoped that non-perturbative methods would be able to take over in treating the dynamics of the infrared end, including the apparent innexistence of free quarks. The present situation is indeed very satisfactory, as far as the perturbative treatment for the "ultraviolet" (UV) region goes. It has also been successfully extended – beyond high-energy electroweak hadronic interactions, corresponding to the current-quarks aspects of NRQM – to

[*]Wolfson Chair Extraordinary in Theoretical Physics
[†]Also on leave from Centre for Particle Physics, University of Texas, Austin, Texas 78712, USA
[‡]Supported in part by the Science Foundation (Belgrade)
[§]Supported by the Wolfson Chair Extraordinary in Theoretical Physics, Tel-Aviv University.

hadronic strong interactions in both the "hard" and "semi-hard" regimes, using the leading and double-log approximations in perturbative QCD and Reggeized diagram techniques[1].

Nothing of the sort has emerged to date in the "infrared" (IR) frequency antipodes. After almost a quarter of a century, we still have no full proof of colour confinement, beyond Wilson's original lattice calculations. Lattice gauge theory, developed as a directly non-perturbative technique, guaranteed to provide a dynamical answer to any set of initial conditions, suffers nevertheless from inherent limitations, which make it hard to extract definite conclusions in the key issues – and especially with respect to colour confinement). Actual treatments are qualitative and follow a variety of individual algorithms: dual superconductivity, monopole pairs, flux tubes, etc. All introduce notions modelled on other domains in physics, though assuming that the analogues exist within QCD itself. For example, QCD has no monopoles, but it is conjectured that other non-perturbative quantum solutions emulate the effects of monopoles. Another possibility is that the Yang-Mills gauge theory of QCD is insufficient for the understanding of confinement and other soft features and that some additional ingredients have to be added. We have seen such an example at this conference in Gribov's analysis.

Almost right from the beginning, Salam and others[2-5] have used a gravitational algorithm, assuming it to be truly extraneous to QCD – perhaps a true short range component of gravity[4,5]. Since 1990[6], however, we have assumed that the effect derives from QCD itself, as an 'effective' formalism for the IR region. In the last two years, we have been able to provide a mathematical derivation of this assumption[7,8]. In sections 2-4, we present a summary of these proofs. In section 5, we set up the suitable dynamical methodology for this algorithm and review the results to date. We believe they justify pursuing the approach and developing it into a practical technique adaptable to most problems in the IR region and the hadron spectrum.

2. 2. Chromo-diffeomorphisms: the two gluons contribution

The chromogravity algorithm advanced by the authors of this paper is based on the following ansätze: (a) That the QCD contribution of *the exchange of any number of gluons, set in colour neutral combinations, includes a gravity-like component with an "effective" (or pseudo-) diffeomorphism invariance ("chromo-diffeomorphism") and with two-gluon exchange playing the role of a "chromometric" field.* We believe we have proved this ansätze[7,8] and shall include in what follows the main points of the proof. (b) That gluon exchange forces (with the gluons in colour-neutral combinations) and in particular the chromo-gravity component, make an *important contribution to colour confinement, to the hadron spectrum and to inter-hadron interactions in their "softest" region.* This is still a physical conjecture, though we shall provide in its favour several strong arguments based on experiment.

We begin by analysing the two-gluon system. We define an effective chromometric field

$$G_{\mu\nu}(x) = (\kappa)^{-2} \, g_{ab} \, B_\mu^a(x) \, B_\nu^b(x) \tag{1}$$

(κ has the dimensions of mass, $< \mu, \nu, \cdots >$ are Lorentz 4-vector indices, $< a, b, \cdots >$ are $SU(3)$ adjoint representation (octet) indices, g_{ab} is the Cartan metric for the $SU(3)$ octet, B_μ^a is a gluon field).

The gluon $SU(3)_{colour}$ gauge field transforms under an infinitesimal local $SU(3)$ variation according to

$$\delta_\epsilon B_\mu^a = \partial_\mu \epsilon^a + B_\mu^b \{\lambda_b\}_c^a \, \epsilon^c = \partial_\mu \epsilon^a + i \, f_{bc}^a \, B^b \, \epsilon^c \tag{2}$$

(we use the adjoint representation $\{\lambda_b\}_c^a = -i f_{b\,c}^a = i f_{bc}^a$). To deal with the non-perturbative IR region, we expand the gauge field operator around a constant global vacuum solution N_μ^a,

$$\partial_\mu N_\nu^a - \partial_\nu N_\mu^a = i f_{bc}^a \, N_\mu^b \, N_\nu^c \tag{3}$$

$$B_\mu^a = N_\mu^a + A_\mu^a \,. \tag{4}$$

Such a vacuum solution might be of the instanton type, for instance. Consider, e.g. the first nontrivial class, with Pontryagin index $n = 0$. Expand around this classical configuration, working, as always for instantons, in a Euclidean metric (i.e. a tunnelling solution in Minkowski spacetime). At large distances the instanton field is required to approach a constant value

$$g_{ab} \, N_\mu^a \, \partial_\nu \epsilon^b = \partial_\nu (g_{ab} \, N_\mu^a \, \epsilon^b) \tag{5}$$

with the $A_\mu^a(x)$ field representing a fluctuation around the constant value, vanishing at large distances. We construct the constant vacuum solution by mapping $SU(3) \to S^4$, namely directly onto the complete Euclidean manifold, compactified by the addition of a point at infinity.

In the $SU(3)$ Cartan basis, H^1 and H^2 are the diagonal operators spanning the Cartan subalgebra's space; E^{+i} and E^{-i}, $i = 1, \ldots, 3$, are respectively the three raising and three lowering operators of I, U, V spins, using Lipkin's original flavour-inspired terminology. In this basis, the Cartan metric takes the form

$$g_{ab} = \sigma^0 \oplus \sigma^{1I} \oplus \sigma^{1U} \oplus \sigma^{1V} \tag{6}$$

where σ^0 is the 2×2 identity matrix, operating in the Cartan subalgebra subspace. σ^1 is that the Pauli matrix, operating in each of the three non-diagonal subspaces constituted by paired positive and negative roots. We now select for the vacuum solution a gauge

$$N_\mu^{H_1} = \frac{1}{\sqrt{2}} \begin{vmatrix} 1 \\ 0 \\ 0 \\ 0 \end{vmatrix}, \quad N_\mu^{H_2} = \frac{1}{\sqrt{2}} \begin{vmatrix} 1 \\ 0 \\ 0 \\ 0 \end{vmatrix}, \quad N_\mu^{I+} = \frac{1}{\sqrt{2}} \begin{vmatrix} 0 \\ 1 \\ 0 \\ 0 \end{vmatrix}, \quad N_\mu^{I-} = \frac{1}{\sqrt{2}} \begin{vmatrix} 0 \\ 1 \\ 0 \\ 0 \end{vmatrix} \tag{7}$$

$$N_\mu^{U+} = \frac{1}{\sqrt{2}} \begin{vmatrix} 0 \\ 0 \\ 1 \\ 0 \end{vmatrix} , \quad N_\mu^{U-} = \frac{1}{\sqrt{2}} \begin{vmatrix} 0 \\ 0 \\ 1 \\ 0 \end{vmatrix} , \quad N_\mu^{V+} = \frac{1}{\sqrt{2}} \begin{vmatrix} 0 \\ 0 \\ 0 \\ 1 \end{vmatrix} , \quad N_\mu^{V-} = \frac{1}{\sqrt{2}} \begin{vmatrix} 0 \\ 0 \\ 0 \\ 1 \end{vmatrix} \quad (8)$$

and obtain the (flat) Euclidean metric from

$$\eta_{\mu\nu} = N_\mu^a \, g_{ab} \, N_\nu^b \tag{9}$$

writing the two non-perturbative field solutions respectively as 4×8 and 8×4 matrices; the octet 8-dimensionality is contracted by the action of the 8×8 Cartan metric matrix. Note that in the instanton system, such a constant solution would arise from a null solution $B_\mu^a = 0$ through the application of a local gauge transformation involving a gauge function $\epsilon^a(x)$ linear in x,

$$\epsilon^{H_1}(x^\mu) = \frac{x^0}{\sqrt{2}} , \quad \epsilon^{H_2}(x^\mu) = \frac{x^0}{\sqrt{2}} , \quad \epsilon^{I+}(x^\mu) = \frac{x^1}{\sqrt{2}} , \quad \epsilon^{I-}(x^\mu) = \frac{x^1}{\sqrt{2}} \tag{10}$$

$$\epsilon^{U+}(x^\mu) = \frac{x^2}{\sqrt{2}} , \quad \epsilon^{U-}(x^\mu) = \frac{x^2}{\sqrt{2}} , \quad \epsilon^{V+}(x^\mu) = \frac{x^3}{\sqrt{2}} , \quad \epsilon^{V-}(x^\mu) = \frac{x^3}{\sqrt{2}} \tag{11}$$

the N_μ^a then being given by the constants $\partial_\mu \epsilon^a(x)$ and thus defining an instanton with Pontryagin index $n = 1$, i.e. a topologically nontrivial object.

In what follows, we preserve the definition (5) and the gauge (6); should a gauge transformation adjoin an x-dependent variation, we choose to include it in the $A_\mu^a(x)$ component of (4). Returning to the "curved" pseudo-metric (1), we can now replace κ by the "flat" density,

$$G_{\mu\nu} = \frac{g_{ab} \, B_\mu^a \, B_\nu^b}{[\det(g_{ab} \, N_\mu^a \, N_\nu^b)]^{1/4}} \tag{12}$$

thus yielding a non-singular dimensionless Euclido-Riemannian "metric". Its colour-$SU(3)$ infinitesimal gauge variation is given by

$$\begin{aligned}
\delta_\epsilon G_{\mu\nu} &= \delta_\epsilon \left\{ g_{ab} (N_\mu^a + A_\mu^a)(N_\nu^b + A_\nu^b) \right\} \\
&= g_{ab}(\partial_\mu \epsilon^a \, N_\nu^b + N_\mu^a \, \partial_\nu \epsilon^b + \partial_\mu \epsilon^a \, A_\nu^b + A_\mu^a \, \partial_\nu \epsilon^b) \\
&\quad + i g_{ab} \left\{ f_{cd}^a \, B_\mu^c \, \epsilon^d \, B_\nu^b + f_{cd}^b \, B_\mu^a \, B_\nu^c \, \epsilon^d \right\}
\end{aligned} \tag{13}$$

The last bracket vanishes, since it represents the homogeneous $SU(3)$ transformation of the $SU(3)$ scalar expression in (1)

$$i f_{bcd} \left(B_\mu^b \, B_\nu^c + B_\mu^c \, B_\nu^b \right) \epsilon^d \tag{14}$$

(or, more technically, due to the total antisymmetry of f_{abc} in a compact group). With N_μ^a, N_ν^b, representing constant fields, we rewrite the terms in which they appear as a new infinitesimal variation,

$$\xi_\mu = \eta_{ab} \, \epsilon^a \, N_\mu^b . \tag{15}$$

We note that the expansion (4)-(5) implies that *at these distances, any Gauss theorem field-fluxes will only involve the N_μ^a constant component,* whereas the $A_\mu^a(x)$ "fluctuation" will not contribute. As a result, when integrating by parts the terms in A_μ^a, A_ν^b in eq. (13), we get

$$g_{ab} \left(\epsilon^a \, \partial_\mu A_\nu^b + \partial_\nu A_\mu^a \, \epsilon^b \right) , \tag{16}$$

an expression whose Fourier transform vanishes for $k \to 0$, i.e. in the infrared sector. We shall return and provide a generalized definition of this *"IR limit"*. Meanwhile, as a result, we can write in this limit,

$$\delta_\xi G_{\mu\nu} = \partial_\mu \xi_\nu + \partial_\nu \xi_\mu = \partial_\mu (\xi^\sigma \, G_{\sigma\nu}) + \partial_\nu (\xi^\sigma \, G_{\mu\sigma}) \tag{17}$$

where we have changed over to the ξ^σ variable of (15), and where we can reidentify δ_ξ as a variation under a formal diffeomorphism of the R^4 manifold. Equation (17) simulates the infinitesimal variation of a "world tensor" $G_{\mu\nu}$ under Einstein's covariance group, $x^\sigma \to x^\sigma + \xi^\sigma$. Therefore, ξ^σ has to be defined as a contravariant vector; $G_{\mu\nu}$ of (1) is invertible, thanks to the constant part N_μ^a, in (4), using a Taylor expansion to evaluate the inverse $G^{\mu\nu}(x)$. Note that as the μ, ν indices are "true" Lorentz indices, acted upon by the physical Lorentz group, the manifold has to be Riemannian: only Riemannian manifolds – with or without torsion – have tangents with orthogonal or pseudo-orthogonal symmetry. Thus

$$D_\sigma G_{\mu\nu} = 0 . \tag{18}$$

This Riemannian condition can also be regarded as a set of 4 constraints on the 10 components (with spins $J = 2, 1, 0, 0.$) of the chromometric 'field'. It leaves as 'physical quanta' (in the chromogravity approximation) only $J = 2$ and $J = 0$. These should be construed as collective modes, somewhat like *phonons* in condensed matter. We shall apply this algorithm extensively, both for hadrons and for nuclei (see section 5). Returning to our proof, we now evaluate the commutator of two such chromo-diffeomorphic variations,

$$[\delta_{\xi_1}, \delta_{\xi_2}] G_{\mu\nu} = \delta_{\xi_3} G_{\mu\nu} , \tag{19}$$

and verify that

$$\xi_{3\mu} := (\partial_\nu \xi_{1\mu}) \, \xi_2^\nu + (\partial_\mu \xi_{1\nu}) \, \xi_2^\nu - (\partial_\nu \xi_{2\mu}) \, \xi_1^\nu - (\partial_\mu \xi_{2\nu}) \, \xi_1^\nu \tag{20}$$

indeed closes on the covariance group's commutation relations.

In section 3, the definition of our "IR limit", which we based on the vanishing of the 4-momenta of the 'fluctuating fields' A_μ^a, A_ν^b in (16) – after an integration by parts in which only the constant fields N_μ^a contribute to the surface terms – will be extended so as to include similar terms with vanishing momenta in all many-gluon zero-colour exchanges. This can be taken as an operational definition, sufficient for our general purpose. To gain some additional insight, however, we remind the reader that such an IR approximation of QCD can also be thought of as the first step, the "zeroth approximation", of a *strong coupling* regime – in terms of a "small parameter" representing the number of "hard", or nonsoft, virtual quanta held in the evaluation of any physical quantity. We can write a generic IR state, carrying 4-momentum k, as follows:

$$|\phi_{IR},\ k\rangle = \sum_{m=1}^{\infty} f_m(k_1, k_2, \ldots, k_m)\delta_{k, k_1+k_2+\cdots+k_m}|k_1 k_2 \ldots k_m\rangle \qquad (21)$$

where $|k_1 k_2 \ldots k_m\rangle$ represents a state of m soft gluons ($k_i \approx 0$, $i = 1, 2, \ldots, m$). Integrating by parts (with surface terms again appearing only for the constant parts N_μ^a), the matrix elements of the terms in A_μ^a, A_ν^b become in this IR approximation

$$\langle \phi'_{IR}, k'|g_{ab}(\epsilon^a \partial_\mu A_\nu^b + \partial_\nu A_\mu^a \epsilon^b)|\phi_{IR},\ k\rangle, \qquad (22)$$

an expression that is proportional to the soft 1-gluon momentum, and that vanishes for $k \to 0$, i.e. in the infrared sector. As a result, when changing over to the ξ^σ variable of (15) and reidentifying δ_ξ as a variation under a formal R^4 diffeomorphism, we get (17). For the sake of completeness, we note that in general one has to consider expressions of the following form

$$\langle \phi'_{IR}, k'|O(A_\mu^a, \partial_\nu A_\mu^a)\delta_\epsilon G_{\mu\nu}|\phi_{IR},\ k\rangle. \qquad (23)$$

We evaluate such expressions, in this IR approximation, by inserting a complete set of states, and retaining only the *soft* virtual quanta. It is explained in Ref. 9 that by making use of the Fradkin representation[10] for relevant Green's functions one has a continuous family of "soft", or IR, approximations which maintain gauge invariance. Thus, one finds a consistent gauge-invariant (strong coupling) IR approximation with dressed gluon propagators which incorporate the iteration of all relevant quark bubbles, each carrying all possible internal, soft-gluon lines.

The consistency of this IR approximation requires one to consider only those QCD variations that connect IR gluon configurations *mutually*. Let us consider the expression for the $A = B - N$ variation. It follows from Eq. (2) that $\delta_\epsilon A_\mu^a = \partial_\mu \epsilon^a + i f^a_{bc} A_\mu^b \epsilon^c$ The left-hand side of this expression is a difference between two soft gluons, implying that the IR matrix elements of its partial derivative are soft. Thus, we find the following "IR constraint" on the QCD gauge parameters:

$$\langle \phi'_{IR}, k'|\partial_\rho \partial_\mu \epsilon^a + i f^a_{bc} A_\mu^b \partial_\rho \epsilon^c|\phi_{IR},\ k\rangle \approx 0 . \qquad (24)$$

3. n-gluon fields

Our treatment is nonperturbative; we use, however, the formal expansion provided by the generating functional of Green's functions, for a classification of the contributions making up the overall non-quark component. In this, although our formalism is meant for the IR regime, we parallel the on-mass-shell and high-energy considerations[11], which had led Harari and Freund to a precise definition of the s-channel processes composing (in the finite energy sum rules) the t-channel Pomeranchuk trajectory: elastic scattering, as represented by duality Harari-Rosner diagrams[12] in which no resonance is formed and no Susskind "rubber-bands" can be stretched into the t-channel. These duality features have been incorporated in QCD[12], so that we now have a coherent picture of the gluon component's contribution 'all the way' from the IR region to high-energy scattering. We shall return to this overlap with dual models in section 5.

The expansion we use involves all possible colour-singlet configurations of gluon fields. We rearrange the sum by lumping together contributions from n-gluon irreducible parts, $n = 2, 3, \ldots, \infty$ and with the same Lorentz quantum numbers. Thus, QCD "gluon-made" operators, which mutually connect various hadron states, are characterized by colour-singlet quanta. The corresponding colour-singlet n-gluon field operator has the following form

$$G^{(n)}_{\mu_1 \mu_2 \cdots \mu_n} = d^{(n)}_{a_1 a_2 \cdots a_n} B^{a_1}_{\mu_1} B^{a_2}_{\mu_2} \cdots B^{a_n}_{\mu_n} , \tag{25}$$

where

$$d^{(2)}_{a_1 a_2} = g_{a_1 a_2},$$
$$d^{(3)}_{a_1 a_2 a_3} = d_{a_1 a_2 a_3},$$
$$d^{(n)}_{a_1 a_2 \cdots a_n} = d_{a_1 a_2 b_1} g^{b_1 c_1} d_{c_1 b_2 a_3} \cdots g^{b_{n-4} c_{n-4}} d_{c_{n-4} b_{n-3} a_{n-2}} g^{b_{n-3} c_{n-3}} d_{c_{n-3} a_{n-1} a_n}, \quad n > 3, \tag{26}$$

B^a_μ is the dressed gluon field, $g_{a_1 a_2}$ is the $SU(3)$ Cartan metric, and $d_{a_1 a_2 a_3}$ is the $SU(3)$ totally symmetric $8 \times 8 \times 8 \to 1$ tensor. It was shown by Biedenharn[13] that the set of all $d^{(n)}_{a_1 a_2 \cdots a_n}$ tensors, $n = 1, 2, \ldots$, can be used to form, together with the group generators, a basis of all $SU(3)$-invariant operators. In this case all such higher-rank operators can be expressed in terms of two invariant operators. In our case, the set of all $G^{(n)}_{\mu_1 \mu_2 \cdots \mu_n}$ operators, $n = 1, 2, \ldots$, forms the basis of a vector space of colourless, purely gluonic configurations. Moreover, in our case, in contradistinction to the ordinary group theoretical situation, these field operators are also all functionally independent.

The QCD variation of the $G^{(n)}_{\mu_1 \mu_2 \cdots \mu_n}$ field is given by

$$\delta_\epsilon G^{(n)}_{\mu_1 \mu_2 \cdots \mu_n} = d^{(n)}_{a_1 a_2 \cdots a_n} \left(\partial_{\mu_1} \epsilon^{a_1} B^{a_2}_{\mu_2} \cdots B^{a_n}_{\mu_n} + B^{a_1}_{\mu_1} \partial_{\mu_2} \epsilon^{a_2} \cdots B^{a_n}_{\mu_n} + \cdots + B^{a_1}_{\mu_1} B^{a_2}_{\mu_2} \cdots \partial_{\mu_n} \epsilon^{a_n} \right) + \tag{27}$$

$$d^{(n)}_{a_1 a_2 \cdots a_n} \left(g^{a_1 r} f_{rst} B^s_{\mu_1} B^{a_2}_{\mu_2} \cdots B^{a_n}_{\mu_n} + g^{a_2 r} f_{rst} B^{a_1}_{\mu_1} B^s_{\mu_2} \cdots B^{a_n}_{\mu_n} + \cdots g^{a_n r} f_{rst} B^{a_1}_{\mu_1} B^{a_2}_{\mu_2} \cdots B^s_{\mu_n} \right) \epsilon^t \tag{28}$$

Here again the homogeneous terms vanish as before due to the fact that $d^{(n)}_{a_1 a_2 \cdots a_n}$ is totally symmetric. Applying the decomposition (4) we can now rewrite the n-gluon configuration transformation as

$$\delta_\epsilon G^{(n)}_{\mu_1\mu_2\cdots\mu_n} =$$

$$d^{(n)}_{a_1a_2\cdots a_n}\left(\partial_{\mu_1}\epsilon^{a_1}N^{a_2}_{\mu_2}\cdots N^{a_n}_{\mu_n} + N^{a_1}_{\mu_1}\partial_{\mu_2}\epsilon^{a_2}\cdots N^{a_n}_{\mu_n} + \cdots + N^{a_1}_{\mu_1}N^{a_2}_{\mu_2}\cdots\partial_{\mu_n}\epsilon^{a_n}\right) +$$

$$d^{(n)}_{a_1a_2\cdots a_n}\left(A^{a_1}_{\mu_1}\partial_{\mu_2}\epsilon^{a_2}N^{a_3}_{\mu_3}\cdots N^{a_n}_{\mu_n} + \cdots + A^{a_1}_{\mu_1}N^{a_2}_{\mu_2}N^{a_3}_{\mu_3}\cdots\partial_{\mu_n}\epsilon^{a_n}\right) + (1\leftrightarrow i=2,\ldots,n) +$$

$$d^{(n)}_{a_1a_2\cdots a_n}\left(\partial_{\mu_1}\epsilon^{a_1}A^{a_2}_{\mu_2}A^{a_3}_{\mu_3}N^{a_4}_{\mu_4}\cdots N^{a_n}_{\mu_n} + \cdots + A^{a_1}_{\mu_1}A^{a_2}_{\mu_2}N^{a_3}_{\mu_3}\cdots N^{a_{n-1}}_{\mu_{n-1}}\partial_{\mu_n}\epsilon^{a_n}\right) + \cdots +$$

$$d^{(n)}_{a_1a_2\cdots a_n}\left(\partial_{\mu_1}\epsilon^{a_1}A^{a_2}_{\mu_2}\cdots A^{a_n}_{\mu_n} + A^{a_1}_{\mu_1}\partial_{\mu_2}\epsilon^{a_2}\cdots A^{a_n}_{\mu_n} + \cdots + A^{a_1}_{\mu_1}A^{a_2}_{\mu_2}\cdots\partial_{\mu_n}\epsilon^{a_n}\right).$$

As to the terms in $A^{a_i}_{\mu_i}$, $i=1,2,\ldots n$, the considerations we mentioned prior to the integration by parts in (16) hold here. Thus, applying integration by parts here too, we get

$$-d^{(n)}_{a_1a_2\cdots a_n}\left(\partial_{\mu_2}(A^{a_1}_{\mu_1})\epsilon^{a_2}\cdots N^{a_n}_{\mu_n} + \cdots + \partial_{\mu_n}(A^{a_1}_{\mu_1})N^{a_2}_{\mu_2}\cdots\epsilon^{a_n}\right) - (1\leftrightarrow i=2,\ldots,n)$$

$$-d^{(n)}_{a_1a_2\cdots a_n}\left(\epsilon^{a_1}\partial_{\mu_1}(A^{a_2}_{\mu_2}A^{a_3}_{\mu_3})N^{a_4}_{\mu_4}\cdots N^{a_n}_{\mu_n} + \cdots + \partial_{\mu_n}(A^{a_1}_{\mu_1}A^{a_2}_{\mu_2})N^{a_3}_{\mu_3}\cdots N^{a_{n-1}}_{\mu_{n-1}}\epsilon^{a_n}\right) - \cdots$$

$$-d^{(n)}_{a_1a_2\cdots a_n}\left(\epsilon^{a_1}\partial_{\mu_1}(A^{a_2}_{\mu_2}\cdots A^{a_n}_{\mu_n}) + \epsilon^{a_2}\partial_{\mu_2}(A^{a_1}_{\mu_1}\cdots A^{a_n}_{\mu_n}) + \cdots + \epsilon^{a_n}\partial_{\mu_n}(A^{a_1}_{\mu_1}A^{a_2}_{\mu_2}\cdots)\right).$$

But taking Fourier transforms – i.e. the matrix elements for these gluon fluctuations – we find that *these terms are precisely those that vanish in our definition of an IR region*, as discussed in the previous section. The terms involving the constant connections $N^{a_i}_{\mu_i}$, $i=1,2,\ldots,n$, can be rewritten in terms of effective pseudo-diffeomorphisms:

$$\delta_\epsilon G^{(n)}_{\mu_1\mu_2\cdots\mu_n} = \partial_{\{\mu_1}\xi^{(n-1)}_{\mu_2\mu_3\cdots\mu_n\}} \equiv \delta_\xi G^{(n)}_{\mu_1\mu_2\cdots\mu_n}\ , \tag{29}$$

where $\{\mu_1\mu_2\cdots\mu_n\}$ denotes symmetrization of indices, and

$$\xi^{(n-1)}_{\mu_1\mu_2\cdots\mu_{n-1}} \equiv d^{(n)}_{a_1a_2\cdots a_n}N^{a_1}_{\mu_1}N^{a_2}_{\mu_2}\cdots N^{a_{n-1}}_{\mu_{n-1}}\epsilon^{a_n}\ , \tag{30}$$

generalizing our results as derived for $G^{(2)}_{\mu\nu} = G_{\mu\nu}[\det(g_{ab}N^a_\mu N^b_\nu)]^{1/4}$. Note that the essential point we have demonstrated is that *the set of colourless many-gluon exchanges includes a gravity-like component*. True, we have no way – at this stage – of evaluating the quantitative weight of this contribution from the parameters of QCD. What therefore remains as a *working hypothesis* is the assignment of gravity-like observational features (including those features which had caused Salam and others to believe in a strong gravity element, extraneous to QCD) to this *chromogravity* component's contribution and the derivation of the chromogravity quantitative parameters from this phenomenological approach.

A subsequent application of two $SU(3)$-induced variations implies

$$[\delta_{\epsilon_1},\ \delta_{\epsilon_2}]G^{(n)}_{\mu_1\mu_2\cdots\mu_n} = \delta_{\epsilon_3}G^{(n)}_{\mu_1\mu_2\cdots\mu_n}\ , \quad \text{i.e.}\quad [\delta_{\xi_1},\ \delta_{\xi_2}]G^{(n)}_{\mu_1\mu_2\cdots\mu_n} = \delta_{\xi_3}G^{(n)}_{\mu_1\mu_2\cdots\mu_n} \tag{31}$$

generalizing the $n = 2$ case in (19), i.e. an infinitesimal nonlinear realization of the $Diff(4, R)$ group in the space of fields $\left\{ G^{(n)}_{\mu_1 \mu_2 \cdots \mu_n} \mid n = 2, 3, \ldots \right\}$.

4. The infinite algebra of the $L^{(m)}$ operators

Let us consider an ∞-dimensional vector space over the field operators $\left\{ G^{(n)} \mid n = 2, 3, \ldots \right\}$, i.e.

$$V(G^{(2)}, G^{(3)}, \ldots) = V(G^{(2)}_{\mu_1 \mu_2}, G^{(3)}_{\mu_1 \mu_2 \mu_3}, \ldots). \tag{32}$$

We can now define an infinite set of field-dependent operators $\left\{ L^{(m)} \mid m = 0, 1, 2, \ldots \right\}$ as follows

$$
\begin{aligned}
L^{(0)\rho}_{\nu_1} &= d^{(2)}_{a_1 a_2} B^{a_1}_{\nu_1} \frac{\delta}{\delta(g_{a_2 b} B^b_\rho)} \equiv g_{a_1 a_2} B^{a_1}_{\nu_1} \frac{\delta}{\delta(g_{a_2 b} B^b_\rho)} , \\
L^{(1)\rho}_{\nu_1 \nu_2} &= d^{(3)}_{a_1 a_2 a_3} B^{a_1}_{\nu_1} B^{a_2}_{\nu_2} \frac{\delta}{\delta(g_{a_3 b} B^b_\rho)} \equiv d_{a_1 a_2 a_3} B^{a_1}_{\nu_1} B^{a_2}_{\nu_2} \frac{\delta}{\delta(g_{a_3 b} B^b_\rho)} , \\
&\cdots \\
L^{(m)\rho}_{\nu_1 \nu_2 \cdots \nu_{m+1}} &= d^{(m+2)}_{a_1 a_2 \cdots a_{m+2}} B^{a_1}_{\nu_1} B^{a_2}_{\nu_2} \cdots B^{a_{m+1}}_{\nu_{m+1}} \frac{\delta}{\delta(g_{a_{m+2} b} B^b_\rho)} .
\end{aligned}
$$
$$\cdots . \tag{33}$$

The $L^{(0)\rho}_{\nu_1}$ action on the field operators $\left\{ G^{(n)} \mid n = 2, 3, \ldots \right\}$ reads

$$
\begin{aligned}
L^{(0)\rho}_{\nu_1} G^{(2)}_{\mu_1 \mu_2} &= \delta^\rho_{\mu_1} G^{(2)}_{\nu_1 \mu_2} + \delta^\rho_{\mu_2} G^{(2)}_{\mu_1 \nu_1} , \\
L^{(0)\rho}_{\nu_1} G^{(3)}_{\mu_1 \mu_2 \mu_3} &= \delta^\rho_{\mu_1} G^{(3)}_{\nu_1 \mu_2 \mu_3} + \delta^\rho_{\mu_2} G^{(3)}_{\mu_1 \nu_1 \mu_3} + \delta^\rho_{\mu_3} G^{(3)}_{\mu_1 \mu_2 \nu_1} , \\
&\cdots \\
L^{(0)\rho}_{\nu_1} G^{(n)}_{\mu_1 \mu_2 \cdots \mu_n} &= \delta^\rho_{\mu_1} G^{(n)}_{\nu_1 \mu_2 \cdots \mu_n} + \delta^\rho_{\mu_2} G^{(n)}_{\mu_1 \nu_1 \mu_3 \cdots \mu_n} + \cdots + \delta^\rho_{\mu_n} G^{(n)}_{\mu_1 \mu_2 \cdots \mu_{n-1} \nu_1} .
\end{aligned}
$$
$$\cdots . \tag{34}$$

The $L^{(1)\rho}_{\nu_1 \nu_2}$ action on the field operators $\left\{ G^{(n)} \mid n = 2, 3, \ldots \right\}$ reads

$$
\begin{aligned}
L^{(1)\rho}_{\nu_1 \nu_2} G^{(2)}_{\mu_1 \mu_2} &= \delta^\rho_{\mu_1} G^{(3)}_{\nu_1 \nu_2 \mu_2} + \delta^\rho_{\mu_2} G^{(3)}_{\mu_1 \nu_1 \nu_2} , \\
L^{(1)\rho}_{\nu_1 \nu_2} G^{(3)}_{\mu_1 \mu_2 \mu_3} &= \delta^\rho_{\mu_1} G^{(4)}_{\nu_1 \nu_2 \mu_2 \mu_3} + \delta^\rho_{\mu_2} G^{(4)}_{\mu_1 \nu_1 \nu_2 \mu_3} + \delta^\rho_{\mu_3} G^{(4)}_{\mu_1 \mu_2 \nu_1 \nu_2} , \\
&\cdots \\
L^{(1)\rho}_{\nu_1 \nu_2} G^{(n)}_{\mu_1 \mu_2 \cdots \mu_n} &= \delta^\rho_{\mu_1} G^{(n+1)}_{\nu_1 \nu_2 \mu_2 \cdots \mu_n} + \delta^\rho_{\mu_2} G^{(n+1)}_{\mu_1 \nu_1 \nu_2 \mu_3 \cdots \mu_n} + \cdots + \delta^\rho_{\mu_n} G^{(n+1)}_{\mu_1 \mu_2 \cdots \mu_{n-1} \nu_1 \nu_2} ,
\end{aligned}
$$
$$\cdots . \tag{35}$$

In the general case, $L^{(m)\rho}_{\nu_1 \nu_2 \cdots \nu_{m+1}}$, $m = 0, 1, 2, \ldots$ action on the field operators $\left\{ G^{(n)} \mid n = 2, 3, \ldots \right\}$ reads

$$L^{(m)\rho}_{\nu_1\nu_2\cdots\nu_{m+1}} G^{(2)}_{\mu_1\mu_2} = \delta^\rho_{\mu_1} G^{(2+m)}_{\nu_1\nu_2\cdots\nu_{m+1}\mu_2} + \delta^\rho_{\mu_2} G^{(2+m)}_{\mu_1\nu_1\nu_2\cdots\nu_{m+1}} ,$$

$$L^{(m)\rho}_{\nu_1\nu_2\cdots\nu_{m+1}} G^{(3)}_{\mu_1\mu_2\mu_3} = \delta^\rho_{\mu_1} G^{(3+m)}_{\nu_1\nu_2\cdots\nu_{m+1}\mu_2\mu_3} + \delta^\rho_{\mu_2} G^{(3+m)}_{\mu_1\nu_1\nu_2\cdots\nu_{m+1}\mu_3} + \delta^\rho_{\mu_3} G^{(3+m)}_{\mu_1\mu_2\nu_1\nu_2\cdots\nu_{m+1}} ,$$

$$\cdots$$

$$L^{(m)\rho}_{\nu_1\nu_2\cdots\nu_{m+1}} G^{(n)}_{\mu_1\mu_2\cdots\mu_n} = \delta^\rho_{\mu_1} G^{(n+m)}_{\nu_1\nu_2\cdots\nu_{m+1}\mu_2\cdots\mu_n} + \delta^\rho_{\mu_2} G^{(n+m)}_{\mu_1\nu_1\nu_2\cdots\nu_{m+1}\mu_3\cdots\mu_n}$$
$$+ \cdots + \delta^\rho_{\mu_n} G^{(n+m)}_{\mu_1\mu_2\cdots\mu_{n-1}\nu_1\nu_2\cdots\nu_{m+1}} ,$$
$$\cdots . \tag{36}$$

Let us now consider the algebraic structure defined by the $\left\{ L^{(m)} \mid m = 0,1,2,\ldots \right\}$ operators Lie brackets. For the $L^{(0)}$ operators themselves we find

$$[L^{(0)}, L^{(0)}] \subset L^{(0)} , \tag{37}$$

i.e.

$$[L^{(0)\rho_1}_{\nu_1}, L^{(0)\rho_2}_{\sigma_1}] = \delta^{\rho_1}_{\sigma_1} L^{(0)\rho_2}_{\nu_1} - \delta^{\rho_2}_{\nu_1} L^{(0)\rho_1}_{\sigma_1} . \tag{38}$$

In the most general case, for the brackets of $L^{(l)}$ and $L^{(m)}$ we find

$$[L^{(l)}, L^{(m)}] \subset L^{(l+m)}, \tag{39}$$

and more specifically,

$$[L^{(l)\rho_1}_{\nu_1\nu_2\cdots\nu_{l+1}}, L^{(m)\rho_2}_{\sigma_1\sigma_2\cdots\sigma_{m+1}}]$$
$$= \sum_{i=1}^{m+1} \delta^{\rho_1}_{\sigma_i} L^{(l+m)\rho_2}_{\sigma_1\sigma_2\cdots\sigma_{i-1}\nu_1\nu_2\cdots\nu_{l+1}\sigma_{i+1}\cdots\sigma_{m+1}} - \sum_{j=1}^{l+1} \delta^{\rho_2}_{\nu_j} L^{(l+m)\rho_1}_{\nu_1\nu_2\cdots\nu_{j-1}\sigma_1\sigma_2\cdots\sigma_{m+1}\nu_{j+1}\cdots\nu_{l+1}} \tag{40}$$

We have constructed an ∞-component vector space, $V = V(G^{(2)}_{\mu_1\mu_2}, G^{(3)}_{\mu_1\mu_2\mu_3}, \ldots)$, over the n-gluon field operators, as well as the corresponding algebra of homogeneous diffeomorphisms, $diff_0(4, R) = \left\{ L^{(m)\rho}_{\nu_1\nu_2\cdots\nu_{m+1}} \mid m = 0,1,2,\ldots \right\}$; the vector space V is invariant under the action of the $diff_0(4, R)$ algebra.

To understand the grading, we define a dilation-like operator D as a trace of $L^{(0)\rho}_\nu$, i.e.

$$D = L^{(0)\rho}_\rho. \tag{41}$$

This operator commutes with the $L^{(0)\rho}_\nu$ operators,

$$[D, L^{(0)\rho}_\nu] = 0, \tag{42}$$

and belongs to the centre of the $gl(4, R)$ chromogravity subalgebra generated by the $L^{(0)\rho}_\nu$ operators. On account of the chromodilation operator, one can make the following decomposition

$$gl(4, R) = r \oplus sl(4, R), \tag{43}$$

where D corresponds to the subalgebra r, while the basis of the $sl(4, R)$ subalgebra is given by

$$T_\nu^{(0)\rho} = L_\nu^{(0)\rho} - \tfrac{1}{4}\delta_\nu^\rho D. \tag{44}$$

The commutation relation of D with a generic $diff_0(4, R)$ operator $L_{\nu_1 \nu_2 \cdots \nu_{m+1}}^{(m)\rho}$ reads

$$[D, L_{\nu_1 \nu_2 \cdots \nu_{m+1}}^{(m)\rho}] = m L_{\nu_1 \nu_2 \cdots \nu_{m+1}}^{(m)\rho} \tag{45}$$

and thus, the chromodilation operator D provides us with a Z_+ grading. This grading justifies and/or explains the m-label used for the $L_{\nu_1 \nu_2 \cdots \nu_{m+1}}^{(m)\rho}$ operators.

The chromodilation operator D counts the number of single gluon fields in a multi-gluon configuration, as seen from the following commutation relation

$$[D, G_{\mu_1 \mu_2 \cdots \mu_n}^{(n)}] = n G_{\mu_1 \mu_2 \cdots \mu_n}^{(n)}. \tag{46}$$

5. Dynamical machinery

The existence of a gravity-like component within the QCD gauge should be no surprise, knowing that *while the truncated massless sector of the open string reduces to a $J = 1$ Yang-Mills field theory, the same truncation for the closed string reduces to a $J = 2$ gravitational field theory*, since the closed string is the contraction of two open strings. The field theory corollary has even been used by the Bern-Kosower group with their string-generated method, to evaluate ("true") gravitational amplitudes[14]. Also, other approaches have led to similar conclusions: Lunev[15] and Freedman et al.[16], working respectively in $2 + 1$ and 3 dimensions – and with $SU(2)$ instead of $SU(3)$ – have obtained for this particular configuration a spatial-temporal $GL(3, R)$ and Riemannian structure for zero-colour systems (beyond our definition of an IR region involving vanishing frequencies). These derivations can be taken as strengthening our conjecture as to the quantitative importance of the chromogravity contribution in the IR sector.

Having established our conceptual foundation we now formulate it as a dynamical theory. At this stage, we write the chromogravity Lagrangian as an effective expression, modelled on gravity, except that this has to be quantum gravity, with renormalization adding higher-order counter-terms:

$$L = \sqrt{-G}\left(-a R_{\mu\nu} R^{\mu\nu} + b R^2 - c l_G^2 R + l_S^2 \Sigma_{\mu\nu}^\sigma \Sigma_\sigma^{\mu\nu} + l_Q^2 \Delta_{\mu\nu}^\sigma \Delta_\sigma^{\mu\nu} + L_{matter}\right), \tag{47}$$

The first three terms, with dimensionless parameters a, b, c, represent the most general gravitational Lagrangian with terms quadratic in the curvatures, shown by Stelle to be renormalizable, though not unitary[17]. The third term is Einstein's linear Lagrangian: l_G^2 is Newton's constant in "true" gravity; here it will be a similar constant with dimensions of a squared mass or of the inverse of an area; $R_{\mu\nu}$ is the curvature tensor, R the curvature scalar. A term with the 4-index Riemann-Christoffel tensor squared would not be independent, due to the existence of the Gauss-Bonnet invariant. The terms in $\Sigma^\sigma_{\mu\nu}$ (the spin-current tensor) and in $\Delta^\sigma_{\mu\nu}$ (the shear-current tensor) are "effective" contributions which are not present in the fundamental Lagrangian. When using a first-order ("Palatini") formalism, the variables are $G_{\mu\nu}$ (10 components) and $\Gamma^\sigma_{\mu\nu}$ (64 components). The latter can be written[18] as a sum $\Gamma = K + S + Q$, where K, S, Q are respectively the Christoffel connection $K(G, \partial G)$, the torsion (24 components) and the non-metricity tensor $Q = DG$ (40 components). Replacing G, Γ as variables by G, S, Q and considering the fact that as a Riemannian system the curvature does not involve ∂S and ∂Q, the equations of motion for S, Q are algebraic rather than differential. They equate $S = \Sigma$ and $Q = \Delta$. As a result, we get the two "effective" terms in spin-squared and shear-squared. As in Einstein-Cartan theory, torsion and non-metricity here have point-wise contributions (given by the spin and shear currents of the matter Lagrangian) but they do not propagate.

The strong coupling regime of "true" gravity is the high-energy one; in QCD, the strong-coupling region we are representing is the low-energy region. The correspondence is thus $QCD\text{-}soft \sim Chromogravity\text{-}hard$. (Just as we go from magnetic to electric in the superconductivity algorithm.) We thus drop all but the dimension-4 terms, dominating at high energy. We linearize the field by taking $H_{\mu\nu} = G_{\mu\nu} - \eta_{\mu\nu}$, subtracting the flat Minkowski metric. We get the equation of motion (in momentum space),

$$[a(p^2)^2 - 2l_S^{-2}f_S(J \cdot J) - 2l_Q^{-2}f_Q(T \cdot T)]H_{\mu\nu}(p) = 0 \tag{48}$$

where we have taken the homogeneous part only, as we should in evaluating the propagator. Let us analyse the result:

(a) **Confinement as seen in the propagators.** The propagator (the inverse of (48)) has p^{-4}. This is how Stelle's Lagrangian is dimensionally renormalizable; this is also why it is not unitary, as such a term can be written as the difference between two "p^{-2}" poles, i.e. one of them with a wrong sign in its residue. However, in chromogravity, *the p^{-4} term is a blessing! Throughout the first years after the postulation of QCD, theorists were searching for such a term in QCD (unsuccessfully) to prove confinement, of which it should be a feature*[19]. That the model is not unitary has no importance, as we anyhow know that this is just a 'slice' of QCD.

(b) **The Regge behaviour $J \sim m^2$.** The equation of motion will yield mass formulae for *Regge trajectories*. In these formulae, putting the quartic momentum term in (48) as m^4, *we find the behaviour $J \sim m^2$*. This fits the observed structure and is a rather unusual result in Lorentz-invariant theories (the Majorana equation, for instance, has reducing masses for mounting spins).

(c) $SA(4, R)$, $SL(4, R)$ **Classification.** Gravity (and its chromogravity analogue) is a local gauge theory for (passive) diffeomorphisms; the fields are tensors, i.e. representations (finite and non-unitary) of $GL(4, R)$, the linear subgroup, or of its simple subgroup $SL(4, R)$; the diffeomorphisms are represented non-linearly on the linear representations of the latter. $SL(4, R)$ is thus a classification group, very roughly in the same way as $SU(2)_{weak}$ or as $SU(3)_{flavour}$. This would seem not to hold for spinors, since general relativity treats spinors by going over to the tangent manifold and using the Lorentz group's spinor representations; the frame fields e_a^μ (a is a tetrad Lorentz index) then provide the nonlinear realization for the quotient diffeomorphisms/Lorentz. It used to be thought that this is due to the linear groups $SL(n, R)$, $GL(n, R)$ having no spinor representations (or no double-covering). *This was however a mistake, as discovered by one of the authors*[20,21]. The linear groups all possess double-coverings (for $n = 2$ the covering is infinite) as the topology of a group is that of its compact subgroup – here $SO(n)$, with $\overline{SO}(n) = Spin(n)$, $\overline{SO}(3) = Spin(3) = SU(2)$ (the overline denotes the double-covering), $\overline{SO}(4) = Spin(4) = SU(2) \times SU(2)$, $\overline{SO}(5) = Spin(5) = Sp(4)$, $\overline{SO}(6) = Spin(6) = SU(4)$ etc. We have characterized the systematics of the *metalinear* groups (i.e. the covering groups of the linear groups) in[22]. The linear unitary representations for the $n = 3$ case have been constructed[23] and so have the simpler ones (multiplicity-free) for $n = 4$ [24]. We have constructed *manifields*, i.e. infinite-component fields behaving as de-unitarized representations of $\overline{SL}(4, R)$ [24], and written the equations for *world-spinors* (fermionic) and for it infinitensors (bosonic)[25–28]. At the same time, the hadron Hilbert space should be given by representations of the double-covered affine group $\overline{A}(4, R) = R^4 \times \overline{SL}(4, R)$, a semidirect product like the Poincaré group, with the Lorentz group $\overline{SO}(1, 3) = SL(2, C)$ replaced by the double-covering of $SL(4, R)$. The unitary irreducible representations of $\overline{A}(4, R)$ and of $\overline{SA}(4, R)$ are determined by the relevant "little group" (stability subgroup); the most relevant ones are those for which the little group is $\overline{SL}(3, R)$. We have evaluated the Casimir invariants of $\overline{SA}(4, R)$ [29] to verify if there are no kinematical constraints on the representations as hadron Regge trajectories (such as the Dashen/Gell-Mann "angular condition" for the representations of the algebra of local current densities) and found that precisely for the $\overline{SL}(3, R)$ little group there are none. Note that we had already suggested (from phenomenology) characterizing Regge trajectories by the unitary infinite representations of $SL(3, R)$ long ago[30]. At that time, however, we did not know about spinorial representations, which we worked out for this group some years later[31], still without seeing the conclusion relating to curved-space spinors[20,21].

Returning to chromogravity, *we have thus suggested*[32,33] *that the hadrons be classified by $\overline{SL}(4, R)$ for the fields and by $\overline{SA}(4, R)$ – and in practice by the little group $\overline{SL}(3, R)$ for the particle Hilbert space.*

(d) **Confinement as seen through the algebra.** The above fitting little group $\overline{SL}(3, R)$ is *the invariance group of a volume!* Thus the algebra tells us that the states we describe by *these excitation bands represent the pulsations and deformational vibrations of a fixed volume 'bag'.* The dynamical approximation of confined colour as built into the Bag Model comes here "for free".

(e) **Quadrupolar excitations.** In nuclear physics, it has been found[34] that deformed nuclei realize $SU(3)$, $\overline{E}(3)$, $\overline{SL}(3, R)$, according to whether the mechanical structure makes the quadrupolar operators close on a compact algebra, commute, or close on a non-compact algebra, respectively. In hadron physics, we had conjectured[30] that Regge trajectories are algebraically generated by the time-derivatives of the gravitational quadrupoles (i.e. their pulsation rates) which indeed close on $SL(3, R)$. Could the importance of quadrupoles be due, beyond their kinematical implications, to their being coupled to chromogravity, the strongest hadronic force?

(f) The $J = 2$ and $J = 0$ quanta in nuclei (the IBM model). We saw in (18) and its conclusions that chromogravity involves two quasi-degenerate (pseudo-) quanta (like excitons etc. in condensed matter). With chromogravity producing the longest-range component (see (h) below), $G_{\mu\nu}$ being "massless" thanks to the chromo-diffeomorphisms gauge, it could very well excite the two roughly degenerate ground-state $J = 2, 0$ excitations in even-even nuclei above a closed shell. These are excitations of pairs of protons or of neutrons. Kinematically, these would couple to $J = 0$. We have suggested[35] that they then absorb the 'chromogravitons' and make the observed states. This can be evaluated, e.g. using a Bethe-Salpeter calculation[36]. These $J = 2, 0$ six states have been assigned by Arima and Iachello[37] to the representation **6** of $SU(6)$ and serve as the phenomenological foundation of the successful Interacting Boson Model (IBM) in nuclear physics. We may thus have a derivation of the otherwise phenomenological identification, aside from a reason for these (otherwise unexplained) excitations.

(g) The $J = 2$ and $J = 0$ quanta generating respectively Regge sequences and constituent quarks. In hadrons, we can regard the Regge sequence (a representation of $\overline{SL}(3, R)$ as explained in (c)) as being generated by the self-energy diagrams of the $J = 2$ quanta, thereby raising the spin by two units at each step. As to the $J = 0$, it is natural to think of it as generating self-mass, i.e. turning a "3-current-quarks" state into a "3-constituent-quarks" one. As the two quanta should be roughly degenerate, the implication is an approximate equality between the slope-interval (about 2 GeV^2 for $\Delta J = 2$, for the nucleon trajectory) and the intercept (i.e. the $mass^2$ of 3-constituent quarks, about 1 GeV^2.

(h) Fitting overlap with dual models First, we should emphasize that $G_{\mu\nu}(x)$ of (1) *is not the di-gluon* "glueball", yet another massive state in the hadron spectrum, which lattice calculations put somewhere between 1.5 and 2.5 GeV[38]. We are dealing here with a force emulating gravity, a force component whose long range is protected by the pseudo-diffeomorphisms gauge, within limits set by its emergence at the colourless and vanishing frequencies end of the spectrum only. Our analysis proves that the non-quark-matter (i.e. "vacuum" or gluonic) component of the inter-hadronic chromodynamical interaction *in the IR limit* represents a gravity-like contribution. In the analytically continued S-matrix, there is a pole at $J = 2$, $M^2 = 0$, emerging in that limit – i.e. an effective contribution simulating such a pole. *This explains the emergence of the $J = 2$ pole in closed strings, within the context of dual models, i.e. effectively of the Pomeranchuk trajectory, in its Harari-Freund dual structure*[11] *and fits the paradigm behind the application of QCD to that region*[12]. However, this is not the only feature in the $J = 2$ sector of the analytical S-matrix, since gluons also make up a $J = 2$ glueball, as mentioned above (1.5-2.5 GeV); there is thus possible mixing between these two gluonic structures. In addition, we have the f^0 meson at $J = 2$, $M = 1.27$ GeV (Salam's candidate "strong graviton") and the $f^{0'}$ at $J = 2$, $M = 1.525$ GeV; the latter two states are identified as mainly quark/antiquark compounds. We conjecture that all of these mix, effectively removing the Pomeron "quasi-pole" from its "chromogravity" $M = 0$ long-range status, reducing the intercept of the trajectory to $J = 1$ or thereabouts. Should further experiments show that the intercept indeed occurs precisely at $J = 1$, we would consider this as an indication of the presence of yet another gauge-like contribution, which we have not identified. Returning to the QCD realization of the Pomeron, we note that our extension here, of a notion defined in the IR region, into the domain of high-energy scattering, goes over smoothly into representations derived for this regime directly from perturbative QCD[1].

6. Acknowledgements

YN would like to thank CERN and its Theory Division for its hospitality while this article was worked on.

7. References

1. L.V. Gribov, E.M. Levin and M.G. Ryskin, *Phys. Rep.* **100** (1983) 1-150. See Chapter 3 in particular.
2. A. Salam, in *Five Decades of Weak Interactions*, *Ann. NY Acad. Sci.* **294** (1977) 12; C.J. Isham, A. Salam and J. Strathdee, *Phys. Rev.* **D8** (1973) 2600 and **D9** (1974) 1702; C. Sivaram and K. Sinha, *Phys. Rep.* **51** (1979) 11.
3. F.W. Hehl, Y. Ne'eman, J. Nitzsch and P. v.d. Heyde, *Phys. Lett.* **B78** (1978) 102.
4. Y. Ne'eman and Dj. Šijački, *Ann. Phys. (NY)* **120** (1979) 292.
5. P. Caldirola, M. Pavšič and E. Recami, *Nuovo Cimento* B (1978) 205; *Phys. Lett.* **A66** (1978) 9; *Lett. Nuovo Cimento* **24** (1979) 565.
6. Dj. Šijački and Y. Ne'eman, *Phys. Lett.* **B247** (1990) 571.
7. Y. Ne'eman and Dj. Šijački, *Phys. Lett.* **B276** (1992) 173.
8. Y. Ne'eman and Dj. Šijački, Tel Aviv preprint TAUP N232-94, to be published.
9. H.M. Fried, *Phys. Rev.* **D27** (1983) 2956.
10. E.S. Fradkin, *Nucl. Phys.* **76** (1966) 588.
11. a. P.G.O. Freund, *Phys. Rev. Lett.* **20** (1968) 235; H. Harari, *Phys. Rev. Lett.* **20** (1968) 1395; 11b. H. Harari, *Phys. Rev. Lett.* **22** (1969) 562; J. L. Rosner, *Phys. Rev. Lett.* **22** (1969) 689.
12. L.V. Gribov, *Phys. Rep.* **100** (1983) 3.
13. L.C. Biedenharn, *J. Math. Phys.* **4** (1963) 436.
14. Z. Bern, D.C. Dunbar and T. Shimada, *String-Based Methods in Perturbative Gravity*, preprint IASSNS-HEP-93-31 (May 1993).
15. F.A. Lunev, *Phys. Lett.* **B295** (1992) 99.
16. D.Z. Freedman, P.E. Haagensen, K. Johnson and J.I. Latorre, *The Hidden Spatial Geometry of Non-Abelian Gauge Theories*, Cambridge preprint MIT CTP-2238 (August 1993); M. Bauer, D.Z. Freedman and P.E. Haagensen, preprint CERN-TH. 7238/94.
17. K.S. Stelle, *Phys. Rev.* **D16** (1977) 953; *Gen. Relat. Grav.* **9** (1978) 353.

18. F.W. Hehl, P. v.d. Heyde, G.D. Kerlick and J.M. Nester, *Rev. Mod. Phys.* **48** (1976) 393.
19. J. Kiskis, *Phys. Rev.* **D11** (1975) 2178; G.B. West, *Phys. Lett.* **B115** (1982) 468.
20. Y. Ne'eman, *Proc. Nat. Acad. Sci. USA* **74** (1977) 4157.
21. Y. Ne'eman, *Ann. Inst. H. Poincaré* **28** (1978) 639.
22. Y. Ne'eman and Dj. Šijački, *Int. J. Mod. Phys.* **A2** (1987) 1655.
23. Dj. Šijački, *J. Math. Phys.* **16** (1975) 298.
24. Dj. Šijački and Y. Ne'eman, *J. Math. Phys.* **26** (1985) 2457.
25. J. Mickelsson, *Commun. Math. Phys.* **88** (1983) 551; A. Cant and Y. Ne'eman, *J. Math. Phys.* **26** (1985) 3180.
26. Y. Ne'eman and Dj. Šijački, *Phys. Lett.* **B157** (1985) 275.
27. Y. Ne'eman, in *"Spinors in Physics and Geometry"*, G. Furlan and A. Trautman, eds., (World Sci. Pub., Singapore, 1987), pp. 313-346.
28. F.W. Hehl, J.D. McCrea, E.W. Mielke and Y. Ne'eman, *"Metric Affine Gravity, etc."*, to be pub. in *Phys. Reports*.
29. J. Lemke, Y. Ne'eman and J. Pecina-Cruz, *J. Math. Phys.* **33** (1992) 2656.
30. Y. Dothan, M. Gell-Mann and Y. Ne'eman, *Phys. Lett.* **17** (1965) 148.
31. D.W. Joseph and Y. Ne'eman, Univ. of Nebraska preprint (1969); L. Weaver and L.C. Biedenharn, *Phys. Lett.* **B32** (1972) 326.
32. Y. Ne'eman and Dj. Šijački, *Phys. Lett.* **B157** (1985) 267.
33. Y. Ne'eman and Dj. Šijački, *Phys. Rev.* **D37** (1988) 3267; Dj. Šijački and Y. Ne'eman, *Phys. Rev.* **D47** (1993) 4133.
34. S. Goshen and H. J. Lipkin, *Ann. Phys. (NY)* **6** (1959) 301; G. Rosensteel and D.J. Rowe, *Phys. Rev. Lett.* **47** (1981) 223; J.P. Draayer and K.J. Weeks, *Phys. Rev. Lett.* **51** (1983) 1422.
35. Dj. Šijački and Y. Ne'eman, *Phys. Lett.* **250** (1990) 1.
36. Y. Ne'eman and Dj. Šijački, *"QCD and the Nuclear Physics IBM"* to be published in *Proc. Int. Conf. on "Perspectives for the Interacting Boson Model"*, Padua (1994), A. Vitturi, ed.
37. A. Arima and F. Iachello, *Phys. Rev. Lett.* **35** (1975) 1069.
38. See for example Y. Liang et al., *Phys. Lett.* **B307** (1993) 375 and references 4-9 quoted therein.

THE THEORY OF QUARK CONFINEMENT

V. GRIBOV

Inst. Theoretische KernPhysik, Universitaet Bonn
Nussallee 14-16, 53115 Bonn, Germany

ABSTRACT

Properties of the quark spectrum and Green functions of confined quarks were discussed in the framework of the theory in which light quarks are responsible for confinement.

EXPERIMENTAL STUDIES OF JET HADRONISATION

WITH THE OPAL DETECTOR AT LEP

JAMES R. LETTS

Department of Physics, University Of California, Riverside
Riverside, CA 92521, USA

ABSTRACT

Measurements of baryon and meson production rates and momentum spectra made with the OPAL detector at LEP are presented and compared with the predictions of Monte Carlo models of jet hadronisation. While the momentum spectrum of charged pions is found to agree well with the Monte Carlos, baryons have a softer momentum spectrum and charged kaons a harder one than predicted by models. No tuning of the models studied could reproduce simultaneously all of the measured baryon rates. Comparisons of the data are also made with the predictions of a QCD-based calculation. A study of rapidity correlations between $\Lambda\bar{\Lambda}$ pairs shows the correlations expected from a chain-like particle creation and indicates that simple baryon-antibaryon chains without "popcorn" splitting predict too strong correlations.

1. Introduction

The interaction $e^+e^- \rightarrow Z^0$ and the subsequent decay of the Z^0 to a quark-antiquark pair are well described by the Standard Model of electro-weak interactions. However, what we observe in the detector is not the original quark-antiquark pair, but jets of hadrons (see figure 1). Although the theory underlying the confinement of quarks and gluons into hadrons is thought to be QCD, the non-perturbative nature of calculations has made them difficult to perform. Therefore, we rely on Monte Carlo models of jet hadronisation to describe the measurements we make of hadronic cross sections and correlations between hadrons.

A general assumption of jet hadronisation is a chain-like particle creation with local conservation of quantum numbers. This assumption is used in jet fragmentation models like JETSET[1], which is based on the Lund string fragmentation model [2], and HERWIG[3], which is based on cluster fragmentation[4].

One consequence of a chain-like particle creation is a small rapidity*difference between neighbouring particles. Two different chains for baryon production have been proposed: a chain with a baryon-antibaryon pair produced in sequence, and one in which one or more mesons can appear between the baryon-antibaryon pair, the popcorn mechanism[5].

---

*The rapidity of a particle is defined as $y = \frac{1}{2}\ln(E + p_\parallel)/(E - p_\parallel)$, where E is the energy and $p_\parallel$ is the momentum component parallel to the thrust axis.

2. Measurement of Baryon and Meson Production

With the OPAL detector[6] at LEP, we have recorded more than two million multi-hadronic Z^0 decays. In several publications we have measured the cross sections and momentum spectra of several (non-heavy flavour) particle types, including $\pi^\pm$, $K^\pm$, and $p(\bar{p})$ [7], $K^*(892)^0$ and $\phi(1020)$ [8], $K^*(892)^\pm$ [9], and Λ, Ξ^-, $\Sigma(1385)^\pm$, $\Xi(1530)^0$, and Ω^- [10]. OPAL has also made a study of rapidity correlations in $\Lambda\bar{\Lambda}$ pairs[10]. These measurements and their relevance for studies of jet hadronisation are presented here.

2.1. Strange Baryon Production

Strange baryons can be identified by their weak decays into charged particles. A typical $\Lambda \rightarrow p\pi^-$ event as reconstructed by the OPAL central tracking chambers is shown in figure 2. The Λ can be identified in the presence of a large combinatoric background by reconstructing the decay vertex, which is well displaced from the primary event vertex, as is seen in the lower left of the figure. These candidates are then used to search for the other strange baryon decays: $\Xi^- \rightarrow \Lambda\pi^-$, $\Omega^- \rightarrow \Lambda K^-$, $\Sigma(1385)^\pm \rightarrow \Lambda\pi^\pm$, and $\Xi(1530)^0 \rightarrow \Xi^-\pi^+$.

The cross sections as a function of the scaled energy $x_E(\equiv E_{hadron}/E_{beam})$ are shown in figure 3 for Λ and Ξ^-, compared to the predictions from JETSET, which have been normalised to the observed rate in the data. One can see clearly that JETSET has a too hard momentum spectrum at high x_E for the Λ, and possibly also for the Ξ^-. HERWIG also predicts a similar spectrum, which is too hard.

The origin of this discrepancy is not well understood. A poor modelling of heavy flavour decays into strange baryons might influence the high end of the spectrum, while differences in strangeness or baryon production in quark and gluon jets could influence lower x_E values.

2.2. Charged Hadron Production

The charged hadrons $\pi^\pm$, $K^\pm$, and $p(\bar{p})$ can be measured using the energy loss dE/dx measurement in the OPAL central detector, which provides a good separation of the different particle species[7].

Using a counting method in the low momentum region where the dE/dx of the different types are well separated, and a fitting method in the high momentum region where they are not, OPAL has determined the cross sections for the three charged hadrons. As in the case of the Λ, the proton is found to have a softer momentum spectrum than predicted by JETSET and HERWIG. Since the overall event shapes have been tuned in the Monte Carlos to agree with our data, the $\pi^\pm$ cross section is well described. To compensate the hardness of the baryon cross sections, we find that the $K^\pm$ momentum spectrum is harder in the data than in the Monte Carlos (see figure 4).

3. Comparison of Hadron Rates with Monte Carlo

OPAL has tried to tune the JETSET and HERWIG Monte Carlos to describe measured baryon production rates[10]. While JETSET can be tuned to describe the overall production rates of charged pions and kaons rather well with little additional tuning, systematic discrepancies can be found in the baryon sector. This is illustrated in figure 5, in which the production rates of the octet and decuplet baryons are compared with the JETSET predictions. One can see good qualitative agreement for the octet baryons, while there is a systematic disagreement in the decuplet. The prediction for the Ω^- rate is many standard deviations too low, and that for the $\Sigma(1385)^\pm$ rate too high. Even with large changes in the fragmentation parameters available in JETSET, we could not simultaneously tune more than one of the $\Sigma(1385)^\pm$, $\Xi(1530)^0$, and Ω^- rates. The measurement of the Δ^{++} production rate in the future should complete the picture.

4. Comparison of Cross Section Results with QCD Predictions

QCD calculations based on the modified leading log approximation (MLLA)[11] predict the shape of the $\xi(\equiv \ln(1/x_p))$ distribution for soft gluons, which can be directly compared to the observed hadron spectra under the assumption of local parton hadron duality (LPHD)[12]. The ξ distribution is expected to be approximately Gaussian-shaped, with the peak position shifted to lower values for more massive hadrons[13].

In figure 6, we can see that although the shape of the distributions are approximately Gaussian, there does not appear to be a simple linear relationship between mass and the position of the maximum. This is further illustrated in figure 7, which presents a compilation of measured maximum positions from OPAL for different hadrons[7,8]. Rather than a simple linear relationship, there appears to be a linearity in the baryon measurements taken by themselves, and quite a lot of scatter in the meson values.

However, studies with the JETSET Monte Carlo show that the effects of resonance decays are important and shift the peak positions significantly[8]. In figure 8 are compared the peak positions in JETSET for primary hadrons only and for all hadrons including secondary and resonance decays. One sees that the inclusive $K^\pm$, $K^*(892)$, and $\phi(1020)$ distributions are shifted to much lower peak positions, which is a consequence of hard $c \to s$ decays. The baryons are less affected by this kind of problem, and maintain a more or less linear relation between the position of the maximum and the particle mass.

This effect shows us that we must be careful when interpreting discrepancies between data and Monte Carlo models. A disagreement might be due to the failure of the Monte Carlo to describe some secondary decay, and not a problem with the tuning of the fragmentation parameters or the assumptions of the model. For example, the proton rate is approximately $0.20 - 0.25$ per event too high in the default version of JETSET. Is this a problem with the tuning of the JETSET fragmentation parameters, or is it a consequence of much too large rate for the Δ baryons, which produce a large fraction of the secondary protons? Given the poor description of the other decuplet baryon rates by the models we have studied, this could be an important factor.

5. Rapidity Correlations Between Baryon-antibaryon Pairs

Rapidity correlations between $\Lambda\bar{\Lambda}$ pairs in the same event provide a way to study the dynamics of jet hadronisation. Chain-like models of hadronisation predict that particles produced nearby on the chain will have correlated rapidities. The strength of this correlation will depend on the assumptions made in the models.

The rapidity distribution of the $\bar{\Lambda}$ corrected for background and efficiency is shown in figure 9 for five rapidity intervals of the tagging Λ. The corresponding distributions for pair candidates with the same baryon number are shown in figure 10. Strong rapidity correlations are clearly seen over the entire rapidity range for the pairs with opposite baryon number, whereas the same baryon number pairs show an essentially flat rapidity distribution as expected from uncorrelated production. The observed rapidity correlations show evidence for a chain-like production of baryon-antibaryon pairs, and are qualitatively reproduced by the JETSET and HERWIG Monte Carlos.

To obtain a more quantitative statement about the rapidity correlation strength, the rapidity difference of all $\Lambda\bar{\Lambda}$ pairs is used. The normalised distribution of the rapidity difference, again corrected for efficiency and the uncorrelated background, is shown in figure 11. The distributions become asymmetric for large rapidity differences because of the limited phase space to find both baryons in the same hemisphere.

The distributions obtained from the HERWIG Monte Carlo and the JETSET Monte Carlo with a popcorn parameter of zero are also shown in figure 11. Both models predict smaller rapidity differences than found in the data. The prediction of the UCLA model[14], with a different ansatz to describe the fragmentation and the popcorn mechanism (on average about 1.7 mesons are produced in between the baryon antibaryon pair) is also shown in figure 11. It predicts weaker rapidity correlations than seen in the data.

The shape of the rapidity difference distribution within JETSET is sensitive only to the choice of the popcorn parameter, which enables the production of baryon-meson-antibaryon configurations. Within this model one can compare the observed rapidity correlation strength with different choices for the popcorn parameter. Figure 12 shows the rapidity difference in the data and the JETSET Monte Carlo for 0%, 50% (the default value), 80%, and 95% probability of the production chain baryon-meson-antibaryon. Agreement with the data is obtained for large parameter values, indicating a value of more than 80% for the popcorn probability.

6. Conclusions

Much work has gone into studying jet hadronisation using measurements of particle production cross sections and correlations between baryon-antibaryon pairs. Although general event properties are well described by the Monte Carlos we studied, there are certain details which are not well modelled. We find that the predicted momentum spectra of protons and Λ baryons are too hard in the models, while those of strange mesons are too soft. Furthermore, no tuning of the models studied could reproduce more than one of the measured decuplet baryon rates. Comparison with a QCD-based calculation (MLLA+LPHD) shows good agreement with the shape of the ξ distribution, although the mass dependence of the peak position is obscured by the effects of resonance decays. A study of rapidity differences between $\Lambda\overline{\Lambda}$ pairs showed that the simple production chain baryon-antibaryon predicts too strong rapidity correlations. The observed correlations can best be described by the JETSET Monte Carlo with a dominant production chain of baryon-meson-antibaryon, the popcorn mechanism.

7. References

1. T. Sjöstrand, *Comp. Phys. Comm.* **39** (1986) 347, and CERN-TH.6488;
   T. Sjöstrand and M. Bengtsson, *Comp. Phys. Comm.* **43** (1987) 367.
2. B. Andersson, G. Gustafson, G. Ingleman, and T. Sjöstrand, *Phys. Rep.* **97** (1983) 33.
3. G. Marchesini and B. Webber, *Nucl. Phys.* **B310** (1988) 461.
   G. Marchesini et al., *Comp. Phys. Comm.* **67** (1992) 465.
4. S. Wolfram, in *Proc. 15th Rencontre de Moriond,* ed. J. Tran Thanh (March 1980), and Caltech preprint CALT-68-778 (April 1980).
5. B. Andersson, G. Gustafson, and T. Sjöstrand, *Physica Scripta* **32** (1985) 574.
6. OPAL Collab., M. Ahmet et al., *Nucl. Instrum. Meth.* **A305** (1992) 275.
7. OPAL Collab., R. Akers et al., *Z. Phys.* **C63** (1994) 181.
8. OPAL Collab., P. Acton et al., *Z. Phys.* **C56** (1992) 521.
   OPAL Collab., *Inclusive Strange Vector and Tensor Meson Production in Hadronic Z^0 Decays,* preliminary results presented at ICHEP94 Glasgow.
9. OPAL Collab., P. Acton et al., *Phys. Lett.* **B305** (1993) 407.
10. OPAL Collab., P. Acton et al., *Phys. Lett.* **B291** (1992) 503.
    OPAL Collab., P. Acton et al., *Phys. Lett.* **B305** (1993) 415.
    OPAL Collab., *Strange Baryon Production and Correlations in Hadronic Z^0 Decays,* preliminary results presented at ICHEP94 Glasgow.
11. Y.L. Dokshitzer, V.A. Khoze, and S.I. Troyan, *Z. Phys.* **C27** (1985) 65.
12. D. Amati and G. Veneziano, *Phys. Lett.* **B83** (1979) 87.
    Y.I. Azimov et al., *Phys. Lett.* **B165** (1985) 147.
13. Y.L. Dokshitzer, V.A. Khoze, and S.I. Troyan, *J. Phys. G: Nucl. Part. Phys.* **17** (1991) 1481.
    Y.L. Dokshitzer, V.A. Khoze, and S.I. Troyan, *Z. Phys.* **C55** (1992) 107.
14. S.B. Chun and C.D. Buchanan, UCLA-HEP-92-008 (August 1992).

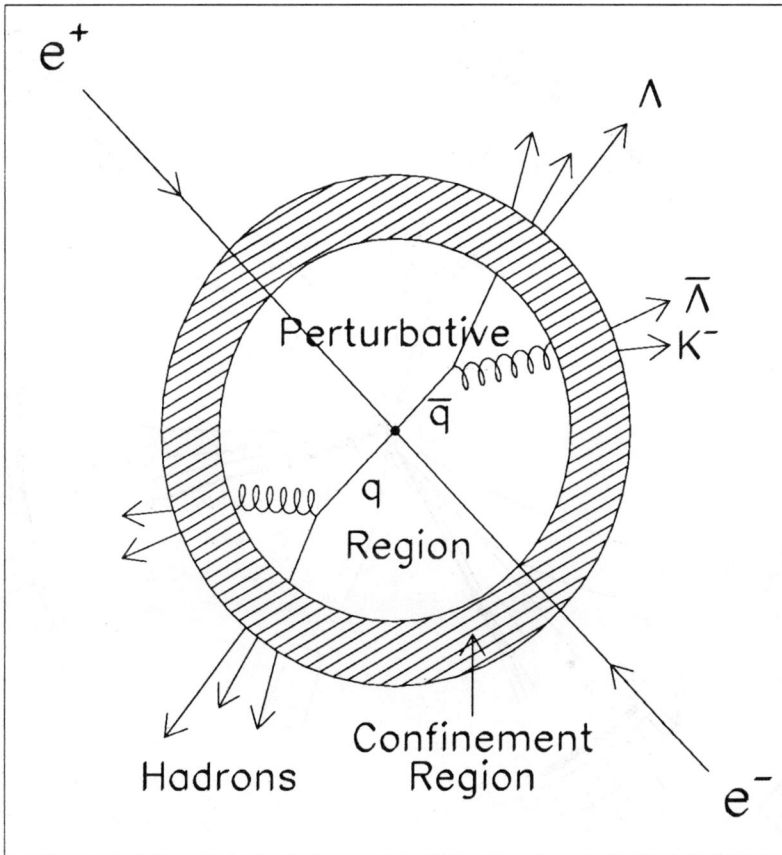

Fig. 1. Schematic diagram of the process $e^+e^- \rightarrow q\bar{q}$ and the subsequent confinement of the quarks and gluons in hadrons.

Fig. 2. An event from OPAL showing the central tracking chambers. In the event are found a Ξ^-, a $\bar{\Lambda}$, and a K^+, which compensates the net strangeness of the two baryons.

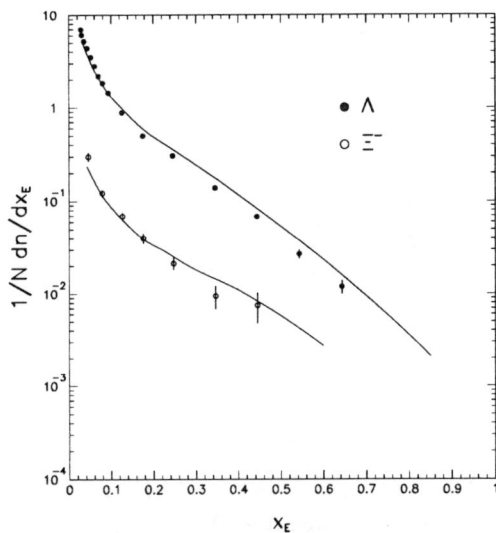

Fig. 3. Differential cross sections for Λ and Ξ^-, compared to the predictions of JETSET, which have been normalised to the rate observed in the data.

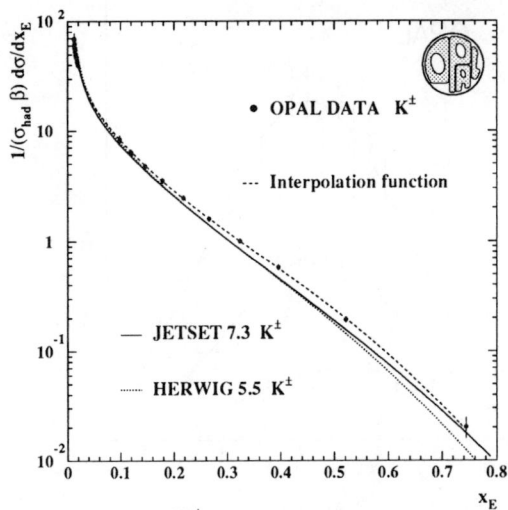

Fig. 4. Differential cross section for $K^\pm$, compared to the predictions from JETSET and HERWIG.

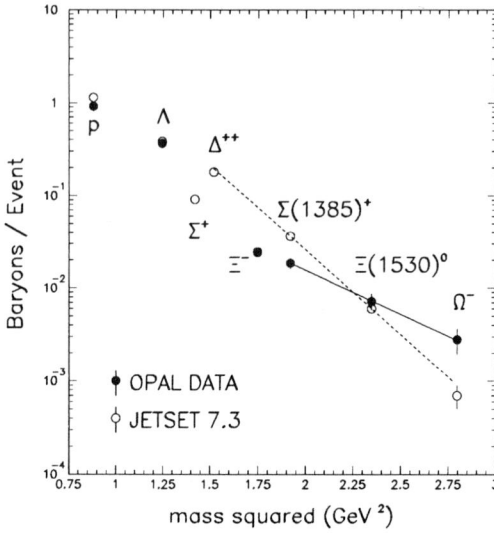

Fig. 5. Baryon rates per hadronic event as measured by OPAL, compared with the predictions from JETSET.

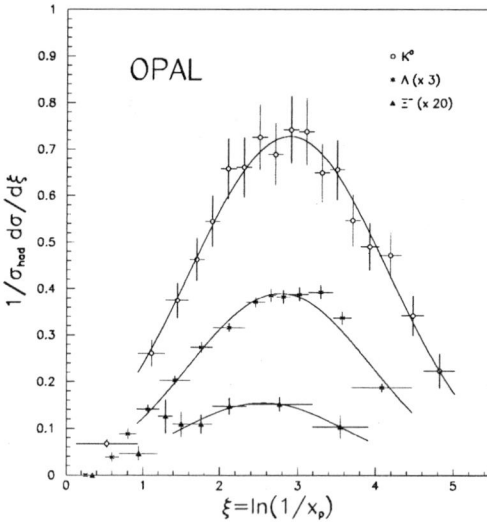

Fig. 6. Differential cross sections for K^0, Λ, and Ξ^-, as a function of $\ln(1/x_p)$.

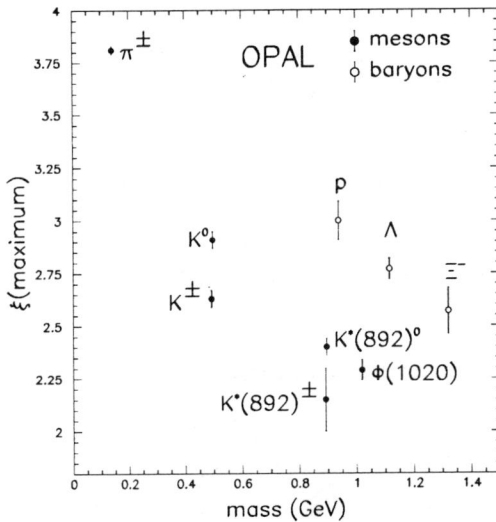

Fig. 7. OPAL measurements of the maximum of the ξ distribution for various particle types, shown as a function of mass.

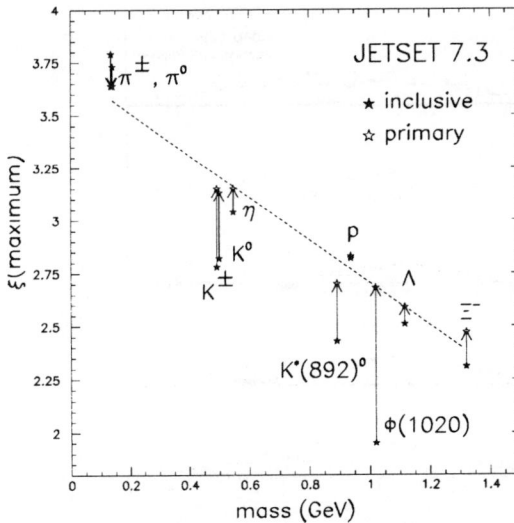

Fig. 8. Maxima of the ξ distribution in JETSET for primary and inclusive (primary plus secondary) hadrons. The large shifts are due to the inclusion of resonance decays.

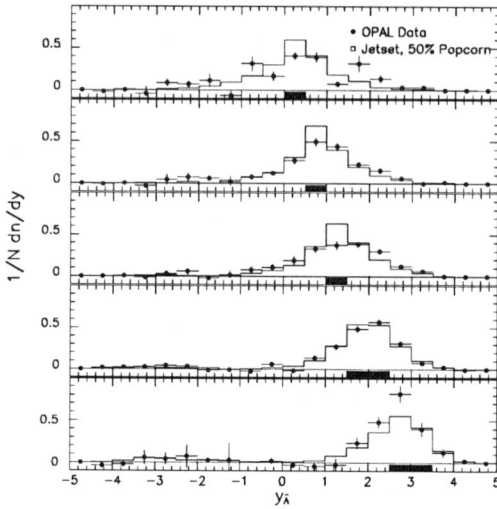

Fig. 9. Rapidity distribution of $\overline{\Lambda}$ for various tagging intervals of Λ's (shown as solid boxes). The uncorrelated background from pair combinations with the same baryon number is subtracted.

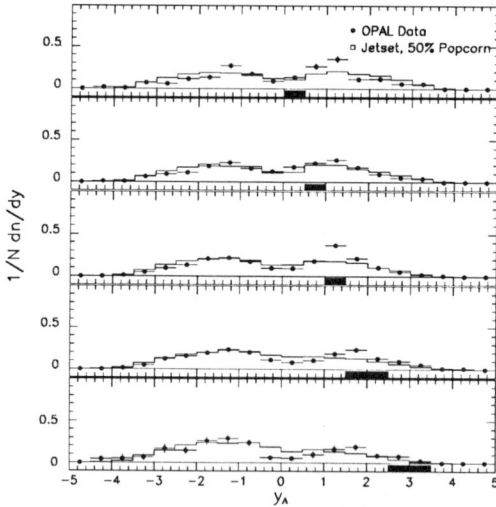

Fig. 10. Rapidity distribution of the second Λ, including background, for various tagging intervals of the first Λ's (shown as solid boxes) for combinations with the same baryon number.

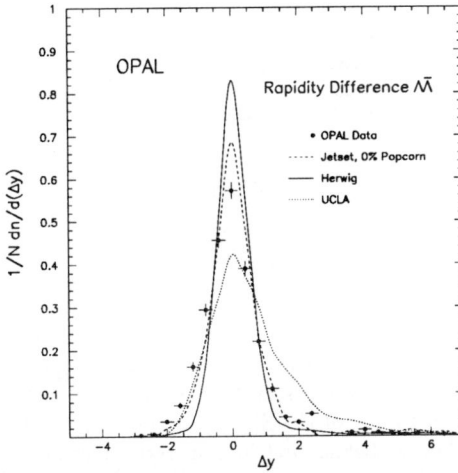

Fig. 11. Rapidity difference, corrected for efficiency, for all $\Lambda\overline{\Lambda}$ pairs (same baryon number pairs are subtracted). The distributions expected from the HERWIG, JETSET without popcorn splitting, and UCLA Monte Carlo models are also shown.

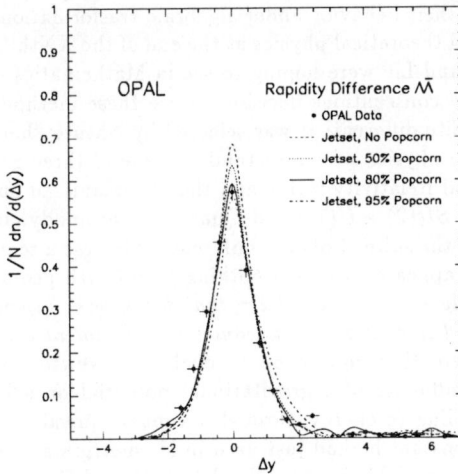

Fig. 12. Rapidity difference, corrected for efficiency, for all $\Lambda\overline{\Lambda}$ pairs (same baryon number pairs are subtracted). The distributions expected from the JETSET Monte Carlo with a popcorn production of 0%, 50% (default), 80% and 95% are also shown.

WHERETO QCD

YUVAL NE'EMAN[*†]

Sackler Faculty of Exact Sciences, Tel-Aviv University
69978 Tel-Aviv, Israel

1. Origins

1.1. Gauge Theories

Twenty years have passed since the adoption of the Standard Model[1] (SM) as a description of the physics of particles and fields – experimentally explored, at the time, up to a few GeV (in the center of mass system). The synthesis included Quantum Chromodynamics (QCD), as a theory of the Strong Interactions[2]. QCD, as also the rest of the SM, is a gauge theory. Let us first look in retrospect at the evolution of these ideas.

In 1872, Felix Klein and Sophus Lie launched the "Erlangen Program". This was a suggestion to use a combination of geometry and algebra, in order to classify the whole of mathematics and discover the inner relationships between the various concepts. Differential Geometry would "carry" the constructs, group theory would categorize them, by testing their behavior under algebraic trasformations.. It is most interesting that fundamental theoretical physics at the end of the XXth Century looks very much like what Klein and Lie were hoping to see in Mathematics – except that whereas they had in mind a conscentious decision to use these methods, the way it happened in Physics was quite different: it was selected by Nature "herself"!

The part of fundamental physics we understand is made of three gauge theories, one classical, namely General Relativity (GR), and the two quantized ones of the SM – the Electroweak (gauging $SU(2) \times U(1)$) and Quantum Chromodynamics (QCD), gauging $SU(3)_{color}$, which is the subject of our conference. GR was a sequel to Special Relativity, a system which appeared to have nothing to do with geometry when it was discovered. *It was only several years later that Minkowski explained that his former student's theory could be regarded as a geometrical statement about the metric of spacetime.* When Einstein then moved on to deal with Newtonian gravity, he found that evaluating the influence of a gravitational potential on a beam of light (using $E = mc^2$, i.e. a coupling to energy, through its mass equivalent), the beam's deflection meant that the constant he had just used in his metric was now no more a constant! This led him (helped by his former schoolmate Marcel Grossmann) to GR. *This dynamical theory is (1) a local gauge theory for the group of Diffeomorphisms – a passive symmetry, except for its Poincaré subgroup (the "orbital" part of the Poincaré*

[*]Wolfson Chair Extraordinary in Theoretical Physics
[†]Also on leave from Centre for Particle Physis, University of Texas, Austin, Texas 78712, USA

group). (2) The latter is locally gauged as an active symmetry, together with the entire Poincaré group, taken as acting on the tangent frames, including its double-covering, i.e. the spin part, (which would not be contained in $Diff(4, R)$) – somehow as an additional local gauge principle (I am describing Einstein's GR as 'completed' after the insertion of spinors through tetrad frames in 1928-29).

This was the first local gauge theory and Einstein worried throughout 1911- 1915 whether this invariance was not too demanding, as a constraint.

Hermann Weyl made in 1919 the first generalization to the geometry of a Fibre Bundle, with a one-parameter Lie group as structure group. At the recent Beersheva conference, Lochlan O'Raifeartaigh gave a beautiful review of the history of gauge theories and showed that Weyl even had the phase (loop) integral, in his paper.[3] He did change the group itself ten years later, from the non-compact R^1 (for the scale function) to the compact $U(1)$, when he saw London's use of that group for the phase transformation of the electron's complex wave function. Multiplication by 'i' did it.. C.N. Yang and H. Mills made the next installment in 1953, generalizing to a non-Abelian group. The applicability to particle physics was not obvious. Nature intervened again, first with the Weak interactions. "She" did it in 1958, through *the identification by R.E. Marshak and E.C.G. Sudarshan of the "V-A" nature of the weak transitions.* This meant that the weak interactions (WI) involve gauge currents and probably gauge couplings. Very soon, Nature intervened again, this time in favor of the description of the Strong Interactions (SI) by a local gauge theory. R. Hofstadter had experimentally probed the electromagnetic structure of the nucleon. He found that the meson cloud surrounding the nucleon contained unexpected resonances, which were shown by Y. Nambu and by W.R. Frazer and J.R. Fulco to represent $J = 1$ mesons, later known respectively as the ω and the ρ. J.J. Sakurai then demonstrated in 1960 that *many features of the SI could be understood in terms of a local gauge theory of $SU(2)_I \times U(1)_Y \times U(1)_B$ ('I' stands for strong-isospin, 'Y' for strong-hypercharge and 'B' for baryon-charge).* This was the first plausible gauge theory of the SI[5], a precursor of QCD.

1.2. $SU(3)_{flavor}$ and quarks

I was very impressed by the Yang-Mills paper when I first came across it in May 1960. Originally, one of the motivations which had drawn me into physics had been the "aesthetics" of GR. Now, after first learning that particle physics was very different, I again had that feeling of beauty conveyed by the combination of geometry with group theory. I was also influenced by Sakurai's paper. My introduction of $SU(3)_{flavor}$[5] was thus entitled "Derivation of the Strong Interactions from a Gauge Invariance". Gell-Mann's 'Eightfold Way' unpublished preprint presented the same gauge theory. We were predicting a vector $8+1$, and they were soon found. They were even coupled 'universally', as necessary for gauge couplings. These couplings notwithstanding, they are now demoted from the lofty status of gauge fields, their proletarian origins as quark-antiquark mesons having been revealed. They had been somewhat suspect from the beginning, because of their masses. This being six years before the

Higgs-Kibble-Anderson mechanism, there were no serious attempts to defend their status. Instead, the universal couplings were explained by a dispersion-relations pole-dominance argument[6]. Pole dominance was one facet of a complicated juxtaposition between SI and WI, to which I shall return.

An octet in $SU(3)$ is a second-rank tensor and within a year I had constructed a candidate fundamental constituent as a triplet with $B = 1/3$, so that 3 of them would make up a baryon[7]. I am dwelling on the subject here because after the Como newspaper headlines, I was asked about the story[8] by many of the participants. Beyond the sketchy Nuovo Cimento article, I developed the idea in a series of lectures at the Weizmann Institute[9]. The problem was that very few people believed in the octet version of $SU(3)_{flavor}$, just because it appeared as too abstract – as compared with the Sakata version, very popular in 1962. Announcing that the octet, in addition, entails the adoption of a fractional electric charges basic model made you sound crazy. I remember, for example, Vicky Weisskopf's visit to Israel in that year. I took him together with Gertrude Goldhaber on a trip to Galilee. We talked of $SU(3)$ on the way – and he was skeptical and completely uninterested. I even mentioned that if one were to look at the Sakata-like fundamental representation, it would have fractional charges. It only made things worse.. Things changed completely early in 1964, after the discovery of the Ω^- hyperon, which we had predicted[10]. Gell-Mann's paper on the quark model appeared just then, in the wake of the discovery, so did Zweig's preprint[11]. Weisskopf, who in his skepticism about $SU(3)$) had given low priority to the Ω^- search at CERN, now called for a massive 'hunt of the quark'.. Later on he even developed a nice simple way of explaining the model.

The identification of static $SU(6)$ by Gürsey, Radicati, Sakita[12] and Zweig gave a great boost to the quark model. In the next two years (1964-66) a tremendous wealth of good predictions strengthened what has since been known as the Non-Relativistic Quark Model (NRQM). In addition, a 'naive' approach also provided good results for the high-energy asymptotic region. I think two ratios of 3:2 predicted by $U(6) \times U(6)_\beta$ symbolize the type of surprising results we were witnessing – the ratio between the magnetic moments of the proton and the neutron (nobody had ever tried dividing the two numbers by each other before..) and the ratio between the $\sigma(NN) : \sigma(N\pi)$ asymptotic cross-sections[13].

And yet quarks were still not regarded as 'legitimate' until 1969, when the deep-inelastic elctron-nucleon scattering experiments at SLAC – using the parton model formalism developed by R.P. Feynman and J.D. Bjorken[14] – showed that the photon probes were indeed finding dot-charges (later even found fitting the fractional charges of the quark model). This became the parton quark model. Some time earlier the results for the hadron spectrum and the high energy asymptotics had been abstracted into the constituent quark model and the local current algebra results into the current quark model, related by a transformation of the Foldy-Wouthuysen type[15]. The current quark model was soon further extended to almost the entire $U(12)$ system.

2. Difficulties Leading to the QCD Ansatz

2.1. Problems with quarks and their resolution in the Han-Nambu model

All of this suffered from the following difficulties: (a) Fractional electric charges. (b) Anomalous spin-statistics correlations, especially after the successes of $SU(6)$, in which the baryon ground state is assigned to the symmetric **56**. (c) A phenomenological restriction of the hadron spectrum only to zero-triality representations of $SU(3)_{flavor}$.

The introduction of another $SU(3)$ by M. Han and Y. Nambu[16] was an attempt to resolve all three issues at the same time: (a) by having 3 different sets of 3 quarks with integer charges, set so that the fractional charges come out as the averages; (b) antisymmetry in the new $SU(3)_{HN}$ assignments (including the singlet as lowest state) (c) three quark or quark-antiquark structures make singlets and are thereby dynamically saturated. Some of the answers provided by this model were incorporated in QCD (though not the integer charges systematics).

2.2. Problems with the broken symmetry

The first impressive successes relating to SI amplitudes or mass formulae also ushered in a mystery. I remember, for example, Syd Drell's visit to Israel in 1962. He was also somewhat cool with respect to $SU(3)_{flavor}$, but for a good theoretical reason: how could the non-perturbative SI so nicely obey these perturbative predictions? Some time before the discovery of the Ω^-, R.J. Oakes and C.N. Yang[17] showed concretely how such predictions had to fail. They were right in principle, except that their actual sample calculation was wrong, due to their having selected an s-wave model, as Oakes recently reminisced. Many were quick to point to their error, but the problem was a true one. I think I was first to provide the answer that was later incorporated in QCD. In an article entitled 'The Fifth Interaction"[18], I suggested that we were confusing two different interactions when discussing the hadron spectrum, etc.. On the one hand, there is the 'true' SI, non-perturbative, responsible for such features as Regge sequences, nuclear binding or high energy scattering. On the other hand, the $SU(3)_{flavor}$-breaking interaction (which I called Fifth - a name later 'recycled' and applied to a hypothetical medium range component of gravity) is a perturbative force, responsible for the mass of the 's' quark and probably also the muon's. The successes of the supposed SI perturbative formulae were really in the same class as those for electromagnetic mass differences, magnetic moments, etc..

QCD incorporates this solution by selecting the second $SU(3)$, that of the *color charges* so that $[SU(3)_{color}, SU(3)_{flavor}] = 0$. *This was not possible with the Han-Nambu group*. The breaking of $SU(3)_{flavor}$ is assigned to the input masses, with the 'u,d' masses being of the order of the electron mass size, as against the 's' (and μ) with masses of the order of $100 MeV$. In GUT language, the Fifth is part of the Higgs sector, or of the 'horizontal' generations-building mechanism.

2.3. Problematics with overlapping gauges between SI and WI

This third area started out as a great advantage and an important tool, as long as the SI groups were restricted to global symmetries. Once they became considered as candidate local gauge symmetries, the entire concept became problematic.

The first such overlap was the *conserved vector current* (CVC)[19] discovered by Gershtein and Zeldovich. They noticed that the WI coupling for beta-decay is almost equal to that of muon decay – i.e. *there is no SI renormalization in the decay of the neutron, although this would have been expected to happen.* The explanation is a *generalized equivalence principle,* modeled on Einstein's. In the latter, the coupling to the energy-momentum tensor is universal (for states at rest, this is Newton's coupling to masses, G), again notwithstanding the actual composition of that matter and its other interactions (e.g. as tested by the Eötvös experiment). This is explained by the fact that the gravitational field is coupled to a conserved quantity a la Emmy Noether, a symmetry obeyed by all interactions, including the SI. The energy-momentum tensor is indeed the 'current-density' of the translations, in applying Noether's theorem to the Poincaré group. For CVC, it means that the weak gauge field ($W^{\pm}$, in SM parlance) is coupled to a Noether current of the SI symmetry. Such equivalence principles were identified for most of the $U(12)$ algebra of the current components (I made a 'hobby' of identifying these[20]). Summarizing, *generalized equivalence relates WI or EMI universality to global symmetries of the SI.*

The *dominance* ansatz is complementary. In the above cases, we were explaining the lack of renormalization (or the universality) of a non-strong transition, as resulting from its gauge-current's coinciding with a Noether-conserved current of the SI. In 'dominance', one explains the (also unexpected) universality of the SI transition, such as the couplings of the $\omega^0, \phi^0, \rho, K^*$, in terms of the given (gauge-generated) universality of the WI or QED. The SI pseudo gauge field appears as a dominating pole in the dispersion analysis for the relevant WI or QED current of the hadrons. The photon's coupling to the hadron electric current thus generates the quasi- universal coupling of the ρ^0 and for a combination of ω^0, ϕ^0; the WI current coupled to the W^- does it for the ρ^-, etc.. PCAC, discovered[21] by M. Goldberger and S. Treiman, is also such a dominance postulate: the SI coupling of pions (generalized to the entire 0^- multiplet) to hadron matter is 'universal' due to these mesons' dominating the dispersion relations for the divergence of the axial-vector Noether current, coupled to the WI gauge field. Both cases were described accordingly in[6]. Summarizing – a dominance ansatz justifies a SI universal coupling by linking it to a WI or EMI gauge-universality.

It is the third possibility which is problematic, when both SI and WI are assumed to couple universally, both as a result of local gauges. For instance, we would have the ρ (as a fundamental gauge field) and the W *coupling to the same current (up to a Cabibbo rotation).* The same current – coupled with different strengths to two gauge bosons with the same quantum numbers but different masses.. This situation is hard to justify. I had noted the problem in 1962 and had tried to suggest an answer, based on a hierarchical structure[22]. Steve Weinberg later pointed out that quantum field theory would cause under such conditions an enhancement of parity non-conservation, which is not seen. Clearly, the conceptual structure and experimental fit of the

SI, as reproduced by gauging $SU(3)_{flavor}$, could only be taken as generated by the dominance argument, not as a quantum gauge field theory. In the sixties, this was not considered important, with an effective taboo declared on quantum field theory anyhow, by the charismatic leadership of the time.

3. Quantum Chromodynamics Twenty Years Later

3.1. Birth of the Theory

As I just noted, field theory had been declared wrong and uninteresting since 1958. Still practiced in odd places (where appointments did not depend on recommendations from that charismatic leadership), the supposedly sick theory was quietly recovering, in the hands of such as R.P. Feynman, B. DeWitt, L.D. Faddeev and V.M. Popov, M. Veltman. In 1971, G. 't Hooft completed the task[23]. Within a few months the 'stepmother' (known as 'S-matrix theory') was abandoned, relativistic quantum (gauge) field theory (RQGFT) taking over everywhere. Two years later, H.D. Politzer, D.J. Gross and F. Wilczek and 't Hooft himself[24] were discovering asymptotic freedom (AF), as a specific feature of Yang-Mills theories. It fitted a picture in which quarks are very weakly coupled at high energy – the "UV end" or the 'hard' sector – fitting the quark parton model and current quark situations, including the SLAC experiments. There was some hope that this would be extendable to the "hard-soft" or semi-hard sector, where the model had been experimentally validated in predictions based on the incorporation of notions such as the impulse approximation (in the context of additive quark amplitudes in high energy scattering). At the same time it hinted at the possibility of confinement, i.e. a strengthening of the coupling at longer distances, perhaps a physical IR divergence. H. Fritzsch and M. Gell-Mann, S. Weinberg made such suggestions[2]. This resembled the 1965 Han-Nambu gauge theory[16], except that $SU(3)_{color}$ was now made to commute with $SU(3)_{flavor}$, i.e. keeping the fractional charges, as a price for getting rid both of the overlap and the broken flavor symmetry problems.

3.2. Present Status

At the Aachen workshop "QCD – 20 years later", Harald Fritzsch reviewed the situation, mainly at the perturbative end. The main parameter of the theory (aside from the Λ renormalization group scale) has been measured at various energies and is very well known, e.g. $\alpha_s(M_Z^2) = 0.12$ from a large number of different experiments. Altogether these perturbative sectors fit very well in a large variety of experimental results – electron-positron annihilation into created hadron matter (thresholds and the R ratio), deep-inelastic e-N scattering – with precise density functions and scaling violations fitting QCD predictions, 'gluonic' jets, etc..

The situation at the non-perturbative end is less encouraging. On the one hand, we would like to have (like in QED, when we calculate the binding energy of positronium) the possibility, using the values of α_s and the UV masses of the 'u,d' quarks as input, of calculating the masses of the proton and neutron (or of a constituent quark), that of the pion, and the $N\pi$ coupling $(g^2/4\pi) = 14$.. We would also like the calculation to tell us about the existence of a Regge sequence of excited states, their composition and the slope of that trajectory in the Chew-Frautschi plot. In brief – *reproduce the hadron spectrum and hadron physics from QCD*. The other issue in this sector is *color confinement*. We would like to have a mathematical derivation leading to a proof of color confinement, with precise indications with respect to deconfinement transitions, if they exist.

3.3. Lattice Gauge Theory; Light Front Physics

This is the main versatile tool the theory disposes of, in the non- perturbative region, sometime almost comparable to a perturbative calculation. It is thus a combination between a regulator technique (the lattice spacing playing the role of a cut-off), a necessary stage even in a perturbative calculation, and a method for the direct computation of the generating functional of Green's functions. It was amply utilized in this meeting: a check on guesses from Quantum Field Theory (as in Gliozzi's study of the quantum behavior of the QCD purported flux-tubes, applying notions from the known physics of fluid interfaces) or in a potential calculation as in Prosperi's study of the binding of hadrons – or in predicting new states (mirror particles) as done by Creutz, evaluating the gluon propagator's non-perturbative behavior (Stella, Parrinello), studying the spontaneous breaking of chiral symmetry (Faber), getting the interquark potential (Schilling), calculating decay amplitudes (Richards), high energy scattering (Dosch), etc.. The main weakness is in the method's approximate nature, without the mitigation arising from the optional 'customer decision' about the precision worth achieving, which is in the nature of the perturbation series. Aside from the aspects inherent in the lattice technique, there is the non-relativistic approximation applied to most calculations. Nevertheless, the program of evaluating the nucleon and pion masses and couplings has produced encouraging results[25]. It appears especially hard to construct and evaluate excitations higher than the first – even though it has even proved possible in some cases to get an evaluation of the slope of the trajectory.

There is, however, another possibility (which was not much touched upon in our meeting) – namely "Light Front Physics"[26,27]. Already when working with local current algebra in the sixties, it was hoped that by using the infinite-momentum frame, one could expect to preserve the perturbative advantages of the UV end, while at the same time probing the structure of the spectrum, generally approachable through the IR end (and often forcing our hand in selecting a non-relativistic regime). The "Light Front" kinematical set up is close to that infinite momentum limit and K. Wilson, still using lattice gauge theory as the computational tool, hopes to improve on the results in "constructing" the hadrons and evaluating their couplings, working in l.f.p.[27]. Presentations at the conference were few (e.g. Dubin's) and I refer the reader to S. Brodsky's review and to Wilson's presentation of his program. In particular, the hope is to have a better insight into the transition between current (or parton) quarks and the constituent quark model. One feature which this approach will bring is the use of "constituent" (massive) gluons, which were not envisaged in the pre-QCD quark model studies.

4. Confinement

4.1. Present status

This is where the theory appears to have fared less well than was hoped for, twenty years ago. I have compared the first twenty years of QCD with the first twenty years (1928-48) of QED, *prior to the renormalization breakthrough of 1946-48.* Many calculations were performed in QED before 1948, thanks to a variety of schemes and approximations, each designed to cope with one specific aspect of the phenomena. After 1948, any process could be calculated, probably even just by a computer, with an appropriate programming of (perturbative) QED. QED is basically a weak coupling area, and its one big problem came from the UV divergences. Once this was resolved, QED became the most perfect and precise theory of (quantum) physics. QCD is a strong coupling theory and its key issue is color confinement. One of these days we shall see – I hope – a methodology which will reproduce confinement and all related phenomena, at will. Meanwhile, like in pre-1948 QED, *we apply different external algorithms, all based on the idea that something of the same nature must occur within QCD itself, except that we have no direct (or sufficient) knowledge of these internal realizations of the algorithms.*

4.2. The Dual-Superconductivity and related algorithms (flux-tubes)

This Meissner-effect inspired model has been the prime algorithm, since the earliest days, put forward by Nambu, Mandelstam, 't Hooft, Parisi and others, also related to the *hadron string*-inspired (Nielsen-Olesen) flux. We have had very interesting expositions from De Giacomo, Gliozzi, Zachariasen, Schlichter and others. Taking, for example, Zachariasen's exposition, the agent were monopoles – assuming that QCD, though it does not have straightforward monopole solutions, will sometime be shown to produce "effective" monopoles. Di Giacomo gave a derivation, using vortice structures more directly related to QCD. Some very recent results of Witten with Sieberg and Vafa[28–30] represent an important advance in the growth of the credibility of this algorithm. They use supersymmetric Yang-Mills theory (with its N=4 non-renormalization features), but this time N=2 (also a 'miraculous' theory in several respects). The theory's supermultiplets contain dyons, i.e. monopoles (and charges) which do not arise as 'soliton' solutions of the equation, just representations of the supergroup. The authors manage to produce a mathematical proof, according to which the monopoles make a condensate (like the artificial monopoles in Zachariasen's presentation) and the $SU(2)$ charges are thereby confined. The advance is thus *in having now a particle-physics exact theory in which the dual-superconductivity algorithm is realized.* We are still not at the stage of knowing what precisely are the condensates in QCD made of and how, but we have a model closer than a condensed matter analog, a vindication of the idea in principle. Other contributions to these approaches at the conference were made by Schlichter, Olejnik, Owen, Zach (in 'compact QED' as a model). A subject related to confinement is chiral symmetry and its *spontaneous breaking*. Here we have had interesting contributions (Sasaki, Suganuma, Achasov, Gogohia, Nagy, Simonov, etc.). Some studies were interesting in fixing the parameters of the confining structure (Leonidov, Khoze, Petronzio, etc.)

4.3. Potentials

These calculations were given a heavenly boost when the Lord, in his graciousness, had us discover the 'c' quark, only a short time after the launching of QCD in 1974. With 'c' (and since then even more so 'b,t') the non-relativistic approximations which were unjustifiable for 'u,d' and even for 's', rightfully gained the status of fairly decent approximations of QCD.. We had some nice applications here by Prosperi, Schilling, Koerner, Yndurain, Dong, Hussain, Sharma, Olsson (related also to the flux tube treatment), Rosina, Ball, Pastrone, Badalian, Uzikov, etc.. Some presentations used combined algorithms, e.g. Page (with the flux-tube), De Fazio, Pestov, Scorletti, etc.. A related topic consists in applying the Bethe-Salpeter equation; we heard such applications from Prosperi, Fulcher, Schoeberl, Owen, Williams, Bijtebier and a related discussion by Mourad. Toernqvist has calculated meson 'molecules'. N. Byers' treatment of the e^+e^- collisions (thresholds) threw some new light on a well studied region.

4.4. Gravity-inspired algorithms

Even before QCD, Salam and his collaborators[31] had suggested a *strong gravity* force, based in the sixties on the $f^0(1,250;2^+)$ mixing with the graviton. In the seventies and after the advent of QCD, they constructed models[32] in which such forces could arise, e.g. gauging $SL(6,C)$, the tensor completion of the product of $SL(2,C)_{Lorentz} \times SU(3)_{color}$. I also partook in such attempts (independently), using enlarged versions of gravity, extensions which could arise from quantum gravity counterterms. One model used Poincaré-gauge gravity[33]; in another, we tried affine gravity[34], later pursued on its own in the Quantum Gravity program, abandoning the assumption of SI contributions. Other groups also tried this approach[35].

More recently, we have (Dj. Sijacki and myself) found such a gravity-like component *inside QCD*. This then becomes an *internal algorithm*. (See my contribution in this conference and[36−39]). One practical result is a classification of the hadron spectrum[40−41], based on $\overline{SL}(4,R)$ manifields (infinite-component fields[42]) and an $\overline{SA}(4,R)$ particle Hilbert space (the overline stands for the double-coverings). Other approaches pointing to a spatial (gravity-like) aspect of Yang-Mills theory, for zero-color systems, have recently been tried by Freedman and coll.[43], and by Lunev[44], working in three dimensions and with $SU(2)$. Note that the Freedman et al. paper uses Dirac's Hamiltonian quantization. Considering that this approach has recently scored points in Quantum Gravity (the Ashtekar program[45]), it might be interesting to try it in Yang-Mills theory.

4.5. Other approaches

Scalar field models, Sigma models, Skyrme-Witten models, Scalar QCD - all of these have been useful in the conceptual development of the theory, in understanding its relationship with current algebra and the previous particle physics mode based on flavor. For want of something better it was certainly a useful occupation (like the effort expanded in S-matrix theory in the sixties), though perhaps sometimes diverting too much of the prime research effort into somewhat remotely connected channels. In this conference, we did not have much of these approaches (except for Ferrando's paper on QCD_2).

The Bag Models. These served their purpose as a check on the effects of confinement. It does not seem there is much further to be done with this type of 'model' (literally).

The Gribov approach: In V. Gribov's analysis, color confinement requires more than QCD proper. He finds that two other inputs are needed: light quarks (as measured against the QCD scale) and a spontaneously broken chiral symmetry, with the light-quark made pion as Goldstone particle. In this picture, the QCD vacuum is unconfined. It might be possible to test this approach against the usual credo of confinement for color as such, whatever the matter field parameters.

Returning to our meeting, it was refreshing to listen also to experiments (Berat, Popa), and good phenomenological calculations (Rosina, Cocolicchio).

5. Acknowledgements

I think I speak for all participants in expressing our sincere thanks to Prof. Prosperi, Prof. Di Giacomo and the other organisers (not the least Dr. N. Brambilla) and members of the International Committee, in having initiated such a meeting, worked out an excellent program – and selected such a marvelous site! On another matter: at the meeting, I added to this final talk a view of Physics at the end of the XXth century, following Kelvin's of the XIXth. I have not included this material here and refer the interested reader (especially for the quotes from Kelvin) to[46].

6. References

1. S. Weinberg, *Phys. Rev.*D11 (1975) 3583.
2. H. Fritzsch and M. Gell-Mann, in *Proc. XVI ICHEP (Chicago 1972)*, v.2, p. 135; S. Weinberg, *Phys. Rev. Lett.* 31 (1973) 494.
3. L. O'Raifeartaigh, in *Mathematical Physics towards the XXI Century (Proc.*

198

Int. Conf. Beersheva 1993), R. Sen and A. Gersten, eds., Ben-Gurion Univ. of the Negev Pub. (1994).

4. J.J. Sakurai, *Ann. Phys. (NY)* **11** (1960) 1.

5. Y. Ne'eman, *Nucl. Phys.* **26** (1961) 222; M. Gell-Mann, Caltech report CTSL20 (1961) unpublished.

6. M. Gell-Mann, *Phys. Rev.* **125** (1962) 1067.

7. H. Goldberg and Y. Ne'eman, *Nuo. Cim.* **27** (1963) 1.

8. Y. Ne'eman, in *Symmetries in Physics (1600-1980)*, M.G. Doncel et al. eds., Univ. Aut. d. Barcelona and World Scientific (Singapore) Pub. (1987), pp. 501-538.

9. H.J. Lipkin, *Phys. Rep.* **8C** (1973) 173; A. Pickering, *Constructing Quarks*, Univ. of Chicago Press (1984).

10. G. Goldhaber, "The Encounter in the Bus", in *From SU(3) to Gravity*, E. Gotsman and G. Tauber, eds., Cambridge Uni. Press (1985), p. 103. M. Gell-Mann in *Proc. XI ICHEP (CERN 1962)* p. 805.

11. M. Gell-Mann, *Phys. Lett.* **8** (1963) 214. G. Zweig, CERN reports TH401,402 (unpublished) 1964.

12. F. Gursey and L.A. Radicati, *Phys. Rev. Lett.* **13** (1964) 173; B. Sakita, *Phys. Rev.* **136** (1964) B1756. G. Zweig, in *Symmetries in Elementary Particle Physics*, A. Zichichi ed., Acad.Pr. (1965), p. 192.

13. E.M. Levin and L.L. Frankfurt, *JETP Lett.* **2** (1965) 105.

14. R.P. Feynman, *Phys. Rev. Lett.* **23** (1969) 1415. J.D. Bjorken, *Phys. Rev.* **179** (1969) 1547.

15. H.J. Melosh, *Phys. Rev.* **D9** (1974) 1095.

16. M. Han and Y. Nambu, *Phys. Rev.* **B139** (1965) 1006.

17. R.J. Oakes and C.N. Yang, *Phys. Rev. Lett.* **11** (1963) 174.

18. Y. Ne'eman, *Phys. Rev.* **134** (1964) B1355.

19. S.S. Gershtein and Y.B. Zeldovich, *JETP* **29** (1955) 698.

20. Y. Ne'eman and V.T.N. Reddy, *Nucl. Phys.* **B84** (1975) 221.

21. M.L. Goldberger and S.B. Treiman, *Phys. Rev.* **110** (1958) 1178.

22. Y. Ne'eman, *Nucl Phys* **30** (1962) 347.

23. G. 't Hooft, *Nucl. Phys.* **B33** (1971) 173.

24. D.J. Gross and F. Wilszek, *Phys. Rev. Lett.* **30** (1973) 1323; H.D. Politzer, *Phys. Rev. Lett.* **30** (1973) 1346. G. 't Hooft, unpublished.

25. F.J. Yndurain, *The Theory of Quark and Gluon Interactions*, Springer Verlag, Berlin-Heidelberg etc. (2nd edition, 1993).

26. S.J. Brodsky, G. McCartor, H.C. Pauli and S.S. Pinsky, *Particle World* **3** (1993) 109.

27. K.G. Wilson, T.S. Walhout, A. Harindranath, W-M. Zhang, R.J. Perry and S.D. Glazek, preprint hep-th/9401153 (1994).

28. N. Seiberg and E. Witten, preprint hep-th/9407087 (Rutgers RU-94-52, Ins. Ad. Stu. IAS-94-43)

29. C. Vafa and E. Witten, preprint hep-th/9408074 (Harvard HUTP-94/A017, Ins. Ad. Stu. IAS-94-54)

30. N. Seiberg and E. Witten, preprint hep-th/9408099 (Rutgers RU-94-60, Ins. Ad. Stu. IAS-94-55)

31. R. Delbourgo, A. Salam and J. Strathdee, *Nuo. Cim.* **49** (1967) 593.

32. A. Salam and J. Strathdee, *Phys. Rev.* **D8** (1978) 4598.

33. F.W. Hehl, Y. Ne'eman, J. Nitzsch and P. Von der Heyde, *Phys. Lett.* **78B** (1978) 102.

34. Y. Ne'eman and Dj. Sijacki, *Ann. Phys. (NY)* **120** (1979) 292.

35. P. Caldirola, M. Pavsic, E. Recami, *Nuo. Cim.* **B48** (1978) 205.

36. Dj. Sijacki and Y. Ne'eman, *Phys. Lett.* **B247** (1990) 571.

37. Y. Ne'eman and Dj. Sijacki, *Phys. Lett.* **B276** (1992) 173.

38. Y. Ne'eman and Dj. Sijacki, Tel-Aviv University preprint TAUP N232-94.

39. Dj. Sijacki and Y. Ne'eman, *Phys. Lett.* **B250** (1990) 1.

40. Y. Ne'eman and Dj. Sijacki, *Phys. Rev.* **D37** (1988) 3267.

41. Dj. Sijacki and Y. Ne'eman, *Phys. Rev.* **D47** (1988) 4133.

42. Y. Ne'eman and Dj. Sijacki, *Intern. J. Mod. Phys.* **A2** (1987) 1655; Dj. Sijacki and Y. Ne'eman, *J. Math. Phys.* **26** (1985) 2457.

43. D.Z. Freedman, P.E. Haagensen, K. Johnson and J.I. Latorre, *"The Hidden Spatial Geometry of Non-Abelian Gauge Theories"*, unpub; M. Bauer, D.Z. Freedman and P. Haagensen, CERN preprint TH-7238/94.

44. F.A. Lunev, *Phys. Lett.* **B295** (1992) 99.

45. A. Ashtekar, *Phys. Rev. Lett.* **57** (1986) 2244; C. Rovelli and L. Smolin *Phys. Rev. Lett.* **61** (1989) 155.

46. Y. Ne'eman, in *Mathematical Physics towards the XXI Century (Proc. Int. Conf. Beersheva 1993)*, R. Sen and A. Gersten eds., Ben-Gurion University of the Negev Pub., pp. 59-73.

SECTION A

Vacuum structure and nonperturbative effects;
flux tube configurations and strings;
lattice theory and numerical simulations.

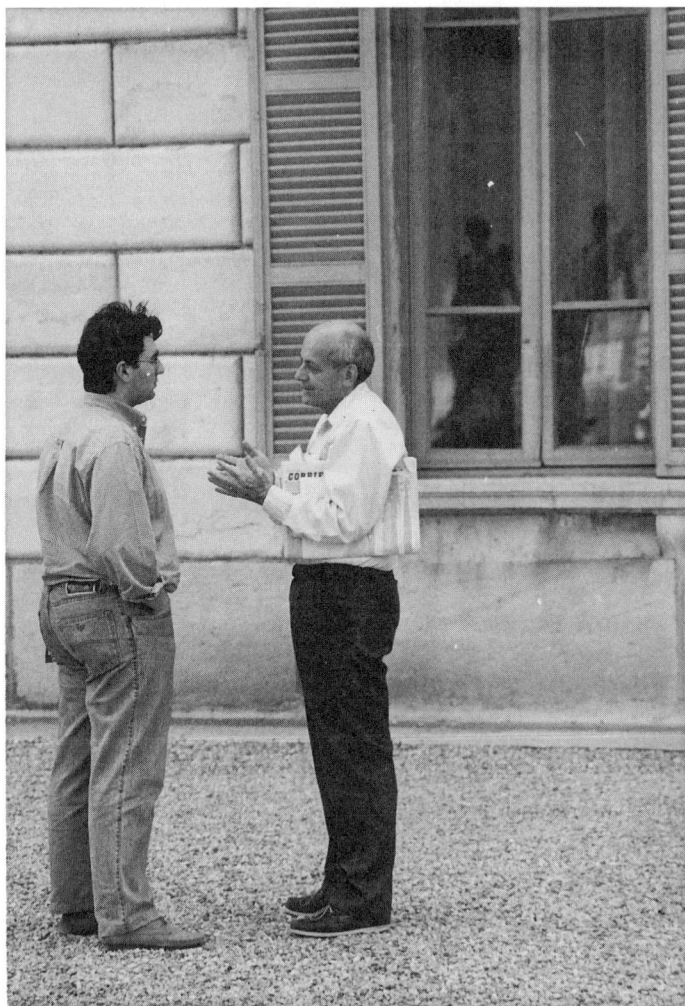

NON-PERTURBATIVE QCD TREATMENT OF
HIGH ENERGY HADRON–HADRON SCATTERING

H.G. DOSCH

Institut für Theoretische Physik der Universität Heidelberg
D69120 Heidelberg, Philosophenweg 16, Fed. Rep. Germany

ERASMO FERREIRA

Departamento de Fisica, Univ. Fed. do Rio de Janeiro
Cidade Universitária, 21919 Rio de Janeiro, Brazil

ABSTRACT

The model of the stochastic vacuum is applied in order to evaluate soft high energy hadron-hadron scattering. The parameters determined from hadron spectroscopy and lattice gauge calculations allow to calculate high energy total and elastic cross sections. The same mechanism which leads to confinement gives rise to a string-string interaction in high energy scattering.

1. The Model of the Stochastic Vacuum

First I shall shortly review a simple model for non-perturbative QCD contributions which has been introduced by Yuri Simonov and myself several years ago [1] [2] [3].

This model, called the model of the stochastic vacuum (MSV) is based on the idea that the low frequency contributions to the functional integral in quantum field theoretical Greens functions can be approximated by a simple stochastic process. Let me express that idea in a somewhat more formal language. Let

$$\prod \mathcal{D}\tilde{\phi}_i(K)e^{-S[\phi]} \tag{1}$$

be the measure of a quantum field theory with fields $\tilde{\phi}_i(K)$ in Euclidean momentum space (coordinates in capital letters indicate a Euclidean space time continuum). The measure is split into low and high frequency parts

$$\prod \mathcal{D}\tilde{\phi}_i(K) = \prod_{|K|\leq\mu} \mathcal{D}\tilde{\phi}_i(K) \prod_{|K|>\mu} \mathcal{D}\tilde{\phi}_i(K). \tag{2}$$

In an asymptotically free theory like QCD the integration over the high frequency components can be accounted for by perturbation theory. Little is known analytically about the functional measure at low frequencies. In a non-Abelian gauge theory like QCD most probably one has to introduce genuine nonperturbative ingredients. Since nature has managed to regularize the infra-red problems of QCD (after all we do observe hadrons) it is reasonable to approximate the contributions of the low frequency components of the fields by a simple stochastic process. One can show [1] [2] that already the requirement of a convergent linked cluster expansion of that process leads to linear confinement. Much more stringent, but also phenomenologically more useful is the assumption, that the low frequency contributions to QCD can be approximated reasonably well by a Gaussian stochastic process, i.e. be described by one correlator.

In order to define a gauge invariant correlator, we introduce the gluon field strength tensor $\mathbf{G}_{\mu\nu}(X, W)$ whose color content has been transported to some fixed reference point W:

$$\mathbf{G}_{\mu\nu}(X, W) = \phi^{-1}(X, W)\mathbf{G}_{\mu\nu}(X)\phi(X, W). \tag{3}$$

with

$$\phi(X, W) = P \exp[ig_s \mathbf{A}_\mu(X + \lambda(W - X))(W - X)_\mu] \tag{4}$$

and

$$\mathbf{G}_{\mu\nu}(X) = \sum_{C=1}^{N_C^2-1} G_{\mu\nu}^C(X)\tau_C \tag{5}$$

Under a local gauge transformation all the parallel transported field strength tensors transform with the gauge transformation at the point W:

$$\mathbf{G}_{\mu\nu}(X, W) \rightarrow \mathbf{U}(W)\mathbf{G}_{\mu\nu}(X, W)\mathbf{U}^*(W). \tag{6}$$

We make now the crucial asumption that the correlator of two field strength tensors does not depend on the reference point W and thus are led to the most general form of the correlator [2]:

$$< g^2 G_{\mu\nu}^C(X, W)G_{\rho\lambda}^D(X', W) >_B = \frac{\delta^{CD} < g^2 GG >}{12(N_C^2 - 1)}((\delta_{\mu\rho}\delta_{\nu\lambda} - \delta_{\mu\lambda}\delta_{\nu\rho})D(Z^2) \cdot \kappa)$$

$$+ [(\frac{1}{2}\frac{\partial}{\partial Z_\mu}(Z_\rho\delta_{\nu\lambda} - Z_\lambda\delta_{\nu\rho}) + \frac{1}{2}\frac{\partial}{\partial Z_\nu}(Z_\lambda\delta_{\mu\rho} - Z_\rho\delta_{\mu\lambda}))D_1(Z^2)(1 - \kappa)]). \tag{7}$$

where $Z = X - X'$ and $< g^2 GG >$ is the vacuum expectation value of the operator $g^2 G_{\mu\nu}(0)G_{\mu\nu}(0)$. Evaluating the expectation value of a Wegner-Wilson loop with the Gaussian measure defined by Eq. (7) we otain the area law and the string tension ρ:

$$\rho = \frac{\pi}{48N_C} < g^2 GG > \kappa \int_0^\infty D(Z^2)dZ^2. \tag{8}$$

It is important to note that the string tension is proportional to κ, the correlator $D_1(Z^2)$ of Eq. (7) does not contribute to it. In an Abelian gauge theory the homogeneous Maxwell equations

$$\partial_\rho \epsilon_{\rho\mu\nu\sigma} G_{\mu\nu} = 0 \tag{9}$$

imply for Eq. (7) that $\kappa = 0$, i.e. we have no area law (and hence no confinement) in an Abelian gauge theory without monopoles.

After the importance of the correlator $D(Z^2)$ has been stressed in ref. [2] it has been calculated on the lattice by Di Giacomo and Panagopoulos [4] using the cooling technique. Their results showed unambiguously that this tensor structure is present and even dominant with a value $\kappa \approx 0.74$.

2. Soft High Energy Scattering

The correlator in Eq. (7), specifying the Gaussian process that approximates non-perturbative effects of QCD, is the starting point for an evaluation of observables in soft high-energy scattering, i. e. high energy scattering with small momentum transfer. In this application the basic entity is the expectation value of Wegner-Wilson loops. The model of the stochastic vacuum is even pushed further, since there the leading term is an expectation value of a product of four field strength tensors, which is in the Gaussian approximation reduced to the product of two-point correlators by means of factorization

In a general analysis on soft high energy scattering Nachtmann [5] has evaluated the quark-quark scattering amplitude using the eikonal approximation for the interaction of the quarks with the gluon field. In a first step, we follow the same approach, and consider the scattering amplitude of a single quark in an external field A_μ. If the energy of the quark is very high and the background field has only a limited frequency range, the quark moves on an approximately straigth light-like line and the eikonal approximation can be applied.

Along its path Γ, the quark picks up the eikonal phase $\mathbf{V}$ (which here is a unitary $N_c \times N_c$ matrix)

$$\mathbf{V} = P \exp[-ig \int_\Gamma \mathbf{A}_\mu(z) \, dz^\mu] \ . \tag{10}$$

The phase factor for an antiquark is obtained by complex conjugation.

From the scattering amplitudes for single quarks in the background field, one obtains the non-perturbative quark-quark scattering amplitude by functional integration, with respect to the background field, of the product of the two scattering amplitudes. More specifically, consider two quarks travelling along the light-like paths Γ_1 and Γ_2 given by

$$\Gamma_1 = (x^0, \vec{b}/2, x^3 = x^0) \text{ and } \Gamma_2 = (x^0, -\vec{b}/2, x^3 = -x^0) \ , \tag{11}$$

corresponding to quarks moving with velocity of light in opposite directions, with an impact vector $\vec{b}$ in the $x^1 - x^2$-plane (referred to in the following as the transverse plane). Let $V_{1,2}(\pm\vec{b}/2)$ be the phases picked up by the quarks along these paths

$$\mathbf{V}_{1,2}(\pm\vec{b}/2) = P \exp\left[-ig \int_{\Gamma_{1,2}} \mathbf{A}_\mu(z) \, dz^\mu\right] \ . \tag{12}$$

Then according to ref. [5] the scattering amplitude for two quarks with momenta p_1, p_2 and colours c_1, c_2 leading to two quarks of momenta p_3, p_4 and colours c_3 c_4 is given by

$$T_{c_3 c_4 ; c_1 c_2}(s,t) = \bar{u}(p_3) \, \gamma^\mu(p_1) \, \bar{u}(p_2) \, \gamma_\mu u(p_2) \, \mathcal{V} \ , \tag{13}$$

where

$$\mathcal{V} = i \langle Z_\psi^{-2} \rangle_A < \int d^2 \vec{b} \, e^{-i\vec{q}\cdot\vec{b}} \{ [\mathbf{V}_1(-\frac{\vec{b}}{2}) - 1]_{c_3 c_1} [\mathbf{V}_2(+\frac{\vec{b}}{2}) - 1]_{c_4 c_2} \} >_A . \tag{14}$$

Here $<>_A$ denotes functional integration over the gluon field; $\vec{q}$ is the momentum transfer $(p_1 - p_3)$ projected on the transverse plane. Of course the approximation makes sense only if $|\vec{q}| \ll |\vec{p}|$. The quantity Z_ψ is the fermion wave-function renormalization constant in the eikonal approximation, given by

$$Z_\psi[A] = \frac{1}{N_c} \text{tr}[V^1(0)] = \frac{1}{N_c} \text{tr}[V^2(0)] . \tag{15}$$

The subtraction of the unit operator from the phase-matrices V is due to the transition from the S to the T-operator.

In the limit of high energies we have helicity conservation

$$\bar{u}(p_3) \, \gamma^\mu u(p_1) \, \bar{u}(p_4) \, \gamma_\mu u(p_2) \xrightarrow{s\to\infty} 2s \delta_{\lambda_1 \lambda_3} \delta_{\lambda_4 \lambda_2} . \tag{16}$$

where λ_i are the helicities of the quarks and $s = (p_1 + p_2)^2$. In the following we can thus ignore the spin degrees of freedom.

The scattering amplitude Eq. (13) is explicitly gauge dependent. But we know that, in meson-meson scattering, for each quark there is an antiquark moving on a nearly parallel line. Furthermore, the meson must be a colour singlet state under local gauge transformations. To construct such a colourless state we have to parallel-transport the colour content from the quark to the antiquark (or vice versa) . Since this parallel-transport of the colours is made by a Schwinger string $\phi(x_q, x_{\bar{q}})$ (see Eq. (4)), we obtain for the meson a Wilson loop whose light-like sides are formed by the quark and antiquark paths, and front ends by the Schwinger strings (see Fig. 1). The direction of the path of an antiquark is effectively the opposite of that of a quark, so that the loop has a well defined internal direction. The resulting loop-loop amplitude is now specified not only by the impact parameter, but also by the transverse extension vectors $\vec{R}_i$ of the loops. We thus introduce the loop-loop scattering amplitude

$$J(\vec{b}, \vec{R}_1, \vec{R}_2) = \frac{1}{< \frac{1}{N_c} \text{tr} \, W_1(0, \vec{R}_1) >_A < \frac{1}{N_c} \text{tr} \, W_2(0, \vec{R}_2) >_A}$$

$$< \frac{1}{N_c} \text{tr}[W_2(-\frac{\vec{b}}{2}, \vec{R}_2) - 1] \frac{1}{N_c} \text{tr}[W_1(-\frac{\vec{b}}{2}, \vec{R}_1) - 1] >_A , \tag{17}$$

where the closed loop ∂S_1 of the W-loop $W_1(-\vec{b}/2, \vec{R}_1)$ is a rectangle whose long sides are formed by the quark path $\Gamma_1^q = (x_0, \vec{b}/2 + \vec{R}_1, x_3 = x_0)$ and the antiquark path $\Gamma_1^{\bar{q}} = (x_0, \vec{b}/2 - \vec{R}_1, x_3 = x_0)$ and whose front sides are formed by lines from $(T, \vec{b}/2 + \vec{R}_1, T)$ to $(T, \vec{b}/2 - \vec{R}_1, T)$ for large positive and negative T (we will then take the limit $T \to \pm\infty$). $W_2(\vec{b}/2, \vec{R}_2)$ is constructed analogously. The first factor

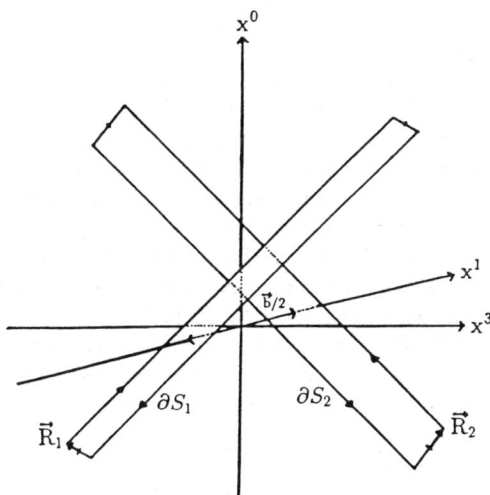

Fig. 1. Wegner-Wilson loops formed by the paths of quarks and antiquarks inside two mesons. The impact vector $\vec{b}$ is the distance vector between the middle lines of the two loops. $\vec{R}_1$ and $\vec{R}_2$ are the vectors in the transverse plane from the middle lines to the quark lines of meson 1 and 2 respectively. The front lines of the loops guarantee that the mesons behave as singlets under local gauge transformations.

in Eq. (17) is the loop renormalization that replaces the quark field renormalization in Eq. (14).

Our next aim is to perform the functional integration over the gluon field by applying the model of the stochastic vacuum discussed in section 1. Since the correlator is given in terms of the parallel-transported field tensor $F_{\mu\nu}(x, w)$, we have first to express the line integrals $\int A_\mu dz^\mu$ through integrals over the field tensor. This is done with the help of the non-Abelian Stokes-theorem [6][3]. Since the reference point W in the correlator Eq. (7) must be the same for both fields we have to choose a common reference point for both traces in the product

$$< \text{tr}[W_1(-\frac{\vec{b}}{2}, \vec{R}_1) - 1] \; \text{tr} \; [W_2(\frac{\vec{b}}{2}, \vec{R}_2) - 1] >_A \; . \tag{18}$$

We choose the point W in the most symmetric way and then the surface emerging from the loop ∂S_1 is formed by the sliding sides of a pyramid with the loop ∂S_1 as basis and the point with coordinates w as apex; the same holds for ∂S_2.

Before it can be applied to high-energy scattering, the model of the stochastic vacuum must be translated from Euclidean space-time, in which it is naturally formulated, to the Minkowski continuum. Unfortunately we cannot go the other way and continue Eq. (17) to the Euclidean continuum, which would be the safe way from the point of view of the functional integration. However the Wilson loops occurring in Eq. (17) have light-like sides which would shrink to a point if continued to a space time continuum with Euclidean metric. We think that this is a serious obstacle for all attempts to evaluate soft high-energy amplitudes numerically on a lattice.

Since we cannot adapt the scattering amplitude to the Euclidean world, we have to go the other way and adapt the model of the stochastic vacuum to the Minkowski world. We are fully aware that this is by no means a trivial step and, pending a better analytical understanding of non-perturbative effects, we have to let the experiment decide on its justification. Thus we must translate the correlation function (7) to the Minkowski world. This is obvious for the tensor structure, where we just substitute $\delta_{\mu\nu}$ by $-g_{\mu\nu}$, etc, but simple choices for the correlation functions like $\exp(-z^2/a^2)$ or $\exp(-\sqrt{z^2}/a)$ cannot be continued in a meaningful way. Therefore, we must look for correlation functions $D(z^2/a^2)$, $D_1(z^2/a^2)$ which fall off for negative values of z^2 (corresponding to Euclidean distances), and whose Fourier transforms exist in Minkowski metric, since these will enter essentially in the scattering amplitudes. We write for the correlator, in terms of the Fourier transforms

$$\langle g^2 G_{\mu\nu}^A(x,w)\, G_{\rho\sigma}^B(y,w)\rangle_A = \frac{\delta^{AB}}{N_c^2 - 1}\,\frac{1}{12}\langle g^2 GG\rangle$$

$$\cdot \int \frac{d^4 k}{(2\pi)^4}\, e^{-ik\frac{x-y}{\lambda}}\{(g_{\mu\nu}g_{\nu\sigma} - g_{\mu\sigma}g_{\nu\rho})\,\kappa i\widetilde{D}(k^2)$$

$$+\ (-g_{\nu\sigma}k_\mu k_\rho + g_{\nu\rho}k_\mu k_\sigma - g_{\mu\rho}k_\nu k_\sigma + g_{\mu\sigma}k_\nu k_\rho)(1 - \kappa)\,i\widetilde{D}_1(k^2)\}\,, \qquad (19)$$

where

$$\widetilde{D}(k^2) = \int d^4 z D(z^2/\lambda^2)\, e^{ikz/\lambda} \quad \text{and} \quad i\widetilde{D}_1(k^2) = \frac{d}{d(k^2)}\int d^4 z D_1(z^2/\lambda^2)\, e^{ikz/\lambda}. \qquad (20)$$

After this choice is made, all functional integrations can be performed, in principle. The quantities $W_{1,2}$ in Eqs. (17) can be expressed as surface integrals by applying the non-Abelian Stokes theorem. The exponential being expanded, the expectation value can be calculated according to Eq. (19) and the factorization in a Gaussian process. I shall not describe the details of the rather tedious evaluation but rather quote some of the principal results [7] [8] [9]. In an Abelian model (i.e. $\kappa = 0$, only the structure D_1 present) the "scattering amplitude" $J(b, R_1, R_2)$ is independent of R_1 and R_2 as soon as $|R_i| \gg \lambda$, the correlation length of the correlator. The typically non-Abelian structure D, however, leads to an amplitude which increases with R_1 and R_2 even if $|R_i| \gg \lambda$. This can be interpreted in the following way:

In an Abelian theory we have only a quark-(anti)quark interaction, whereas in a non-Abelian theory the same correlator which causes confinement gives also rise to a (non-perturbative) string-string interaction.

From the string-string amplitude one constructs a hadron-hadron scattering amplitude by averaging over a transversal wave function $\psi_H(\vec{R})$, which for a meson is simply choosen as a Gaussian

$$\psi_H(\vec{R}) = \sqrt{\frac{2}{\pi}}\frac{1}{S_H}e^{-R^2/S_H^2}$$

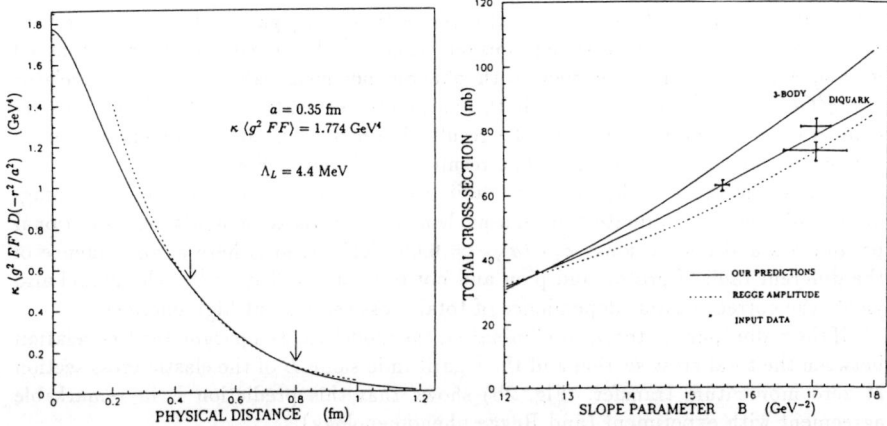

Fig. 2. a) Correlator used in the model of the stochastic vacuum for high energy scattering [8] (solid line) and the one from the lattice calculation [4] (dashed line) with $\Lambda_{Lattice} = 4.4$ MeV. The arrows indicate the range inside which lattice data exist. b)Relation between the total cross section and B, the logarithmic slope of the elastic cross section at zero momentum transfer for proton-antiproton scattering. Stars are experimental results [11]. Solid line: prediction of ref. [8]. Dashed line: relation obtained from the Regge amplitude [12] with Pomeron intercept $\alpha(0) = 1.0808$ and slope $\alpha'(0) = 0.25$.

For the confining correlator the following form is chosen [7] [9]

$$D((z^2/\lambda^2)) = \int \frac{d^4k}{(2\pi)^4} e^{ikz/\lambda} \tilde{D}(k^2)\tilde{D}(k^2) = \frac{27\pi^4}{4 \cdot \left(k^2 - \frac{9\pi^2}{64}\right)^4}$$

A consistent choice of the parameters of the model, the gluon condensate and the correlation length and the hadron radius is [9]:

$$\lambda = 0.35 \text{ fm}, \kappa < g^2 GG >= 1.7 \text{ GeV}^4, \; S_P = 0.84 \text{ fm}. \tag{21}$$

These parameters yield the right total cross section and logarithmic slope of proton-(anti-)proton scattering, a string tension (slope of the linear confining potential) $\sigma = 0.18 \text{ GeV}^2$ in full agreement with phenomenological values and the correlator $\kappa < g^2 GG > D(Z^2)$ obtained with the parameter set (21) is in the Euclidean region in excellent agreement with the result obtained from lattice calculations from DiGiacomo and Panagopoulos [4]. The result is displayed in Fig.2a.

Taking the ratio of the hadronic radii S_H equal to the ratio of the electromagnetic radii one can evaluate the pion-nucleon total cross section without additional parameters and one obtains: $\sigma_{\pi N}/\sigma_{NN} = 0.66$. This ratio is here a consequence of the different radii of proton and pion and not of quark additivity [10]. The model also yields the correct flavour dependence of total cross sections at high energies [9].

If the radius parameter S_H is eliminated, the model yields a parameter-free relation between the total cross section and the logarithmic slope B of the elastic cross section at zero momentum transfer. Fig. 2b) shows that this prediction is in remarkable agreement with experiment (and Regge phenomenology).

Concluding one can remark, that the model of the stochastic vacuum does not only yield a good description of soft high energy scattering, but also relates the parameters of hadron spectroscopy to high energy scattering and its fundamental ingredient, the correlator Eq. (19), is in excellent agreement with results from lattice gauge theory.

3. Acknowledgements

It i a pleasure to thank the organizers of the conference and especially Dr. Nora Brambilla for creating such a nice and stimulating atmosphere.

1. Dosch, H.G., *Phys. Lett.* **B190** (1987) 555
2. Dosch, H.G., Yu.A. Simonov, *Phys. Lett.* **B205** (1988) 339
3. Simonov, Yu. A. *Yad. Fiz.* **48** (1988) 1381 and **50** (1988) 213
4. DiGiacomo, A., H. Panagopoulos, *Phys. Lett.* **B285** (1992) 133
5. Nachtmann, O., *Ann. Phys. N.Y.* **209** (1991) 436
6. Bralic, N.E. *Phys. Rev.* **D22** (1980) 3090
7. Krämer,A., and H.G.Dosch, *Phys. Lett.* **B272** (1991) 114
8. Dosch, H.G., E.Ferreira, *Phys. Lett.* B 318 (1993) 197
9. Dosch, H.G., E.Ferreira, A. Krämer, *Phys. Rev.* **D**(1994), in print
10. E.M.Levin and L.L.Frankfurt, *JETP Lett.* **2** (1965) 65

11. see contributions of A. Bueno, R. Rubin and P. Giromini in **H.M. Fried** *et al.* (Ed.) *ELASTIC AND DIFFRACTIVE SCATTERING* (Vth Blois Workshop), Singapore 1994

12. A.Donnachie and P.V.Landshoff, *Phys. Lett.* **B296** (1992) 227

A LATTICE STUDY OF THE GLUON PROPAGATOR, IN THE LANDAU GAUGE[*]

P. MARENZONI[a], G. MARTINELLI[b], N. STELLA[c], M. TESTA[b]

[a] Dip. di Ingegneria dell'Informazione
Università di Parma, Viale delle Scienze, 43100 Parma, Italy
[b] Dip. di Fisica, Università degli Studi di Roma "La Sapienza" and
INFN, Sezione di Roma, P.le A. Moro 2, 00185 Rome, Italy.
[c] Physics Department, "The University", S09 5NH Highfield, Southampton, U.K.

ABSTRACT

We present the results of two high-statistics studies of the gluon propagator in the Landau gauge, at $\beta = 6.0$, on different lattice volumes. The dependence of the propagator on the momenta is well described by the expression $G(k^2) = \left[M^2 + Z \cdot k^2 (k^2/\Lambda^2)^\eta \right]^{-1}$. We obtain a precise determination of $\eta = 0.532(12)$, and verify that M^2 does not vanish in the infinite volume limit.

The non-perturbative investigation of the behavior of the basic fields of the QCD Lagrangian is crucial to shed light on the mechanism of confinement and can be achieved through numerical lattice computations of the Gluon Propagator[1,2] [GP]. The Euclidean GP in the Landau Gauge is:

$$D_{\mu\nu}(k) = \int d^4 x \text{Tr}\langle A_\mu(x) A_\nu(0) \rangle e^{-ikx} = G(k^2) \left(\delta_{\mu\nu} - \frac{k_\mu k_\nu}{k^2} \right), \qquad (1)$$

where $\mu, \nu = 1, \ldots, 4$ and the trace is intended over color indices.

Recently, there has been much effort in trying to obtain a non perturbative form for $G(k^2)$, both from analytic[3-7] and numerical[8-11] analyses. With the present studies, we investigate the non-perturbative form of $G(k^2)$ and its behaviour in the infinite volume limit.

In tab.1, the parameters of our simulations are summarized. The gauge fields have been generated with a Hybrid Monte Carlo algorithm[12]. The Landau gauge-fixing has been performed and checked carefully, being this a crucial point when dealing with gauge-dependent quantities. The fluctuation left-over after gauge-fixing $\langle \partial_\mu A_\mu(x) \rangle_{\text{Latt}} \leq 10^{-6}$, are absolutely negligible[1], with respect to the statistical errors. This can be checked since the condition $\partial_\mu A_\mu(x) = 0$ implies $\partial_t A_0(\vec{0}, t) = 0$. One can define the correlation $\langle A_0(t) A_0(0) \rangle$ at zero momentum, and study its time derivative. We have shown[1,2] that it is zero within errors, and the gluon field $A_0(\vec{0}, t)$ is constant at the level of 0.008% on each individual configuration. These results demonstrate that the present gauge-fixing procedure is the most effective, among those implemented in the literature[9-11].

Accordingly to eqn.1, we compute 2-pt functions of the gluon field, defined in term of the link variable $U_\mu(x)$ as $A_\mu(x) = [U_\mu(x) - U_\mu^\dagger(x)]/2i$, which, using spectral

---

[*]Talk presented by N.Stella.

β	# confs.	Volume	$\partial_\mu A_\mu$
6.0	1000	$16^3 \times 32$	$< 10^{-6}$
6.0	500	$24^3 \times 48$	$< 10^{-6}$

Table 1: Summary of the parameters of our simulations.

decomposition and translation invariance, can be written as

$$D(t, \vec{k}) = \sum_{\vec{x}} \mathrm{Tr}\langle A_j(x) A_j(0)\rangle e^{i\vec{k}\cdot\vec{x}} = \sum_{|i\rangle} \frac{|\langle A_j(0)|i\rangle|^2}{\mathcal{N}_i} e^{-E_i t}, \qquad j = 1, \ldots, 3, \qquad (2)$$

where the sum is over the states which couple to the gluon field and E_i is the energy of the state $|i\rangle$. From eq. (2), the effective energy, defined as

$$\omega_{\mathrm{eff}}(t, \vec{k}) = \log \frac{D(t, \vec{k})}{D(t+1, \vec{k})}. \qquad (3)$$

should be a decreasing function of the time, for any value of the momentum $\vec{k}$. In fig. 1, we show that $\omega_{\mathrm{eff}}(t, \vec{k})$ is increasing with time, for all the momentum combinations considered. Hence, it is impossible[1,8] to fit the GP to a sum of single particle pole function, neither if physical states ($\mathcal{N}_i > 0$) nor "ghost" ($\mathcal{N}_i < 0$) are considered. This is an unacceptable feature for the propagator of a physical particle. To avoid systematic uncertainties due to the presence of infrared and ultraviolet cut-offs on the lattice, one can study the GP in the intermediate region of momenta. We find that $G(k^2)$ is well described by the following modeling function

$$G(k^2) = \frac{1}{M^2 + Z k^2 \left(\frac{k^2}{\Lambda^2}\right)^\eta} \xrightarrow{\text{on the lattice}} \frac{1}{M_L^2 + Z_L (k^2)^{1+\eta}}, \qquad (4)$$

which depends on the parameters η, M_L^2, and Z_L. The stability and the quality of the fits have been checked in different ways[1,2]. On the two volumes, we find (see fig. 2)

$$V = 16^3 \times 32 \begin{cases} M_L^2 = 2.8(1) \times 10^{-3} \\ Z_L = 9.01(4) \times 10^{-2} \\ \eta = 0.56(6) \\ \chi^2_{\mathrm{ndof}} = 1.5 \end{cases} \qquad V = 24^3 \times 48 \begin{cases} V = 24^3 \times 48 \\ M_L^2 = 4.46(9) \times 10^{-3} \\ Z_L = 0.102(1) \\ \eta = 0.532(12) \\ \chi^2_{\mathrm{ndof}} = 1.08 \end{cases}$$

$$(5)$$

We observe that the finite volume has a very small effect in the value of the anomalous dimension, provided that the range of k^2 is large enough. Indeed, we obtain a fairly accurate determination of η. The two determinations of M^2 are inconsistent, and M^2 increases with the volume. This feature rules out the hypothesis that the non-zero value of M^2 is merely due to finite volume effects. If this were the case, we would expect M^2 to scale roughly as $1/L^2$. It is possible to try a first, very crude extrapolation of the value of M^2 to the infinite volume limit. Using

$$M^2(V) = M^2(V = \infty) + \mathrm{cost} \frac{1}{\sqrt{V}} \qquad \text{we get} \quad M^2(V = \infty) = 6.202(8) \times 10^{-3}. \qquad (6)$$

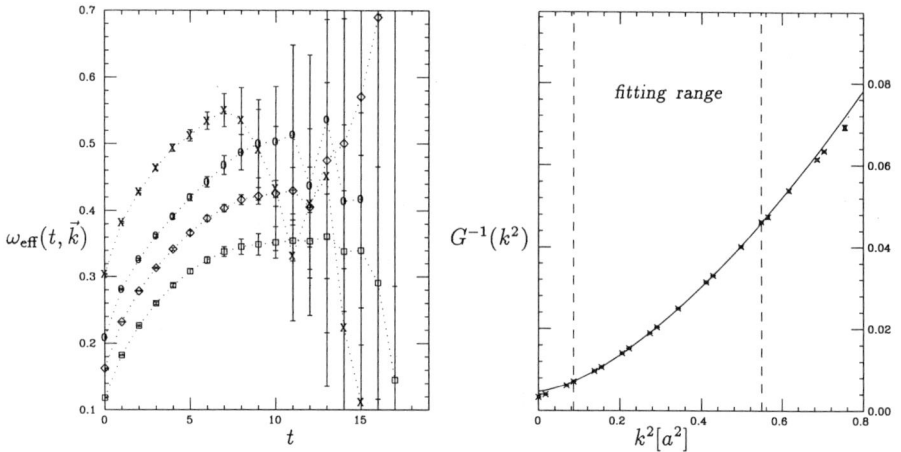

Figure 1: Effective energy for gluon 2-pt functions. The four curves correspond to the following momenta: $\Box : \vec{k} = \vec{0}$, $\Diamond : \vec{k} = (2\pi/24, 0, 0)$, x: $\vec{k} = (2\pi/24, 2\pi/24, 0)$ and $0 : \vec{k} = (4\pi/24, 0, 0)$. Figure 2: Best fit of the propagator in momentum space to the function 5. The figure corresponds to the case $V = 24^3 \times 48$.

Using $a^{-1} \sim 2\text{GeV}$, as determined by several simulations at $\beta = 6.0$, we obtain

$$M_{\text{phys}}^2 \simeq (160 \text{ MeV})^2 \simeq (\Lambda_{QCD})^2. \tag{7}$$

An interpretation of this result, in connection with colour confinement is, at present, absent. However, we stress that it is fundamental to understand the behaviour of both η and M^2 in the continuum limit.

Acknowledgments

We are indebted to the Thinking Machines Corporation for allowing us to perform these simulations. G.M. and M.T. acknowledge the partial support os MURST, Italy. N.S. thanks the Noopolis-Sovena Foundation for financial support.

1. P.Marenzoni, G.Martinelli, N.Stella, M.Testa, *Phys.Lett.* **B318** (93) 511.
2. P.Marenzoni, G.Martinelli, N.Stella, *SHEP prep.* 93/94-31.
3. V.N. Gribov, *Nucl.Phys.* **B139** (78) 19.
4. N.Brown, M.Pennington, *Phys.Rev.* **D38** (88) 2266.
5. N.Brown, M.Pennington, *Phys.Rev.* **D39** (89) 2723.
6. M.Stingl, *Phys.Rev.* **D34** (86) 3863.
7. D.Zwanziger, *Nucl.Phys.* **B378** (92) 525.
8. J.E. Mandula, M. Ogilvie, *Phys.Lett.* **B185** (87) 127.
9. R.Gupta *et al*, *Phys.Rev.* **D36** (87) 2813.
10. C.Bernard, C. Parrinello, A. Soni, *Nucl.Phys.* *B30* **Proc. Suppl.** (92) 535.
11. C.Bernard, C.Parrinello, A. Soni, *Phys.Rev.* **D49** (94) 1585.
12. P.Marenzoni, G.Pugnetti, P.Rossi, *Phys.Lett.* **B315** (93) 152.

THE RUNNING QCD COUPLING
FROM THE LATTICE TRIPLE GLUON VERTEX

DAVID HENTY and CLAUDIO PARRINELLO

Department of Physics and Astronomy, University of Edinburgh
Edinburgh EH9 3JZ, Scotland

ABSTRACT

We give preliminary results on a lattice determination of the QCD running coupling. The coupling is computed by calculating the relevant renormalization constants from the lattice three-gluon vertex and gluon propagator. For momenta larger than 2 GeV, the coupling is found to run according to the 2-loop asymptotic formula. In our preliminary results the influence of lattice artifacts appears negligible. Further work is in progress to fully investigate this point.

1. Introduction

The running coupling $\alpha_s(q)$, where q is a momentum scale, is a fundamental QCD quantity providing the link between the low and high-energy properties of the theory. Given a renormalization scheme, $\alpha_s(q)$ can be measured experimentally for a wide range of momenta. A precise determination of $\alpha_s(q)$ (or equivalently of the scale Λ determining the rate at which α_s runs) is very important as it would fix the values of various parameters in the Standard Model, providing bounds on new physics.

Given the difficulty of the experimental measurements, computing the QCD running coupling is a major challenge for the lattice community. Many different methods have been devised in this respect, [1,2,3,4], but our method differs by adopting a definition of the coupling which directly involves the fundamental QCD degrees of freedom.

2. The Method

The method [5] aims to compute the renormalized running coupling $g(q)$ (where $\alpha = g^2/4\pi$) by calculating the three-gluon vertex function $\Gamma^{(3)}_{\alpha\beta\gamma}(q)$, and the gluon propagator $G^{(2)}_{\mu\nu}(q)$, in a fixed gauge on a lattice with spacing a. The vertex renormalization constant, $Z_V(qa)$, can be defined at any fixed momentum scale q by choosing appropriate kinematics to project out the part of Γ proportional to the tree-level vertex. In particular, one evaluates $\Gamma^{(3)}_{\alpha\beta\gamma}$ at the asymmetric point defined by

$$\alpha = \gamma \neq \beta, \qquad p_1 = -p_3 = p_\beta, \quad p_2 = 0. \tag{1}$$

Here, $p_\beta = q\,\hat{e}_\beta$, where $\hat{e}_\beta$ is the unit vector in the β direction. In the limit $a \to 0$ it can be written as

$$\Gamma^{(3)}_{\alpha\beta\alpha}(-p_\beta, 0, p_\beta) = 2\, Z_V^{-1}(qa)\, g_0(a)\, q. \tag{2}$$

Analogously, the wavefunction renormalization constant $Z_A(qa)$ is defined from the gluon propagator by

$$G_{\mu\nu}^{(2)}(q) = T_{\mu\nu}(q)Z_A(qa)\frac{1}{q^2}, \tag{3}$$

where $T_{\mu\nu}$ is the usual transverse projector (we choose to work in Landau gauge). The renormalized coupling is finally obtained via

$$g(q) = Z_A^{3/2}(qa)\, Z_V^{-1}(qa)\, g_0(a). \tag{4}$$

3. Computational Procedure

We have generated three data sets, each comprising 150 gluon configurations, on a 16^4 lattice. Each set corresponds to a different value of the parameter $\beta = 6/g_0^2(a)$. We choose $\beta = 6.0$, 6.2 and 6.4, corresponding to an ultraviolet cutoff $a^{-1} = 1.94$, 2.72 and 3.62 GeV respectively [2].

As an initial step, we evaluate the gluon propagator in momentum space and compute Z_A from Eq. (3). Next, we compute the complete three-point function $G^{(3)}$ of the gluon field in momentum space. We then obtain Γ, the amputated vertex function, by dividing $G^{(3)}$ by the propagators corresponding to the momenta on the external legs. This yields Z_V from Eq. (2). Finally, $g(q)$ is obtained from Eq. (4).

All the above quantities are calculated after numerically transforming the gauge fields to Landau gauge. To achieve a good signal to noise ratio, we find it necessary to reduce $\partial_\mu A_\mu$ to zero to the level of machine precision. To the same end, we also fully exploit the translational invariance and symmetries of Eq. (2).

4. Results

We plot the running coupling versus momentum in Fig.(1). It is pleasing to note that we obtain a very clear signal despite the complicated numerical procedure. The important physical question is whether there exists a range of momenta for which the coupling that we measure runs according to the 2-loop perturbative expression

$$g^2(q) = \left[b_0\ln(q^2/\Lambda^2) + \frac{b_1}{b_0}\ln\ln(q^2/\Lambda^2)\right]^{-1}. \tag{5}$$

where $b_0 = 11/16\pi^2$, $b_1 = 102/(16\pi^2)^2$ and Λ is the QCD scale parameter (in this study we are neglecting sea-quark effects). To answer this question, and to obtain an estimate for Λ, we compute Λ as a function of the measured values of $g(q)$ according to the formula

$$\Lambda = q\exp\left(-\frac{1}{2b_0g^2(q)}\right)\left[b_0g^2(q)\right]^{-\frac{b_1}{2b_0^2}}. \tag{6}$$

If the coupling is running according to Eq. (5), then we expect Λ defined from the above equation to be constant. We plot Λ versus q in Fig.(1), and see that for $q > 2$

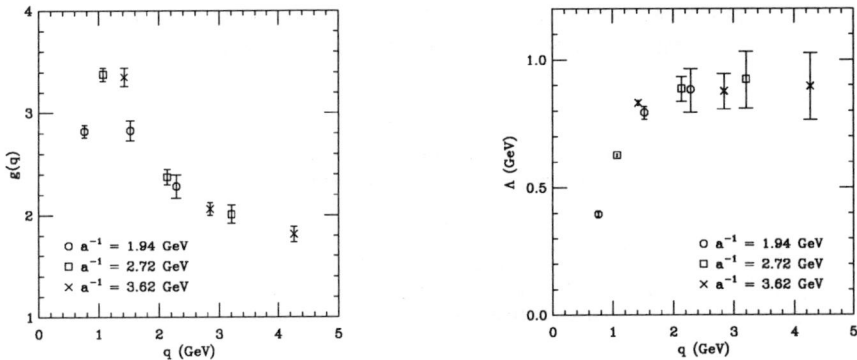

Figure 1: Running coupling (left) and Λ parameter (right) vs. momentum.

GeV the data are consistent with a constant value for the QCD scale parameter. As this range of q contains results obtained with different values of a^{-1}, this provides the important check that this value for Λ is independent of the ultraviolet cutoff. Within errors, our data appear to be free of lattice artifacts.

5. Conclusions

We have shown that a non-perturbative determination of the QCD running coupling can be obtained from first principles by a lattice study of the triple gluon vertex. Our numerical results have to be further checked for the occurrence of lattice artifacts. Eventually, our values for the coupling $g(q)$ (or equivalently our value for Λ) will be converted to predictions for the corresponding quantities in the $\overline{MS}$ scheme via a perturbative matching.

6. Acknowledgements

This work was carried out within the UKQCD collaboration on the Connection Machine 200 at the University of Edinburgh. The authors acknowledge the support of PPARC through grant GR/J 21347 (CP) and a Personal Fellowship (DH).

7. References

1. S. P. Booth et al., *Phys. Lett.* **B294** (1992) 385.
2. G. S. Bali and K. Schilling, *Phys. Rev.* **D47** (1993) 661.
3. A. X. El-Khadra et al., *Phys. Rev. Lett.* **69** (1992) 729.
4. M. Lüscher et al., *Nucl. Phys.* **B413** (1994) 481.
5. C. Parrinello, preprint 94/543, University of Edinburgh, to appear in Phys. Rev. **D** (Rapid Communications).

SPONTANEOUS CHIRAL SYMMETRY BREAKING AND
THE MESON CLOUD WITHIN LATTICE QCD

M. FABER, M. SCHALER

Institut für Kernphysik, Technische Universität Wien
Wiedner Hauptstr. 8-10, A-1040 Vienna, Austria

and

H. GAUSTERER

Institut für Theoretische Physik, Universität Graz
Universitätsplatz 5, A-8010 Graz, Austria

ABSTRACT

We investigate chiral symmetry breaking locally around a static quark in the frame-work of lattice QCD by calculating the correlation of the Polyakov loop with the chiral order parameter. We observe a restoration of chiral symmetry in the vicinity of the color source with a different mechnism in both phases of QCD. In the hadronic phase the chiral condensate is screened by a pion cloud, and the effect is more pronounced for lighter dynamical quarks. Above the deconfinement phase transition we find that the Polyakov loop is surrounded by a cloud of scalar $\sigma-$mesons.

1. Theory

To investigate the breaking of chiral symmetry locally around static quarks we calculate the chiral condensate in a system containing a static source at finite temperature. The Polyakov loop $L(\vec{r})$ describes the time propagation of a static quark [1]. We obtain insight into the local chiral structure in the vicinity of the static quark by computing the correlations between the Polyakov loop L and the order parameter of chiral symmetry $\bar{\psi}\psi$ in the form $< L(0)\bar{\psi}\psi(\vec{r}) >$. This observable can be written as a path-integral

$$< L(0)\bar{\psi}\psi(\vec{r}) >= \frac{\int \mathcal{D}[U]\mathcal{D}[\bar{\psi},\psi]\, L(0)\bar{\psi}\psi(\vec{r})\, e^{-S(U,\bar{\psi},\psi)}}{\int \mathcal{D}[U]\mathcal{D}[\bar{\psi},\psi]\, e^{-S(U,\bar{\psi},\psi)}}. \tag{1}$$

The functional integration extends over all degrees of freedom of the gauge fields U and the fermionic fields $\bar{\psi}$ and ψ. The integral (1) is calculated numerically in a Monte Carlo simulation. For the lattice regularized action of the gauge fields we use Wilson's plaquette action. As far as the fermionic degrees of freedom are concerned we employ the Kogut-Susskind formulation. This method retains for $m = 0$ a chiral $U(1)_V \otimes \bar{U}(1)_A$ symmetry of the chiral flavor group $SU(n_f)_V \otimes SU(n_f)_A$ in continuum space-time. The spontaneous breaking of the $\bar{U}(1)_A$ symmetry leads to the appearance of a pseudoscalar Goldstone boson [2].

Table 1. Amplitude α, screening mass μ and asymptotic constant γ for the chiral condensate around a static quark according to Eq. 2 for various dynamical quark masses m.

m	α	μ	γ
$\beta = 4.9$			
0.1	0.117	1.806	0.999
0.05	0.108	1.322	0.999
0.025	0.102	0.947	0.997
0.0125	0.118	0.863	0.998
$\beta = 5.2$			
0.1	0.047	0.924	0.998
0.05	0.052	1.087	0.998
0.025	0.067	1.323	0.998
0.0125	0.066	1.321	0.998

2. Numerical Simulations

The system was simulated on an $8^3 \times 4$ lattice with inverse gluonic coupling constants $\beta = 4.9$ below the deconfinement phase transition and $\beta = 5.2$ for the system in the quark-gluon plasma phase. We used the Hybrid Monte Carlo algorithm. The quark mass was decreased from $ma = 0.1$ to $ma = 0.0125$ in both phases, and the number of flavors in the Kogut-Susskind action was $n_f = 4$. The lattice constant a corresponds to approximately 0.3 fm giving for the lowest quark mass $m \approx 8$ MeV.

As a result, we observe that the chiral condensate is lowered in the vicinity of a static source. This suppression decreases with increasing quark mass in the hadronic phase, whereas the effect is nearly mass independent and less pronounced above the deconfinement phase transition [3]. Taking $\bar\psi\psi$ as the order parameter of chiral symmetry breaking we find that chiral symmetry is locally restored around static quarks, which is in agreement with the basic ideas of bag models.

The data has been fitted with an exponential function taking lattice anisotropies into account

$$\frac{< L(0)\bar\psi\psi(\vec{r}) >}{< L >< \bar\psi\psi >} = \gamma - \alpha\, e^{-\mu r}, \qquad (2)$$

and the systematics extracted from our analysis is summarized in Table 1. In the chirally broken phase we obtain an increase of the screening mass μ with higher quark mass. We see a linear behaviour $\mu^2 \sim m$ for $m \geq 0.025$, and therefore the Gell-Mann$-$Oakes$-$Renner relation $m_\pi^2 \sim m < \bar\psi\psi >$ with the pion mass m_π tells us that the chiral condensate around a static quark source is screened by pions. For the lightest dynamical quarks ($m = 0.0125$) we are faced with the finite-size effect of the pion as the Goldstone boson of chiral symmetry breaking [4]. These observations suggest that the Polyakov loop L is surrounded by a pion cloud, and it interacts with the scalar density $\bar{\psi}\psi$ via two-pion exchange. The appearance of a two-pion exchange reflects the fact that $\bar{\psi}\psi$ is a scalar operator, which can only be saturated by two pseudoscalar pions in order to fulfill parity conservation.

The correlation $< L(0)\bar{\psi}\psi(r) >$ measures the screening of the Polyakov loop by a pion cloud because a Polyakov loop represents an infinitely heavy quark Q in a color triplet state. Due to the $Z(3)$-symmetry of the QCD vacuum in the confined phase a single color source has infinite energy and therefore cannot exist. The fermionic determinant provides the possibilities to screen the color field of the infinitely heavy quark Q by means of a light antiquark $\bar{q}$ or two light quarks qq [5]. On the other hand, $\bar{\psi}\psi$ represents a colorless operator, which can create or destroy a particle with the quantum numbers of the σ-meson. The energetically most favorable interaction between the two colorless states (the heavy-light system and the chiral condensate) is via the exchange of almost massless Goldstone bosons.

Above the deconfinement phase transition ($\beta = 5.2$) chiral symmetry is realized through parity doubling, and the pion and its scalar parity partner, the σ−meson, are degenerate. Our data in Table 1 and recent lattice simulations [6] lead to the conclusion that in the plasma phase $\bar{\psi}\psi$ interacts with L via the exchange of a scalar σ−meson [3]. This screening mechanism includes parity conservation trivially.

3. Conclusions

We have shown that for light dynamical quarks the chiral condensate is suppressed in the vicinity of a static quark, whereas the screening mechanism is different in the confinement and the deconfinement phase. In the hadronic phase the static quark is surrounded by a pion cloud, and the chiral condensate is screened due to two-pion exchange. In the quark-gluon plasma phase chiral symmetry is realized through parity doubling of the hadron spectrum. The chiral condensate and the Polyakov loop interact via σ−exchange, and the local restoration of chiral symmetry is only weakly mass dependent.

References

1. L. D. McLerran and B. Svetitsky, *Phys. Rev.* **D24** (1981) 450.
2. J. B. Kogut, M. Stone, H. W. Wyld, S. H. Shenker, J. Shigemitsu and D. K. Sinclair, *Nucl. Phys.* **B225** (1983) 326.
3. M. Faber, H. Gausterer and M. Schaler, *Phys. Lett.* **B317** (1993) 1021.
4. T. Jolicoeur and A. Morel, *Nucl. Phys.* **B262** (1985) 627.
5. M. Faber, O. Borisenko, S. Mashkevich and G. Zinovjev, *Fresh Look on Triality*, TU Vienna preprint IK-TUW-Preprint 9308401 (1993).
6. A. Gocksch, *Phys. Rev. Lett.* **67** (1991) 1701.

FLUX TUBES AND TOPOLOGICAL CHARGES
IN FINITE-TEMPERATURE LATTICE QCD[†]

M. FABER, H. MARKUM, Š. OLEJNÍK* and W. SAKULER

Institut für Kernphysik, Technische Universität Wien, A-1040 Vienna, Austria

ABSTRACT

Topological properties of the vacuum were investigated in finite-temperature lattice QCD without and with dynamical quarks using a local operator of topological charge density and a variant of cooling. Below the deconfinement temperature evidence for flux tube formation was observed in the distribution of topological charge density around a static $Q\bar{Q}$ pair. With dynamical quarks, the flux tube breaks when the distance between the static quark and antiquark increases.

1. Introduction

Many phenomenological models of strong interactions motivated by QCD share a set of simple assumptions: confinement of quarks, antiquarks and gluons; nontriviality of the QCD ground state; existence of a flux tube between Q and $\bar{Q}$ with constant energy density per unit length; existence of a phase transition temperature T_c above which confinement and flux tubes disappear.

Lattice QCD is a powerful tool to test the above assumptions nonperturbatively from first principles. The existence of flux tubes has been proven in numerous lattice studies both indirectly, by showing that the $Q\bar{Q}$ potential rises linearly for large $Q\bar{Q}$ separations, and directly, by measuring correlations of plaquettes with the Wilson loop or a pair of Polyakov lines.[1,2]

Colour confinement is believed to be due to the structure of the QCD vacuum, which in turn should be a complicated superposition of quark and gauge fields with nontrivial topology. Then it is natural to expect that the existence of the flux tube between a $Q\bar{Q}$ pair will also be reflected in quantities characterizing topological properties of vacuum configurations containing a $Q\bar{Q}$ pair. The aim of the present contribution is to show that the existence and basic properties of flux tubes can be inferred from distributions of the topological charge density around a static $Q\bar{Q}$ pair.

2. Topological charge on a lattice and cooling

The topology of a continuum gauge field configuration is characterized globally by the (integer) value of the topological charge Q:

[†]Presented by Š. Olejník. Work supported in part by BMWF.

*Permanent address: Institute of Physics, Slov. Acad. Sci., SK–842 28 Bratislava, Slovak Republic.

$$Q = \int d^4x \, q(x) = \frac{g^2}{64\pi^2} \int d^4x \, \epsilon^{\mu\nu\rho\sigma} \, F^a_{\mu\nu}(x) \, F^a_{\rho\sigma}(x), \tag{1}$$

where $q(x)$ is the topological charge density. A natural choice for defining Q on a lattice is to replace the integral in Eq. 1 by a sum over all lattice sites, $Q_L = \sum_x q_L(x)$, with some lattice operator of topological charge density $q_L(x)$ which in the naive continuum limit converges to $q(x)$. Here we will restrict ourselves to results which we obtained using the *plaquette* definition[3]:

$$q_L(x) = -\frac{1}{2^4 32\pi^2} \sum_{\mu\nu\rho\sigma=\pm1}^{\pm4} \tilde{\epsilon}_{\mu\nu\rho\sigma} \, \mathrm{Tr} \, [U_{\mu\nu}(x) \, U_{\rho\sigma}(x)], \tag{2}$$

with $U_{\mu\nu}$ being a plaquette with (μ,ν)-orientation.* Employing this definition, a problem immediately arises: topological charges of lattice configurations are *non*-integer and in general rather small. The origin of this problem was clarified by Campostrini *et al.*[5]: lattice and continuum versions of the theory are different renormalized quantum field theories which differ from one another by finite renormalization factors, *e. g.* $q_L(x) = a^4 \, Z(\beta) \, q(x) + O(a^6)$. $Z(\beta)$ is rather small for the range of β values used in Monte Carlo simulations.

A way to get rid of renormalization constants while preserving physical information contained in lattice configurations is the *cooling* method.[6,7] One assumes that small local changes of a configuration do not change its global topological properties. We applied the variant of cooling suggested by Hoek *et al.*[7]

3. Results

Our results come from three runs, each of them performed on an $8^3 \times 4$ lattice using the Metropolis algorithm. In the first two runs we evaluated path integrals in pure SU(3) gauge theory with standard Wilson action for $\beta = 5.6$ (confinement phase) and 5.8 (deconfinement phase). The third run was done for QCD at $\beta = 5.2$ with SU(3) Wilson action for gluons and 3 flavours of Kogut–Susskind quarks of equal mass $ma = 0.1$ (confinement phase again). We made 100k iterations in each run and measured our observables after every 50th sweep. Each recorded configuration was subjected to 50 cooling steps.

In all simulations, cooling quickly removed the renormalization factor $Z(\beta)$ and the measured values of Q_L converged (close) to integer values. We then observed substantial differences between lattice configurations below and above the deconfinement phase transition. The $Q_L = 0$ sector dominates in both phases, but in the confinement phase almost 70% of configurations have non-zero charge and the distribution of charges is approximately Gaussian (in both pure QCD and full QCD), while in the deconfinement phase $Q_L \neq 0$ appears only very rarely.

The most interesting result comes from measuring $\langle L(0)L^\dagger(d) \, q_L^2(r)\rangle$, the correlation function which is related to (the square of) the topological charge density around

---

*Similar results were found with the *hypercube* definition of Di Vecchia *et al.*[3], see Ref. 4.

Pure gauge QCD, $\beta = 5.6$	Pure gauge QCD, $\beta = 5.8$	Full 3-flavour QCD, $\beta = 5.2$, $ma = 0.1$
(a) $d = 3$	(b) $d = 3$	(c) $d = 3$
(d) $d = 4$	(e) $d = 4$	(f) $d = 4$

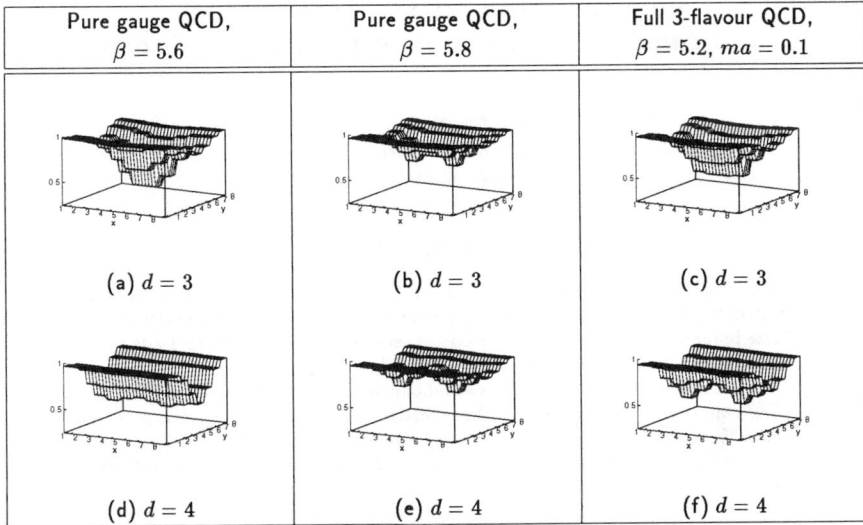

Fig. 1. The square of the topological charge density around a static quark-antiquark pair, $\langle L(0)L^\dagger(d)\, q^2(r)\rangle/(\langle L(0)L^\dagger(d)\rangle\langle q^2\rangle)$, at distances 3 and 4, after 5 cooling steps.

a static $Q\bar{Q}$ pair with separation d (L denotes the usual Polyakov line). In both phases we see a suppression of the topological charge density around static charges, consistent with phenomenological models. However, there is an essential difference between the phases which is illustrated in Fig. 1. In the deconfinement phase the suppression exists only in the vicinity of static sources (1b,e), while in the confinement phase it occurs in the whole flux tube between Q and $\bar{Q}$, both without (1a,d) and with dynamical quarks (1c). In the pure gauge case the flux tube is most clearly visible at the largest separation, $d = 4$ (1d). In full QCD for $d = 4$ (1f) we see an indication of flux tube breaking due to creation of a virtual $q\bar{q}$ pair. A similar effect was observed earlier also in colour field distributions around a static $Q\bar{Q}$ system.[8]

4. References

1. For a review see A. Di Giacomo, *Acta Phys. Pol.* **B25** (1994) 215.
2. C. Schlichter, *this Conference*.
3. P. Di Vecchia *et al.*, *Nucl. Phys.* **B192** (1981) 392.
4. M. Faber *et al.*, *Phys. Lett.* **B334** (1994) 145.
5. M. Campostrini *et al.*, *Phys. Lett.* **B212** (1988) 206.
6. M. Campostrini *et al.*, *Nucl. Phys.* **B329** (1990) 683.
7. J. Hoek *et al.*, *Nucl. Phys.* **B288** (1987) 589.
8. W. Feilmair and H. Markum, *Nucl. Phys.* **B370** (1992) 299.

MONOPOLES AROUND ELECTRIC CHARGES
AND THE CONFINEMENT POTENTIAL IN COMPACT QED

M. ZACH, M. FABER, W. KAINZ and P. SKALA

Institut für Kernphysik, Technische Universität Wien,
A-1040 Wien, Austria

ABSTRACT

The confinement in compact QED is known to be related to magnetic monopoles. Electric flux lines align the monopole currents, which is seen in lattice calculations. These induced currents on the other side expel the electric flux, which leads to the formation of a flux tube between positive and negative charges. This scenario can be described by the dual version of Maxwell–London equations, including a dual Dirac string connecting the charges. In this effective model one can derive a potential with a linearly rising term. In order to get agreement with lattice simulations of compact QED at finite temperature we have to take into account fluctuations of the string.

1. Introduction

It was suggested many years ago that confinement could be understood in analogy to superconductivity[1]. In the dual superconductor picture dynamically generated topological excitations provide the screening supercurrents[2]. Compact U(1) gauge theory contains magnetic monopoles in addition to photons. At strong coupling compact U(1) has a confining phase, therefore it can be regarded as a prototype of a confining theory. It has been demonstrated via numerical simulations that the monopole density is high in the confinement phase of compact U(1), whereas it decreases exponentially above the phase transition to the Coulomb phase[3].

We examine the monopole condensate consisting of fluctuating current loops and find a suppression around electric charges and in flux tubes. In regions of electric fields these loops are partly aligned and form a persistent current distribution[4].

2. Lattice Calculations in Compact QED

Our simulations are done on an Euclidean $8^3 \times 4$ lattice with periodic boundary conditions. The U(1) gauge degrees of freedom are expressed by means of the link variables $U_\mu(x) = exp[i\theta_\mu(x)]$. We use the standard Wilson action

$$S = \beta \sum_{plaquettes} [1 - \cos\theta_{\mu\nu}(x)], \tag{1}$$

where $\beta = \frac{1}{e^2}$ and $exp[i\theta_{\mu\nu}(x)]$ is an oriented product of gauge variables around an elementary plaquette. The static charges are represented by Polyakov loops $L(\vec{r})$. So we can calculate observables around a static $q\bar{q}$-pair by means of the correlation function $\langle L(0)L^{\dagger}(d)\mathcal{O}(x)\rangle$. $\mathcal{O}(x)$ stands both for $\sin\theta_{i4}$, which measures the electric flux through a space–time–plaquette (using this definition Gauss' law for electric charges turns out to be fulfilled[4]) and for the magnetic current $\vec{J}_M$ defined by the dual versions of Maxwell's equations and composed of the operators measuring the electric and magnetic field.

We measure the distribution of $\vec{E}$ and $\vec{J}_M$ in the confinement phase ($\beta = 0.96$) for various distances of charges (from 1 to 4 lattice spacings). As expected, the monopole currents behave like a coil, squeezing the electric field into a narrow flux tube. We show a profile of $\vec{E}$ and $\vec{J}_M$ for a distance $2a$ of charges in a plane perpendicular to the $q\bar{q}$-axis (see fig.1).

3. An Effective Model

If the dual superconductor hypothesis is correct, the field and current distributions should obey the coupled Maxwell–London equations

$$\mathrm{rot}\,\vec{E}(\vec{r}) = -\vec{J}_M(\vec{r}), \tag{2}$$

$$\vec{E}(\vec{r}) = \lambda^2 \mathrm{rot}\,\vec{J}_M(\vec{r}) + \vec{\mathcal{E}}(\vec{r}), \tag{3}$$

where we have introduced a dual Dirac string connecting the charges $+Q, -Q$

$$\vec{\mathcal{E}}(\vec{r}) = Q \int_l d\vec{R} \ \ \delta^{(3)}(\vec{r} - \vec{R}), \tag{4}$$

This scenario was introduced by Dirac to include magnetically charged particles and adapted by Nambu[5] for explaining confinement. One can calculate analytically $\vec{E}$ and $\vec{J}_M$ and the total energy of the electric field[6]. If we take l to be a straight line which is energetically most favorable we get

$$W = \frac{1}{2} \int d^3 r \vec{E}^2(\vec{r}) = \frac{Q^2}{8\pi\lambda^2}d - \frac{Q^2}{4\pi}\frac{e^{-\frac{d}{\lambda}}}{d} + const., \tag{5}$$

a linearly rising and a Yukawa potential. The free parameter λ introduced in (3) is the London penetration depth of the electric field. For a straight string of infinite length $\vec{E}$ and $\vec{J}_M$ take a simple form, and have been used by Singh et al.[4] for comparison to lattice QED results.

Our compact QED calculations contain thermal fluctuations, therefore we include fluctuations of the dual Dirac string in our effective model. This is done by solving equations (2–3) numerically on an 8^3 lattice for all possible strings up to a certain length, and by calculating the thermal expectation values of $\vec{E}$ and $\vec{J}_M$. If we fit the parameter λ (we find $\lambda = 0.163a$ for $\beta = 0.96$), we get very good agreement between

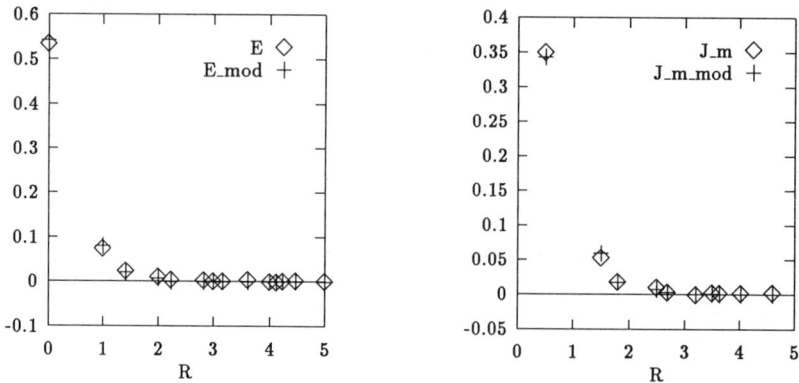

Fig. 1. Profile of electric field and monopole current: QED calculations (E, J_m) compared to effective model with $\lambda = 0.163a$ (E_mod, J_m_mod)

the effective model and compact QED (see fig.1) on the whole lattice and for each distance of charges ($\chi^2 \approx 1$). The free energy as a function of distance, however, is about 40 % smaller than the potential from QED correlations. This may be due to the "time" dependent fluctuations of the string as well as to the suppression of the monopole condensate in the flux tube.

4. Conclusions

We have shown that the dual superconductor model is appropriate for describing the field and current distributions of compact QED. Agreement is only achieved if fluctuations of the string are taken into account. Further, we found a suppression of the monopole condensate in the regions of electric fields, which should give a further contribution to the string energy and possibly remove the remaining discrepancy between compact QED and the effective model.

5. References

1. H. B. Nielsen and P. Olesen, *Nucl. Phys.* **B61** (1973) 45.
2. S. Mandelstam, *Phys. Rep.* **23C** (1976) 245.
3. T. A. DeGrand and D. Toussaint, *Phys. Rev.* **D22** (1980) 2478 .
4. V. Singh, R. H. Haymaker, D. A. Browne, *Phys. Rev.* **D47** (1993) 1715 .
5. Y. Nambu, *Phys. Rev.* **D10** (1974) 4262.
6. M. Faber, A. N. Ivanov, W. Kainz, M. Zach, *A Classical Effective Model for the Formation of a Gluon String*, IK-TUW-Preprint 9404401

CONFINING FORCES AND STRING FORMATION FROM THE LATTICE

G.S. Bali, K. Schilling, Ch. Schlichter, and A. Wachter
Fachbereich Physik, Bergische Universität,
D 42097 Wuppertal, Germany
and
Forschungszentrum Jülich, HLRZ, D 52425 Jülich, Germany

ABSTRACT

We show the running coupling as derived from the SU(3) $Q\bar{Q}$ potential and discuss preliminary results on spin dependent heavy quark potentials from high statistics lattice simulations of SU(2) gauge theory. The precision suffices to study scaling properties and lattice artifacts (at short distances). We identify the Coulomb like short range interaction as a mixed vector-scalar exchange. We measure flux tube formation between a static $Q\bar{Q}$ pair over physical distances up to 2 fm, with a spatial resolution as small as .05 fm. Consistency with the string picture is found for separation larger than about 1 fm, with a half width of the profile of approximately .7 fm.

1. The Lattice Approach

Lattice gauge theory provides a powerful tool to compute non-perturbative properties of strong interactions from the basic QCD Lagrangian, such as the confining force between a quark-antiquark pair. The central potential has been the object of lattice investigations, ever since the seminal work of M. Creutz[1] back in 1979.

Over the last years considerable progress has been accumulated, *both* in computational methods and in the computing power, in particular through the advent of high performance parallel computing devices. In this way, the static central potential, V_0, could be computed — from improved Wilson loops — with statistical accuracy of a few per mille in $SU(3)$ gauge theory, on large lattices with high resolution[2]. This level of precision allowed to carry out a direct analysis of the short range interaction in terms of the running coupling[3,4] $\alpha_{q\bar{q}}$ in this theory, as we demonstrate[5] in fig. 1, in units of the string tension $K \approx (440 \text{ MeV})^2$. The full curves correspond to the two-loop running coupling with $\Lambda_R = .66(4)\sqrt{K}$. The dotted curves are expected from a funnel potential.

The computation of the spin dependent potentials presents a much greater challenge to large scale computing, as it requires the summation over a variety of connected plaquette-Wilson loop operators[6] with large time extent. The first pioneering attempts in this direction[7,8] have been made in the mid eighties, and have provided evidence – within the statistical and systematic errors of the period – in favour of the common prejudice that the long range confining force is related to scalar exchange, while the short range Coulombic part is predominantly vector like in nature.

Another, not less demanding problem is the computation of the energy and action density distributions around static sources that could shed light on the mechanism

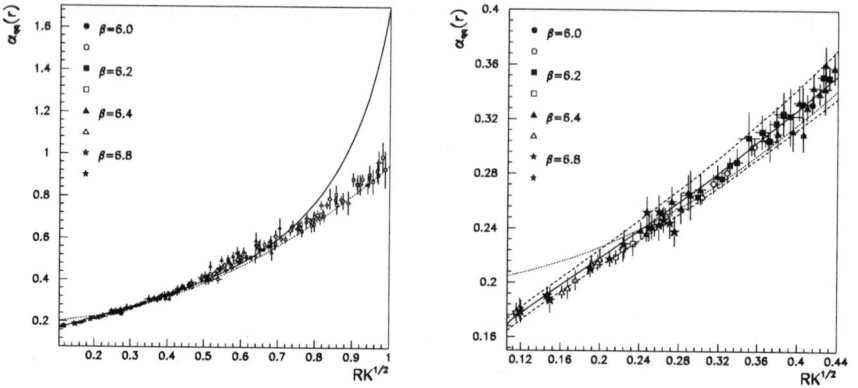

Fig. 1. (a) The running coupling from the interquark force. (b) Same as fig. 1a for $1/r <$ 1 GeV.

of string formation. Technically speaking, these distributions are extracted from the disconnected correlation between a plaquette and a Wilson loop whose time extent, T, should be asymptotically large such as to exhibit ground state dominance. In order to study string formation, one needs to realize spatial separations beyond one fermi; in this respect, previous lattice studies were hampered either by too coarse lattice resolution or by too small source separations[9,10].

In order to produce decent signal to noise ratios from these lattice measurements, it is crucial to ensure early T-asymptotics in the observed correlators. This can be achieved by a suitable "smearing" of the spatial parts in the Wilson loops[11,2,4]. In this contribution we shall present an outline of the physics results from our recent high statistics[12,13] investigations, where we applied such smearing techniques. For the sake of statistics, we base our work on $SU(2)$ gauge theory as it is easier to handle on the computer and has a vacuum structure similar to the $SU(3)$ case.

2. Spin Dependent Potentials

A $1/m$ expansion of the heavy quark potential in the equal mass case yields the Breit-Fermi type expression (assuming spin independence of the leading order potential, V_0),

$$V_{SD}(r) = V_0(r) + \frac{1}{m^2}\left(V_{LS}(r) + V_{SS}(r) + \text{spin independent corrections}\right) + \cdots \quad (1)$$

with

$$V_{LS}(r) = \frac{\vec{L}_1\vec{s}_1 - \vec{L}_2\vec{s}_2}{r}\left(\frac{V_0'(r)}{2} + V_1'(r)\right) + \frac{\vec{L}_1\vec{s}_2 - \vec{L}_2\vec{s}_1}{r}V_2'(r) \quad , \quad (2)$$

$$V_{SS}(r) = \left((\hat{r}\vec{s}_1)(\hat{r}\vec{s}_2) - \frac{\vec{s}_1\vec{s}_2}{3}\right)V_3(r) + \frac{\vec{s}_1\vec{s}_2}{3}V_4(r) \quad , \quad (3)$$

where $\vec{L}_i = \vec{r} \times \vec{p}_i$. For Dirac fermions explicit expectation values have been associated to the spin-orbit and spin-spin "potentials" V_1', V_2' and V_3, V_4, which can be computed in form of "eared" Wilson loops on the lattice[6,14]. These potentials are related to scalar (S), vector (V) and pseudo-scalar (P) exchange contributions in the following way:

$$V_0(r) = S(r) + V(r) \tag{4}$$

$$V_3(r) = \frac{V'(r) - P'(r)}{r} - (V''(r) - P''(r)) \tag{5}$$

$$V_4(r) = 2\nabla^2 V(r) + \nabla^2 P(r) \quad . \tag{6}$$

In addition, relativistic invariance leads to the Gromes relation[15]

$$V_0' = V_2' - V_1' \quad . \tag{7}$$

The central potential, V_0, has a very clear linear confining part (see fig. 2a) which must be of scalar exchange type as a vector exchange can have at most a logarithmic asymptotic r dependence[16]. The lattice potentials V_1^L to V_4^L undergo multiplicative renormalizations in respect to their continuum counterparts, while V_0 does not. We have performed a non-perturbative renormalization of these quantities in the manner suggested by Michael et al.[17]. In fig. 2b we compare the Gromes combination $V_2' - V_1'$ (in units of the string tension, K) to the force, derived from the fit to the central potential as given in fig. 2a. Two data sets for the β-values 2.74 and 2.96 (lattice resolutions .04 fm and .02 fm) are combined: we find good scaling behaviour and agreement with the Gromes relation outside of the region of lattice artifacts. This demonstrates the success of Michael's renormalization prescription.

We confirm the second spin-orbit potential (see fig. 3b) to be definitely of short range nature, leaving little room for a scalar contribution to this potential. The first spin-orbit potential, V_1', is smaller by a factor five and more noisy (see fig. 3a). Our lattice resolution enables us to establish an attractive short range contribution that can be well fitted to a Coulomb $(1/R^2)$ form, in addition to the constant long-range term, that is in agreement with the string tension, K.

In principle, the potentials V_0, V_3, and V_4 allow for a determination of S, V, and P. At this stage, we will assume P to vanish and V_1 to be pure scalar (which induces the equalities $V_2 = V$ and $V_1 = -S$). This leads to a prediction for V_3, according to eq. (5), which can be checked in fig. 4a: we find reasonable agreement between the data sets and the predicted curve, $V_2'/R - V_2''$, the deviations being qualitatively understood from tree level lattice perturbation theory.

The remaining spin-spin potential, V_4, exhibits oscillatory behaviour as a lattice artifact (see fig. 4b) and can largely be understood as a δ-contribution, according to

$$V_2 = -\frac{c}{r} \Rightarrow V_4 = 2\nabla^2 V_2 = 8\pi c \delta_L^3(\vec{R}) \quad , \tag{8}$$

where δ_L is the appropriate lattice δ-function for the chosen (symmetric) ear combination of our lattice operator. This expectation is indicated by open squares in the

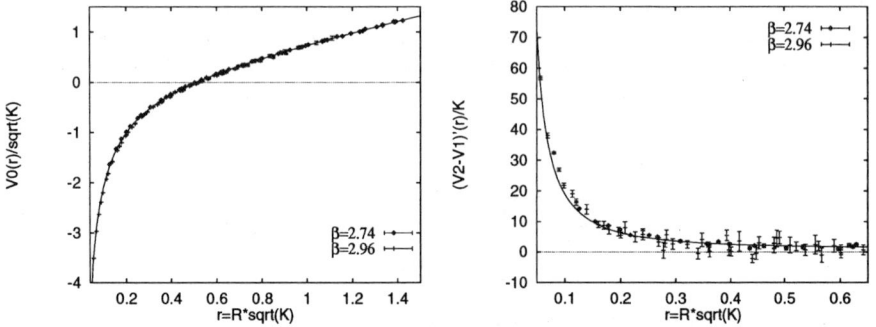

Fig. 2. (a) Central potential with a funnel-type fit curve. (b) Comparison of $V_2' - V_1'$ (data) with the (fitted) central force (curve) (Gromes relation).

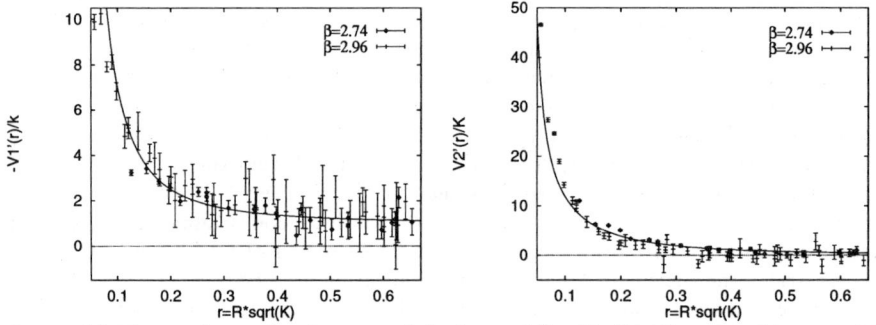

Fig. 3. (a) V_1', together with a fit curve of the form $a/r^2 + K$. (b) The spin-orbit potential V_2'. The curve corresponds to the central force plus the (fitted) V_1'.

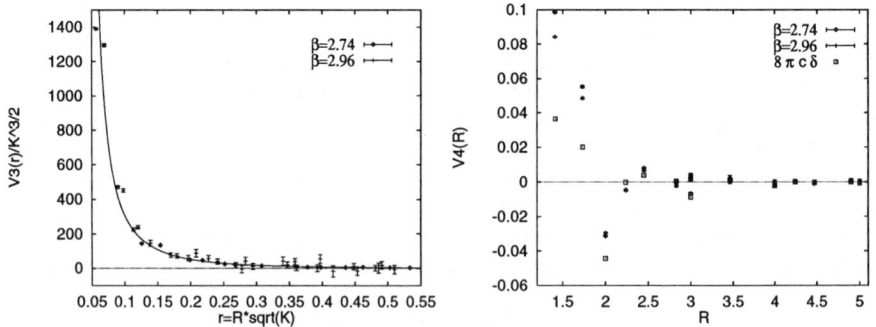

Fig. 4. (a) Comparison of V_3 with $V_2'/r - V_2''$ (curve). (b) V_4 for the two β-values in lattice units (points with error bars). The squares indicate the appropriate lattice δ function with the coefficient, c, taken from the V_2' analysis.

figure. Notice, that we have plotted V_4 in lattice units on both axes! At $R \leq 2$ we observe an additional $1/R^4$ contribution.

In conclusion, we find a consistent picture when identifying $-V_1$ (V_2) with the scalar (vector) exchange contribution, other exchange types being negligible to order $1/m^2$. Apart from the linear large distance part, we observe a short distance Coulomb like scalar contribution from the first spin-orbit potential. It appears that the Coulomb part of the central potential splits up into a vector/scalar ratio in between 3/1 and 4/1.

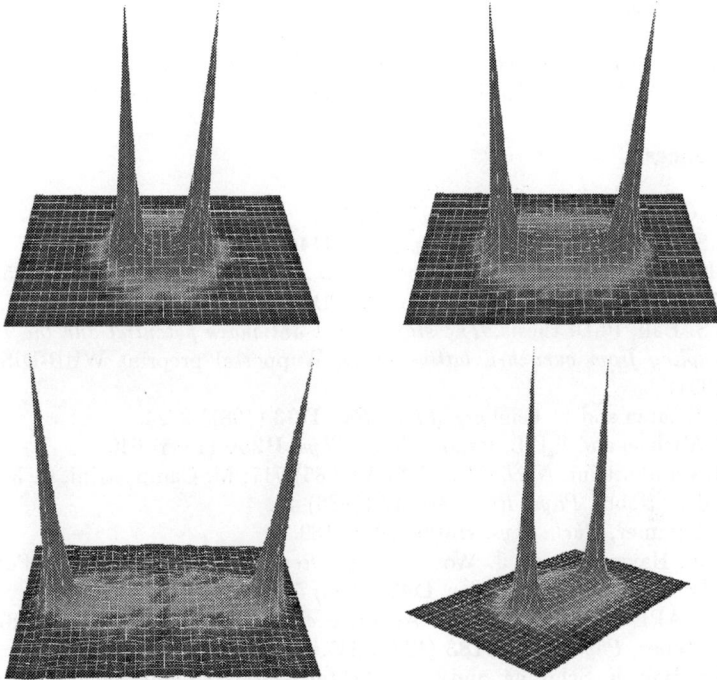

Fig. 5. The action density distribution for various quark separations, measured at $\beta = 2.5$.

3. String Formation

We are able to map out the action density distributions around static SU(2) sources with separations, r, ranging up to 2 fm[12]. Figs. 5a to 5c illustrate the formation of an elongated string by varying the source separation from .6 over 1.1 to 2 fm, the largest distance corresponding to 24 lattice spacings at $\beta = 2.5$. Fig. 5d contains the information of fig. 5b, but observed from a different angle of view, in order to exhibit the width of the string. Notice, that the plots represent the data without any interpolation. The brightness has been computed from the relative statistical errors[*]

We observe a linear widening of the width of the flux tube with the quark separation, r, up to $r \approx .7$ fm. After saturation, from $r \approx 1$ fm onwards, the width remains rather constant. The transverse profile is in agreement with both, a dipole type and a Gaussian shape, within statistical accuracy. We find a plateau value, $\rho \approx .7$ fm, for the half width. The corresponding mean squared radius, δ, is found to be within the range $.5$ fm $< \delta < .8$ fm.

4. Acknowledgements

This project was supported by DFG (grant Schi 257/3-1) and by EU (grant SC1*-CT91-0642). We thank HLRZ for the computer time on their CM5 system.

5. References

1. M. Creutz, *Phys. Rev.* **D21** (1980) 2308.
2. G.S. Bali and K. Schilling, *Phys. Rev.* **D46** (1992) 2636.
3. UKQCD collaboration: S.P. Booth *et al.*, *Phys. Lett.* **B294** (1992) 385.
4. G.S. Bali and K. Schilling, *Phys. Rev.* **D47** (1993) 661.
5. G.S. Bali, PhD. thesis, *The static quark-antiquark potential and the running coupling from quenched lattice QCD*, Wuppertal preprint WUB-DIS 94-01 (1994).
6. E. Eichten and F. Feinberg, *Phys. Rev.* **D23** (1981) 2724.
7. C. Michael and P.E.L. Rakow, *Nucl. Phys.* **B256** (1985) 640.
8. M. Campostrini, *Nucl. Phys.* **B256** (1985) 717; M. Campostrini, C. Moriarty and C. Rebbi, *Phys. Rev. Lett.* **57** (1986) 44.
9. R. Sommer, *Nucl. Phys.* **B306** (1988) 180.
10. R.W. Haymaker and J. Wosiek, *Phys. Rev.* **D43** (1990) 1991; Y. Peng and R.W. Haymaker, *Phys. Rev.* **D47** (1993) 5104.
11. The APE Collaboration: M. Albanese *et al.*, *Phys. Lett.* **B192** (1987) 163; M. Teper, *Phys. Lett.* **B183** (1987) 345.
12. G.S. Bali, K. Schilling, and C. Schlichter, *Observing long colour flux tubes in SU(2) lattice gauge theory*, CERN preprint CERN-TH.7413/94 (1994).
13. G.S. Bali, K. Schilling, and A. Wachter, *to be published*.
14. D. Gromes, *Z. Phys.* **C22** (1984) 265.
15. D. Gromes, *Z. Phys.* **C26** (1984) 401.
16. D. Gromes, *Phys. Lett.* **B202** (1988) 262; D. Gromes, W. Lucha, and F.F. Schöberl, *Phys. Rep.* **200** (1991) 127.
17. A. Huntley and C. Michael, *Nucl. Phys.* **B286** (1987) 211.

---

[*]A data base of colour images of the flux tubes and related quantities can be accessed via anonymous ftp from wpts0.physik.uni-wuppertal.de. The (compressed) .rgb and .ps files can be found in the directory pub/colorflux.

LATTICE RESULTS ON THE DECAYS OF HEAVY-LIGHT MESONS

UKQCD COLLABORATION - Presented by D G Richards

Department of Physics and Astronomy, The University of Edinburgh
Edinburgh EH9 3JZ, Scotland

ABSTRACT

We present results for the leptonic and semi-leptonic decay matrix elements of heavy-light mesons obtained in quenched QCD, using an $O(a)$-improved fermion action. We find substantial non-scaling corrections to the leptonic decay constants: of order 30% for f_D and of order 10% for f_B. We compute the Isgur-Wise function by studying semi-leptonic decays of the form $P \rightarrow P'l\bar{\nu}$, where P and P' are heavy-light pseudoscalar mesons. Finally, we present results for the leading-order matrix element for the decay $B \rightarrow K^*\gamma$. We find results for the on-shell form factor $T_1(q^2=0)$ to be consistent (in the Standard Model) with the recent CLEO experimental branching ratio.

1. Introduction

The leptonic and semi-leptonic decays of heavy-light pseudoscalar mesons provide a crucial laboratory for the determination of the parameters of the Standard Model. However, a necessary prerequisite is a quantitative understanding of strong-interaction effects, characterised by the pseudoscalar decay constants, f_P, in the case of leptonic decays, and by the semi-leptonic form factors in the case of the semi-leptonic decays. Lattice QCD enables an *ab initio* investigation of these effects. This talk provides a compendium of the *UKQCD Collaboration* results on heavy-light physics. We conclude this introduction with a brief description of the simulation[1].

Our principal results are obtained from a simulation in the quenched approximation on a $24^3 \times 48$ lattice at $\beta = 6.2$, using 60 gauge field configurations. The quark propagators are computed using an $O(a)$-improved Wilson fermion action[2], with "rotated" operators[3]. We find an inverse lattice spacing $a^{-1} = 2.73(5)\text{GeV}$, obtained by measuring the string tension.

2. Leptonic Decays

The finite lattice spacing employed in our simulations precludes the study of the B-meson directly, since it would be subject to uncontrolled discretisation errors. Instead, we chose to investigate the decays of heavy-light mesons containing a quark with a mass around that of the charm quark, and to extrapolate our results to the physical b-quark mass. In this study we have used four values of the heavy-quark mass. To perform the extrapolation, we appeal to the Heavy-Quark Effective Theory; in the limit of exact heavy-quark symmetry, the quantity

$$\hat{\Phi}(M_P) \equiv (\alpha_s(M_P)/\alpha_s(M_B))^{2/\beta_0} Z_A^{-1} f_P \sqrt{M_P}, \tag{1}$$

is a constant, where M_P is the pseudoscalar mass, and Z_A the matching coefficient. To quantify the non-scaling effects in f_P, we fit Eq. (1) to linear and quadratic functions of $1/M_P$. The fit reveals $O(1/M_P)$ corrections to f_P of approximately 10% at M_B and 30% at M_D, in agreement with previous analyses, and we obtain for our best results[4]:

$$
\begin{aligned}
f_D &= 185 \; {}^{+}_{-} \, {}^{4}_{3}(\text{stat}) \; {}^{+}_{-} \, {}^{42}_{7}(\text{syst}) \text{ MeV} \\
f_B &= 160 \; {}^{+}_{-} \, {}^{6}_{6} \; {}^{+}_{-} \, {}^{53}_{19} \text{ MeV} \\
\frac{f_{D_s}}{f_D} &= 1.18 \; {}^{+}_{-} \, {}^{2}_{2}, \frac{f_{B_s}}{f_B} = 1.22 \; {}^{+}_{-} \, {}^{4}_{3}.
\end{aligned}
\tag{2}
$$

3. Semi-leptonic Decays

In the limit of exact heavy-quark symmetry, the six form factors for the decays $B \to D$ and $B \to D^*$ are given by the universal Isgur-Wise function $\xi(\omega = v \cdot v')$, where v and v' are the four-velocities of the ingoing and outgoing mesons. This function is absolutely normalised: $\xi(1) = 1$. We obtain $\xi(\omega)$ by studying decays of the form $P \to P'l\bar{\nu}$ where P and P' are heavy-light pseudoscalar mesons. These processes are described by matrix elements of the heavy-quark vector current $\bar{Q}'\gamma^\mu Q$, which can be decomposed in terms of two form factors, $h^+_{Q \to Q'}(\omega)$ and $h^-_{Q \to Q'}(\omega)$,

$$
\frac{\langle P'(\vec{p}')|\bar{Q}'\gamma^\mu Q|P(\vec{p})\rangle}{\sqrt{M_P M_{P'}}} = (v + v')^\mu h^+_{Q \to Q'}(\omega) + (v - v')^\mu h^-_{Q \to Q'}(\omega).
\tag{3}
$$

For heavy quarks of finite mass, there are corrections relating the form factors to $\xi(\omega)$. For the case of $h^+(\omega)$ they are parametrised as

$$
h^+(\omega) = (1 + \beta^+(\omega) + \gamma^+(\omega))\xi(\omega) ,
\tag{4}
$$

where $\beta^+(\omega)$ and $\gamma^+(\omega)$ contain the radiative and non-scaling corrections respectively. We obtain $\xi(\omega)$ from $h^+(\omega)$, correcting for $\beta^+(\omega)$[5]. We combine data at different values of the heavy-quark mass in our analysis, and find no evidence for non-scaling effects in our data.

We fit the measured $\xi(\omega)$ to $s\xi_{NR}(\omega)$ where

$$
\xi_{NR}(\omega) \equiv \frac{2}{\omega + 1} \exp\left(-(2\rho^2 - 1)\frac{\omega - 1}{\omega + 1}\right)
\tag{5}
$$

and s accounts for the uncertainty in the normalisation of our data. ρ^2 is the slope of the Isgur-Wise function at zero recoil, for which we obtain

$$
\rho^2 = 0.9 \; {}^{+}_{-} \, {}^{2}_{2}(\text{stat}) \; {}^{+}_{-} \, {}^{4}_{2}(\text{syst}).
\tag{6}
$$

The experimentally-measured decay rate is $dB(\bar{B} \to D^*l\bar{\nu})/d\omega$; we can use our determination of the Isgur-Wise function to perform a single-parameter fit to the data

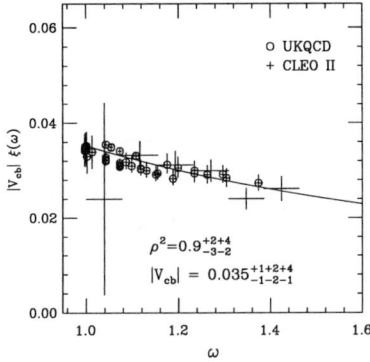

Figure 1: Least-χ^2 fits to CLEO II experimental data for $| V_{cb} | \xi(\omega)$, using the parametrisation $\xi_{NR}(\omega)$ with ρ^2 given by equation Eq. (6). The first error on $| V_{cb} |$ is experimental, the second from lattice statistical uncertainties, and the third from the lattice systematic errors on ρ^2.

to extract V_{cb}. This is illustrated for CLEO-II data[6] in Fig. 1. Note that, though radiative corrections at $\omega = 1$ have been taken into account in the data, we otherwise assume exact heavy-quark symmetry in relating $\xi(\omega)$ to the experimentally measured form factors.

4. Rare Decays

Excluding QCD contributions, the free quark decay $b \rightarrow s\gamma$ in the Standard Model proceeds via the "penguin" diagrams. Because it does not occur at tree-level, it provides a particularly subtle test of the model. The charm and top quarks dominate, because the up quark contribution to the loop is suppressed by the small CKM factor $|V_{ub}V_{us}^*|$. The exclusive decay $B \rightarrow K^*\gamma$ has recently been measured by CLEO[7], who obtained $BR(B \rightarrow K^*\gamma) = (4.5 \pm 1.5 \pm 0.9) \times 10^{-5}$. The phenomenological importance of this process has encouraged the lattice study of the decay matrix elements[8]. The hadronisation ratio may be related to a single on-shell form factor: $T_1(q^2 = 0)$. In Fig. 2 we show $\hat{T}_1(q^2 = 0)$, defined by

$$\hat{T}_1 \equiv T_1(q^2 = 0) \left(\frac{M_P}{M_B}\right)^{3/2} \left(\frac{\alpha_s(M_P)}{\alpha_s(M_B)}\right)^{2/\beta_0} \tag{7}$$

which, assuming pole-dominance, scales in the limit of exact heavy-quark symmetry, together with linear and quadratic fits in $1/M_P$. Also shown is the "experimental" determination of the form-factor, using $m_t = 170$ GeV and $\tau_B = 1.5$ ps. Alternatively,

234

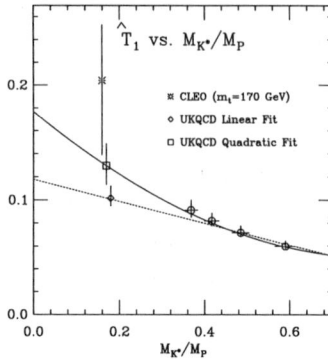

Figure 2: The scaling quantity $\hat{T}_1(q^2 = 0)$ as defined in the text. The dashed and solid lines are linear and quadratic fits in $1/M_P$ respectively. Also shown is the "experimental" determination using the CLEO data.

using our determination of $T_1(0)$, we obtain

$$BR(B \to K^*\gamma) = \left(1.5 \pm 0.6\,(\text{stat.})\,^{+9}_{-8}\,(\text{sys.})\right) \times 10^{-5} \qquad (8)$$

consistent with the experimental branching ratio above.

5. Acknowledgements

This research was supported by the UK Science and Engineering Research Council under grants GR/G 32779 and GR/J 21347, by the University of Edinburgh and by Meiko Limited. We are grateful to Edinburgh University Computing Service and, in particular, to Mike Brown for maintenance of service on the Meiko i860 Computing Surface. DGR acknowledges the support of the Particle Physics and Astronomy Research Council through an Advanced Fellowship.

1. UKQCD Collaboration, C. Allton *et al.*, Nucl. Phys. **B407** (1993) 331.
2. B. Sheikholeslami and R. Wohlert, Nucl. Phys. **B259** (1985) 572.
3. G. Heatlie *et al.*, Nucl. Phys. **B352** (1991) 266.
4. UKQCD Collaboration, R. M. Baxter *et al.*, Phys. Rev. **D49** (1994) 1594.
5. M. Neubert, Phys. Rev. **D46** (1992) 2212.
6. "Semi-leptonic B Decays", S. Stone, in "B Decays", ed. S. Stone, 2nd Edition (World Scientific 1994).
7. CLEO Collaboration, R. Ammar *et al.*, Phys. Rev. Lett. **71** (1993) 674.
8. UKQCD Collaboration, K. C. Bowler *et al.*, Phys. Rev. Lett. **72** (1994) 1398; and Edinburgh Preprint 94/544, Southampton Preprint 93/94-29.
9. C. Bernard, P. Hsieh, and A. Soni, Phys. Rev. Lett. **72** (1994) 1402.

ON THE RELATION BETWEEN 2D QUANTUM CHROMODYNAMICS

AND HIGHER DIMENSIONAL GAUGE THEORIES

A. FERRANDO, A. JARAMILLO and V. VENTO

Departament de Física Teòrica and I.F.I.C. Centre Mixt Universitat de València – C.S.I.C.
E-46100 Burjassot (València), Spain.

ABSTRACT

Quantum Chromodynamics on a strip of width L can be used as a tool to study the connection between 2D and 3D QCD. We analyze pure gauge QCD in the small width regime. The $L \to 0$ limit turns out to be non-trivial. The glueball spectrum and the string tension can be calculated analytically. The evolution of these observables with L shows a different behavior than in the abelian theory. We propose a 3D confinement mechanism based on the continuity and relative smoothness of the dimensional transition between 2D and 3D physics. Finite temperature lattice results give support to the previous mechanism.

1. Non-abelian gauge dynamics on narrow strips

We develop a continuous procedure to lower the dimensionality of a gauge theory. Our starting point is the pure gauge QCD_3 action. We compactify the transversal degrees of freedom by requiring periodic boundary conditions on the gauge field. The great advantage of the previous procedure is that it yields an effective 2D theory mimicking 3D physics[1]. After the choice of the strip version of the so called static gauge $\partial_2 A_2 = 0$, the theory can be written in terms of the gauge field Fourier modes. They are 2D fields. The resulting action is still invariant under 2D *longitudinal* gauge transformations. These transformations are compatible with the "static" gauge condition just implemented. The theory on the strip contains besides the conventional QCD_2 (2D) gluons an infinite set of covariant (non-gauge) interacting fields $(\phi, \{V_\alpha^n, \alpha = 0,1; n \in Z, n \neq 0\})$ belonging to the adjoint representation of the color group.

The 2D gluon field is the zero mode of the *longitudinal* components of the gauge field $(A_\alpha, \alpha = 0,1)$. The ϕ field is a massless gauge-covariant degree of freedom. It is the remnant of the gauged-away *transversal* field A_2. This scalar field ϕ couples to the 2D gluons through the covariant derivative to preserve the 2D *longitudinal* gauge invariance of the strip action. The V_n fields of the non abelian strip theory self-couple with couplings which are not arbitrary, but determined by the 3D gauge invariance, lost after the implementation of the gauge fixing condition. The 3D gauge invariance also defines originally the coupling of the V_n fields with the other covariant non-gauge particle of the reduced action, the scalar field ϕ.

At a given width L and coupling constant g_3 we can characterize our theory by fixing the 2D gauge coupling constant $g_2 = g_3/L^{1/2}$ and the dimensionless width $\epsilon = g_3 L^{1/2}$. The non-zero mode fields V_n become infinitely heavy ($m_n \sim 1/\epsilon$) in the $\epsilon \to 0$ limit and a good description of the system can be given in terms of the lighter degrees of freedom. That is in terms of the remnant *classically* massless degrees of freedom, the 2D gluon and the scalar field ϕ. The scalar field ϕ undergoes a mass renormalization process which provides it with mass. Unlike the 2D gluons, whose masslessness is protected by the 2D gauge invariance, there is no custodial symmetry for the scalar field. The integration of the heavy modes turns out to be non-trivial. They decouple eventually. However vacuum V_n fluctuations in the form of one-loop contributions provide the scalar field ϕ with a very heavy mass for small widths. As a consequence the action describing dynamics on very narrow strips (small ϵ's) is given by,

$$S_\epsilon = \int d^2x \left\{ \frac{1}{2} f_{\alpha\beta}^2 + (D_\alpha \phi)^2 + \mu_\phi^2(\epsilon)\phi^2 \right\}, \quad \mu_\phi^2(\epsilon) = \frac{g_2^2 N}{\pi} + \frac{g_2^2 N}{2\pi} \ln\left(\frac{1}{\epsilon^2}\right) \qquad (1)$$

The low lying glueball spectrum ($\phi\phi$ and $V_n V_n^*$ bound states) can be calculated analytically using a one-dimensional Schrödinger equation. This is possible because both the ϕ and V fields are very heavy for small ϵ's. The glueball spectrum is given in terms of the renormalized ϕ and V mass μ_R, the string tension $\sigma_\phi = \frac{1}{2} g_2^2 C_N$ (C_N is a color invariant constant) and the zeroes of the Airy function $-\varepsilon_n{}^{1,2}$,

$$M_n(\epsilon) = 2\mu_R(\epsilon) + \varepsilon_n \left[\frac{\sigma_\phi^2}{\mu_R(\epsilon)} \right]^{1/3} \qquad (2)$$

2. Confinement mechanism

The theory on a narrow strip is still confining. The integration of the non-abelian scalar field in Eq. (1) influences the 2D gauge dynamics by dressing the kinetical gluon term. Confinement is thus linear because is identical to that of pure gauge QCD_2. The only difference comes from the appearance of a ϵ-dependent $q\bar{q}$ string tension arising from the renormalization of the 2D coupling constant,

$$\sigma_q(\epsilon) = \frac{6\, g_2^2\, c_N\, N\, \log(\epsilon^{-2})}{1 + 12\, N\, \log(\epsilon^{-2})} \qquad (3)$$

The 2D gluonic long distance dominance in the Wilson loop is also present at any finite width[1]. The role of the transversal fields, V_n and ϕ, is to dress the *longitudinal* effective interaction driven by the 2D gluons. The string tension can be obtained from the 2D gluon effective action just by looking at the constant ($Z(\epsilon)$) dressing the gluon kinetic term ($f_{\alpha\beta}^2$). There is a simple relation between them, $\sigma_q(\epsilon) = (1/2)Z(\epsilon)c_N$.

In this context the string tension given by Eq. (3) is just the leading order approximation in a perturbative expansion in the dimensionless width ϵ. Exactly the same happens to the glueball spectrum previously presented Eq. (2). Both equations show a flow in the dimensionless width ϵ. This flow goes beyond the perturbative regime when $\epsilon \geq 1$. The prevalence of 2D gluon dynamics at long distances on the strip ensures that the dominant contribution to the static $q\bar{q}$ energy is linear $\sigma_q(\epsilon)R$. Certainly the behavior of the string tension in QCD on the strip is highly non-trivial, as the low ϵ result Eq. (3) already exhibits . Moreover we can outline the behavior of the ϵ flux beyond the perturbative regime. This behavior will define a mechanism describing the dimensional transition between 2D and 3D gauge QCD.

The mechanism we propose is characterized by the following two properties:

1. The flow of the string tension with ϵ is continuous and different from zero. That is, one can extrapolate the perturbative result in Eq. (3) to the $\epsilon \geq 1$ regime in a continuous way providing a non-zero value for the QCD_3 string tension in the $\epsilon \to \infty$ limit.

2. There is a stabilization of the ϵ flow at a critical width L^*. The string tension and the glueball spectrum tend quickly to a constant, their QCD_3 value, above the critical point

There is a strong evidence supporting the previous mechanism arising from finite temperature lattice calculations[3]. It is known in FT gauge theories that the *spatial* Wilson loop is *not* an order parameter for confinement. There is no reason for the spatial string tension to be different from zero even in the deconfined phase. This qualitative picture, continuity and non-zero value of the spatial string tension for all T, is completely confirmed by the lattice simulation. When properly interpreted in the string formalism the *spatial* string tension becomes the ordinary time string tension. Thus lattice results support the first property of the confinement mechanism. Lattice results also show that the spatial string tension is flat below T_c reaching its zero temperature value. This gives an evidence for the second property of the dimensional transition mechanism. The critical length L^* has a nice interpretation in the strip framework. It can be understood mathematically as an IR fixed point of the scaling transformations of the strip width. Physically the interpretation of L^* as the flux tube width arises in a natural way. If physics is frozen at L^*, the only relevant configurations to the functional generator are confined on a strip of width L^*.

3. References

1. A. Ferrando and A. Jaramillo, *University of Valencia Preprint* FTUV/94-41, IFIC/94-36.
   A. Ferrando, A. Jaramillo, V. Vento, work in progress.
2. E. D'Hoker, *Nucl. Phys.* **B201** (1982) 401.
3. M. Teper, *Phys. Lett.* **B311** (1993) 223.
   G.S. Bali, J. Fingberg, U.M. Heller, F. Karsch, K. Schilling, *Wuppertal Preprint* HLRZ-93-43, June 1993.
   M. Caselle, R. Fiore, F. Gliozzi, P. Guaita and S. Vinti, *Nucl. Phys.* **B422** (1994) 397.

COLOR CONFINEMENT AND DYNAMICAL EFFECT OF LIGHT QUARKS
IN THE DUAL GINZBURG-LANDAU THEORY

HIDEO SUGANUMA

RIKEN, Wako, Saitama 351-01, Japan

and

SHOICHI SASAKI and HIROSHI TOKI

RCNP, Osaka University, Ibaraki, Osaka 567, Japan

ABSTRACT

We study nonperturbative features of QCD using the dual Ginzburg-Landau theory. The color confinement is realized through the dual Higgs mechanism, which is brought by QCD-monopole condensation. We investigate the infrared screening effect on the color confinement due to the light-quark pair creation. By solving the Schwinger-Dyson equation, we find that the dynamical chiral-symmetry breaking is largely brought by the confining force.

1. Dual Higgs Mechanism for Color Confinement

We study color confinement and dynamical effects of light quarks in the dual Ginzburg-Landau (DGL) theory.[1,2] Because of the asymptotic freedom, QCD in the infrared region exhibits the nonperturbative features like the color confinement and the dynamical chiral-symmetry breaking (DχSB). The color confinement is characterized by the vanishing of the color dielectric constant and squeezing of the color electric flux, so that it has been regarded as the dual version of the superconductor using the duality of gauge theories. In this analogy, the confinement is brought by the dual Meissner effect originated from QCD-monopole condensation, which corresponds to Cooper-pair condensation in the superconductivity. As for the appearance of monopoles in QCD, 't Hooft[3] proposed an interesting idea of the abelian gauge fixing, which is defined by the diagonalization of a suitable gauge dependent variable. In this gauge, QCD is reduced into an abelian gauge theory with magnetic monopoles, which appear from the hedgehog-like configuration corresponding to the nontrivial homotopy class on the nonabelian manifold, $\pi_2(\mathrm{SU}(N_c)/\mathrm{U}(1)^{N_c-1}) = Z_\infty^{N_c-1}$.

We compare the QCD vacuum with the superconductor in terms of the abelian gauge fixing. In the superconductor, there are two kinds of degrees of freedom, the gauge field (photon) and the matter field corresponding to the electron and the metallic lattice, which provide the Cooper pair. On the other hand, there is only the gauge field in the pure gauge QCD, and therefore it seems difficult to find the analogous point between these two systems. However, in the abelian gauge, the diagonal part and the off-diagonal part of gluons play different roles. While the diagonal part behaves as the gauge field, the off-diagonal part behaves as the charged matter and provides QCD-monopoles, whose condensation leads to the dual Higgs mechanism, mass generation of the dual gauge field. Thus, QCD can be regarded as a similar system to the dual superconductor in the abelian gauge. Recent studies[4] using the lattice gauge theory have reported many evidences on the abelian dominance scheme and monopole condensation for the color confinement.

2. Dual Ginzburg-Landau Theory and Quark Confinement Potential

The DGL theory is considered as an infrared effective theory of QCD in the abelian gauge. Its Lagrangian in the Zwanziger form is described by the diagonal gluon $\vec{A}_\mu$ and the dual gauge field $\vec{B}_\mu$,[1,2]

$$\mathcal{L}_{\mathrm{DGL}} = -\frac{1}{n^2}[n \cdot (\partial \wedge \vec{A})]^\nu [n \cdot^* (\partial \wedge \vec{B})]_\nu - \frac{1}{2n^2}[n \cdot (\partial \wedge \vec{A})]^2 - \frac{1}{2n^2}[n \cdot (\partial \wedge \vec{B})]^2$$
$$+ \sum_{\alpha=1}^{3} \left[|(i\partial_\mu - g\vec{\epsilon}_\alpha \cdot \vec{B}_\mu)\chi_\alpha|^2 - \lambda(|\chi_\alpha|^2 - v^2)^2 \right] \tag{1}$$

apart from the quark sector. (The notations are the same as those in Refs.1 and 2.) The self-interaction of the QCD-monopole field χ_α is introduced phenomenologically like the Ginzburg-Landau theory in the superconductivity. There is the dual gauge symmetry corresponding to the local phase invariance of the QCD-monopole field χ_α as well as the residual gauge symmetry embedded in $SU(3)_c$. When QCD-monopoles are condensed, the dual Higgs mechanism occurs accompanying mass generation of the dual gauge field $\vec{B}_\mu$ and the spontaneous breaking of the dual gauge symmetry. On the other hand, the residual gauge symmetry is never broken in this process.

In this framework, the Dirac condition $eg = 4\pi$ for the dual gauge coupling constant g is naturally derived in the same way as in the Grand Unified Theory.[1,3] In view of the renormalization group, the DGL theory is not asymptotically free in terms of g similar to the scalar QED. Hence, asymptotic freedom is expected for the gauge coupling constant e owing to the Dirac condition. Thus, the DGL theory qualitatively shows asymptotic freedom in terms of e,[1] which seems a desirable feature for an effective theory of QCD.

First, we investigate the Q-$\bar{Q}$ system in the quenched level using the DGL theory. The Q-$\bar{Q}$ static potential includes the Yukawa and linear part,[1]

$$V(r) = -\frac{\vec{Q}^2}{4\pi} \cdot \frac{e^{-m_B r}}{r} + kr, \quad k = \frac{\vec{Q}^2 m_B^2}{8\pi} \ln(\frac{m_B^2 + m_\chi^2}{m_B^2}), \tag{2}$$

where $\vec{Q}$ denotes the color electric charge of the color source. Here, m_B is the mass of the dual gauge field $\vec{B}_\mu$, whose inverse corresponds to the cylindrical radius of the flux tube. The expression of the string tension k is quite similar to the energy per unit length of the Abrikosov vortex in the type-II superconductor.[1]

3. Quark Pair Creation, Infrared Screening Effect and DχSB

Next, we consider the dynamical effect of light quarks, which should be taken into account for the study of DχSB. A long hadron string can be cut through the light q-$\bar{q}$ pair creation, which is estimated by using the Schwinger formula.[1] This provides the screening effect on the long-range part of the confinement potential, as is observed in the lattice QCD with light dynamical quarks.[5] Taking account of such an infrared screening effect, we introduce the corresponding infrared cutoff ε to the gluon propagator as[1]

$$D_{\mu\nu}^{sc} = -\frac{1}{k^2}\left\{g_{\mu\nu} + (\alpha_e - 1)\frac{k_\mu k_\nu}{k^2}\right\} - \frac{1}{k^2}\frac{m_B^2}{k^2 - m_B^2} \cdot \frac{\epsilon^\lambda{}_{\mu\alpha\beta}\epsilon_{\lambda\nu\gamma\delta}n^\alpha n^\gamma k^\beta k^\delta}{(n \cdot k)^2 + \varepsilon^2} \tag{3}$$

without breaking the residual gauge symmetry. Here, we have introduced the infrared cutoff to the non-local factor $\frac{1}{(n\cdot k)^2}$, which provides the strong and long-range correlation as the origin of the confinement potential. Using this gluon propagator, we obtain a compact formula for the screened quark potential,[1]

$$V_{\text{linear}}^{sc}(r) = k \cdot \frac{1 - e^{-\varepsilon r}}{\varepsilon} \tag{4}$$

apart from the Yukawa part. This screened potential certainly exhibits the saturation for the longer distance than ε^{-1}.

Finally, we investigate DχSB in the DGL theory. We use the Schwinger Dyson (SD) equation with the gluon propagator including the nonperturbative effects on the confinement and the infrared screening. We find that QCD-monopole condensation largely contributes to the dynamical generation of the quark mass.[1] As the physical interpretation, the dual Higgs mechanism leads to the confining force (a strong and long-range attractive force) between the q-$\bar{q}$ pair with opposite color charges, which promotes q-$\bar{q}$ pair condensation similarly in the Nambu-Jona-Lasinio model.

In conclusion, we have studied nonperturbative features of QCD using the dual Ginzburg-Landau theory. The confinement potential has been reproduced through the dual Higgs mechanism, which is brought by QCD-monopole condensation. We have investigated the infrared screening effect on the linear confinement potential due to the light-quark pair creation, and obtained a compact formula for the screened potential. DχSB have been also studied in terms of the SD equation. We have found that DχSB is largely brought by the confining force between the light q-$\bar{q}$ pair.

One of the authors (H.S.) is supported by the Special Researchers' Basic Science Program at RIKEN.

REFERENCES

1. H. Suganuma, S. Sasaki and H. Toki, to be published in *Nucl. Phys.* **B**.
2. T. Suzuki, *Prog. Theor. Phys.* **80** (1988) 929 ; **81** (1989) 752.
   S. Maedan and T. Suzuki, *Prog. Theor. Phys.* **81** (1989) 229.
3. G. 't Hooft, *Nucl. Phys.* **B190** (1981) 455.
4. A. S. Kronfeld, G. Schierholz and U.-J. Wiese, *Nucl. Phys.* **B293** (1987) 461.
   T. Suzuki and I. Yotsuyanagi, *Phys. Rev.* **D42** (1990) 4257.
5. W. Bürger, M. Faber, H. Markum, M. Müller, *Phys. Rev.* **D47** (1993) 3034.

D. Zwanziger, PR D3, 880 (1971).

DYNAMICAL CHIRAL-SYMMETRY BREAKING
IN THE DUAL GINZBURG-LANDAU THEORY

SHOICHI SASAKI

RCNP, Osaka University, Ibaraki, Osaka 567, Japan

and

HIDEO SUGANUMA

RIKEN, Wako, Saitama 351-01, Japan

and

HIROSHI TOKI

Tokyo Metropolitan University, Hachiohji, Tokyo 192-03, Japan

ABSTRACT

We study the effect of QCD-monopole condensation on the dynamical chiral-symmetry breaking by using the dual Ginzburg-Landau theory of QCD. We formulate the Schwinger-Dyson equation and solve it numerically. The large enhancement is found for the chiral-symmetry breaking due to QCD-monopole condensation, which suggests the close relation between the color confinement and the chiral-symmetry breaking.

1. Introduction

The lattice QCD simulation showed a remarkable coincidence between the deconfinement phase transition and the chiral-symmetry restoration.[1] It is therefore natural to think the existence of a strong correlation between confinement and dynamical chiral-symmetry breaking (DχSB). In order to understand these non-perturbative properties, we use the dual Ginzburg-Landau theory[2] as an infrared effective theory of QCD based on the dual Higgs mechanism. According to 't Hooft's conjecture,[3] QCD is reduced into an abelian gauge theory with color-magnetic monopoles in the abelian gauge, which is assumed to be the relevant gauge for the study of the color confinement in terms of the dual Meissner effect or monopole condensation. Indeed, the recent lattice QCD simulations reported the evidence of abelian dominance and the important role of QCD-monopole condensation for the color confinement.[1,4]

In our previous works, we found that a linear confinement potential is derived between the static quark-antiquark system as the results of QCD-monopole condensation in the dual Ginzburg-Landau theory.[5] In this paper, we study the role of QCD-monopole condensation on DχSB by investigating the Schwinger-Dyson equation in this theory.[5,6]

2. Dynamical Chiral-Symmetry Breaking

The Schwinger-Dyson (SD) equation in the rainbow approximation is given by

$$S_q^{-1}(p) = i\not{p} + \int \frac{d^4k}{(2\pi)^4} \, \vec{Q}^2 \gamma_\mu \, S_q(k) \, \gamma_\nu \, D_{\mu\nu}^{SC}(p-k) \; ; \; \vec{Q}^2 = C_F \frac{e^2}{N_c+1} \quad (1)$$

in the Euclidian metric after the Wick rotation in the k_0-plane. Here, e is the gauge coupling constant and $C_F = \frac{N_c^2-1}{2N_c}$ for $SU(N_c)$. We take a simple form for the quark propagator, $S_q^{-1}(p) = i\not{p} - M(p^2)$. By taking the trace of Eq.(1), an integral equation for the dynamical quark mass $M(p^2)$ is obtained. Taking the angular average with respect to the direction of the Dirac string, the trace of the gluon propagator in the Landau gauge is obtained as

$$D_\mu^{SC\ \mu}(k) = \frac{1}{k^2} + \frac{2}{k^2+m_B^2} + \frac{1}{\varepsilon} \cdot \frac{4}{\varepsilon+\sqrt{\varepsilon^2+k^2}} \left(\frac{m_B^2-\varepsilon^2}{k^2+m_B^2} + \frac{\varepsilon^2}{k^2} \right), \quad (2)$$

which includes the mass of the dual gauge field, m_B, and the infrared cut-off ε. The mass m_B is generated through the dual Higgs mechanism when QCD-monopoles are condensed.[2,5,6] The infrared cut-off parameter ε is introduced for the infrared screening effect of the strong and long-range correlation due to the q-$\bar{q}$ pair creation.[5,6] As for the gauge coupling constant, we use the Higashijima-Miransky approximation, $\vec{Q}^2 = 4\pi C_F \cdot \alpha_s^{\text{eff}}(\max\{p^2,k^2\})$, as the hybrid type of the running coupling,[7]

$$\alpha_s^{\text{eff}}(p^2) = \frac{12\pi}{(11N_c-2N_f)\ln\left[(p^2+p_c^2)/\Lambda_{\text{QCD}}^2\right]} \quad (3)$$

where p_c approximately divides the momentum scale into the infrared region and the ultraviolet region, $p_c^2 = \Lambda_{\text{QCD}}^2 \exp[\frac{48\pi^2}{e^2} \cdot \frac{N_c+1}{11N_c-2N_f}]$.[5,6] Then, we get the final expression for the SD equation after performing explicitly the angle integrations,[6]

$$\begin{aligned}
M(p^2) = C_F \int_0^\infty \frac{dk^2}{4\pi} \, \alpha_s^{\text{eff}}(\max\{p^2,k^2\}) & \frac{M(k^2)}{k^2+M^2(k^2)} \left(\frac{k^2}{\max\{k^2,p^2\}} \right. \\
& + \frac{4k^2}{k^2+p^2+m_B^2+\sqrt{(k^2+p^2+m_B^2)^2-4k^2p^2}} \\
& + \frac{8k^2}{\pi\varepsilon} \int_0^\pi d\theta \frac{\sin\theta^2}{\varepsilon+\sqrt{k^2+p^2+\varepsilon^2-2kp\cos\theta}} \\
& \left. \times \left[\frac{m_B^2-\varepsilon^2}{k^2+p^2+m_B^2-2kp\cos\theta} + \frac{\varepsilon^2}{k^2+p^2-2kp\cos\theta} \right] \right), \quad (4)
\end{aligned}$$

which is reduced to the usual one of the QCD-like theory in the ultraviolet limit.[7,8]

We solve the SD equation, Eq.(4), numerically.[6] We show in Fig.1 the dynamical quark mass $M(p^2)$ for several values of m_B, which is proportional to the QCD-monopole condensate. The quark mass $M(p^2)$ increases with m_B, which means that QCD-monopole condensation provides a large contribution to DχSB.[5,6] (Only a trivial solution exists for small $m_B \leq 0.2$GeV).

We also estimate the quark condensate $\langle \bar{q}q \rangle$ and the pion decay constant f_π under one parameter set : $e = 5.5$, $m_B = 0.5\text{GeV}$ and $\varepsilon = 85\text{MeV}$. The quark confinement potential (the string tension $k \simeq 1\text{GeV}/\text{fm}$) and flux tube radius $\simeq 0.4\text{fm}$ are reproduced with these parameters.[5] Here, the QCD scale parameter is fixed at $\Lambda_{\text{QCD}} = 200\text{MeV}$. We find the dynamical quark mass $M(0) \simeq 354\text{MeV}$, the quark condensate $\langle \bar{q}q \rangle = -(222\text{MeV})^3$ and the pion decay constant $f_\pi = 88\text{MeV}$.[6] Thus, the dual Ginzburg-Landau theory seems to give good results on both the quark confining potential and DχSB semi-quantitatively.

In conclusion, we find that DχSB is enhanced by QCD-monopole condensation, which leads to the color confinement. This result indicates the evidence for the close relation between the color confinement and DχSB through QCD-monopole condensation in the dual Ginzburg-Landau theory.

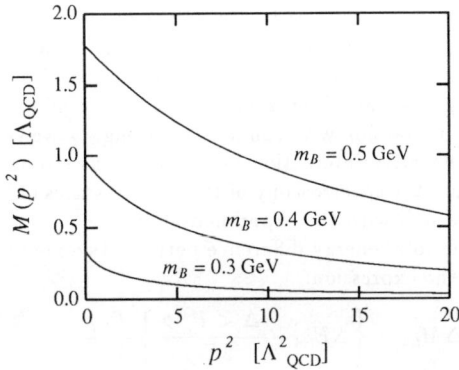

Fig. 1. : The dynamical quark mass $M(p^2)$ as a function of the Euclidian momentum squared p^2 for $m_B = 0.3, 0.4$ and 0.5GeV. The other parameters are $e = 5.5$ and $\varepsilon = 85\text{MeV}$, while the QCD scale parameter is fixed at $\Lambda_{\text{QCD}} = 200\text{MeV}$.

REFERENCES

1. For Instance, papers in *Proc. of the symposium on Lattice '93, Nucl. Phys.* **B** (Proc. Suppl.) **34** (1994) 1.

2. T. Suzuki, *Prog. Theor. Phys.* **80** (1988) 929 ; **81** (1989) 752.

3. G. 't Hooft, *Nucl. Phys.* **B190** (1981) 455.

4. A. S. Kronfeld, G. Schierholz and U.-J. Wiese, *Nucl. Phys.* **B293** (1987) 461.

5. H. Suganuma, S. Sasaki and H. Toki, to be published in *Nucl. Phys.* **B**.

6. S. Sasaki, H. Suganuma and H. Toki, RIKEN preprint AF-NP-172 (1994).

7. K. Higashijima, *Prog. Theor. Phys. Suppl.* **104** (1991) 1 and references therein.

8. A. Barducci, R. Casalbuoni, S. De Curtis, D. Dominici and R. Gatto, *Phys. Rev.* **D38** (1988) 238.
   K.-I. Aoki, M. Bando, T. Kugo, M. G. Mitchard and H. Nakatani, *Prog. Theor. Phys.* **84** (1990) 683.

ELECTROMAGNETIC HADRON MASS DIFFERENCES
AND ESTIMATION OF ISOTOPIC SYMMETRY VIOLATION
PARAMETERS OF QCD VACUUM FROM QUARK MODEL.

A. E. DOROKHOV

Joint Institute for Nuclear Research
Bogoliubov Theoretical Laboratory
Dubna, Moscow region, 141980, Russia

In the present note, we shall consider the isospin mass splittings of low - lying hadrons and obtain estimations of the isospin violation in quark masses and condensates. To this end, we shall use the version of the bag model based on the idea that the interaction of hadron constituents with background vacuum fields in the bag plays the dominant role [1]. It has been shown that the spin - dependent forces are determined by the interaction of quarks with instantons (short - range vacuum fluctuation) while the stability (confinement forces) is due to their interaction with condensates (long - wave vacuum fluctuation). This nonperturbative interaction between quarks strongly depending on quark masses defines the spectroscopy of the ground states of hadrons. The results obtained agree well with the experimental ones[2].

Let us consider the total energy difference between two members of a multiplet. It is given by the expression:

$$\Delta M_{tot} = \left\{ \Delta E_{tot} - \frac{\Delta < P^2 >}{2E_{tot}} \right\} \frac{E_{tot}}{M_{tot}} \qquad (1)$$

with $\Delta E_{tot} = \Delta E_{kin} + \Delta E_{em} + \Delta E_{strong}$, and $\Delta E_{strong} = \Delta E_{vac} + \Delta E_{inst} + \Delta E_{gl}$, where M_{tot} is a hadron mass, $M_{tot}^2 = E_{tot}^2 - < P^2 >$, with center - of - mass motion correction $< P^2 >$ taken into account [3], ΔE_{gl} is due to the QCD hyperfine interaction of quarks inside a bag [4], and ΔE_{vac} and ΔE_{inst} are due to the interaction of quarks with vacuum fields.

The contributions, ΔE_{em}, ΔE_{kin}, ΔE_{gl} ΔE_{vac} and ΔE_{inst}, are discussed in detail in [2]. The results of calculations are presented in Table I. The numbers correspond to the parameters of the QCD vacuum:

$$< 0|\bar{U}U|0 > = -(221 \ MeV)^3, < 0|\frac{\alpha_s}{\pi}G^2|0 > = 0.031 \ GeV^4, \quad \rho_c^2 = 1 \ GeV^{-2},$$

the differences of u and d quark masses $(m_u = 5 \ MeV)$ and their condensates

$$m_d - m_u = 3.5 \ \text{MeV}, \gamma = \frac{< \bar{D}D > - < \bar{U}U >}{< \bar{U}U >} = -2 \cdot 10^{-3}. \qquad (2)$$

In our calculations we take the values $m_s = 200 \ MeV$, $\alpha_s = 0.4$ and $\delta = \frac{< \bar{S}S > - < \bar{U}U >}{< \bar{U}U >} = -0.1$.

There are two exceptional EMD combinations which depend only on the electromagnetic term ΔE_{em} thus being sensitive only to the bag radius. They are the $I = 2$ part of the $\Sigma-$ and the π mass differences: $\Sigma^+ + \Sigma^- - 2\Sigma^0 = 1.71 \pm 0.15\ MeV$, $\pi^\pm - \pi^0 = 4.59 \pm 0.05\ MeV$ (this is valid for $\rho^\pm - \rho^0$, too, but its experimental value is not well defined). The first one is satisfied in the range of bag radii $R = 5 \sim 6\ GeV^{-1}$, which confirms a good and self - consistent description of the mass scales of the baryon octet and splittings within it. From Table I we see that the bag stability radius, $R = 5.6\ GeV^{-1}$, belongs to this interval and $\Sigma^+ + \Sigma^- - 2\Sigma^0 = 1.71\ MeV$ in fine agreement with experimental value. As to the pion, due to large negative instanton and c.m. energy contributions, it has no radius of bag stability. This is a signal of the Goldstone nature of the pion in our model.

Given a typical bag radius and due to the absence of instanton contribution, the $p - n$ mass difference $(p - n = 1.3\ MeV)$ is mainly defined by the sum of the electromagnetic energy term, ΔE_{em}, and the kinetic energy term, ΔE_{kin}. The latter directly depends on the $u - d$ quark mass difference. Within some uncertainties in the definition of the model parameters, the value of Δm_q is grouped around $3.5\ MeV$.

Usually, in the bag model there is the problem with the description of the $I = 1$ part of octet splitting. In fact, the bag radius and Δm_q fixed as above, it is impossible to saturate the Coleman - Glashow relation (CG), $p - n + \Xi^0 - \Xi^- = \Sigma^+ - \Sigma^-$, only with the kinetic energy, ΔE_{kin}, and electromagnetic energy, ΔE_{em}, contributions. That is, $CG_0^{theor} = \Delta E_{kin}^{CG} + \Delta E_{em}^{CG} + ... \approx -4.5\ MeV$ while the experimental value is about $CG^{ex} \approx -8\ MeV$. It is important to stress that the color - magnetic energy contribution could not save the situation with CG even for the large constant $\alpha_s^{MIT} = 2.2$ [4].

The $I = 1$ part of the $\Sigma-$ and $\Xi-$ mass differences is essentially due to the instanton contribution that is proportional to the number of strange quarks. It is important that these splittings are of the same order as the $u-$, $d-$ quark mass differences. Then one has: $\Delta E_{inst}^\Sigma = \dfrac{E_{inst}^{0s}}{m_0^*}\Delta m_q$, and for typical values for $E_{inst}^\Sigma \approx -70\ MeV$ it follows that the effective quark mass m_0^* is of the same order as E_{inst}^Σ. This ratio requires quite a small value for $m_0^* \sim < \bar{Q}Q > \rho_c^2$ and a large value for a gluon condensate $E_{inst} \sim < 0|\frac{\alpha_s}{\pi}G^2|0 > / < \bar{Q}Q >^2$. The Coleman - Glashow relation is satisfied by each contribution separately because, as noted above, the bag radii for N, Σ, Ξ are well equal. As to the absolute value of the left and right sides of this relation, the instanton contribution is very important. ¿From Table I we find $p - n + \Xi^0 - \Xi^- = -8.30\ MeV$ $(-7.7 \pm 0.6\ MeV)_{exp}$ and $\Sigma^+ - \Sigma^- = -8.06\ MeV$ $(-8.07 \pm 0.09\ MeV)$ in excellent agreement with experiment values.

Thus, it is shown that the instanton plays the key role in the saturation of the CG relation between octet baryon states. In this case, as in the case of the dynamical explanation of the Okubo - Zweig - Iizuka rule, the gluon exchange contributions are very small and, therefore, from the magnitudes of these effects we can clearly judge on the strength of the instanton induced interaction.

The contributions related to the condensate difference are not large, and are poor fixed from hadron mass differences. From our analysis we can define only the lower bound of γ: $0 > \gamma > -0.003$. With a precision of the model and data it is difficult to expect for more.

In summary, we conclude from our results for isospin mass hadron differences that $m_d - m_u = 3.5 MeV$ and $< \bar{D}D > - < \bar{U}U > = -(0 \sim 3) \cdot 10^{-3} < \bar{U}U >$. It would be interesting to consider the D and D^* isospin mass differences in the framework of the quark model with QCD vacuum induced interaction.

The author is very thankful to Professor A.W. Thomas and members of the theoretical seminar of Adelaide University and N.N. Achasov, S.B. Gerasimov, N.I. Kochelev, E. Rodionov for stimulating discussions.

1. A. E. Dorokhov, N. I. Kochelev *Sov. J. Nucl. Phys.* **52** (1990) 135 (214); In Proc. Int. Conf. *Quarks-86, Tbilisi*, p. 392, Moscow, 1986.
2. A. E. Dorokhov, *Nucl. Phys.*A (1994) to be appeared, JINR preprint E2-94-231, Dubna.
3. J. Donoghue, K. Johnson, *Phys. Rev.* **D21** (1980) 1975.
4. R.P. Bickershtaff, A.W. Thomas *Phys. Rev.* **D25** (1982) 1869.

Table I. The electromagnetic mass differences of Hadrons (MeV).

Particles	R	M	ΔM_{kin}	ΔM_{EM}	ΔM_{gl}	ΔM_{vac}	ΔM_{inst}	ΔM_{tot}	ΔM_{exp}
P-n	5.60	940	-1.06	0.55	0.04	-0.15	0	-1.23	-1.2933± 0 0(
$\Sigma^+ - \Sigma^0$	5.40	1230	-1.19	0.55	-0.01	-0.13	-1.49	-3.18	-3.18 ±0.17
$\Sigma^0 - \Sigma^-$			-1.19	-1.16	-0.01	-0.13	-1.49	-4.89	-4.89 ±0.08
$\Xi^0 - \Xi^-$	5.30	1330	-1.22	-1.21	-0.07	-0.13	-3.16	-7.07	-6.4 ±0.6
$\Delta^{++} - \Delta^0$	6.40	1240	-2.56	2.51	0.08	-0.48	0	-1.75	-2.70 ±0.30
$\Delta^+ - \Delta^0$			-1.28	0.48	0.04	-0.24	0	-1.61	
$\Delta^0 - \Delta^-$			-1.28	-0.97	0.04	-0.24	0	-2.90	
$\Sigma^{*+} - \Sigma^{*-}$	6.65	1380	-2.66	-0.50	0.08	-0.56	0	-4.89	-4.4 ±0.5
$\Sigma^{*0} - \Sigma^{*-}$			-1.33	-0.94	0.04	-0.28	0	-3.03	-3.5 ±1.2
$\Xi^{*0} - \Xi^{*-}$	6.70	1510	-1.37	-0.96	0.03	-0.29	0	-3.18	-3.2 ±0.6
$K^+ - K^0$	5.70	700	-0.87	0.56	-0.11	-0.16	-3.34	-5.40	-4.024 ±0.03
$\rho^\pm - \rho^0$	6.00	780	0.00	0.77	0	0	0	0.75	-0.3 ±2.2
$K^{*+} - K^{*0}$	6.00	890	-1.07	0.53	0.03	-0.20	0	-1.19	-6.7 ±1.2

SHOCK WAVES AND THE VACUUM STRUCTURE OF GAUGE THEORIES

M. MARTELLINI and M. ZENI

Dip. di Fisica, Univ. di Milano, Via Celoria, 16
I.N.F.N., Sezioni di Pavia e Milano
Milano, 20133, ITALY

and

A. SAGNOTTI

Dip. di Fisica, Univ. di Roma "Tor Vergata", Via della Ricerca Scientifica, 1
I.N.F.N., Sezione di Roma "Tor Vergata"
Roma, 00133, ITALY

ABSTRACT

In Yang-Mills theory massless point sources lead naturally to shock-wave configurations. Their magnetic counterparts endow the vacuum of the four-dimensional compact abelian model with a Coulomb-gas behaviour whose physical implications are briefly discussed.

The current semiclassical picture of the vacuum in gauge theories rests, to a large extent, on the known solutions of the Yang-Mills field equations[1]. It is common wisdom that attaining a detailed understanding of this vacuum is a major challenge for Quantum Field Theory, as well as a crucial step in assessing its actual role in High-Energy Physics. In this talk we consider a class of rather simple shock-wave solutions of the field equations with massless point sources. In the compact abelian gauge theory, their euclidean counterparts exhibit rather neatly a phase transition[2], thus providing a simple explicit realization of the standard picture[3].

Let us begin by considering the Yang-Mills field equation $D^\mu F_{\mu\nu} = 4\pi j_\nu$ with the massless point source

$$j_\nu^a = q \, I^a \, \delta_\nu^u \, \delta(u) \, \delta^2(r) \qquad , \tag{1}$$

where u and v are light-cone coordinates ($u = \frac{x^0 - x^3}{\sqrt{2}}$; $v = \frac{x^0 + x^3}{\sqrt{2}}$), r is a space-like coordinate vector orthogonal to u and v, and I^a labels the color charge of the point particle. The resulting classical solution,

$$A^a_{\ \mu} = -2 \, q \, I^a \, \delta_\mu^u \, \delta(u) \, \log r \quad , \tag{2}$$

where r denotes the length of r, may be obtained by inspection, or by a simple extension of the "cut and paste" procedure[4] used to generate a similar type of gravitational wave. The electric and magnetic field strengths associated to eq. (2),

$$E_i = \sqrt{2} \, q \, I^a \, \delta(u) \, \frac{r^i}{r^2} \qquad , \tag{3}$$

$$B_i = -\sqrt{2} \, q \, I^a \, \delta(u) \, \varepsilon^{ij} \, \frac{r^j}{r^2} \qquad (i = 1, 2) \qquad ,$$

may also be obtained as singular boosts of a static Coulomb field in the limit where the velocity $v/c \to 1$ [5]. The "two-dimensional" shape of the shock wave applies to arbitrary superpositions of comoving massless point-like currents and may be ascribed to the relativistic contraction of the fields in the longitudinal direction.

In the quantum theory, the topological character of the field configuration space plays a crucial role in determining the nature of the correct degrees of freedom. In the prototype example for this type of phenomenon, the 2D XY model, the fundamental field is an angular variable $\theta(r)$ and the naive elementary excitations, long-wavelength spin waves, must be supplemented with genuinely new ones, vortices. From the mathematical viewpoint, vortices are singular field configurations that result in effective (quantized) violations of the "Bianchi identity" for $d\theta$. These excitations are responsible for the Kosterlitz-Thouless (KT) phase transition[6] that separates the weak and strong coupling regimes of the model.

Vortices are actually relevant in a number of different models with periodic field configuration spaces, most notably in lattice gauge theories, where they may be associated with magnetic monopoles. Again, in continuum formulations they manifest themselves as singular field configurations that violate the relevant "Bianchi identities". A number of reasons call for a periodic formulation of abelian gauge theories, most notably the fact that in unified models abelian gauge fields typically emerge from the spontaneous breaking of non-abelian gauge symmetries[3].

For compact 3D QED Polyakov[7] has shown that the periodicity results in a vacuum filled with a plasma of monopoles and in a mass gap for the gauge fields, as well as in a confinement phenomenon at all scales. For compact 4D QED a quantitative discussion of the role of vortices has been hampered by the string-like nature of the corresponding excitations, monopole current loops. Still, the basic argument is rather simple. It is best exhibited starting from the Villain[8] form of the action

$$S = \frac{1}{4g^2} \sum_{x,\alpha\beta} (F_{\alpha\beta}(x) + 2\pi n_{\alpha\beta}(x))^2 \quad , \tag{4}$$

where the gauge potential is an angular variable and the $n_{\alpha\beta}(x)$ are matrices of integers associated to the lattice plaquettes that may be related to integer-valued currents upon integration over elementary cubes $\square^\mu$ of the lattice,

$$\oint_{\partial\square^\mu} n_{\alpha\beta}(x) = j^\mu \quad . \tag{5}$$

The Berezinsky decomposition $n(x)_{\alpha\beta} = \partial_\alpha n_\beta - \partial_\beta n_\alpha + \varepsilon_{\alpha\beta\gamma\delta}\, \partial_\gamma \varphi_\delta$ exhibits the longitudinal part of n, as well as the vector field φ_μ, whose source is the current of eq. (5) and whose (dual) field strength adds to the usual one. Since the total field strength

$$F'_{\alpha\beta} = F_{\alpha\beta} + 2\pi n_{\alpha\beta} = F_{\alpha\beta} + \ldots + \frac{1}{2}\, \varepsilon_{\alpha\beta\gamma\delta}\, \mathcal{F}_{\gamma\delta} \tag{6}$$

violates the Bianchi identity of the original gauge field, the current in eq. (5) is of *magnetic* origin.

An elegant lattice description of this model, allowing for electric current (Wilson) loops as well, was presented in [12]. The resulting picture involves interacting monopole and charge loops, but it is difficult to turn it into a quantitative analysis, since in this case the effective sine-Gordon description of 3D QED[7] should be replaced by a theory of monopole loops. Although a proper description of this theory is likely to be complicated, the intuition gained from ordinary string theory[9] suggests a possible way of gaining quantitative insight into the problem. This may be associated to "straight" current loops, and conceivably to massless ones near the phase transition, that according to numerical estimates appears to be of second order[2].

In the remainder of this talk we would like to show how the truncation of the monopole strings to these "zero modes" accounts both for the phase transition and for Wilson's area law[11] in a rather neat fashion. To this end, we need the duals of the fields in eq. (3). They may be derived from a "magnetic" analogue of the vector potential of eq. (2) or, alternatively, from the ordinary potential

$$A_\mu = (0, 0, A_i) , \qquad where \qquad A_i = q_{(m)}\theta(u) \, \varepsilon^{ij} \, \frac{r^j}{r^2} \qquad (i, j = 1, 2) \quad , \qquad (7)$$

whose Dirac string

$$B_3 = 2\pi q_{(m)}\theta(u) \, \delta^2(r) \qquad (8)$$

would be ineffective in eq. (4) upon suitable quantization of the monopole charge, a single Dirac quantum corresponding to $q_{(m)} \pm 1$. Interestingly, the potential in eq. (7) plays an important role in knot theory[10].

In computing the action for a pair of waves, it is convenient to resort to a covariant notation, that has the further advantage of illuminating the geometry of the space-like planes where the interactions take place. A "straight" current in a generic direction may be written

$$j_\mu(x) = \frac{1}{2} \, q_{(m)}k_\mu \int d\tau \delta^4[x^\alpha - x_0^\alpha - k^\alpha \tau] \quad , \qquad (9)$$

where in the massless case k_μ is a null vector. The total action for a pair (ij) of shock waves is then

$$S_{ij} = \frac{\pi q_{(m)i}q_{(m)j}}{g^2} \, \log(x_{ij} \, \Pi \, x_{ij}) \quad , \qquad (10)$$

where x_{ij} is the distance between the two lines and Π is a projector onto the space-like plane orthogonal to the two wave vectors. This result essentially holds in the Wick rotated case as well, where the calculation requires a suitable extension of $\delta(x)$ to the complex plane, rather interesting in its own right. The logarithmic interaction closely parallels the state of affairs for the XY model and is just enough to compete with the point-like entropy of these configurations, thus displaying a KT-like phase transition, while the divergent self interactions require globally neutral sets of monopole currents. Interestingly, a transition of this type would be predicted by the Migdal-Kadanoff approximation[13], known to become less accurate as the space dimensionality is increased, consistenly with our neglect of higher extended excitations. The naive estimate of the transition temperature from eq. (10) is rather amusing, since it yields $2g^2_c = \pi/2$ for a pair of fundamental Dirac monopoles with a KT-like measure d^4x, to be compared with the loop-space estimate for the full gas of monopole strings[12,2], $2g^2_c \approx 1.57$!

Finally, the area law[11] may be anticipated by comparing the effective KT structure of our "straight-line" vacuum to the Coulomb-gas picture of the three-dimensional model of ref.[7]. One may then arrive at an effective Sine-Gordon dynamics to infer that, above the critical temperature T_c, double layers of monopole lines form around the Wilson loop, thus implying the area law and confinement. One may also envisage a similar analysis of black-hole physics[4], given the formal similarity between the logarithm of the Wilson loop and the Bekenstein entropy. More details, including the explicit invariant measure for the moduli, will be presented elsewhere[14].

Acknowledgements

It is a pleasure to acknowledge stimulating discussions with T. Banks and A. Di Giacomo. We also benefitted from the kind hospitality of the Physics Departments of the University of Milan and of the University of Rome "Tor Vergata" and of the Centre de Physique Theorique of the Ecole Polytechnique. This work was supported in part by E.E.C. Grants CHRX-CT93-0340 and CHRX-CT92-0035.

References

1. For a review, see A. Actor, *Rev. Mod. Phys.* **51** (1979) 461.
2. For a review, see J.B. Kogut, in *Recent Advances in Field Theory and Statistical Mechanics*, Proc. Les Houches 1982, eds. J.B. Zuber and R. Stora (North-Holland, Amsterdam 1984).
3. G. 't Hooft, *Proc. 1975 EPS Conference*, ed. A. Zichichi (Compositori, Bologna, 1976);
   S. Mandelstam, *Phys. Rep.* **23C** (1976) 245;
   A.M. Polyakov, *Gauge Fields and Strings* (Harwood, New York, 1987).
4. T. Dray and G. 't Hooft, *Nucl. Phys.* **B253** (1985) 173.
5. R. Jackiw, D. Kabat and M. Ortiz, *Phys. Lett.* **B277** (1992) 148.
6. J.M. Kosterlitz and D.J. Thouless, *J. Phys.* **C6** (1973) 1181;
   J.M. Kosterlitz, *J. Phys.* **C7** (1974) 1046.
7. A.M. Polyakov, *Phys. Lett.* **B59** (1975) 79.
8. J. Villain, *J. Physique* **36** (1975) 581.
9. M.B. Green, J.H. Schwarz and E. Witten, *Superstring Theory* (Cambridge University Press, Cambridge, 1987).
10. T. Khono, *Adv. Studies in Pure Math.* **16** (1988) 255;
    E. Guadagnini, M. Martellini e M. Mintchev, *Nucl. Phys.* **B336** (1990) 581.
11. K.G. Wilson, *Phys. Rev.* **D10** (1974) 2445.
12. T. Banks, R. Myerson and J. Kogut, *Nucl. Phys.* **B129** (1977) 493;
    A. Ukawa, P. Windey and A.H. Guth, *Phys. Rev.* **D21** (1980) 1013.
13. A.A. Migdal, *Sov. Phys. JETP* **42** (1975) 413;
    L.P. Kadanoff, *Rev. Mod. Phys.* **49** (1977) 267.
14. M. Martellini, A. Sagnotti and M. Zeni, in preparation.

THE STABILITY OF NON-ABELIAN FLUX TUBES AND THEIR LENGTH

E.I. GUENDELMAN
D.A. OWEN
Physics Department, Ben Gurion University of the Negev
Beer Sheva, Israel

A. LEONIDOV
Physics Department, Bielefeld University, Germany
and
Lebedev Physics Institute, Moscow, Russia

ABSTRACT

We investigated the effect of external field boundary conditions of a background chromomagnetic field on the instability (discovered by Nielsen and Olesen) for the vacuum of a non-Abelian field theory. We find that the vacuum is neither stabilized by a one-dimensional nor a two dimensional cutoff of this magnetic field. However, the vacuum in the presence of flux tubes whose length is restricted to be under a critical value, L_0, is stable. There is a tendency for flux tubes of lengths greater than L_0 to spontaneously fragment into segments each of which is smaller than L_0. This corresponds to a dual picture which allows stable electric flux tubes.

1. Introduction

It had been found in 1977 by Savvidy[1] that the infrared singularities lead to an instability of the "trivial" vacuum (i.e. the vacuum in which all fields have zero expectation value) towards the formation of a nontrivial magnetic condensate. On the other hand, Nielsen and Olesen[2] found that even this vacuum cannot be the true one since it is intrinsically unstable as well. They demonstrated this by examining the field equations in the context of an SU(2) Yang-Mills theory and showing that the fluctuations about the magnetic condensate contain exponentially growing modes.

In the discussion of Nielsen and Olesen,[2] in which they found the vacuum instability, they assumed a constant magnetic field. This, of course, assumes that the magnetic field extends over all space and is not confined to any localized region. In the later part of the paper of Ambjørn, et. al.[3] it was conjectured that the instability disappears when the background field develops a domain structure (spaghetti vacuum). However, the proof that such a structure for the vacuum would be stable, was lacking. The results of this present work, makes such a vacuum structure less plausible since we have shown that infinitely long, background filaments, taken individually, are intrinsically unstable. We examine the stability associated with a variety of external chromo-magnetic field configurations which we discuss below. More details may be found in our recent publication.[4]

2. Instability as a Function of Background Field Configuration

We examine the effect on the vacuum stability when boundary conditions are imposed on the background magnetic field. What follows will be in the context of *pure* SU(2) non-Abelian gauge theory. The Lagrangian is therefore given by

$$\mathcal{L} = -\frac{1}{4} [G^a_{\mu\nu}]^2 \tag{1}$$

where

$$G^a_{\mu\nu} = \partial_\mu A^{(a)}_\nu - \partial_\nu A^{(a)}_\mu + g\varepsilon^{abc} A^{(b)}_\mu A^{(c)}_\nu \tag{2}$$

We define a W-boson field by $W^\pm_\mu = \frac{1}{\sqrt{2}} [A^{(1)}_\mu \pm (-iA^{(2)}_\mu)]$ and the "photon" by $A_\mu \equiv A^{(3)}_\mu$.

In terms of $W^\pm$ and A_μ, $\mathcal{L}$ becomes

$$\mathcal{L} = -\frac{1}{4} F^{\mu\nu}F_{\mu\nu} - \frac{1}{2} |D_\mu W^+_\nu - D_\nu W^+_\mu|^2 + \frac{1}{2} g \, F_{\mu\nu} W^-_\rho \, S^{\mu\nu\rho\sigma} W^+_\sigma$$

$$-\frac{1}{4} g^2 \, W^-_\rho \, S^{\mu\nu\rho\sigma} W^+_\sigma W^- \alpha \, S_{\mu\nu\alpha\beta} W^{+\beta} \tag{3}$$

$$F_{\mu\nu} = \partial_\mu A_\nu - \partial_\nu A_\mu; \quad D_\mu = \partial_\mu + igA_\mu \quad \text{and} \quad S_{\mu\nu\rho\sigma} = -i(g_{\mu\rho}g_{\nu\sigma} - g_{\mu\sigma}g_{\nu\rho}) \tag{4}$$

The equations of motion for the W-field in the presence of an external field A_μ resulting from (3) are (neglecting the terms cubic in $W_\mu \equiv W^-_\mu$ and W^+_μ)

$$[D_\lambda D^\lambda g^{\sigma\nu} - D_\lambda D^\sigma g^{\lambda\nu} + igF^{\sigma\nu}]W_\nu = 0 \tag{5}$$

If we assume the background Landau gauge condition:[2] $(\partial^\nu + igA^\nu)W_\nu = 0$, Eq. (5) becomes

$$[(\partial_\lambda + igA_\lambda)(\partial^\lambda + igA^\lambda)g^{\sigma\nu} + 2igF^{\sigma\nu}]W_\nu = 0 \tag{6}$$

If in addition we consider the external field to be a constant "color magnetic field"[2], we find the result found by Nielsen and Olesen that the square of the energy eigenvalue, E^2 can take on negative eigenvalues. This leads exponentially growing modes which is results in the vacuum instability. Physically, the anomalous magnetic moment of the "non-Abelian" W-field (gluons for QCD) lowers the E^2 of the W-modes in the constant magnetic field in such a way as to overcome the zero point kinetic energy so that the total zero point energy is imaginary.

2.1 One-Dimensional Cutoff

We considered a constant magnetic field along the z-axis such that in a perpendicular direction, say $x \equiv x^1$, the field goes to zero for $x < -L$ and for $x > L$ for some given $L > 0$ where L is a constant. Eq. (6) reduces to a Schrödinger-like equation[4]

$$\left(-\frac{d^2}{dx^2} + V(x)\right)u_\mu(x) = E^2 u_\mu(x)$$

$$\tag{7}$$

No matter how small we take the magnetic field to be, it can be shown that Eq. (7) has at least one negative eigenvalue (i.e. bound state) and thus for that state, $E^2 < 0$ and the vacuum instability remains.

2.2 Cylindrically Symmetric Configuration

(a) Quantized Magnetic Flux

We considered a constant magnetic flux along the z-axis within a cylinder bounded by the surface $\rho_0 = \sqrt{x^2 + y^2}$. The solution of Eq. (6) for this case the standard confluent hypergeometric function for $\rho < \rho_0$ and a Bessel function for $\rho > \rho_0$. Here, too, independent of how small the external magnetic field is taken to be, E^2 is negative for the lowest energy state and hence, the instability remains.

(b) Shielded Magnetic Flux Cylinder

In this case, we assume that the vector potential is confined in the cylinder and so it vanishes for $\rho > \rho_0$. Here, too, we find from analyzing Eq. (6) that independent of the field strength or the value of ρ_0, there is an instability associated with the lowest state.

3. Instability and the Geometry of Flux Tubes

If we now consider a flux tube of finite length, and the flux quantization condition[4], Eq. (6) becomes

$$\frac{d^2 U_\mu^{(E)}}{d\rho^2} - \left(\frac{3}{4\rho^2} + k_z^2\right) U_\mu^{(E)} = -E^2 U_\mu^{(E)}; \qquad (\rho > \rho_0) \qquad (8)$$

$$\frac{d^2 U_\mu^{(E)}}{d\rho^2} + \left(\frac{1}{4\rho^2} - \frac{g^2 \rho^2 H^2}{4} - k_z^2 \pm 2gH\right) U_\mu^{(E)} = -E^2 U_\mu^{(E)}; \qquad (\rho < \rho_0) \qquad (9)$$

where H is the magnitude of the magnetic field and $\hbar k_z$ the momentum of the W-field which has a minimum value which is $\frac{\hbar}{L}$,(L is the length of the flux tube) given by the uncertainty principle. If the flux tube is short enough, i.e. for $k_z^2 > 2gH$ or $(\frac{1}{L})^2 > 2gH$ the above equations (8) and (9) have no bound state solutions and the flux tubes are stable. However, if they are longer than $L_0 = \sqrt{\frac{1}{2gH}}$ then the above equations have a bound state solution and the flux tubes are unstable. The boundaries of this configuration correspond to magnetic monopoles. To make contact with strings connecting quarks and antiquarks, the above picture has be regarded as representing a dual model for QCD.

4. Confinement and Duality in QCD

It is useful to relate our results to the duality considerations of both 't Hooft[5] and Mandelstam[6] which studied the connection between the stability of electric and magnetic flux tubes. Our result that magnetic flux tubes greater than a critical length are inherently unstable is consistent with only one of the possibilities considered by 't Hooft, namely the existence of stable flux tubes in the electric domain. Our result can also be important in the context of QCD vacuum building using the abelian chromomagnetic monopoles.[7] An isolated abelian monopole has an infinitely long Dirac string attached to it and is therefore unstable. This also means that the flux tubes connecting monopole and antimonopole can not be too long, which should lead to a peak in the distribution of the flux tube length in the monopole-antimonopole configurations.

1. G. Savvidy, *Phys. Lett. B71*, (1977) 133.
2. N. Nielsen and P. Olesen, *Nucl. Phys.* **B144** (1978) 376.
3. J. Ambjørn, N.K. Nielsen and P. Olesen, *Nucl. Phys.* **B152** (1979) 75.
4. E. I. Guendelman, D. A. Owen and A. Leonidov, *Int. J. Mod. Phys.* **A8** (1993) 4745.
5. G. 't Hooft, *Nucl. Phys.* **B153** (1979) 141.
6. S. Mandelstam, *Phys. Rev.* **D19** (1979) 2391.
7. K. Yee. Monopoles and Quark Confinement: Introduction and Overview. Preprint LSUHEP-022094, hep-ph/9404363 (1994).

TOWARDS REALIZATION OF CHIRAL SYMMETRY

N. N. ACHASOV

Laboratory of Theoretical Physics, Institute for Mathematics
630090 Novosibirsk 90 RUSSIA

ABSTRACT

It is shown that the Goldstone boson associated with the spontaneous breakdown of chiral symmetry can reproduce the axial–anomaly pole in the single-loop approximation only in a peculiar "conspiracy" with a massless pseudovector boson.

The classical axial anomaly problem is still in the focus of attention of theorists.

Recently I published[1] the analytic form of invariant amplitudes free of kinematical singularities describing the triangle graph: (axial-vector current) $\to q\bar{q} \to \gamma(k_1)\gamma(k_2)$ at $k_1^2 = 0$, $k_2^2 \neq 0$. I showed that the axial–anomaly pole emerges only in the massless–fermion limit and only for real photons, $k_2^2 = 0$, in contrast to an opinion which can be found sometimes.

In the present paper the problem of spontaneous breaking of chiral symmetry in connection with the results published in[1] is analyzed.

As it is known, the amplitude describing the transition: (axial-vector current) * $\to q\bar{q} \to$ (conserved current) $\times$ (conserved current) reads as follows

$$T_{\alpha\beta\mu} = \sum_i A_i t_{\alpha\beta\mu}^i = A_1 k_1^\sigma \epsilon_{\sigma\alpha\beta\mu} + A_2 k_2^\sigma \epsilon_{\sigma\alpha\beta\mu} + A_3 k_{1\beta} k_1^\delta k_2^\sigma \epsilon_{\delta\sigma\alpha\mu} +$$
$$+ A_4 k_{2\beta} k_1^\delta k_2^\sigma \epsilon_{\delta\sigma\alpha\mu} + A_5 k_{1\alpha} k_1^\delta k_2^\sigma \epsilon_{\delta\sigma\beta\mu} + A_6 k_{2\alpha} k_1^\delta k_2^\sigma \epsilon_{\delta\sigma\beta\mu} \tag{1}$$

Local gauge invariance (vector current conservation),

$$k_1^\alpha T_{\alpha\beta\mu} = k_2^\beta T_{\alpha\beta\mu} = 0, \tag{2}$$

is ensured by the following constraints:

$$A_1 = k_2^2 A_4 + (k_1 k_2) A_3, \qquad A_2 = k_1^2 A_5 + (k_1 k_2) A_6. \tag{3}$$

Besides that

$$A_3(k_1, k_2) = -A_6(k_2, k_1), \qquad A_4(k_1, k_2) = -A_5(k_2, k_1). \tag{4}$$

The invariant amplitudes A_3, A_4, A_5 and A_6 are free of kinematical singularities and well-defined. They can be calculated in analytic form if $k_1^2 = 0$ (or $k_2^2 = 0$)[1].

In the chiral limit $(m_q \to 0)$ at $k_1^2 = 0$ in the single-loop approximation[1]:

$$A_1 = \frac{1}{4\pi^2}\left\{\frac{Q^2}{Q^2 - W^2}\ln\frac{Q^2}{W^2} + 1\right\},$$

$$A_2 = \frac{1}{4\pi^2}\left\{\frac{Q^2}{Q^2 - W^2}\ln\frac{Q^2}{W^2} - 1\right\},$$

$$A_3 = -A_6 = -\frac{1}{2\pi^2} \cdot \frac{1}{Q^2 - W^2}\left\{\frac{Q^2}{Q^2 - W^2}\ln\frac{Q^2}{W^2} - 1\right\}, \tag{5}$$

---

*Here the third component of the isovector axial-vector current $j_{5\mu} = \bar{u}\gamma_\mu\gamma_5 u - \bar{d}\gamma_\mu\gamma_5 d$, which has not the gluon axial anomaly is meant. It is convenient to omit the factor $2N_c e^2(Q_u^2 - Q_d^2)$.

which are correct for $0 < Q^2 = -E^2 = -k_2^2$ and $0 < W^2 = -M^2 = -(k_1 + k_2)^2$.

In the other regions of M^2 and E^2 the analytic continuation is carried out in the following manner.

$$i) \qquad 0 < -Q^2 = E^2 : \qquad \ln Q^2 \to -i\pi + \ln E^2;$$

$$ii) \qquad 0 < -W^2 = M^2 : \qquad \ln \frac{1}{W^2} \to i\pi + \ln \frac{1}{M^2}. \tag{6}$$

So, in the massless–fermion limit on the physical sheets of the amplitudes A_i there are cuts over $0 \leq E^2 < \infty$ and $0 \leq M^2 < \infty$ if $k_2^2 \neq 0$. Only when $Q^2 \to 0$, a pole arises in A_3 and A_6 at $M^2 = 0$:

$$A_3 = -A_6 = \frac{2}{M^2} A_1 = -\frac{2}{M^2} A_2 = \frac{1}{2\pi^2} \cdot \frac{1}{M^2}. \tag{7}$$

Let us allow for the contribution of the massless pseudoscalar Goldstone boson in the axial-vector channel:

$$T_{\alpha\beta\mu}^G = \frac{f_\pi g_{\pi\gamma\gamma}}{M^2} (k_1 + k_2)_\mu \, k_1^\delta k_2^\sigma \epsilon_{\delta\sigma\beta\alpha} \tag{8}$$

Using the equations

$$k_{1\mu} k_1^\delta k_2^\sigma \epsilon_{\delta\sigma\beta\alpha} = k_1^2 k_2^\sigma \epsilon_{\sigma\alpha\beta\mu} - (k_1 k_2) k_1^\sigma \epsilon_{\sigma\alpha\beta\mu} - k_{1\beta} k_1^\delta k_2^\sigma \epsilon_{\delta\sigma\alpha\mu} - k_{1\alpha} k_1^\delta k_2^\sigma \epsilon_{\delta\sigma\beta\mu},$$

$$k_{2\mu} k_1^\delta k_2^\sigma \epsilon_{\delta\sigma\beta\alpha} = (k_1 k_2) k_2^\sigma \epsilon_{\sigma\alpha\beta\mu} - k_2^2 k_1^\sigma \epsilon_{\sigma\alpha\beta\mu} - k_{2\beta} k_1^\delta k_2^\sigma \epsilon_{\delta\sigma\alpha\mu} - k_{2\alpha} k_2^\delta k_1^\sigma \epsilon_{\delta\sigma\beta\mu}, \tag{9}$$

one passes to the form of Eq.(1):

$$A_1^G = -\frac{f_\pi g_{\pi\gamma\gamma}}{M^2} \left[(k_1 k_2) + k_2^2 \right] = -\frac{f_\pi g_{\pi\gamma\gamma}}{2M^2} \left(M^2 + k_2^2 - k_1^2 \right),$$

$$A_2^G = \frac{f_\pi g_{\pi\gamma\gamma}}{M^2} \left[(k_1 k_2) + k_1^2 \right] = \frac{f_\pi g_{\pi\gamma\gamma}}{2M^2} \left(M^2 + k_1^2 - k_2^2 \right),$$

$$A_3^G = A_4^G = -A_5^G = -A_6^G = -\frac{f_\pi g_{\pi\gamma\gamma}}{M^2}. \tag{10}$$

As seen the amplitudes A_i^G contain the Goldstone pole not only at $k_1^2 = k_2^2 = 0$.

Consequently, the Goldstone boson, associated with the spontaneous breakdown of chiral symmetry, cannot reproduce the axial anomaly pole[2]. However, there is a price to pay for introducing the Goldstone boson[2]. To understand how high the price is, we identify the Goldstone pole with the one from axial anomaly at $k_1^2 = k_2^2 = 0$. Comparison of Eq.(10) with Eqs.(5) and (7) yields

$$f_\pi g_{\pi\gamma\gamma} = -\frac{1}{2\pi^2}. \tag{11}$$

Now let us consider the differences between the amplitudes Eq.(5) and Eq.(10) $(\overline{A}_i = A_i - A_i^G$ at $k_i^2 = 0)$, taking into account Eq.(11):

$$\overline{A}_1 = \overline{A}_2 = \frac{1}{4\pi^2} Q^2 \left\{ \frac{1}{Q^2 + M^2} \ln \frac{Q^2}{-M^2} + \frac{1}{M^2} \right\},$$

$$\overline{A}_6 = -\overline{A}_3 = \frac{1}{2\pi^2} \cdot \frac{Q^2}{Q^2 + M^2} \left\{ \frac{1}{Q^2 + M^2} \ln \frac{Q^2}{-M^2} + \frac{1}{M^2} \right\}. \tag{12}$$

It can be seen from Eq.(12) that the invariant amplitudes $\overline{A}_i$ have a pole at $M^2 = 0$ when $k_2^2 \neq 0$. This pole looks like the contribution of a massless boson.

One can find easily that the amplitude

$$\overline{T}_{\alpha\beta\mu} = \sum_i \overline{A}_i t^i_{\alpha\beta\mu} \tag{13}$$

is transverse in the axial-vector channel:

$$(k_1 + k_2)^\mu \overline{T}_{\alpha\beta\mu} = (\overline{A}_1 - \overline{A}_2) k_1^\delta k_2^\sigma \epsilon_{\delta\sigma\alpha\beta} = 0, \tag{14}$$

that is, only pseudovector intermediate states (1^+) are possible in the axial-vector channel of the amplitude $\overline{T}_{\alpha\beta\mu}$. Consequently, the amplitude $\overline{T}_{\alpha\beta\mu}$ has a contribution looking like one from a massless pseudovector boson if $k_2^2 \neq 0^2$.

From a formal point of view it has been shown above that in the chiral limit the amplitude of the transition of the isovector axial-vector current in two photons in the single-loop approximation has a pseudovector massless pole in parallel with the pseudoscalar massless one. The pseudovector massless pole vanishes when the photons are real, in full accord with the Landau-Yang theorem. If one of photons is virtual then the pseudovector pole cancels the pseudoscalar one in the amplitudes free of kinematical singularities.

So, in the case of spontaneous breaking of chiral symmetry, the massless pseudoscalar Goldstone boson can reproduce the axial–anomaly pole in the single-loop approximation only in the peculiar "conspiracy" with a massless pseudovector boson.

Does this result take place for the "dressed" amplitude? That is the problem.

Nonperturbative effects could "give" the massless pseudovector meson a mass and convert it into the $a_1(1260)$ meson, that would be an elegant solution of the problem.

Remind that the linear realization of chiral symmetry through massless nucleons meets with a crucial theoretical opposition. So, if the "dressed" amplitude $T_{\alpha\beta\mu}$ has the axial–anomaly pole only for the real photons then there is such a dilemma: either a conspiracy between the massless pseudoscalar meson and the massless pseudovector meson or the linear realization of chiral symmetry by the massless quarks, i.e., the "liberation" (deconfinement) of the quarks in the chiral limit.

I am grateful to the organizers of the International Conference QUARK CONFINEMENT AND THE HADRON SPECTRUM and to the International Science Foundation (Grant 1478 1) for giving me a chance to give this talk. This visit was supported partly by the Russian Foundation for Fundamental Sciences, Grant 94-02-05 188.

1. N. N. Achasov, *Phys. Lett.* **B287** (1992) 213.
2. N. N. Achasov, *ZhETF* **103** (1993) 11 [*JETP* **76** (1993) 5].

NONPERTURBATIVE QCD EFFECTS IN EXCLUSIVE REACTIONS

N. G. STEFANIS

Institut für Theoretische Physik II, Ruhr-Universität Bochum
D-44780 Bochum, Germany

ABSTRACT

The transverse-momentum dependence (Sudakov and confinement-size effects) of the proton and neutron magnetic form factor in the spacelike region is critically studied, using a whole spectrum of model distribution amplitudes for the nucleon obtained on the basis of QCD sum rules.

1. Introduction

Recent theoretical developments [1,2,3] concerning the incorporation of gluonic radiative corrections (Sudakov effects) in the calculation of exclusive reactions, like hadron form factors, enable rigorous considerations of the kinematic endpoint regions rendering the perturbative contribution self-consistent, albeit considerably reduced relative to previous results [4,5,6] within the standard convolution scheme [7]. The characteristic feature underlying the method, i.e., the "modified convolution scheme" (MCS), is resummation of Sudakov logarithms according to renormalization-group (RG) techniques [8], giving exclusive amplitudes an explicit dependence on transverse momenta. However, the validity of this approach, although quite general, becomes obscured when several transverse scales are involved, and calls for a careful infrared (IR) regularization. Indeed, we have recently shown [9] that Li's analysis [3] of the proton form factor in the spacelike region is seriously flawed, giving rise to infinite expressions due to uncompensated singularities of the one-loop perturbative expression of the effective coupling constant. In order to avoid infinities, while still using the $\overline{MS}$ scheme, one has to set all transverse interquark separations inside the nucleon equal and use the maximum one as a common IR cut-off ("MAX" prescription) [9]. Results within such an approach will be presented below.

2. Nucleon form factors within the modified convolution scheme

The basic feature of the MCS becomes apparent by explicitly showing one of the Fourier-transformed hard-scattering amplitudes entering the calculation of the proton form factor ($C_F = 4/3$)

$$\hat{T}_1 = \frac{8}{3} C_F \, \alpha_s(t_{11})\alpha_s(t_{12}) K_0\left(\sqrt{(1-x_1)(1-x_1')}Qb_1\right) K_0\left(\sqrt{x_2 x_2'}Qb_2\right), \quad (1)$$

where K_0 is the Macdonald function and b_l denotes the length of the corresponding interquark transverse separation vector. The t_{ji} are defined as the maximum scale of either the longitudinal momentum $\propto Q$ or the inverse transverse separation $\propto 1/b_l$, appearing in the argument of K_0. They are associated with the virtualities of the exchanged gluons. Imposing our IR prescription, we set

$$\tilde{b} \equiv \max\{b_1, b_2, b_3\} = \tilde{b}_1 = \tilde{b}_2 = \tilde{b}_3 \qquad (2)$$

with the consequence that the Sudakov form factors always cancel the α_s-singularities (see [9] for details). The physical nature of this treatment is that soft gluons with wavelengths larger than the maximum transverse interquark separation cannot resolve single color charges and probe the nucleon rather as a whole, i.e., as a colorless object. This approach not only suffices to protect the amplitudes from becoming singular, but it also yields form-factor results which are rather insensitive to distances of order $1/\Lambda_{QCD}$, thus providing saturation. On the contrary, long-distance dominance in the perturbative contribution would invalidate the consistency of factorization. The IR cut-off scales $1/\tilde{b}_l$ enter also the exponents of the Sudakov form factors and mark the interface between genuine nonperturbative momenta—implicitly taken care of in the proton wave function—and those due to soft gluons which have explicitly been accounted for in the Sudakov form factor. Obviously, the IR cut-off serves at the same time as the gliding factorization scale μ_F to be used in the evolution of the nucleon wave function.

It is apparent that, besides such theoretical prerequisites, concrete applications become computationally useful only when explicit and reliable models for the valence-quark structure of the nucleon are available. The nucleon distribution amplitude $\Phi(x_i, \mu_F^2)$ contains the complicated bound-state dynamics and hence must be determined by nonperturbative techniques. Its momentum-scale dependence is RG-controlled, meaning that $\Phi(x_i, Q^2) = \Phi_{as}(x_i) \sum_{n=0}^{\infty} B_n \tilde{\Phi}_n(x_i)[\alpha_s(Q^2)/\alpha_s(\mu_F^2)]^{\gamma_n}$, where $\{\tilde{\Phi}_n\}_0^{\infty}$ are orthonormalized eigenfunctions of the interaction kernel of the evolution equation [7] expressed in terms of Appell polynomials and $\Phi_{as}(x_i) = 120 x_1 x_2 x_3$ is the asymptotic form of the nucleon distribution amplitude. The eigenvalues γ_n coincide with the anomalous dimensions of multiplicatively renormalizable $I_{1/2}$ baryonic operators of twist three. A basis including a total of 55 eigenfunctions (polynomial order 9) together with the associated normalization coefficients and anomalous dimensions has been constructed in [10]. The expansion coefficients B_n, which contain the nonperturbative input, can be analytically determined [10] in terms of the moments, $\Phi^{(i0j)}(\mu_F^2) = \int_0^1 [dx] x_1^i x_2^0 x_3^j \Phi(x_i, \mu_F^2)$, of the nucleon distribution amplitude [10]:

$$\frac{B_n(\mu_F^2)}{\sqrt{N_n}} = \frac{\sqrt{N_n}}{120} \sum_{i,j=0}^{\infty} a_{ij}^n \, \Phi^{(i0j)}(\mu_F^2), \qquad (3)$$

where the projection coefficients a_{ij}^n are calculable to any polynomial order. The moments, in turn, are computed via a short-distance operator product expansion in a (spacelike) momentum interval, in which quark-hadron duality is valid (see, for instance, [4,11]). Scanning the parameter space of the Appell decomposition coefficients B_n, a series of solutions to the sum rules [11] has been determined [6]. These solutions constitute a finite *orbit* in the $(B_4, |G_M^n|/G_M^p)$-space. The lower part of the orbit is associated with the heterotic solution [5], corresponding to the smallest $|G_M^n|/G_M^p$ value still compatible with the sum-rule constraints. Its upper region controls COZ-type [11] amplitudes. In Fig. 1 we present theoretical results for the proton [9] and the neutron [13] magnetic form factor in the form of a shadowed strip which corresponds to the orbit of the sum-rules solutions. The calculations have been performed under the proviso

260

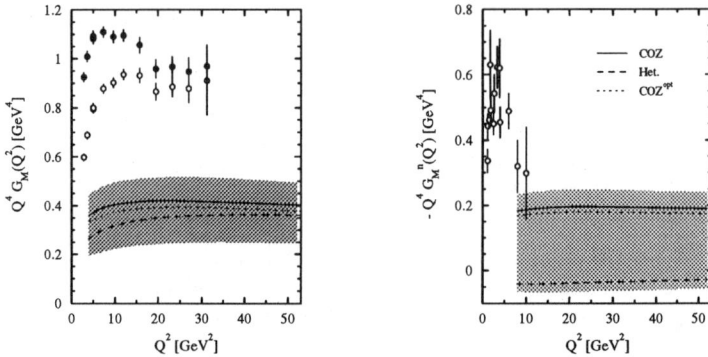

Figure 1: Proton and neutron magnetic form factor including the intrinsic $k_\perp$-dependence vs. Q^2 in comparison with data [14]. The open circles denote F_1^p data.

of the "MAX" prescription for the Sudakov form factors, including evolution and normalizing the wave functions to unity. The intrinsic $k_\perp$-dependence, accounting for confinement-size effects, has been parametrized by a Gaussian distribution [12].

3. Summary and Conclusions

The inclusion of transverse-momentum effects in the calculation of form factors improves the self-consistency of the perturbative calculation at the expense of a reduced magnitude. This calls for sizeable higher perturbative corrections (large K-factor) and/or soft contributions, e.g., higher twists, at accessible momentum transfers.

1. J. Botts and G. Sterman, *Nucl. Phys.* **B325** (1989) 62.
2. H.-N. Li and G. Sterman, *Nucl. Phys.* **B381** (1992) 129.
3. H.-N. Li, *Phys. Rev.* **D48** (1993) 4243.
4. N. G. Stefanis, *Phys. Rev.* **D40** (1989) 2305; **D44** (1991) 1616 (E).
5. N. G. Stefanis and M. Bergmann, *Phys. Rev.* **D47** (1993) R3685.
6. M. Bergmann and N. G. Stefanis, *Phys. Rev.* **D48** (1993) R2990; *Phys. Lett.* **B325** (1994) 183.
7. G. P. Lepage and S. J. Brodsky, *Phys. Rev.* **D22** (1980) 2157.
8. J. C. Collins, D. E. Soper, and G. Sterman, in *Perturbative Quantum Chromodynamics*, ed. A. H. Mueller (World Scientific, 1989).
9. J. Bolz, R. Jakob, P. Kroll, M. Bergmann, and N. G. Stefanis, Bochum preprint RUB-TPII-01/94 (May 1994), to be published in *Z. Phys.* C.
10. N. G. Stefanis and M. Bergmann, in *Workshop on Exclusive Reactions at High Momentum Transfer*, ed. C. E. Carlson, P. Stoler, and M. Taiuti, Elba, Italy, 24-26 June, 1993, (World Scientific, Singapore, 1994).
11. V. L. Chernyak, A. A. Ogloblin, and I. R. Zhitnitsky, *Z. Phys.* **C42** (1989) 569.
12. R. Jakob and P. Kroll, *Phys. Lett.* **B315** (1993) 463; **B319** (1993) 545(E).
13. J. Bolz, R. Jakob, P. Kroll, M. Bergmann, and N. G. Stefanis, Bochum preprint RUB-TPII-04/94 (July 1994), to be published in *Phys. Lett.* **B**.
14. A. F. Sill et al., *Phys. Rev.* **D48** (1993) 29; P. Bosted et al., *Phys. Rev. Lett.* **68** (1992) 3841.

POLARIZATION PHENOMENA AND CONFINEMENT FORCES

M.G.RYSKIN

PNPI, Gatchina, S.Petersburg, 188350, Russia

ABSTRACT

The simple model which explains the polarization as a result of the interaction with the colour-magnetic field at large distances is considered. It agrees with experiment. Therefore studying the polarization effects one can get the information about the colour-magnetic field in the confinement region.

1. Introduction

It is well known that the polarization effects come mainly from the large distances, where the effective masses of the quarks are close to the costituent quark masses and the QCD coupling α_s is not small. It is impossible to explain the large polarizations (or asymmetry) in the framework of pert.QCD as the imaginary part of the apmlitude (which is needed to get the nonzero polarization $\mathbf{P} = \frac{Im(A_{spin\ flip} \cdot A_{n.f.}^*)}{(|A_{n.f.}^2| + |A_{s.f.}^2|)}$) is proportional to the small coupling α_s and the current mass of the quark (which flips the spin at the small distances) is also extremely ($< 10 MeV$) small. Thus being originated at large distances the polarization effects should reflects some properties of the confinement mechinery.

In the present talk I'll discuss very simple model in which the polarization(asymmetry) is caused by the interaction with the colour magnetic field at the periphery of the colour tube. In spite of the oversiplification, the comparision of the model with the data in not too bad.

2. Asymmetry

To describe the idea, let us consider the asymmetry of the particle produced in the inelastic collision with the polarized target.

After a collision and exchange of a coloured gluon (colour charge exchange) a colour flux tube (string) connects the colliding hadrons. In the simplest case one takes into account only the electric colour field in this tube just as in one-dimensional electrodynamics (the Shwinger model). However this solution is unstable. As was shown in[1], a colour-magnetic field arises around the tube (just like around a wire carrying a current) and interacts with the quark colour-magnetic moment μ. The interaction energy is $\mathbf{H} \cdot \mu$. Since the tube is directed along the longitudinal axis, the nonuniform colour-magnetic field transfers to a quark with spin up an additional transverse momentum δq_t pointing to the right and to a quark with spin down – one pointing to the left.

262

The typical strength of the colour-magnetic field is[1] $H \simeq \sqrt{\alpha_s(R_c)}/1.6R_c^2$, where R_c is the distance at which the constant α_s begins to change sharply. If we use the instanton model of the QCD vacuum[2-4], the additional transverse momentum transferred by the colour-magnetic field to a polarized quark will be *

$$\delta q_t \simeq H\mu \frac{\alpha_s C_2}{2m_q \cdot 1.6R_c^2} \approx 100\,MeV \qquad (1)$$

(here we put $R_c = 1/(400\,MeV)$; $\alpha_s = 1/2$ and $m_q = 330\,MeV$).
Since the change of momentum δq_t is small in compare with the average secondary-hadron transverse momenta $< q_{t\pi} > \simeq 400MeV$, the polarized-quark emission asymmetry A can be found from the expression

$$A = \frac{d\sigma/d^3q_\uparrow - d\sigma/d^3q_\downarrow}{d\sigma/d^3q_\uparrow + d\sigma/d^3q_\downarrow} = \frac{d\sigma(q_t + \delta q_t) - d\sigma(q_t - \delta q_t)}{d\sigma(q_t + \delta q_t) + d\sigma(q_t - \delta q_t)}$$

$$= \delta q_t \frac{\partial}{\partial q_t}\left(\frac{d\sigma}{d^3q}\right)\left(\frac{d\sigma}{d^3q}\right)^{-1}, \qquad (2)$$

which makes it possible to calculate the asymmetries (or polarizations) on the basis of the dependence of the inclusive cross section $\frac{d\sigma}{d^3q}$ on the transverse momentum.

It is interesting to note that Eq. (2) closely matches the caculations in the low-order pert. QCD and, in this sense, is the correct continuation to large q_t (see ref.[5]).

3. Handedness

The same idea may be used to describe the origin of the jet handednes[6] – the conception proposed more than 15 years ago:
the decay products of a jet generated by a polarized parton should reveal the correlation – $\epsilon^{\alpha\beta\gamma\delta}q_{1\alpha}q_{2\beta}q_{3\gamma}s_\delta \neq 0$ [7]; where: q_i are the momenta of the secondary hadrons (which form the jet) and s_δ is the spin vector of the initial parton (quark or gluon).
At the stage of jet fragmentation, when the colour-electric potential becomes strong enough, it produces a new quark-antiquark pair, breaking the colour string. At the begining the transverse (with respect to the jet axis) momenta of these new partons are balanced ($\mathbf{q'}_{t2} = -\mathbf{q'}_{t3}$) but moving in the colour-magnetic field $\mathbf{H}\|\mathbf{s}$, originated by the initial polarized quark, the partons turn around and get an additional momenta δq_t. As they have an opposite colour charge and opposite dirrections (in transverse plane), the momenta δq_{t2} and δq_{t3} acquire the same sign. (In the simplest symmetric case $\delta q_{t2} = \delta q_{t3} = \delta q_t$.) Thus one gets the system of 3 acoplanar axes (momenta q_1, q_2, q_3).

As was estimated in [6] the value of δq_t in this case is very close to the previous (Eq. (1)) one ($\delta q_t \simeq 90\,MeV$).

---

*Of course, this estimate gives only the order of magnetude of δq_t, its accuracy being $\sim 50\%$.

The typical scale of the effect is proportional to the ratio $\delta q_t / \bar{q}_t$. On the parton level it is large enough ($\sim 10\%$ for mean $\bar{q}_t \sim 400\,MeV$) but, of course, it is whashed out crucially after the hadronization and the decays of resonances. If one selects the pion with rather large transverse momentum (say $q_t > 1 GeV$) the role of the resonances decays will be suppressed and we hope to observe the effect of the order of $\delta q_t / 2 q_t \sim 4 - 5\%$.

4. Conclusion

The qualitative agreement of the model with the data (see ref.[8]) says us that the proposed picture is not too bad. Therefore studying the polarization effects one can extract some nontrivial information about the colour-magnetic field in the coloure tube (which provide the hadronization and the confinement of the initial quark) and the polarization of the partons (quarks) inside the hadron wave function. In other words one can measure the expectation value of the non local operator

$$T = < p, s_t | \bar{\Psi}(0) \gamma^+ \int dy_2^- \epsilon_{\sigma\rho\alpha\beta} s_t^\sigma n^\alpha \bar{n}^\beta F^{\rho+}(y_2^-) \Psi(y_1^-) | p, s_t >$$

that combines the quark field with a single gluonic field strength[9].

5. Acknowledgements

I would like to thank the organizers and especially G.Marchesini for the very good conference and interesting discussions.

6. References

1. A. B. Migdal and S. B. Khokhlachev, *JETP Lett.* **41** (1985) 194.
2. R. D. Callan, R. Dashen and D. J. Gross, *Phys. Rev.* **D17** (1978) 2717; **D20** (1979) 3279.
3. R. D. Carlitz and D. B. Creamer, *Ann. Phys. (N.Y.)* **118** (1979) 429.
4. D. I. D'yakonov and Yu. V. Petrov, *Sov. Phys. JETP* **62** (1985) 204.
5. M. G. Ryskin, *Sov. J. Nucl. Phys.* **46** (1987) 337.
6. M. G. Ryskin, *Phys. Lett.* **B319** (1993) 346.
7. O. Nachtmann, *Nucl. Phys.* **B127** (1977) 314; A. V. Efremov, *Sov. J. Nucl. Phys.* **28** (1978) 83; A. V. Efremov, L. Mankiewicz and N. A. Tornquist CERN-TH 6430/92.
8. M. G. Ryskin *Sov. J. Nucl. Phys.* **48** (1989) 708.
9. A. V. Efremov and O. V. Teryaev, *Sov. J. Nucl. Phys.* **36** (1982) 140; **39** (1984) 962;
   J. Qiu and G. Sterman, *Phys. Rev. Lett.* **67** (1991) 2264; *Nucl. Phys.* **B378** (1992) 52.

NON–STANDARD QUARK PROPAGATORS FROM
CONFORMAL FIELD THEORY

RAINER DICK

School of Natural Sciences, Institute for Advanced Study
Olden Lane, Princeton, NJ 08540, USA
and
Department of Physics, University of Munich
Theresienstr. 37, 80333 Munich, Germany

ABSTRACT

It is shown that Weyl spinors in Minkowski space are isomorphic to primary fields of half–integer conformal weights. This yields representations of fermionic 2–point functions in terms of correlators of primary fields with a factorized transformation behavior under the Lorentz group. I employ this observation to determine the general structure of the corresponding Lorentz covariant correlators by methods similar to the methods employed in conformal field theory to determine 2– and 3–point functions of primary fields. In particular, there appear chiral symmetry breaking terms which resemble fermionic 2–point functions of 2D CFT up to a function of the product of momenta.

1. Introduction

Despite many efforts and impressive results, the infrared limit of QCD still provides many puzzles and represents one of the great challenges of theoretical physics today. Characteristic features of this sector are confinement and chiral symmetry breaking, and it has been frequently argued that chiral symmetry breaking may appear as a consequence of confinement. Well established methods to study these subjects have been devised in lattice gauge theory and chiral perturbation theory, and through approaches based on Bethe–Salpeter equations or on Schwinger–Dyson equations augmented by Slavnov–Taylor identities. Most of our present understanding of low energy QCD is in one or another way related to these methods, and many surveys along with interesting new developments have been presented at this conference. Particularly interesting developments arise when established methods are combined with new ideas, like e.g. the inclusion of a 5th dimension as a technical tool to restore chiral symmetry on the lattice[1], the explicit construction of monopole creation operators in order to provide further evidence for the dual superconductivity picture of confinement[2], or systematic derivations of qqq potentials from Wilson loops[3]. Other recent and also very promising developments have evolved in heavy quark effective theory, see[4,5] and references therein, and through path integral hadronization[6,7].

Here I would like to point out that notions and methods which seem pretty far away from bound state calculations in QCD like two–dimensional conformal field theory may have the potential to add useful observations and results to the subject of chiral symmetry breaking. What I will describe is essentially a group theoretical investigation of correlators of massless fermions, which becomes amenable by identifying the non–vanishing components of a Weyl spinor with local representatives of primary fields on a sphere in momentum space.

The relevance of methods of 2D field theory in certain kinematical regimes or large N expansions has been noticed in many places. In particular, I should mention the string based calculations of gluon scattering amplitudes[8]. Recent constructions of effective 2D field theories to describe high energy scattering in QCD were given in Refs.[9,10].

The physical motivation for the present work results from the well known observation that light mesons and especially pions can be understood as Goldstone bosons of spontaneously broken chiral symmetry, with their masses resulting from explicit chiral symmetry breaking through quark masses and a non–vanishing condensate (for a well written survey of this, see the corresponding chapters in Ref.[11]). Therefore, it seems worthwhile to address the problem, whether chiral symmetry breaking may appear in an initially massless theory, which acquires massterms through a transmutation of terms which are due to breaking of translational invariance. The breaking of translational invariance then would be expected as a consequence of confinement.

Motivated by this picture, I would like to point out that Eq. (2) displayed below provides a general form of a fermion propagator in the free massless limit which complies with Lorentz covariance and contains chiral symmetry breaking terms.

2. Correlators

The construction proceeds through the following steps[12]: First we prove that massless spinors in 3+1 dimensions are isomorphic to local representations of primary fields of half–integer weights under Lorentz transformations. More specifically, let $z(\vec{p})$ denote stereographic coordinates in momentum space:

$$z = \frac{p_+}{|\vec{p}| - p_3}$$

The entries of a spinor of negative helicity appear as local representations $\psi(z, \bar{z}, E)$ of primary fields with respect to a conformal atlas, and transform under holomorphic transformations according to

$$\psi'(z', \bar{z}', E') = \psi(z, \bar{z}, E) \left(\frac{\partial z'}{\partial z}\right)^{-\frac{1}{2}} \tag{1}$$

Special cases of this transformation behavior imply the Weyl equation for massless fermions. Lorentz transformations induce via $SL(2, \mathbf{C})$ holomorphic transformations of spheres in momentum space, and the resulting transformation behavior of left handed spinors agrees with the equation above.

As a consequence, the correlation functions of massless fermions in Minkowski space can be written as a sum over correlation functions which exhibit a factorized transformation behavior under the Lorentz group. The factorized transformation behavior in turn makes it possible to determine the structure of the correlation functions from Lorentz covariance by methods similar to the techniques employed to calculate 2– and 3–point functions in 2D conformal field theory: Choose a generating set of the symmetry group and solve the corresponding covariance conditions. The on–shell correlation function then turns out to be fixed up to two functions f_1, f_2 of single arguments[12]:

$$\langle \Psi(\vec{p}\,)\overline{\Psi}(\vec{p}\,') \rangle = \qquad (2)$$

$$\begin{pmatrix} 0 & 1 \\ 0 & 0 \end{pmatrix} \otimes \begin{pmatrix} \bar{z}z' & \bar{z} \\ z' & 1 \end{pmatrix} \langle \phi(\vec{p}\,)\phi^+(\vec{p}\,') \rangle + \begin{pmatrix} 0 & 0 \\ 1 & 0 \end{pmatrix} \otimes \begin{pmatrix} 1 & -\bar{z}' \\ -z & z\bar{z}' \end{pmatrix} \langle \psi(\vec{p}\,)\psi^+(\vec{p}\,') \rangle$$

$$+ \begin{pmatrix} 1 & 0 \\ 0 & 0 \end{pmatrix} \otimes \begin{pmatrix} \bar{z} & -\bar{z}\bar{z}' \\ 1 & -\bar{z}' \end{pmatrix} \langle \phi(\vec{p}\,)\psi^+(\vec{p}\,') \rangle + \begin{pmatrix} 0 & 0 \\ 0 & 1 \end{pmatrix} \otimes \begin{pmatrix} z' & 1 \\ -zz' & -z \end{pmatrix} \langle \psi(\vec{p}\,)\phi^+(\vec{p}\,') \rangle$$

$$\langle \psi(\vec{p}_1)\psi^+(\vec{p}_2) \rangle = \langle \phi(\vec{p}_2)\phi^+(\vec{p}_1) \rangle = f_1\left(\frac{|\vec{p}_1|}{|\vec{p}_2|} \right) \frac{1 + z_1\bar{z}_2}{\sqrt{|\vec{p}_1||\vec{p}_2|}} \delta_{z\bar{z}}(z_1 - z_2) \qquad (3)$$

$$\langle \psi(\vec{p}_1)\phi^+(\vec{p}_2) \rangle = \overline{\langle \phi(\vec{p}_2)\psi^+(\vec{p}_1) \rangle} = \frac{1}{z_1 - z_2} f_2\left(|\vec{p}_1||\vec{p}_2| \frac{(z_1 - z_2)(\bar{z}_1 - \bar{z}_2)}{(1 + z_1\bar{z}_1)(1 + z_2\bar{z}_2)} \right) \qquad (4)$$

where the decomposition of the correlator into correlators for the fields ψ, ϕ essentially goes back to a helicity expansion of the spinor Ψ, and the trivial dependence on color and flavor indices has been suppressed.

The form of the single correlation functions in Eqs. (3,4) is fixed from covariance with respect to the proper orthochronous Lorentz group, while the relations between different correlation functions are fixed by invariance with respect to parity, time reversal, or charge conjugation.

The terms containing the function f_1 preserve chiral symmetry: They do not contribute to a chiral condensate and anticommute with γ_5. The consistency of the result is expressed by the fact that these terms contain a δ–function which restricts the correlator to parallel momenta. On the other hand, the terms containing the function f_2 break chiral symmetry, and therefore must also account for breaking of translational invariance. This is in agreement with Eq. (4), because the right hand side of this equation cannot accomodate for a δ–function in the external momenta.

The unperturbed result for the on–shell correlation

$$\langle \psi(\vec{p}\,)\overline{\psi}(\vec{p}\,') \rangle = -2p \cdot \gamma |\vec{p}| \delta(\vec{p} - \vec{p}\,')$$

is recovered from Eqs. (2–4) for $f_1(x) = \delta(x - 1)$, $f_2 = 0$.

Off–shell extensions of Eq. (2) can be inferred from the requirement to yield the same propagator in configuration space:

$$S(x, x') = \frac{1}{(2\pi)^4} \int d^4p \int d^4p' \exp(ip \cdot x) S(p, p') \exp(-ip' \cdot x')$$

$$= \frac{\Theta(t-t')}{(2\pi)^3} \int \frac{d^3\vec{p}}{2|\vec{p}|} \int \frac{d^3\vec{p}''}{2|\vec{p}''|} \exp(\mathrm{i}p \cdot x)\mathrm{i}\langle\Psi(\vec{p})\overline{\Psi}(\vec{p}'')\rangle \exp(-\mathrm{i}p' \cdot x')$$

$$- \frac{\Theta(t'-t)}{(2\pi)^3} \int \frac{d^3\vec{p}}{2|\vec{p}|} \int \frac{d^3\vec{p}''}{2|\vec{p}''|} \exp(-\mathrm{i}p \cdot x)\mathrm{i}\langle\Psi(\vec{p})\overline{\Psi}(\vec{p}'')\rangle \exp(\mathrm{i}p' \cdot x')$$

thus fixing the structure up to the 2 functions f_1, f_2, which parametrize the chiral symmetry preserving and violating parts.

3. Acknowledgements

I would like to thank Hermann Nicolai and Julius Wess for helpful discussions at various stages of this work.

4. References

1. M. Creutz and I. Horváth, *Surface States and Chiral Symmetry on the Lattice*, BNL–60062 (February 1994), and contribution by M. Creutz to this workshop.
2. L. Del Debbio, A. Di Giacomo and G. Paffuti, *Detecting Dual Superconductivity in the Ground State of Gauge Theory*, IFUP–TH–16–94, and contribution by A. Di Giacomo to this workshop.
3. N. Brambilla, P. Consoli and G.M. Prosperi, *A Consistent Derivation of the Quark–Antiquark and Three–Quark Potentials in a Wilson Loop Context*, IFUM 452/FT (December 1993), and contribution by N. Brambilla and G.M. Prosperi to this workshop.
4. F. Hussain, J.G. Körner, K. Schilcher, G. Thompson and Y.L. Wu, *Phys. Lett.* **B249** (1990) 295, and contributions to this workshop by J.G. Körner and F. Hussain.
5. J.G. Körner, M. Krämer and D. Pirjol, *Heavy Baryons*, in *Progr. Part. Nucl. Phys.* **33** (1994) 787.
6. H. Reinhardt, *Phys. Lett.* **B244** (1990) 316; *Phys. Lett.* **B248** (1990) 365.
7. D. Ebert, H. Reinhardt and M.K. Volkov, *Effective Hadron Theory of QCD*, in *Progr. Part. Nucl. Phys.* **33** (1994) 1.
8. D.A. Kosower, B.-H. Lee and V.P. Nair, *Phys. Lett.* **B201** (1988) 85.
9. L.N. Lipatov, *Nucl. Phys.* **B365** (1991) 614; *Phys. Lett.* **B309** (1993) 394.
10. E. Verlinde and H. Verlinde, *QCD at High Energies and Two–Dimensional Field Theory*, IASSNS–HEP–92/30, PUPT–1319 (February 1993).
11. F.J. Ynduráin, *The Theory of Quark and Gluon Interactions* (Springer, Berlin–Heidelberg–New York, 1993).
12. R. Dick, *Half–Differentials and Fermion Propagators*, LMU–TPW 94–6 (June 1994), to appear in *Rev. Math. Phys.*; *On Chiral Symmetry Breaking with Massless Fermions*, LMU–TPW 94–9 (July 1994), to appear in *Mod. Phys. Lett. A*.

SECTION B

Interquark potential and quarkonia;
Bethe-Salpeter equation;
relativistic equations.

ON THREE-QUARK ANOMALOUS DIMENSIONS

M. BERGMANN

Institut für Theoretische Physik II, Ruhr-Universität Bochum
D-44780 Bochum, Germany

and

N. G. STEFANIS

Institut für Theoretische Physik II, Ruhr-Universität Bochum
D-44780 Bochum, Germany

ABSTRACT

The anomalous dimensions of three-quark operators are calculated by diagonalizing the one-gluon exchange kernel of the evolution equation of nucleon distribution amplitudes. This is done within a symmetrized basis of Appell polynomials for polynomial degrees $\gg 1$. Technically, this is accomplished by a combination of analytical and numerical algorithms. The calculated eigenvalues form a degenerate system whose weight seems to follow an empirical logarithmic law.

1. Introduction

A key clue to the dynamics of exclusive reactions is the renormalization-group behavior of the hadron distribution amplitude (DA) Φ, controlled by an evolution equation, derived by Brodsky and Lepage [1]. Specifically, the momentum dependence of the nucleon DA for the valence-quark Fock state $\Phi(x_i, Q^2)$ is given by

$$\left\{ Q^2 \frac{\partial}{\partial Q^2} + \frac{3C_F}{2\beta} \right\} \Phi(x_i, Q^2) = \frac{C_B}{\beta} \int_0^1 [dy] \, V(x_i, y_i) \, \Phi(y_i, Q^2). \tag{1}$$

$[\int_0^1 \equiv \int_0^1 dx_1 \int_0^{1-x_1} dx_2 \int_0^1 dx_3 \delta(1 - x_1 - x_2 - x_3)]$. Here C_B and C_F are the Casimir operators of the fundamental and adjoint representation of $SU(3)$, respectively, and β is the Gell-Mann and Low function. The leading-order expression for the integral kernel V has been calculated in [1].

2. Higher-order eigenfunctions of the nucleon evolution equation

It was noticed in [2] that $\left[\hat{P}_{13}, \int_0^1 [dy] \, V(x_i, y_i) \right] = 0$ (where $\hat{P}_{13}$ is the permutation operator applied to quarks 1 and 3). Thus it is more appropriate to introduce a symmetrized basis of Appell polynomials $\tilde{\mathcal{F}}_{mn}(x_1, x_3) = (1/2) \, (\mathcal{F}_{mn}(x_1, x_3) \pm \mathcal{F}_{nm}(x_1, x_3))$ for $(m \geq n \, / \, m < n)$, in terms of which the eigenfunctions $\tilde{\Phi}_n(x_i)$ can be explicitly represented. Within this particular basis, the evolution kernel becomes block-diagonal with respect to different polynomial orders $m + n$ and also within a particular order for different symmetry classes with respect to $\hat{P}_{13}$. A typical feature of the Appell polynomials is that their number is quadratically increasing with the polynomial order. This renders the analytic diagonalization of the evolution equation increasingly complicated. Use of the symmetrized basis allows for an analytic diagonalization up to order 7. This improves previous calculations [3] which were limited to order 3. The eigenfunctions $\tilde{\Phi}_n(x_i)$ are represented as a finite series of ordinary polynomials, e.g., in powers of x_1, and x_3

$$\tilde{\Phi}_n(x_i) = \frac{1}{\sqrt{N_n}} \sum_{ij} a_{ij}^n \, x_1^i x_3^j \tag{2}$$

[with expansion coefficients a_{ij}^n and normalization factors N_n, given in [2,4]]. It is precisely this peculiar property of the eigenfunctions, which allows one to express them in terms of their *own* moments [4] (as it is known from Taylor expansions).

Another important property of the eigenfunctions of the evolution equation is that they form a commutative algebra analogous to angular momentum eigenfunctions obeying the triangle relation:

$$\tilde{\Phi}_k(x_i)\,\tilde{\Phi}_n(x_i) = \sum_{l=0}^{\infty} F_{kn}^l \tilde{\Phi}_l(x_i) \quad \text{with} \quad |\mathcal{O}(k) - \mathcal{O}(n)| \leq \mathcal{O}(l) \leq \mathcal{O}(k) + \mathcal{O}(n), \tag{3}$$

where $\mathcal{O}(k)$ denotes the polynomial order of eigenfunction $\tilde{\Phi}_k(x_i)$. The structure coefficients [2] F_{kn}^l of the group are calculated via

$$F_{kn}^l = N_l \int_0^1 [dx]\, x_1 x_3 (1 - x_1 - x_3)\, \tilde{\Phi}_k(x_i)\tilde{\Phi}_n(x_i)\tilde{\Phi}_l(x_i), \tag{4}$$

with the particularly important case $F_{kk}^0 = \frac{N_0}{N_k}$. The utility of this method has been shown in [4] in studying the evolution behavior of a more complicated ansatz for the nucleon DA of the Brodsky-Lepage-Huang-Mackenzie-type which incorporates higher-order Appell series contributions. In the present contribution we focus on the large-order behavior of the three-quark anomalous dimensions γ_n. These are related to the eigenvalues of evolution equation and are explicitly given by $\gamma_n = -\left(\frac{3}{2}\frac{C_F}{\beta} + \eta_n \frac{C_B}{\beta}\right)$, where η_n are the zeros of the characteristic polynomial that diagonalizes the evolution kernel. These anomalous dimensions belong to multiplicatively renormalizable three-quark operators appearing in the operator product expansion [5]. Our analytical results up to order 7 are given in [2,4]. The emerging pattern of anomalous dimensions up to this order seems to follow a power-law behavior empirically fitted (dashed line in Fig. 1) to $\gamma_n = 0.37\,\mathcal{O}(n)^{0.565}$. However, in order to determine the asymptotic behavior of the anomalous dimension pattern, one has to take into account the eigenvalues of much higher orders. This becomes possible by combining analytic and high-precision numerical algorithms [2]. Our new results are compiled in Fig. 1.

3. Discussion and conclusions

Interpreting the above results, one observes that the multiplet structure already indicated in [4] is also confirmed for higher orders and exhibits an inherent symmetry of

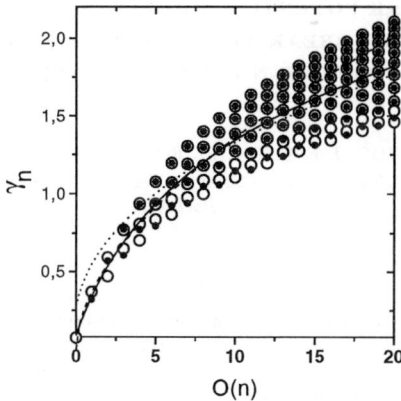

Figure 1: Pattern of three-quark anomalous dimensions γ_n as functions of the polynomial order $\mathcal{O}(n)$. Open (full) circles denote eigenvalues of symmetric (antisymmetric) eigenfunctions $\tilde{\Phi}_n$.

the spectrum of eigenfunctions, not appreciated previously. Comparison of symmetric and antisymmetric eigenvalues effects a strong degeneracy which increases with the polynomial order. This makes it apparent that a separation of symmetry classes is a necessary prerequisite to obtain results for higher orders. Furthermore, the resulting pattern indicates that the asymptotic behavior of both symmetry classes may be identical. It is also important to emphasize that the power-law behavior of eigenvalues suggested in [4] on the basis of lower-order Appell polynomials is not confirmed if higher-order terms are included [c.f. (dotted curve) $\gamma_n = 0.52 \, \mathcal{O}(n)^{0.417}$]. Indeed, up to order 20 the spectrum of γ_n-eigenvalues is better described by a logarithmic behavior of the form (solid line): $\gamma_n = 1.25 \, log^{1.32} (\mathcal{O}(n) + 1.37)$. In order to accurately separate power law from logarithmic behavior, one has to extend the calculation, at least, to polynomial order 100. Such a program is in progress. These estimates are of particular importance to check against results extracted from Wilson-loop calculations [6] which also predict a logarithmic behavior of such anomalous dimensions.

4. Acknowledgements

M.B. thanks the organizers of the conference for financial support from EEC funds.

1. G.P. Lepage and S.J. Brodsky, *Phys. Rev.* **D22** (1980) 2157.
2. M. Bergmann, Ph.D thesis, Bochum University, 1994.
3. K. Tesima, *Phys. Lett.* **110B** (1982) 319.
4. M. Bergmann and N. G. Stefanis, *Workshop on Exclusive Reactions at High Momentum Transfer*, Elba, Italy, June 24-26, 1993, C. E. Carlson, P. Stoler, and M. Taiuti (Eds.), World Scientific, Singapore, 1994.
5. M. Peskin, *Phys. Lett.* **88B** (1979) 128.
6. G. P. Korchemsky and G. Marchesini, *Nucl. Phys.* **B406** (1993) 225.

CONFINING QUARK POTENTIAL MODELS COMPATIBLE WITH CHIRAL SYMMETRY BREAKING AND LORENTZ INVARIANCE ARE TESTED IN K-N EXOTIC SCATTERING

P. J. de A. BICUDO, J. E. F. T. RIBEIRO, J. M. R. RODRIGUES

Departamento de Física, Instituto Superior Técnico, Av. Rovisco Pais
1096 Lisboa, Portugal

ABSTRACT

We study the general constraints in quark models due to confinement, chiral symmetry breaking and Lorentz invariance. We verify the models present the litterature and find that most fail to obey our constraints. Then we test the surviving models in K-N exotic scattering. We explain why naive quark models fail to explain it, and show that consistent quark models may succeed. Results are good for the Orsay model.

1. General constraints of consistent quark models

Quarks are widely accepted to exist not only as the partons of deep inelastic scattering but also as the constituents of hadrons. Although gluons are not directly present in quark models, a set of interesting theoretical problems -that QCD itself would not avoid- remains to be solved consistently. If we assume that hadrons are constituted of quarks, then the main properties -and problems- of hadronic physics are confinement, chiral symmetry breaking (χSB) and strong interactions. Confinement is supposed to induce infrared divergences that yield an arbitrarily large mass for quarks while the hadronic spectrum must remain finite. Chiral symmetry breaking has to verify Ward identities, and coexist with confinement. Lorentz invariance is necessary when we ask the quark model to describe hadronic interactions and decays.

1.1. Confinement

Confinement is related to quarks being unobserved individualy at low energies. This fact and the spectrum of hadrons where radial excitations have a nearly constant splitting suggest that confinement corresponds to a strong effective interaction between quarks. Thus we get the first two constraints that quark models should obey,

(a) unbound monotonously growing potential,

(b) infrared divergent potential where color singlets turn out to be finite while colored states have infinite energy

1.2. Chiral symmetry breaking

χSB is present both in the hadronic spectrum where the parity multiplets don't have the same mass, and in the sucesses of PCAC. In the limit of a vanishing quark mass, models must have the properties,

(c) there are solutions with a dynamical mass for quarks

(d) axial and vector Ward identities are verified

(e) the energy of the χSB vacuum is minimum

1.3. Lorentz invariance

The first striking evidence of hadronic physics is their strong interactions, where resonances are strongly coupled to their decay channels that usually are relativistic. In order to reproduce also these effects quark models must either be explicitly Lorentz invariant, or one must know how to boost the bound states. We get,

(f) the model is at least aproximately Lorentz invariant

(g) it is invariant and soluble in Minkowsky space

1.4. Literature

In table 1 we apply constraints (a) $\rightarrow$ (g) to references[1] $\rightarrow$[7] that include the quark models with χSB that we found in the literature. For instance, for a confining potential with a Dirac structure corresponding to the exchange of a scalar s , a pseudoscalar p, a vector v , an axial vector a and a tensor t , we get the conditions,

$$s - p = 0 \quad , \qquad s + p + 6t = 0 \quad , \qquad 4v - 4a = -2v - 2a \tag{1}$$

Table1.Confinement, chiral symmetry breaking and Lorentz invariance.

Ref.	(a) (b)	(c) (d) (e)	(f) (g)	Ref.	(a) (b)	(c) (d) (e)	(f) (g)
1	N N	Y Y Y	Y Y	5	Y N	Y Y Y	Y N
2	Y y	Y Y N	Y N	6	Y N	Y N N	Y Y
3	Y Y	Y Y Y	Y N	7	Y Y	Y Y Y	Y Y
4	Y Y	Y Y Y	Y N	8	Y Y	Y Y Y	Y Y

2. K-N exotic scattering

The $K-N$ exotic scattering is a crucial test for quark models. So far the interquark potentials that reproduce sucessfully the spectroscopy of hadrons, predict a cross section wich is much larger than the experimental one[9]. We now apply this stringent quantitative test to the *new* quark models, with χSB, that survived the general test of the previous section. Thus we study references [3,4] and their covarantized versions of references [7,8] that consist respectively of a quadratic and linear confining potentials.

2.1. Resonating group method

Starting from the Schwinger-Dyson equations with a q-q efective interaction, we get the Resonating Group Method (RGM) equations. In turn, from RGM it is straightforward[10] to deduce the associated Schrödinger equation for the KN system where the effective potential is derived from geometrical overlap kernels. Finally we obtain that the Isospin dependent KN scattering lengths a_I are simply given by,

$$a_I = \frac{1}{\alpha} \frac{4\pi\mu v\Omega_I}{\alpha^2 + 4\sqrt{\pi}\mu v\Omega_I} \quad , \qquad \Omega_{0,1} = \frac{1}{3}, \frac{7}{6} \tag{2}$$

274

where μ is the reduced mass of the KN system, α is the inverse radius of the bare nucleon, v is the hyperfine splitting of the bare baryon pair $\Delta - N$ and Ω_I is an overlap factor that depends of the flavour channel. We show in Fig. (1) the α as a function of v when we take for a_1 the experimental value of $0.32 Fermi$.

2.2. Naive models versus χSB

In a naive model, the nucleon is a 3 quark system, and v and α are fixed respectively by the experimental $\Delta - N$ mass difference and the inverse electromagnetic radius of the nucleon. It can be seen in Eq. (2) and in Fig. (1) that this yields a much too large scattering length a_I.

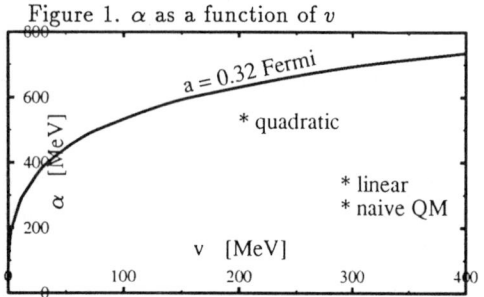

Figure 1. α as a function of v

The experimental scattering length can only be reproduced for a smaller v and a larger α. This is possible in consistent models where the N is coupled to the channel $N\pi$. In models with spontaneous χSB, where the π has a small mass, these coupled channel effects are strong. In this case [11] most of the $\Delta - N$ mass splitting is due to coupled channel effects and the hyperfine v is smaller. Also the bare mass of N is significantly larger than the actual mass because the coupled channels contribute with a negative mass shift of the order of 50% of the N mass, and in the same way α is increased.

In Fig. (1) we show the results of the quadratic confining potential of Refs.[3,7] and the linear potential of Refs.[4,8]. Results are good only for the quadratic case, and for the first time such a close agreement with experiment is reached. This reinforces a recent prediction of that model [12].

3. References

1. Y. Nambu and G. Jona-Lasinio, *Phys. Rev.* **122** (1961) 345; **124** (1961) 246.
2. H. Pagels *Phys. Rev.* **D14** (1976) 2747; **D15** (1977) 2991.
3. A. le Yaouanc, L. Oliver, S. Ono, O. Pene and J.-C. Raynal *Phys. Rev. Lett.* **50** (1983) 87;**54** (1985) 506; *Phys. Rev.* **D29** (1984) 1233; **D31** (1985) 137; *Phys. Lett.* **134B** (1984) 249.
4. S. Adler and A. Davis *Nuc. Phys.* **B224** (1984) 469.
5. Y. Dai, Z. Huang and D. Liu *Phys. Rev.* **D43** (1991) 1717.
6. F. Gross and J. Milana *Phys. Rev.* **D43** (1991) 969; **D45** (1992) 2401.
7. R. Horvat, D. Kekez, D. Klabucar, D. Palle *Phys. Rev.* **D44** (1991) 1584.
8. R. Horvat, D. Kekez, D. Palle, D. Klabucar *Zag. Prep.* ZTF-93-9-R (1993).
9. J. Ribeiro *Zeit. Phys.* **C5** (1980) 27; *Phys. Rev.* **D25** (1982) 2046.
10. P. Bicudo, J. Ribeiro *Zeit. Phys.* **C38** (1988) 453.
11. A. W. Thomas and G. A. Miller, *Phys. Rev.* **D43**, (1991) 288.
12. P. Bicudo, *Phys. Rev. Lett.* **72** (1994) 1600.

THE TWO-FERMION RELATIVISTIC WAVE EQUATIONS

OF CONSTRAINT THEORY

J. MOURAD*

*Laboratoire de Physique Thórique et Hautes Energies†,
Université Paris XI, Bât. 211,
F-91405 Orsay Cedex, France*

and

H. SAZDJIAN

*Division de Physique Théorique‡, Institut de Physique Nucléaire,
Université Paris XI,
F-91406 Orsay Cedex, France*

ABSTRACT

The two manifestly covariant relativistic wave equations of Constraint Theory for two-fermion systems are presented. The compatibility condition of the two equations eliminates the relative energy variable and leads to a three-dimensional internal dynamics, besides the spin degrees of freedom. These equations, relative to a sixteen-component wave function, can be reduced to a single Pauli-Schrödinger type equation for a four-component wave function.

The use of the manifestly covariant formalism with constraints[1] in the two-body problem[2,3] leads to a Poincaré invariant description of the dynamics of the system with the correct number of degrees of freedom. Furthermore, the potentials that appear in the corresponding wave equations are calculable, in perturbation theory, in terms of the kernel of the Bethe-Salpeter equation[4], and therefore allow one to deal with field theoretic problems.

For a system of two spin-1/2 particles, composed of one fermion of mass m_1 and one antifermion of mass m_2, say, Constraint Theory imposes two independent wave equations, which are generalizations of the Dirac equation satisfied by each particle in the free case. These wave equations have not a unique form and can be modified by wave function transformations, but a convenient form, where several properties can easily be read off, is the following one[3] :

$$\left(\gamma_1 . p_1 - m_1 \right) \tilde{\Psi} = \left(- \gamma_2 . p_2 + m_2 \right) \tilde{V} \tilde{\Psi} , \tag{1a}$$

$$\left(- \gamma_2 . p_2 - m_2 \right) \tilde{\Psi} = \left(\gamma_1 . p_1 + m_1 \right) \tilde{V} \tilde{\Psi} . \tag{1b}$$

*Present address : Laboratoire de modèles de Physique Mathématique, Parc de Grandmont, Université de Tours, 37200 Tours, France.
†Laboratoire associé au CNRS.
‡Unité de Recherche des Universités Paris 11 et Paris 6 associée au CNRS.

Here, $\tilde{\Psi}$ is a sixteen-component spinor wave function of rank two and is represented as a 4×4 matrix :

$$\tilde{\Psi} = \tilde{\Psi}_{\alpha_1,\alpha_2}(x_1, x_2) \quad (\alpha_1, \alpha_2 = 1, \ldots, 4) , \tag{2}$$

where $\alpha_1(\alpha_2)$ refers to the spinor index of particle 1(2). γ_1 is the Dirac matrix γ acting in the subspace of the spinor of particle 1 (index α_1); it acts on $\tilde{\Psi}$ from the left. γ_2 is the Dirac matrix acting in the subspace of the spinor of particle 2 (index α_2); it acts on $\tilde{\Psi}$ from the right. In Eqs. (1) p_1 and p_2 represent the momentum operators of particles 1 and 2, respectively. $\tilde{V}$ is a Poincaré invariant potential.

Equations (1) must be compatible among themselves. This imposes restrictions on the wave function and the potential. One finds for the wave function the constraint:

$$\left[(p_1^2 - p_2^2) - (m_1^2 - m_2^2) \right] \tilde{\Psi} = 0 , \tag{3}$$

which allows one to eliminate the relative energy variable in a covariant form. For eigenfunctions of the total momentum operator P, the solution of Eq. (3) is :

$$\tilde{\Psi} = e^{-iP.X} e^{-i(m_1^2 - m_2^2)P.x/(2P^2)} \tilde{\psi}(x^T) , \tag{4}$$

where we have used notations from the following definitions :

$$P = p_1 + p_2 , \quad p = \frac{1}{2}(p_1 - p_2) ,$$
$$X = \frac{1}{2}(x_1 + x_2) , \quad x = x_1 - x_2 . \tag{5}$$

We also define transverse and longitudinal components of four-vectors with respect to the total momentum P :

$$q_\mu^T = q_\mu - \frac{(q.P)}{P^2}P_\mu , \quad q_\mu^L = (q.\hat{P})\hat{P}_\mu , \quad \hat{P}_\mu = P_\mu/\sqrt{P^2} ,$$
$$q_L = q.\hat{P} , \quad P_L = \sqrt{P^2} . \tag{6}$$

This decomposition is manifestly covariant. In the c.m. frame the transverse components reduce to the three spacelike components, while the longitudinal component reduces to the timelike component of the corresponding four-vector. (Note that $x^{T2} = -\mathbf{x}^2$ in the c.m. frame.)

For the potential, the constraint is:

$$\left[p_1^2 - p_2^2 , \tilde{V} \right] \tilde{\Psi} = 0 . \tag{7}$$

By noticing that $p_1^2 - p_2^2 = 2P_L p_L$, this equation is satisfied by demanding that $\tilde{V}$, which is Poincaré invariant, be independent of the longitudinal relative coordinate x_L:

$$\tilde{V} = \tilde{V}(x^T, P_L, p^T, \gamma_1, \gamma_2) . \tag{8}$$

No other constraints are found. Equations (4) and (8) show that the internal dynamics of the system is three-dimensional, besides the spin degrees of freedom, described by the three-dimensional transverse coordinate x^T. Because of the constraint (3), the longitudinal components of the momentum operators can be replaced by their eigenvalues determined in terms of P_L [Eq. (6)] and the masses :

$$p_{1L} = \frac{P_L}{2} + \frac{(m_1^2 - m_2^2)}{2P_L}, \quad p_{2L} = \frac{P_L}{2} - \frac{(m_1^2 - m_2^2)}{2P_L}. \tag{9}$$

In general $\tilde{V}$ is an integral operator in x^T. However, in one-particle exchange diagrams, $\tilde{V}$ turns out to be a function of x^T and P_L, with definite dependences on the γ-matrices. It would also be meaningful to approximate, as a zeroth order approximation, multiparticle exchange contributions by appropriate local functions and thus to use for $\tilde{V}$ quasilocal expressions (in x^T and P_L). Furthermore, in perturbation theory, the c.m. energy (P_L) dependence of the potential takes into account the leading contribution of the nonlocal effects.

In order to solve Eqs. (1), one decomposes the 4×4 matrix wave function on the basis of the matrices 1, γ_L, γ_5 and $\gamma_L\gamma_5$ by defining 2×2 matrix components :

$$\psi = \psi_1 + \gamma_L\psi_2 + \gamma_5\psi_3 + \gamma_L\gamma_5\psi_4. \tag{10}$$

It can be shown[5] that, by eliminating in Eqs. (1) the components ψ_1, ψ_2 and ψ_4 in terms of ψ_3, which is a surviving component in the nonrelativistic limit, one ends up with a Pauli-Schrödinger type equation for ψ_3, where the analysis of the various contributions of the potential becomes simplified. This equation can then be used for the study of the interparticle dynamics for all values of masses and energy, by maintaining, at all stages, the relativistic invariance of the theory. In the limit when the mass of one of the particles tends to infinity, the above equations reduce to the one-body Dirac equation with the static potential created by the heavy particle[5]. For electromagnetic interactions the correct α^4 effects are produced when two-photon exchange contributions are taken into account[5,6].

1. *Constraint's Theory and Relativistic Dynamics*, edited by G. Longhi and L. Lusanna (World Scientific, Singapore, 1987), and references therein.
2. H.W. Crater and P. Van Alstine, *Ann. Phys. (N.Y.)* **148** (1983) 57; *Phys. Rev.* **D36** (1987) 3007.
3. H. Sazdjian, *Phys. Rev.* **D33** (1986) 3401; *J. Math. Phys.* **29** (1988) 1620.
4. H. Sazdjian, in *Extended Objects and Bound Systems*, proceedings of the Karuizawa International Symposium, 1992, edited by O. Hara, S. Ishida and S. Naka (World Scientific, Singapore, 1992), p. 117; *J. Math. Phys.* **28** (1987) 2618.
5. J. Mourad and H. Sazdjian, preprint IPNO/TH 94-5 and LPTHE 94/16, hep-ph/9403232, to appear in *J. Math. Phys.*.
6. H.W. Crater, R.L. Becker, C.Y. Wong and P. Van Alstine, *Phys. Rev.* **D46** (1992) 5117.

ON SOLVING qq̄ BOUND STATE PROBLEM FOR CORNELL POTENTIAL

IN MOMENTUM SPACE

UZIKOV YU.N.

Department of Physics, Kazakh State University, Timiryasev Str., 46
Alma-Ata, Kazakhstan 480121

ABSTRACT

A method for solving integral equation for two-body bound state problem with interaction potential $V = -\alpha/r + br$ is developed in momentum space. A regularization procedure for the potential and polynomial interpolation of wave function are used in order to avoid the singularities caused by linear confinement. The method has been tested on previously solved relativistic and nonrelativistic examples. Good results for the S-states are obtained.

1. Introduction

The possibility of successful description of meson spectra and meson couplings — from the π to the Υ-meson— on the basis of universal $q\bar{q}$–potential motivated by "soft" QCD had been shown some years ago [1]. Relativistic effects both in the kinetic and potential energy operators were found to be very important. However it is unclear at present which features of the potential from Ref. [1] will be present in future theory based on "true" QCD. Indeed, one-gluon exchange plus linear confinement potential is widely used in calculations of meson spectroscopy. The development of methods solving relativistic bound state problem with this potential is of great value. The method suggested in this work is rather simple with enough high accuracy especially in case of pure linear confinement.

2. Mathematical Formalism

We discuss here the following relativistic equation for the $q\bar{q}$ bound-state problem

$$[\sqrt{m_1^2 + \mathbf{p}^2} + \sqrt{m_2^2 + \mathbf{p}^2} - M] \, \psi(\mathbf{p}) + <\mathbf{p}|V|\psi> = 0, \qquad (1)$$

where the interaction operator V has the form of Cornell potential [2]

$$V(r) = -\alpha/r + br + C, \qquad (2)$$

m_i is the i-th quark mass, $<\mathbf{p}|\psi> = \psi(\mathbf{p})$ is the bound state vector, M is the mass of $q\bar{q}$-system, $\mathbf{p}$ is the 3-momentum of a quark in the $q\bar{q}$ c. m. frame. Eq.(1) takes place in so-called Relativistic Quantum Mechanic (RQM) [3] for a system with fixed number of particles (N=2). In case of spinless quarks a QCD basis for this equation with linear confinement potential have been established in Ref.[4]. The main mathematical problem for numerical solving Eq.(1) with potential (2) is the singular behavior of this potential in momentum space. In order to avoid the difficulties arising from these singularities regularization schemes [5,6], elimination technique [7] and square root procedure [8] are used. We use the regularization of the potential $\hat{V}(r, \mu) = \exp(-\mu r)V(r)$ and take the limit $\mu \to 0(+)$ after performing the integration in Eq.(1). For simplicity we discuss only the case $l = 0$. The generalization to the cases of higher partial waves ($l = 1, 2, 3$) is straightforward. In the case $l = 0$ we have for the potential in momentum space $\left(\hat{V}_l(p, q; \mu) = \int_0^\infty j_l(pr)V(r, \mu)j_l(qr)r^2 dr\right)$

$$\hat{V}_0^{conf}(p, q; \mu) = \frac{b}{2pq}\left(\frac{\mu^2 - (p+q)^2}{[(p-q)^2 + \mu^2]^2} - \frac{\mu^2 - (p+q)^2}{[(p+q)^2 + \mu^2]^2}\right). \tag{3}$$

The Exp. (3) has singular behavior at the point $p = q$ in the limit $\mu \to 0$: $V^{conf} \sim \mu^{-2}$ ($V^{Coul} \sim \ln \mu^2$). To avoid the difficulties connected with the term $\sim \mu^{-2}$ we transform the integral operator entering the equation

$$I_0(p, \mu) = \int_0^\infty V_0^{conf}(p, q; \mu)\psi_0(q)q^2 dq. \tag{4}$$

Assuming the relation $q\psi(q) \to 0$ at the limit $q \to \infty$ one can find from (4)

$$I_0(p, \mu) = \frac{b}{2p}\int_0^\infty \left\{\frac{p-q}{(p-q)^2 + \mu^2} + \frac{p+q}{(p+q)^2 + \mu^2}\right\} U_0'(q)dq. \tag{5}$$

Here the definition $U = q\psi(q)$ is introduced. The first term in the integrand of Eq.(5) equals to zero at the point $q = p$ for any $\mu \neq 0$. Therefore the singular behavior of the $< p|V^{Conf}|q >$ potential is absent in the expression for $< p|\hat{V}^{conf}|\psi >$. Taking into account the limit $\mu \to 0$ we find that the integral in Eq.(5) is the principal value one. Thus Eq.(1) with potential (2) in this approach takes the form of integro-differential equation.

When we numerically solve this equation, we restrict the integration to the range $[-B, B]$, where B is enough large momentum. Then we use the mapping $q = Bx$, $x \in [-1, +1]$. For the unknown wave function U_0 the following polynomial interpolation is used in this range

$$U_0(x) = \sum_{i=1}^N C_i \frac{P_N(x)}{(x - x_i)P_N'(x)}, \tag{6}$$

where P_N is the first-kind Legendre polynomial of the N-th order (in our calculations we put $N = 50 \div 60$), x_i is its node. It follows from (6) that unknown coefficients C_i satisfy the condition $U_0(x_i) = C_i$. The advantages of this interpolation are due to the following features: (i) the derivative of U can be presented in analytical form, (ii) the principal value integrals in Exp.(5) over the range $q \in [-B, B]$ can be calculated in analytical form using the following relation $\frac{1}{2}\int_{-1}^{+1} P_N(x)/(t - x)dx = Q_N(t)$ (iii) an integration of the right-hand side of Eq.(6) between the limits $x = -1$ and $x = +1$ gives an exact N-points Gauss quadrature formula with true weightfactors $W_i = -2Q_N(x)/P_N'(x_i)$.

Table 1. The S-state masses of the Hamiltonians $\hat{H} = p^2 + r = \hat{M} - m_1 - m_2$, $(m_1 = m_2 = 1\ GeV)$ and $\hat{H} = p + r = \hat{M}$.

nS	$\hat{H} = p^2 + r = \hat{M} - m_1 - m_2$		$\hat{H} = p + r = \hat{M}$.	
	M,GeV present work	M,GeV from work [2]	M,GeV present work	M,GeV from work [6]
1S	4.3380	4.3381	2.2323	2.2323
2S	6.0879	6.0879	3.3299	3.3297
3S	7.5205	7.5206	4.1662	4.1642
4S	8.7868	8.7867	4.8566	4.8585
5S			5.4669	5.4659

3. Numerical Results

The numerical results are obtained for the nonrelativistic Schrödinger equation with the Hamiltonian $\hat{H} = p^2 + r$, ultrarelativistic system $(m_1 = m_2 = 0)$ with $\hat{H} = p + r$ (Tab.1) and for the ρ-meson states with parameters of the potential (2) from [6] $(m_1 = m_2 = 0.12\ GeV,\ \alpha = 0.468,\ b = 0.2GeV,\ C = -0.486\ GeV)$. The following values are obtained for the low-lying ρ-meson states (in the brackets are presented the corresponding values from Tab. 8 of Ref.[6]): $M(1S) = 0.776(0.770)$ GeV, $M(2S) = 1.486(1.481)$ GeV, $M(3S) = 2.040(2.022)$ GeV. The values of $< r^2 >$, $< p^2 >$, $|\psi(r = 0)|^2$ are calculated for the $\rho-$ meson. The following numbers have been obtained for the $1S-$ state: $|\psi(r = 0)| = 0.55(0.57)$, $< p^2 >= 0.476(0.476)$, $< r^2 >= 3.4(3.5)$; for the $2S-$ state: $|\psi(r = 0)| = 0.70(0.71)$, $< p^2 >= 0.687(0.678)$, $< r^2 >= 5.5(5.7)$.

4. Conclusion

One can see that the results obtained on the basis of our method are in good agreement with other solved examples both for the mass spectrum and wave functions. The next improvement may be reached by combination of our method with the elimination technique [7].

The author would like to thank Professor Yu. A. Simonov for the interest shown in this work and stimulating discussions.

1. S. Godfrey and N. Isgur *Phys.Rev.* **D32** (1985) 189.
2. E. Eichten et al., *Phys. Rev.* **D17** (1978) 3090.
3. F.M. Lev, *Phys. Rev.* **C49** (1994) 383.
4. A. Yu. Dubin, A. B. Kaidalov and Yu. A. Simonov, *Yad. Fiz.* **56** (1993) 213.
5. J. R. Spence and J. P. Vary, *Phys. Rev.* **D35** (1987) 2191.
6. J. L. Basdevant and S. Boukraa, *Z. Phys.* **C28** (1985) 413.
7. J. Norbury, D. E. Kahana and K. M. Maung, *Phys. Rev.* **D47** (1993) 1182.
8. L.Fulcher, in this Conference.

LIGHT-CONE HAMILTONIAN OF THE QCD STRING WITH QUARKS

A.YU.DUBIN

Institute of Theoretical and Experimental Physics
117259, Moscow, B.Cheremushkinskaya 25, Russia

ABSTRACT

The light–cone Hamiltonian of the effective QCD string with quarks is derived from gauge and Lorentz invariant Green function of $q\bar{q}$ system in the confining gluonic fields. Confinement is maintained via the area law asymptotics for the averaged Wilson loop which results in Hamiltonian of quarks connected by the frozen string.

1. Introduction

Recently a new formalism was developed[1,2] to consider large distances nonperturbative dynamics of quark and antiquark in the confining gluonic fields. Starting from the QCD Lagrangian and assuming the minimal area law for the asymptotics of Wilson loop we have obtained the Hamiltonian of the system in the rest frame. Since the area entering Wilson loop can be represented as the worldsheet of the frozen string our Hamiltonian contains explicitly the string energy density determined by the motion of its ends.

In this paper we modify our formalism for consideration of the same system (quark and antiquark interacting via the Wilson loop average with the area law asymptotics) in the light cone (l.c.) frame. Our aim will be to derive on light cone the Hamiltonian of quark and antiquark connected by the rigid string (wave functions of which are directly connected to form factors and structure functions[3] of hadrons).

2. Light Cone Hamiltonian of the system

Our starting point as in[1,2] is the quenched approximation for Green function G of spinless quark and antiquark in the confining gluonic fields[4] in Eulidean space which leads to the following $3D$ effective action of valence quarks connected by the straight line string in Minkowski l.c. variables

$$G = \int D\mu_1 D\mu_2 Dz_\perp D\bar{z}_\perp Dz_- D\bar{z}_- \exp[iA]$$

$$A = K + \bar{K} - \sigma \int d\tau \, d\beta \sqrt{(\dot{w}w')^2 - \dot{w}^2 w'^2} \tag{1}$$

where

$$w_\mu(\tau,\beta) = z_\mu(\tau)\beta + \bar{z}_\mu(\tau)(1-\beta), \quad \dot{w}_\mu = \frac{\partial w_\mu}{\partial \tau}, \quad w'_\mu = \frac{\partial w_\mu}{\partial \beta}, \tag{2}$$

$$z_+(\tau) = \bar{z}_+(\tau) = \tau \tag{3}$$

The quark kinetic terms K, $\bar{K}$ have the form [5]

$$K = \frac{1}{2} \int d\tau \left(\mu_1(\dot{z}_\perp^2 + 2\dot{z}_-) - \frac{m_1^2}{\mu_1} \right), \quad \bar{K} = \frac{1}{2} \int d\tau \left(\mu_2(\dot{\bar{z}}_\perp^2 + 2\dot{\bar{z}}_-) - \frac{m_2^2}{\mu_2} \right) \tag{4}$$

where $\mu_1(\tau), \mu_2(\tau)$ are the functions which should be integrated out in path integral (1) as well as $\vec{z}_\perp, \vec{\bar{z}}_\perp$ and $z_-, \bar{z}_-$.

First we note that our action contains time derivatives of w_μ in the interaction term so that the string carries a finite fraction of the total and relative momentum of the system. To incorporate it explicitly let us introduce in the same way as in ref.[2] the total R_μ and relative r_μ coordinates

$$R_\mu = x(\tau)z_\mu(\tau) + (1 - x(\tau))\bar{z}_\mu(\tau), \quad r_\mu = z_\mu(\tau) - \bar{z}_\mu(\tau) \tag{5}$$

where the parameter $x(\tau)$ will be determined from the condition[2] that $\dot{R}_\mu$ is decoupled from $\dot{r}_\mu$ in the total action (1).

We use the formalism of auxiliary fields[6] in order to write the effective action (1) in a gaussian form which results in the following Hamiltonian[5]

$$H = \frac{1}{2} \left\{ \frac{m_1^2}{\mu_1} + \frac{m_2^2}{\mu_2} + \frac{\vec{p}_\perp^2 - \frac{(\vec{p}_\perp \vec{r}_\perp)^2}{r_\perp^2}}{a_3} + \frac{((\vec{p}_\perp \vec{r}_\perp) + \lambda(P_+ r_-))^2}{\tilde{\mu} r_\perp^2} + \int \frac{\sigma^2}{\nu} d\beta \, r_\perp^2 + \right. \tag{6}$$

$$\left. + \frac{\int \nu d\beta \cdot (P_+ r_-)^2}{P_+(\mu_1 + \mu_2)r_\perp^2} \right\}$$

where $\nu(\tau, \beta)$ is the auxiliary field which should be integrated out in the full path integral representation for the Green function and

$$a_3 = \mu_1(1 - x)^2 + \mu_2 x^2 + \int (\beta - x)^2 \nu \, d\beta$$

$$\tilde{\mu} = \mu_1 \mu_2 / (\mu_1 + \mu_2), \quad \lambda = x - \mu_1 / (\mu_1 + \mu_2) \tag{7}$$

Here we have introduced the total momentum

$$P_+ = p_{1+} + p_{2+} = \mu_1 + \mu_2 + \int \nu \, d\beta, \quad \vec{P}_\perp = 0 \tag{8}$$

and Feynman–Bjorken variable x

$$x = \frac{p_{1+}}{p_{1+} + p_{2+}} = \frac{\mu_1 + \int \beta \nu \, d\beta}{\mu_1 + \mu_2 + \int \nu \, d\beta} \tag{9}$$

which coincides[5] with $x(\tau)$ entering eq. (5). From eqs. (8), (9) it follows that $\nu(\tau,\beta)$ is the density of fraction of total momentum carried by the string while $\mu_i(\tau)$ are fractions of P_+ carried by the corresponding quarks.

One is to express[5] μ_i with the help of eqs. (8), (9) through x, P_+, ν and then substitute $\nu(\tau,\beta)$ by its extremal value[1] ν_{ext} defined by equation

$$\left.\frac{\delta H}{\delta \nu}\right|_{\nu_{ext}} = 0 \tag{10}$$

It results in Hamiltonian depending only on canonically conjugated pairs $\{x,(P_+r_-)\}$, $\{\vec{p}_\perp,\vec{r}_\perp\}$ and Weyl ordering enables to construct the operator of H which will be used to calculate wave functions and formfactors.

3. Discussion

To conclude we derive l.c. Hamiltonian of the effective QCD string with quarks. Due to the noninert nature of the interaction properly defined Feynman–Bjorken variable x and total momentum P_+ involve the contribution from the string. It is important to note that total momentum P_+ and relative distance r_- enter the mass squared operator M^2 only via combination (P_+r_-) which is canonically conjugated to x. This property violates usual assumption that P_+ is decoupled from M^2. Also we stress that M^2 can not be decomposed into a pure kinetic (usually refered to quarks) and pure potential (string) parts, which will result in particular in the dependence of properly defined l.c. orbital momentum on the interaction.

1. Dubin A.Yu., Kaidalov A.B., Simonov Yu.A. *Yad. Fiz.* **56** (1993) 164
   Dubin A.Yu., Kaidalov A.B., Simonov Yu.A. *Phys. Lett.* **B323** (1994) 41.
2. Gubankova E.L., Dubin A.Yu. *Phys. Lett.* **B334** (1994) 180.
   Gubankova E.L., Dubin A.Yu. Submitted to *Phys. Rev. D, HEP–PH 9408278*.
3. Mueller A.H. *Phys. Rep.* **73** (1981) 237,
   Gribov L.V., Levin E.M., Ryskin M.G. *Phys. Rep.* **100** (1983) 1.
4. Yu.A.Simonov Yu.A. *Nucl.Phys.* **B307** (1988) 512,
   Simonov Yu.A. *Phys. Lett.* **226** (1989) 151
5. Dubin A.Yu., Kaidalov A.B., Simonov Yu.A. *Yad. Fiz.* in press,
   HEP–PH 94 08212,
   Dubin A.Yu., Kaidalov A.B., Simonov Yu.A. Submitted to *Phys. Lett.* **B**.
6. Polyakov A.M. Gauge Fields and strings.
   Harwood academic publishers, 1987.

NONPERTURBATIVE INFRARED RENORMALIZATION
AND QUARK CONFINEMENT THEOREM IN QCD

V. GOGOHIA

RMKI, Department of Theoretical Physics, Central Research Institute for Physics,
H-1525, Budapest 114, P. O. B. 49, Hungary

ABSTRACT

Nonperturbative approach to QCD at large distances in the context of the Schwinger-Dyson (SD) equations and corresponding Slavnov-Taylor (ST) identity in the quark sector is presented. Making only one widely accepted assumption that the full gluon propagator becomes infrared (IR) singular like $(q^2)^{-2}$ in an arbitrary covariant gauge, we find three and only three confinement-type solutions for the quark propagator (quark confinement theorem). Two of them vanish after the removal of the IR regularization parameter. The third solution does not depend on this latter parameter, but it has no pole and it implies dynamical chiral symmetry breakdown (DCSB). We also show that multiplication solely by the quark IR renormalization constant would make all the Green's functions IR finite (multiplicative renormalizability). The final forms of the renormalized (IR finite) quark SD equations do not *explicitly* depend on a gauge fixing parameter.

Our aim here is to find (using minimal assumptions) a quark propagator which, on the one hand, should satisfy some of the necessary conceptual requirements and, on the other hand, should make it possible to investigate analytically and calculate numerically any physical quantity from low energy physics in a self-consistent way. Making only one widely accepted phenomenological assumption that the full gluon propagator becomes IR singular like $D_{\mu\nu}(q) \sim (q^2)^{-2}, q^2 \to 0^{1-5}$ in the covariant gauge, precisely such a quark propagator has recently been found within our approach to QCD[5]. Such singular behaviour of the full gluon propagator, on the one hand, satisfies the Wilson criterion of quark confinement-area low [3,6] leading to the derivation of the linear rising quark-antiquark potential at long distances within the Wilson loop approach[7,8], and on the other hand, requires the introduction of a small IR regularization parameter ϵ in order to define the initial SD equations and ST identities in the IR region. Because of this, the quark propagator and other Green's functions become dependent, in general, on this IR regularization parameter ϵ, which is to be set to zero at the end of computation $\epsilon \to 0^+$. By completing our renormalization program, we show explicitly that all Green's functions are IR multiplicative renormalizable (MR).

Let us begin from the exact, unrenormalized SD equation for the quark propagator in momentum space

$$S^{-1}(p) = S_0^{-1}(p) + g^2 C_F \int \frac{d^n q}{(2\pi)^n} \Gamma_\mu(p,q) S(p-q) \gamma_\nu D_{\mu\nu}(q), \qquad (1)$$

where C_F is the eigenvalue of the quadratic Casimir operator in the fundamental representation. Other notions are obvious.

In order to actually define an initial SD Eq.(1) in the IR region (at small momenta) let us apply the gauge-invariant dimensional regularization method[9] in the limit $n = 4 + 2\epsilon$, $\epsilon \to 0^+$, where ϵ is the above mentioned small IR regularization parameter. In what follows we consider the SD equations and the corresponding quark-gluon ST identity in Euclidean space ($d^n q \to i d^n q_E$, $q^2 \to -q_E^2$, $p^2 \to -p_E^2$, but for simplicity the Euclidean subscript will be omitted).

Let us use in the sense of distribution theory the relation[10]

$$(q^2)^{-2+\epsilon} = \frac{\pi^2}{\epsilon} \delta^4(q) + (q^2)_+^{-2} + O(\epsilon), \qquad \epsilon \to 0^+. \tag{2}$$

The above mentioned singularity of the full gluon propagator is a unique initial singularity within our approach to QCD at large distances. All other Green's functions will be considered as regular functions of their arguments.

Substituting Eq.(2) into the SD (1), we obtain the quark propagator expansion in the IR region (in Euclidean space). Within the distribution theory[10] one has

$$\int \frac{d^n q}{(2\pi)^n} \Gamma_\mu(p,q) S(p-q) \gamma_\nu t_{\mu\nu}(q)(q^2)_+^{-2}$$
$$= \int \frac{d^n q}{(2\pi)^n} t_{\mu\nu}(q) \{\Gamma_\mu(p,q) S(p-q) - \Gamma_\mu(p,0) S(p)\} \gamma_\nu (q^2)^{-2}, \tag{3}$$

so, expanding $S(p-q)$ and $\Gamma_\mu(p,q)$ in powers of q and keeping the terms of order q^{-2} (the Coulomb order terms), one finally obtains

$$S^{-1}(p) = S_0^{-1}(p) + \frac{1}{\epsilon} \tilde{g}^2 \Gamma_\mu(p,0) S(p) \gamma_\mu$$
$$+ c_1 \int \frac{d^n q}{(2\pi)^n} t_{\mu\nu}(q)(q^2)^{-2} \left\{ -q^\alpha q^\lambda \Gamma_\mu^\alpha(p,0) S^\lambda(p) + \frac{1}{2} q^\alpha q^\beta \Gamma_\mu^{\alpha\beta}(p,0) S(p) \right\} \gamma_\nu$$
$$+ \int \frac{d^n q}{(2\pi)^n} \left\{ \frac{c_1}{2} t_{\mu\nu}(q)(q^2)^{-2} q^\lambda q^\tau \Gamma_\mu(p,0) S^{\lambda\tau}(p) + c_2 T_{\mu\nu}(q,a)(q^2)^{-1} \Gamma_\mu(p,0) S(p) \right\} \gamma_\nu$$
$$+ O(\epsilon), \quad \epsilon \to 0^+, \tag{4}$$

where $\tilde{g}^2 = C_F \frac{3}{4} g^2 \mu^2 \pi^2 (2\pi)^{-4}$ and $t_{\mu\nu}(q) = [g_{\mu\nu} - \frac{q_\mu q_\nu}{q^2}]$ is the transverse tensor and the tensor $T_{\mu\nu}(q,a) = \beta(a) t_{\mu\nu}(q) + a \frac{q_\mu q_\nu}{q^2}$ explicitly depends on a gauge-fixing parameter a. Here, for example, $\Gamma_\mu^\alpha(p,0) = \left\{ \frac{d\Gamma_\mu(p,q)}{dq^\alpha} \right\}_{q=0}$ and so on.

As mentioned above, all Green's functions become dependent generally on the IR regularization parameter ϵ. Let us introduce the renormalized (IR finite) quark-gluon vertex function at zero momentum transfer and the quark propagator, as

$$\Gamma_\mu(p,0) = Z_1(\epsilon) \bar{\Gamma}_\mu(p,0), \qquad and \qquad S(p) = Z_2(\epsilon) \bar{S}(p), \qquad \epsilon \to 0^+, \tag{5}$$

respectively. Here $Z_i(\epsilon)$ ($i = 1, 2$) are the corresponding IR renormalization constants. $\bar{\Gamma}_\mu(p,0)$ and $\bar{S}(p)$ do not depend on ϵ in the $\epsilon \to 0^+$ limit, i.e. they exist as $\epsilon \to 0^+$.

From Eq.(4) and on account of Eq.(5) a cancellation of nonperturbative IR divergences takes place if and only if (iff)

$$Z_1(\epsilon)Z_2^2(\epsilon) = \epsilon Y_1, \qquad \epsilon \to 0^+. \tag{6}$$

Here and below $Y_i (i = 1, 2, 3, 4.)$ are the arbitrary finite constants which can be set to unity without loosing generality because of the above mentioned MR[5]. This is a quark convergence condition. Because of Eq.(6), the explicitly gauge-dependent terms (the so-called next-to-leading terms) in the SD Eq.(4) become ϵ–order terms. For this reason these noninvariant terms vanish in the $\epsilon \to 0^+$ limit. Obviously, the finite next-to leading terms coming from the IR region and not containing the derivatives of the higher Green's functions, become the terms of order ϵ. For this reason, they vanish in the $\epsilon \to 0^+$ limit as well as the above mentioned finite terms explicitly depending on a gauge fixing parameter. Thus, leading and next-to-leading terms of our expansion for the quark propagator in the IR domain always become not *explicitly* dependent on a gauge-fixing parameter a. It is a general feature of our expansions in the IR region for various Green's functions. The next-to-leading terms coming from the IR region, which remain finite by completing our renormalization program (containing the derivatives of higher Green's functions) should be considered as a good approximation for the intermediate region, which still remains *terra incognita* in QCD. They are of the same Coulomb order of magnitude ($\sim q^{-2}$ in comparison with the initial singularity) as the corrections coming from the UV region. Here and in what follows we will be interested only in the leading terms of the corresponding Green's functions expansions in the IR domain. The renormalization program to remove all nonperturbative IR divergences on a general ground depends, of course, only on these terms.

In the same way should be investigated the SD equation for the ghost self-energy as well as the Slavnov-Taylor (ST) identity for the corresponding quark-gluon vertex function[5]. We are now able to formulate the main result of our approach in terms of the following theorem.

Quark confinement theorem

Let us assume that the full gluon propagator in the covariant gauge becomes an IR singular like $(q^2)^{-2}$ at small momentum ($q^2 \to 0$), then the quark SD equation has three and only three confinement-type solutions for the quark propagator in the IR region. Two of them are IR vanishing solutions after the removal of the IR regularization parameter. The third solution does not depend on the IR regularization parameter at all, but it has no pole on the real axis in the complex plane and it implies DCSB.

Proof

Let us write down the close set of equations for the IR-finite quantities in the quark sector, obtained finally within our approach[5]

$$\bar{S}^{-1}(p) = Z_2(\epsilon)S_0^{-1}(p) + \tilde{g}^2\bar{\Gamma}_\mu(p,0)\bar{S}(p)\gamma_\mu, \tag{7}$$

$$\bar{\Gamma}_\mu(p,0)\left[\epsilon Z_2^{-1}(\epsilon) + \frac{1}{2}\bar{b}(0)\right] = id_\mu\bar{S}^{-1}(p) - \frac{1}{2}\bar{b}(0)\bar{S}(p)\bar{\Gamma}_\mu(p,0)\bar{S}^{-1}(p). \tag{8}$$

Recalling that the quark wave function renormalization constant $Z_2(\epsilon)$ should always be a regular function of the ϵ-parameter, one immediately obtains that there are three and only three different types in the quark propagator behaviour (5) as $\epsilon \to 0^+$. Indeed, from the quark ST identity (8) it follows that $Z_2(\epsilon)$ cannot tend to zero in any case stronger than ϵ in the $\epsilon \to 0^+$ limit. Thus, we obtain the first type of quark propagator behaviour as $\epsilon \to 0^+$, namely it behaves exactly as ϵ for $\epsilon \to 0^+$.

I. $Z_2(\epsilon) = \epsilon$, $\quad S(p) = \epsilon\bar{S}(p)$. The system of Eqs.(7)-(8) in this case was first investigated by Pagels[4] in his pioneering work on nonperturbative QCD.

The second independent type of behaviour of quark propagator can be obtained if the IR renormalization constant $Z_2(\epsilon)$ tends to zero more weakly than ϵ for $\epsilon \to 0^+$, otherwise remaining arbitrary. So that,

II. $\epsilon Z_2^{-1}(\epsilon) = 0$, $\quad S(p) = Z_2(\epsilon)\bar{S}(p)$. Contrary to the previous case, the system of Eqs.(7)-(8) in this case has no regular limit $\bar{b}(0) = 0$.

The third independent solution is the IR finite from the very beginning quark propagator.

III. $S(p) \equiv \bar{S}(p)$ and $Z_2(\epsilon) \equiv Z_2 = 1$ because of the above mentioned MR[5]. The system of Eqs.(7)-(8) in this case always has a chiral symmetry breaking solution $(m_0 = 0, B(t) \neq 0$, dynamical quark mass generation). Any nontrivial solution automatically breaks the γ_5 invariance of the quark propagator and therefore leads to the spontaneous chiral symmetry breakdown. The system III cannot be solved exactly in the general case $(m_0 \neq 0)$, but it is possible to show that the solutions of all three systems cannot have pole-like singularities at any finite point on the real axis in the p^2- complex plane[5]. This is a direct manifestation of quark confinement.

The author would like to thank the organizers of the QUARK CONFINEMENT Conference for the invitation to participate in it. A pleasant atmosphere on the conference at Como is also acknowledged.

1. References

1. S.Mandelstam, *Phys. Rev.* **D20** (1979) 3223
2. M.Baker, J.S.Ball and F.Zachariasen, *Nucl. Phys.* **B186** (1981) 531, 560; *Nucl. Phys.* **B226** (1983) 455
3. M.Bander, *Phys. Rep.* **75** (1981) 205
4. H.Pagels,*Phys. Rev.* **D15** (1977) 2991
5. V.Gogohia, *Int. Jour. Mod. Phys.* **A9** (1994) 759
6. K.Wilson, *Phys. Rev.* **D10** (1974) 2445
7. N.Brambilla and G.M.Prosperi, *Phys. Rev.* **D48** (1993) 2360
8. W.Lucha, F.F.Schoberl and D.Gromes, *Phys. Rep.* **200** (1991) 127-240
9. G.'t Hooft and M.Veltman, *Nucl. Phys.* **B44** (1972) 189
10. J.N.Gelfand and G.E.Shilov, *Generalized Functions* (Academic Press, NY, 1964)

GENERALIZED LADDER BETHE-SALPETER EQUATION
IN MINKOWSKI SPACE

KENSUKE KUSAKA AND ANTHONY G. WILLIAMS

Department of Physics and Mathematical Physics, University of Adelaide

South Australia 5005, Australia

ABSTRACT

The Bethe-Salpeter (BS) equation is solved within a broad class of non-ladder scattering kernels using the perturbation theory integral representation (PTIR). We consider a bound state of two spinless particles. with the formal expression of the full scattering kernel in a $\phi^2\sigma$ scalar model. The resulting BS amplitude is written as a parametric integral of its weight function in this approach. We derive an integral equation for the weight function with a real kernel. In the pure ladder limit we have recovered the usual results obtained in earlier Euclidean treatments.

1. Introduction

In a relativistic field theory two-body bound states are described by the Bethe-Salpeter (BS) amplitude[1]. In order to handle such a singular integral equation the analytic continuation of the relative-energy variable, which is called "Wick rotation", is widely used The ladder BS amplitude is solved as a function of Euclidean relative momentum in the standard approach. If one uses a "dressed" propagator or more complicated kernels in the BS equation, the validity of the Wick rotation becomes highly nontrivial, e.g., almost all the dressed propagator studied previously in the Dyson-Schwinger equation approach contains pathological complex "ghost" poles[2]. It is therefore preferable to formulate and solve the BS equation directly in Minkowski space.

We present a method to solve the BS equation without Wick rotation by making use of the perturbation theory integral representation (PTIR) for the BS amplitude[3]. This integral representation has been studied for a scalar-scalar bound state in the ladder approximation[4]. We extend this method to a wide class of non-ladder kernels.

2. Scalar-Scalar BS Equation

Let us consider a bound state of two spinless particles ϕ having a mass m. They interact each other through the exchange of another spinless particle σ with a mass μ. Let the interaction between ϕ and σ be the Yukawa coupling: $\mathcal{L}_{int} = -g\phi^2\sigma$. The Bethe-Salpeter amplitude $\Phi(p, P)$ for the bound state having the total momentum P and the relative one $2p$ obeys the following equation:

$$[D(P/2 + p)D(P/2 - p)]^{-1}\Phi(p, P) = \int \frac{d^4q}{(2\pi)^4 i} I(p, q; P)\Phi(q, P) \tag{1}$$

where $D(q)$ is the propagator of ϕ-particle and we approximate it with the tree one: $D^0(q) = 1/(m^2 - q^2 - i\epsilon)$. The scattering kernel $I(p, q; P)$ describes the process: $\phi_1\phi_2 \to \phi_3\phi_4$ and the momentum $2p$ and $2q$ are the relative momentum of initial and final states. We consider the following formal expression for the full scattering kernel:

$$I(p, q; P) = \int_0^\infty d\gamma \int_\Delta d\vec{\xi} \frac{\rho_{st}(\gamma, \vec{\xi}; g)}{\gamma - \left[\sum_{i=1}^4 \xi_i q_i^2 + \xi_5 s + \xi_6 t\right] - i\epsilon}$$
$$+ (\text{cyclic permutation of } s, t, u) \tag{2}$$

where q_i is the momentum carried by ϕ_i and s, t and u are Mandelstam variables. The symbol Δ denotes the integral region of 6 dimensionless Feynman parameters ξ_i such that $\Delta \equiv \{\xi_i \,|\, \xi_i \geq 0, \sum \xi_i = 1 (i = 1, \ldots, 6)\}$. The "mass" parameter γ represents a spectrum of the scattering kernel. The function $\rho_{ch}(\gamma, \vec{\xi}; g)$ gives the weight of the spectrum in a different channel; $ch = \{st\}, \{tu\}, \{us\}$. This expression has been derived by Nakanishi and is called the perturbation theory integral representation (PTIR)[3].

3. PTIR for BS amplitude

Let us consider the s-wave bound state for simplicity[*]. We assume that the BS amplitude $\Phi(p, P)$ has an integral representation of the form:

$$\Phi(p, P) = -i \int_{-\infty}^\infty d\alpha \int_{-1}^1 dz \frac{\varphi_n(\alpha, z)}{[m^2 + \alpha - (p^2 + zp \cdot P + P^2/4) - i\epsilon]^{n+2}} \tag{3}$$

where the non-negative integer parameter n is a dummy parameter, since a partial integration with respect to α changes the power. We can utilize this artificial degree of freedom for a numerical study. Substituting the above expression into Eq.(1) together with Eq.(2), we obtain the following integral equation for $\varphi_n(\alpha, z)$ as follows:

$$\varphi_n(\bar{\alpha}, \bar{z}) = \int_{-\infty}^\infty d\alpha \int_{-1}^1 dz \sum_{ch} \int_0^\infty d\gamma \int_\Delta d\vec{\xi} \frac{\rho_{ch}(\gamma, \vec{\xi}; g)}{(4\pi)^2} K_n(\bar{\alpha}, \bar{z}; \alpha, z) \, \varphi_n(\alpha, z), \tag{4}$$

where we have suppressed the dependence on the kernel parameters. Note that the Eq.(4) is frame-independent. The real kernel function $K_n(\bar{\alpha}, \bar{z}; \alpha, z)$ with a fixed kernel parameter set $(\gamma, \vec{\xi})$ has the following structure:

$$K_n(\bar{\alpha}, \bar{z}; \alpha, z) = \frac{\partial}{\partial \bar{\alpha}} (\bar{\alpha}^n \theta(\bar{\alpha})) h_n(\alpha, z) - \theta((\alpha - \omega_1(\bar{\alpha}, \bar{z}, z))(\alpha - \omega_2(\bar{\alpha}, \bar{z}, z)))$$
$$\times \left\{ \frac{g_n(\bar{\alpha}, \bar{z}; \alpha, z)}{\sqrt{(\alpha - \omega_1(\bar{\alpha}, \bar{z}, z))(\alpha - \omega_2(\bar{\alpha}, \bar{z}, z))}} + k_n(\bar{\alpha}, \bar{z}; \alpha, z) \right\} \tag{5}$$

---

[*]Extension to higher partial wave solutions is straightforward.

where $\omega_i(\bar{\alpha}, \bar{z}, z)$, $g_n(\bar{\alpha}, \bar{z}; \alpha, z)$ and $h_n(\alpha, z)$ are regular functions. The function $k_n(\bar{\alpha}, \bar{z}; \alpha, z)$ is also regular in the simple one-σ-exchange ladder kernel, but in general it contains a singularity such as Pf $\cdot 1/(\alpha - \tau(\bar{\alpha}, \bar{z}, z))^n$ where the symbol Pf$\cdot$ stands for Hadamard's finite part and $\tau(\bar{\alpha}, \bar{z}, z)$ is a regular function. Thus the first term of Eq.(5) contains a δ-function singularity at $\bar{\alpha} = 0$, only if $n = 0$. This singularity, independent of $\bar{z}, \alpha$ and z, corresponds to the pole singularity in $\Phi(p, P)$ which comes from the free propagation of two ϕ's. On the other hand, the second term contains a discontinuity due to the step function depending on $\bar{\alpha}, \bar{z}, \alpha$ and z. In addition to the singularity in $k_n(\bar{\alpha}, \bar{z}; \alpha, z)$ as mentioned above, the term has a square root line singularity at the boundary of its support. Since the Hadamard's finite part Pf $\cdot 1/x$ coincides Cauchy's principal value, the kernel function $K_{n=1}(\bar{\alpha}, \bar{z}; \alpha, z)$ is integrable, so that the integral equation (4) with a constant kernel parameter set is numerically tractable by setting the dummy parameter $n = 1$, provided that the weight function $\varphi_1(\alpha, z)$ is differentiable at the singular points of $K_1(\bar{\alpha}, \bar{z}; \alpha, z)$. We have reproduced the known results for the one-σ-exchange ladder kernel by solving Eq.(4) numerically. A case with a general scattering kernel, whose weight function contains a derivative of δ-function, is under investigation.

4. Summary

We have derived a real integral equation of the weight function for the scalar-scalar BS amplitude with a formal expression of the full scattering kernel. We found that our integral equation is numerically tractable for a class of non-ladder scattering kernels. We have verified that our numerical solutions agree with those previously obtained from a Euclidean treatment of the pure ladder limit. This represents a powerful new approach to obtaining solutions of the BSE and a more detailed discussion will appear soon[5].

5. References

1. N. Nakanishi, *Suppl. Prog. Theor. Phys.* **43**, 1 (1969).
2. C.D. Roberts and A.G. Williams, "Dyson-Schwinger equations and their application to hadronic physics" in *Progress in Particle and Nuclear Physics, Vol. 33*, Amand Faessler (Ed.), (Pergamon Press, Oxford, 1994), p. 477.
3. N. Nakanishi, *Graph Theory and Feynman Integrals*, (Gordon and Breach, New York, 1971).
4. N. Nakanishi, *Phys. Rev.* **130**, 1230 (1963); Erratum, *ibid* **131**, 2841 (1963) and references therein.
5. K. Kusaka and A.G. Williams, in preparation.

THE INFINITE MASS LIMIT OF THE IMPROVED SALPETER EQUATION.

J. BIJTEBIER* and J. BROEKAERT†

Theoretische Natuurkunde, Vrije Universiteit Brussel, Pleinlaan 2,
B-1050 Brussels, Belgium.

ABSTRACT

We present a tridimensional reduction of the two-fermion Bethe-Salpeter equation, generalizing Salpeter's equation to non-instantaneous Bethe-Salpeter kernels. In contrast with the original Salpeter equation, our equation has a correct limit when the mass of one fermion goes to infinity.

1. Introduction

If one makes the popular instantaneous approximation in the Bethe-Salpeter (BS) equation, one gets Salpeter's equation. It is well known, however, that Salpeter's equation does not lead to a correct limit when the mass of one of the particles becomes infinite (even when one does not make the ladder approximation). We shall present a tridimensional reduction of the BS equation, which is not based on an instantaneous approximation and leads to a generalization of Salpeter's equation. This improved Salpeter's equation has a correct infinite mass limit.

2. Reduction of the Bethe-Salpeter equation

The BS equation for a two-fermion bound state can be written

$$\chi = G_0 K \chi \tag{1}$$

where the χ is the BS amplitude, while $G_0 = G_{01} G_{02}$ is the product of the propagators

$$G_{0i} = (p_{i0} - h_i + i \epsilon h_i)^{-1} \beta_i \tag{2}$$

with

$$h_i = \vec{\alpha}_i \cdot \vec{p}_i + \beta_i m_i. \tag{3}$$

The BS kernel K is the sum of an infinity of contributions, represented by the irreducible Feynman graphs. Let us first define the total and relative variables:

$$P = p_1 + p_2, \qquad X = \frac{1}{2}(x_1 + x_2), \qquad S = h_1 + h_2, \tag{4}$$

$$p = \frac{1}{2}(p_1 - p_2), \qquad x = x_1 - x_2, \qquad s = \frac{1}{2}(h_1 - h_2), \tag{5}$$

---

*Senior Research Associate at the National Fund for Scientific Research (Belgium).
†Researcher at the Inter-University Institute for Nuclear Sciences (Belgium).

$$\mu = \frac{1}{2\,P_0}\left(h_1^2 - h_2^2\right). \tag{6}$$

If we write the free propagator G_0 in terms of the operators $(P_0 - S)$ and $(p_0 - \mu)$ (that we find respectively in the sum of two free Dirac equations and in the half-difference of the corresponding iterated equations), we can isolate a pole at $P_0 - S = 0$:

$$G_0 = G_\delta + G_R, \qquad G_\delta = -2i\pi\tau\left(P_0 - S + i\,\epsilon\,S\right)^{-1}\delta\left(p_0 - \mu\right)\beta_1\beta_2 \tag{7}$$

with

$$\tau = \Lambda_1^+\Lambda_2^+ - \Lambda_1^-\Lambda_2^-, \qquad \Lambda_i^\pm = \frac{\sqrt{h_i^2} \pm h_i}{2\sqrt{h_i^2}}. \tag{8}$$

If we keep only this pole part G_δ of G_0 in the right-hand side of the BS equation (1), and transport the $G_R\,K\,\chi$ term to the left-hand side, we get the relativistic quantum mechanics equation:

$$\psi = G_\delta\,K_T\,\psi \tag{9}$$

with

$$\psi = \left(1 - G_R\,K\right)\chi, \tag{10}$$

$$K_T = K\left(1 - G_R\,K\right)^{-1} = K + K\,G_R\,K + \cdots. \tag{11}$$

The higher-order terms of the series (11) re-introduce the reducible Feynman graphs into the kernel, but with G_0 replaced by G_R. Multiplying eq.(9) by $(P_0 - S)$ and $(p_0 - \mu)$ gives a pair of coupled Dirac equations. The second of these equations gives the behaviour of ψ with respect to the relative time t. In terms of $\psi' = \psi\,(t = 0)$, the first equation becomes:

$$(P_0 - S)\,\psi' = \tau\,V'\,\psi', \tag{12}$$

$$V' = -2i\pi\int dp_0'\,dp_0\,\delta\left(p_0' - \mu\right)\beta_1\beta_2\,K_T\left(p_0', p_0\right)\delta\left(p_0 - \mu\right). \tag{13}$$

This generalizes Salpeter's equation to a non-instantaneous kernel (when the kernel $K\left(t', t\right)$ can be written as $\delta(t')\,\bar{K}\,\delta(t)$, then V' is simply $-i\,\beta_1\,\beta_2\,\bar{K}$).

The operator τ is $+1, -1, 0$ when applied on eigenstates of h_1, h_2 with positive, negative or mixed eigenvalues respectively. Its presence in eq.(12) allows for the existence of solutions in the $\tau^2 = 1$ subspace. This fact eliminates several difficulties, like the "continuum dissolution" problem when an external potential is present (in our formalism, an external potential can be introduced by simply adding a $V_i\left(\vec{x}_i, \gamma_i\right)$ term to the definition (3) of h_i).

3. Infinite mass limit.

There exists of course an infinity of possible decompositions of G_0 into a G_δ containing the most important singularities of G_0 and a rest. Many existing treatments of the BS equation could be described in the present formalism, with a suitable choice of G_δ. One may for example choose (7), but without the operator τ. This changes nothing (at first order) for the $\tau = +1$ part of the wave function, which is the most important one, but the $\tau = 0$ components could become a source of problems. This choice is well adapted to a situation in which one of the fermions is much heavier than the other one. At the $m_2 \to \infty$ limit, indeed, we get the Dirac equation of the first fermion, with a Coulomb potential already given by the limit of the one-photon exchange term (in QED). The contributions of all higher-order terms cancel mutually at the $m_2 \to \infty$ limit. Sazdjian's reduction (this conference) inspired by the free second-order equations, exhibits the same cancellations. The $m_2 \to \infty$ limit is not so simple with our choice of G_δ, but its physical interpretation is the same. The one-photon exchange diagram gives the Coulomb potential, with a projector Λ_1^+ (coming from our operator τ) at left. Let us compute the $m_2 \to \infty$ limit of the two second-order diagrams (the crossed diagram and the boxed diagram) in three different cases: bound state potential with our choice of G_δ, with or without τ, scattering amplitude. The differences come from the factorizable contribution of the heavy fermion line. This contribution consists in a term coming from the crossed diagram, and one or two terms coming from the boxed diagram (writing G_R as $G_0 - G_\delta$ in the bound state problem, keeping the full G_0 in the scattering problem). We get:

$$\text{our reduction} \qquad (-\omega + i\,\epsilon)^{-1} + (\omega + i\,\epsilon)^{-1} + 2i\pi\,\Lambda_1^+\,\delta(\omega) = -2i\pi\,\Lambda_1^-\,\delta(\omega), \quad (14)$$

$$\text{without } \tau \qquad (-\omega + i\,\epsilon)^{-1} + (\omega + i\,\epsilon)^{-1} + 2i\pi\,\delta(\omega) = 0, \qquad (15)$$

$$\text{scattering} \qquad (-\omega + i\,\epsilon)^{-1} + (\omega + i\,\epsilon)^{-1} = -2i\pi\,\delta(\omega), \qquad (16)$$

where ω is an energy integration variable. The mutual cancellation is complete in the no-τ case. In the scattering case, the two terms combine into the second-order term of the Born series. In our reduction, they combine into the contribution of an intermediate negative energy state. Such a contribution can be obtained by writing Dirac's equation with a Coulomb potential, splitting the wave function into a positive and negative h_1 part and eliminating the negative h_1 part by an iterative procedure. The six third-order "skeleton" diagrams can be combined in a similar way at the $m_2 \to \infty$ limit. We can therefore conclude that the $m_2 \to \infty$ limit of our equation (12) is the equation of the $h_1 > 0$ projection of the wave function of a fermion in a Coulomb potential. The partial (total in (15)) cancellation between the reducible and irreducible graphs, which are neglected in the instantaneous and ladder approximations respectively, indicates that these approximations must either be made together (in a first order calculation) or not at all.

CHIRAL SYMMETRY BREAKING
INDUCED BY A LINEARLY RISING CONFINEMENT POTENTIAL

A. N. IVANOV, N. I. TROITSKAYA
Dept. of. Theor. Phys., State Technical University
195251 St. Petersburg, Russian Federation

and

M. FABER, M. SCHALER
Institut für Kernphysik, Technische Universität Wien, Wiedner Hauptstr. 8-10
A-1040 Vienna, Austria

and

M. NAGY
Institute of Physics, Slovak Academy of Sciences
842 28 Bratislava, Slovakia

ABSTRACT

Chiral symmetry breaking is explored within QCD with a linearly rising confinement potential with the Nambu-Jona-Lasinio (NJL) prescription. The dynamical quark mass is obtained from a gap-equation in terms of the string tension σ. It is shown that the spectra of low-lying quark-antiquark collective excitations coincide with those predicted by the NJL model. In addition vector and axial-vector excitations, whose masses exceed twice the constituent-quark mass, are found stable under decays into quark-antiquark pairs in accordance with colour confinement requirements.

It is well-known that the Nambu-Jona-Lasinio (NJL) model[1,2] realizes a very nice mechanism of spontaneous breaking of chiral symmetry (SBCS). The chiral symmetry breaking, being due to the strong attraction in the quark-antiquark channels produced by the NJL local four-quark interactions, leads to the appearance of quark-antiquark collective excitations with quantum numbers of low-lying mesons[1-4]. The main shortcoming of the NJL model is the absence of confinement.

The experience of lattice QCD[5] and the dual-superconducting picture of QCD[6] shows that confinement in quark-antiquark channels can be described by a linearly rising potential

$$V(\vec{x}_1 - \vec{x}_2) = \sigma|\vec{x}_1 - \vec{x}_2| - 2\sqrt{\sigma}, \tag{1}$$

where $\vec{x}_{1,2}$ are the positions of quark and antiquark, and σ is the string tension.

Following the NJL prescription for the analysis of SBCS in QCD one must consider the four-quark interaction. Of course, the evaluation of such interaction within QCD is a very complicated problem which can hardly be resolved at present. However, for our aim we need the interaction defined at large distances only. In this case we may use Adler's theory of static quark forces[7] and admit that the main contribution to the $\bar{\psi}\psi$-interaction at large distances comes from the interaction between colour quark charges. It gives the following total effective action of quark fields

$$S_{\text{st. QCD}}^{(\bar{\psi}\psi)} = \int d^4 x \; \bar{\psi}(\vec{x},t) i\gamma^0 \frac{\partial}{\partial t} \psi(\vec{x},t) + S_{\text{st. int}}^{(\bar{\psi}\psi)}. \tag{2}$$

The second term describes the interaction of colour charges

$$S_{\text{st. int}}^{(\bar{\psi}\psi)} = -\frac{1}{2} \int d^3 x_1 \; d^3 x_2 \int_{-\infty}^{\infty} dt \; \bar{\psi}(\vec{x}_1,t)\gamma^0 t_c^A \psi(\vec{x}_1,t) V(\vec{x}_1 - \vec{x}_2)\bar{\psi}(\vec{x}_2,t)\gamma^0 t_c^A \psi(\vec{x}_2,t). \tag{3}$$

Here t_c^A $(A = 1,...,N^2 - 1)$ are generators of the $SU(N)_c$ group normalized by $\text{tr}_{\text{C}}(t_c^A t_c^B) = \delta^{AB}/2$.

The action (2) seems to be valid only for heavy quarks, since the kinetic terms of quark fields do not involve coordinate derivatives. The neglect of coordinate derivatives of the light quark fields in (2) can be justified as follows. We assume that the four-quark interaction (3) produces a strong attraction between quarks in the $\bar{\psi}\psi$-channel. By virtue of this tension the quarks are being kept at small relative velocities. This can explain the neglection of coordinate derivatives of light quark fields in the action (2).

In order to analyse SBCS caused by the action (3) we follow the NJL prescription and evaluate the one-quark-loop contribution. As a result we get[8] the constituent-quark mass

$$m = \frac{2\sqrt{\sigma}}{\pi} \tag{4}$$

as the solution of the gap equation at leading order in the large N expansion. From formula (4) at $\sigma = 0.27$ GeV^2 we estimate $m = 0.33$ GeV. Our result agrees with the prediction of the naive quark model[9] and the NJL model[4,3].

The Lorentz covariant part of the four-quark interaction (3) produces four nonets of $\bar{\psi}\psi$-collective excitations with quantum numbers of low-lying scalar, pseudoscalar, vector and axial-vector mesons[8]. Due to the Lorentz non-covariant part of the interaction (3) two nonets of $\bar{\psi}\psi$-collective excitations can be produced. In the chirally invariant hadronic phase the masses of these excitations exceed the SBCS scale $\Lambda_\chi \approx 1$GeV. Therefore they can be regarded as heavy-mass-meson states and, consequently, deleted from the spectra of low-lying $\bar{\psi}\psi$-collective excitations.

The analysis of the mass spectra of low-lying $\bar{\psi}\psi$-collective excitations, performed within the Bethe-Salpeter equation approach in Ref.[1]1, gave results in good agreement with those predicted by the NJL-model with a local four-quark interaction[3,4]. In addition it was shown that the vector and axial-vector $\bar{\psi}\psi$-collective excitations, whose masses exceed twice the constituent-quark mass, are found stable under decays into quark-antiquark pairs which agrees with the confinement requirement. Of course, these mesons can decay into light mesons as usual.

The Lorentz covariant part of the four-quark interaction (3) restores the NJL-local four-quark interaction[8] with equal values of the NJL phenomenological constants. The latter agrees with the results of previous attempts to compute these constants within QCD[12].

References

1. Y. Nambu and G. Jona-Lasinio, *Phys. Rev.* **122** (1961) 345; *Phys. Rev.* **124** (1961) 246.

2. T. Eguchi, *Phys. Rev.* **D14** (1976) 2755; K. Kikkawa, *Progr. Theor. Phys.* **56** (1976) 947; H. Kleinert, in *Proc. Int. Summer School of Subnuclear Physics*, Erice 1976, ed. A. Zichichi.

3. S. Klimt, M. Lutz, U. Vogel and W. Weise, *Nucl. Phys.* **A516** (1990) 429; *Nucl. Phys.* **A516** (1990) 469.

4. A. N. Ivanov, M. Nagy and N. I. Troitskaya, *Int. J. Mod. Phys.* **A7** (1992) 7305; A. N. Ivanov, M. Nagy and N. I. Troitskaya, *Int. J. Mod. Phys.* **A8** (1993) 2027,3425; A. N. Ivanov, *Int. J. Mod. Phys.* **A8** (1993) 853.

5. K. G. Wilson, *Phys. Rev.* **D10** (1974) 2445; T. A. De Grand and D. Toussaint, *Phys. Rev.* **D24** (1981) 466; M. Faber, H. Markum, M. Müller and Š. Olejník, *Phys. Lett.* **B247** (1990) 377; H.-O. Ding, C. F. Baille and G. C. Fox, *Phys. Rev.* **D41** (1990) 2912.

6. M. Baker, J. S. Ball and F. Zachariasen, *Phys. Rev.* **D34** (1986) 3894.

7. S. L. Adler, *Phys. Rev.* **D18** (1978) 411.

8. A. N. Ivanov, N. I. Troitskaya, M. Faber, M. Schaler and M. Nagy, *Nuovo Cimento* **A107** (1994) (in print).

9. J. J. J. Kokkedee, *The Quark Model* (W. A. Benjamin, New York, 1969).

10. Particle Data Group, *Phys. Rev.* **D45** (1992) No.11, part II.

11. A. N. Ivanov, N. I. Troitskaya, M. Faber, M. Schaler and M. Nagy, *Phys. Lett.* **B** (in print).

12. T. Goldman and R. W. Haymaker, *Phys. Rev.* **D24** (1981) 724; R. W. Haymaker and T. Goldman, *Phys. Rev.* **D24** (1981) 743.

PROPORTIONALLY OFF-MASS-SHELL EQUATION

KHIN MAUNG MAUNG

Physics Department, Hampton University
Hampton, Virginia 23668, U.S.A

JOHN W. NORBURY

Department of Physics, University of Wisconsin
La Crosse, Wisconsin 54601, U.S.A

and

DAVID E. KAHANA

Department of Physics, Kent State University
Kent, Ohio 44242, U.S.A

ABSTRACT

A new two-body relativistic equation is presented. The advantage of the proposed equation over existing equations in the literature is that this equation automatically adjusts itself for different mass systems. Besides unitarity and covariance, this equation has a physically meaningful prescription of how the particles go off-mass-shell in the intermediate states.

In the relativistic two-body problem Bethe-Salpeter equation is believed to be the correct two-body equation, and it can be related to the underlying field theory. It is a four-dimensional equation and the kernel of the equation in principle must include all two-particle irreducible diagrams which is in practice impossible. Because of the technical difficulties associated with the full Bathe-Salpeter equation, for practical calculations, approximations are usually made. The most common approximation is to keep the one boson exchange term of the kernel only and solve the resulting ladder equation. Although it is easier to solve, there is an intrinsic problem associated with the ladder Bethe-Salpeter equation. It does not have a one-body limit.[1] That is, it does not reduce to a one-body equation when one of the particle is infinitely massive.

Instead of solving the Bethe-Salpeter equation directly, there exists another way to describe the relativistic two-body problem in a covariant manner. This is to reduce the dimensionality of Bethe-Salpeter equation to three. It was shown by Yaes [2] that there exists infinitely many reduction schemes. All three-dimensional equations appearing in the literature, can be classified into two types. In the first type, one particle is kept on-mass-shell and in the second type both particles are kept equally off-mass-shell, for all intemediate states. The first group of equations keeps one of the particles on-mass-shell and prohibits the negative energy propagation of that particle. Therefore they make good phenomenolgical choices for systems with one very massive particle and one light particle. The second group of equations treats both particles on the same footing and should be good for equal mass systems. But for systems with unequal but comparable masses, there is no general consensus for what type of equation to use.

Therefore, there is always an ambiguity associated with the choice of an equation for practical problems. This fact alone has generated several studies and comparisons of these equations by applying them to different physical problems.[3]

Now consider a system with unequal but comparable masses, such as pionic or kaonic hydrogen or up-antistrange quark bound states or kaon-nucleon scattering problems. Obviously these systems fall between the two extreme limits mentioned above and the the choice of equation is quite arbitrary. Therefore it would be advantagous and aesthetically pleasing to have an equation which would interpolate between the one-body limit and the equal mass limit.

The propagator for our new two-body equation (for two scalar particles) is given by

$$g(k, P) = 2\pi \frac{\delta^{(+)}\left([(aP + k)^2 - m_1^2](\frac{m_1}{m_1+m_2}) - [(bP - k)^2 - m_2^2](\frac{m_2}{m_1+m_2})\right)}{(aP + k)^2 + (bP - k)^2 - m_1^2 - m_2^2 + i\eta} \qquad (1)$$

where k and P are relative and total four-momentum of the two-body system and m_1 and m_2 are the masses of the particles. Therefore $p_1 = aP + k$ is the momentum of particle 1 and $p_2 = bP - k$ is the momentum of particle 2. It was pointed out by Tiemeijer and Tjon [4] that in order to have a one-body limit in three-dimensional equations, one must use Wrightmann-Gordon [5] choice of 'a 'and 'b ',i.e $a = (s + m_1^2 - m_2^2)/2s$ and $b = (s + m_2^2 - m_1^2)/2s$, where s is the total four momentum squared. From the argument of the delta function it can be seen that when $m_1 = m_2$ the particles will be put equally off-mass-shell and the corresponding equation is known as the Todorov [6] equation.When one of the particle becomes infinitely massive, it will be kept on mass-shell and with our choice of relative momentum, we obtain a one-body equation. When the two masses are not equal but comparable, the relative off-mass-shellness of the particles are controlled by this delta function. Therefore no matter what the ratio of the masses are, the propagator automatically adjust itself in a physically meaningful way. Mathematically the propagator proposed above is similar to form to the one proposed by Gross [1] which interpolates between one-body on-mass-shell equation and Todorov [6] equation with the help of a free parameter.

Now in the center of mass, with $P = (W, 0)$, the propagator takes the form

$$g = -2\pi \frac{\delta(k_0 - k_0^{(+)})}{R} \frac{x}{2Wk_0^{(+)} - i\eta} \qquad (2)$$

with

$$k_0^{(+)} \equiv \frac{-W[x(m_1^2 - m_2^2)/W^2 + 1] + R}{2x} \qquad (3)$$

and

$$R \equiv \left[W^2(1 - x^2) - 2x(m_2^2 - m_1^2) + 2x^2(E_1^2(\mathbf{k}) + E_2^2(\mathbf{k})) \right]^{1/2} \tag{4}$$

where $x = (m_1 - m_2)/(m_1 + m_2)$. It can be shown that the factor R produces no singularities and the only singularity in the propagator is the elastic singularity arising from $k_0^{(+)}$, and this singularity arises when the intermediate 3-momentum k equals the on-shell 3-momentum and hence unitarity is satisfied.

In conclusion, we have presented a two-body relativistic equation, which is covariant and unitary. The major advantage of the proposed new equation over existing equations in the literature is that this equation automatically adjusts itself for different mass systems. The novel feature of this equation is that it has a physically meaningful prescription for the off-mass-shell propagation of the particles in the intermediate states. The particles are allowed to go off-shell proportionally to their masses so that when one of the particles becomes infinitely massive, it automatically keeps that particle fully on-mass-shell and the equation can be reduced to a one-body equation. When the particles have equal masses, then the particles are treated symmetrically and the equation reduces to an equation known as the Todorov equation. Since this equation has a physically meaningful prescription for the allowed off-shellness of the intermediate state particles, it will be useful for a very wide variety of problems such as leptonic bound states, light hadronic atoms and $q\bar{q}$ bound state systems where the the ratio of the masses can vary by an order of magnitude.

Acknowledgements

We are very grateful to Professor Franz Gross for his kind advice over the past several years. K.M.M. was supported by NSF grant HRD-9154080.

1. F. Gross, *Phys. Rev. C* **26**,(1982).2203.
2. R. J. Yaes, *Phys. Rev. D* **3**,(1971) 3086.
3. R. M. Woloshyn and A. D. Jackson, *Nucl. Phys.* **64**, (1973)269., J. Frohlich, K. Schwarz and H. F. K. Zingl, *Phys. Rev. C* **27**, (1983)265, A. J. Sommerer, J. R. Spence and J. P. Vary,*Phys. Rev. C* **49**, (1994)513.
4. P. C. Tiemeijer and J. A. Tjon, *Phys. Rev. C* **49**, (1994)494.
5. A. J. MacFarlane,*Rev. Mod. Phys.***34**,(1962)42.
6. I. T. Todorov, *Phys. Rev. D* **10**,(1971)2351.

WEAK COUPLING LIMIT AND GENUINE QCD PREDICTIONS

FOR HEAVY QUARKONIA

WOLFGANG KUMMER

and

WOLFGANG MÖDRITSCH

Institut für Theoretische Physik, Technische Universität Wien
Wiedner Hauptstraße 8-10, A-1040 Wien, Austria

ABSTRACT

Although individual levels of toponium will be unobservable, the top–anti–top system near threshold fulfills all requirements of a rigorous perturbation theory in QCD for weakly bound systems. Corresponding techniques from positronium may thus be transferred successfully to this case. After clarifying the effect of a non-zero width we calculate the $t\bar{t}$ potential to be used for the calculation of e.g. the cross-sections near threshold.

Perturbative quantum field theory by means of the general Bethe–Salpeter (BS) formalism is applicable also for weakly bound systems, described as corrections to a zero order equation for the bound system. The latter is usually taken as the Schrödinger–equation, but experience in positronium has shown that other starting points of perturbation theory, like the Barbieri-Remiddi (BR) equation,[1] offer advantages. That such rigorous methods [2] have not found serious considerations for a long time in quarkonia has to do with the importance of confinement effects in QCD. Parametrizing the latter by a gluon-condensate [3], one finds that for quark masses m up to about 50 GeV nonperturbative effects are of the same order as corrections to the bound states [4], although the situation seems a little less serious using BS-methods[5]. Nevertheless, the possible practical advantages of a rigorous approach have been demonstrated a long time ago, not for level shifts in bound-states, but for perturbative corrections to the decay of quark-antiquark systems. Straightforward applications of perturbative QCD to the annihilation part *alone* of, say, bottom-anti-bottom, with minimal subtraction at a scale $O(2m)$ gave huge a correction $O(10\alpha_s/\pi)^6$. However, in such a decay the bound quark pair is really off-shell. Although from that it is clear that an (off-shell) annihilation part by itself cannot even be gauge independent: The perturbative correction to the wave-function of the decaying state must be added for consistency. In fact, after doing this and performing a careful renormalization procedure at Bohr momentum $O(\alpha_s m)$, appropriate for that system, it could be shown that large corrections tend to compensate in that result, at least in the case of the singlet ground state (0^{-+}) [7].

Now, with a top quark in the range of 150 - 180 GeV [8] for the first time confinement effects are negligible. The large width Γ of the decay $t \to b + W$ makes bound–state 'poles' at real energies unobservable, but at the same time obliterates confinement effects even *above* threshold because the top quark has no time to 'hadronize'. The effect of Γ can be taken into account by simple analytic continuation of the zero order equation to complex energies [9]. Previous work [10] was based upon phenomenological quarkonium potentials. However also BS perturbation theory can be applied to the $t\bar{t}$ Green function G near the poles which move into the unphysical sheet of the complex plane at a distance Γ to the real energy axis. Nevertheless, the residues and hence the wave functions together with the QCD level shifts remain real. Hence they can be used in a straightforward manner to determine the different contributions to a rigorous QCD potential.

After showing that this analytic continuation argument also works for the BR-equation, we determine the different contributions for such a quantity from the (real) energy shifts, including (numerical) $\mathcal{O}(m\alpha_s^4)$–effects. With a generic perturbation H to the Green function G_0 of the zero order equation, the level shifts become [2]

$$\Delta E_n = \langle\langle h_i \rangle\rangle (1 + \langle\langle h_1 \rangle\rangle) + \langle\langle h_0 g_1 h_2 \rangle\rangle + \mathcal{O}(h^3) \tag{1}$$

where h_i and g_i are the expansion coefficients of H, resp. G_0 near the pole $E \sim E_n$ of G_0. The expectation values are *four*–dimensional momentum integrals, taken here with respect to BR wavefunctions. The latter differ by factors produced by relativistic corrections and with p_0 from (normalized) Schrödinger–wavefunctions. Nevertheless, we formulate our final result as a 'potential'. The full formula for the different parts of

$$V = \sum_{i=0}^{2} V_{QCD}^{(i)} + V_{EW} \tag{2}$$

can be found in ref.[11]. (i) refers to the loop order which, of course, is not uniquely correlated with orders of α_s in energy shifts. — $V_{QCD}^{(0)}$ beside the Coulomb exchange contains the $\vec{p}^4$–term and the exchange of a transversal gluon, producing the 'abelian' relativistic corrections $\mathcal{O}(m\alpha_s^4)$. We *do not* include a running coupling constant anywhere because this would mix orders, spoiling even the gauge–independence order by order within any application. Of course, for technical reasons eq.(2) is derived and should be applied in the Coulomb gauge.

$$V_{QCD}^{(1)} = -\frac{33\alpha^2}{8\pi r}(\gamma + \ln \mu r) + \frac{\alpha^2}{4\pi r}\sum_{j=1}^{5}[\mathrm{Ei}(-rm_j e^{\frac{5}{6}}) - \frac{5}{6} + \frac{1}{2}\ln(\frac{\mu^2}{m_j^2} + e^{\frac{5}{3}})] + \frac{9\alpha^2}{8mr^2}$$

consists of vacuum polarization and vertex corrections. The first contain the gluon loop and the loops from fermions. Although the level contributions will be $\mathcal{O}(m\alpha_s^3)$ for the toponium system the mass of the charm and bottom must not be neglected, because they yield *numerical* $\mathcal{O}(m\alpha_s^4)$ corrections. In contrast to the QED case here also the one–loop gluon–splitting vertex yields a potential $\propto \alpha_s^2/mr^2$, important to this order. In the vacuum polarization part of

$$V_{QCD}^{(2)} = c^{(H)}\frac{4\pi\alpha^3}{r}$$

$$-2\frac{\alpha^3}{(16\pi)^2 r}\left\{(33-2n_f)^2[\frac{\pi^2}{6}+2(\gamma+\ln\mu r)^2]+9(102-\frac{38}{3}n_f)(\gamma+\ln\mu r)\right\}$$

we emphasize the importance of *non-leading* logarithms which would only be contained in the usual running coupling constant if a two–loop β–function would be used. Among the 'box' corrections, an H–graph (with the figure H formed by gluons between the two fermion lines) is emphasized as an contribution which gives at least a correction to the Coulomb term. In

$$V_{EW} = -\frac{8}{9}\alpha_{QED}(\mu)\frac{4\pi\alpha}{r} - \sqrt{2}G_F m^2 \frac{e^{-m_H r}}{4\pi r} + \sqrt{2}G_F m_Z^2 a_f^2 \frac{\delta(\vec{r})}{m_Z^2}(7-\frac{11}{3}\vec{S}^2)$$

$$+\sqrt{2}G_F m_Z^2 a_f^2 \frac{e^{-m_Z r}}{2\pi r}[1-\frac{v_f^2}{2a_f^2}-(\vec{S}^2-3\frac{(\vec{S}\vec{r})^2}{r^2})(\frac{1}{m_Z r}+\frac{1}{m_Z^2 r^2})-(\vec{S}^2-\frac{(\vec{S}\vec{r})^2}{r^2})],$$

photon–exchange, Z–exchange and Z–anihilation turn out to be as essential for the $t\bar{t}$–system as the usual relativistic corrections [11].

One of the authors (W.M) thanks the organizers for financial support.This work is supported in part by the Austrian Science Foundation (FWF) in project P10063-PHY within the framework of the EEC- Program "Human Capital and Mobility", Network "Physics at High Energy Colliders", contract CHRX-CT93-0357 (DG 12 COMA).

1. R. Barbieri and E. Remiddi, Nucl.Phys. **B141** (1978) 413.
2. G. Lepage, Phys.Rev.A **16** (1977) 863;
   W. Kummer, Nucl.Phys. **B179** (1981) 365.
3. M.A. Shifman, A.I. Vainshtein, V.I. Zakharov, Nucl.Phys. **B147** (1978) 385, 448.
4. H. Leutwyler, Phys.Lett.B **98** (1981) 447
5. W. Kummer and G. Wirthumer, Phys.Lett.B **150** (1985), 392
6. R. Barbieri, G. Curci, E. d'Emilio, E. Remiddi, Nucl.Phys.B **154** (1979) 535;
   P.B. Mackenzie, G.P. Lepage, Phys.Rev.Lett **47** (1981) 1244.
7. W. Kummer and G. Wirthumer, Nucl. Phys. B **181** 41, err. B **194** (1982) 546.
8. "Evidence for Top Quark Production in $p\bar{p}$ Collisions at $\sqrt{s} = 1.8$ TeV, FERMILAB-PUB-94/116-E;
   J. Ellis, G.L. Fogli and E.Lisi," The Top Quark and Higgs Boson Masses in the Standard Model and the MSSM" CERN-TH-7261-94 (May 1994)
9. V.S. Fadin and V.A. Khoze, Sov.J. Nucl.Phys. **48** (1988) 309.
10. Y. Sumino et al., Phys.Rev. **D47** (1993) 56.
11. W. Kummer and W. Mödritsch, "Rigorous QCD-Potential for the $t\bar{t}$-System at Threshold", prep. TUW–94–14

GAUGE-INDEPENDENT BOUND-STATE CORRECTIONS

TO THE TOPONIUM DECAY WIDTH

WOLFGANG MÖDRITSCH

and

WOLFGANG KUMMER

Institut für Theoretische Physik, Technische Universität Wien
Wiedner Hauptstraße 8-10, A-1040 Wien, Austria

ABSTRACT

Off-shell and relativistic bound-state corrections for the decay $t \rightarrow b + W$ are calculated to $O(\alpha_s^2)$ making full use of the Bethe-Salpeter formalism for weakly bound systems. Thus we are able to take into account all terms to that order in a systematic and straightforward manner. One of the previously not considered contributions cancels precisely gauge dependent terms which appeared in an earlier off-shell calculation. Important cancellations also determine the gauge-independent part.

In calculations e.g. of the cross section $\sigma_{e^+e^- \rightarrow anything}$ near the $t\bar{t}$ threshold [1], it has been realized that a momentum dependent width (e.g. from phase space reduction) could lead to drastic changes [2].

Qualitatively the main physical effects from t decaying inside toponium seem to be quite well understood by now [3]: On the one hand, Γ decreases by time dilatation and by reduction of phase space, caused by the decay below the mass shell. On the other hand, the Coulomb-enhancement induced by the Coulomb interaction of the relativistic b-quark should increase Γ as in the muonium system [4].

The purpose of our present work is to show that existing quantum field theoretical technology, well tested e.g. in the (abelian) positronium case, may be used also here to solve the problem of a gauge-independent "running width", including the qualitative effects listed above. Even the Coulomb enhancement can be taken into account properly in this context, which to the best of our knowledge has not been possible before. We will use the Bethe-Salpeter (BS) approach as described in [5].

The leading imaginary correction to the toponium energy levels is indicated in figure 1 and gives rise to the BS-perturbation kernel

$$H^\Sigma = -i[\Sigma(p) \otimes (\not{p}' - m) + (\not{p} - m) \otimes \Sigma(p')](2\pi)^4 \delta(k - k'), \tag{1}$$

where the second term originates from the same contribution of the $\bar{t}$ quark.

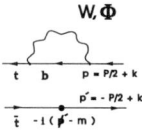 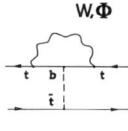 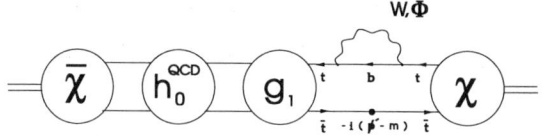

Fig. 1   Fig. 2   Fig. 3

Since we are calculating higher order effects we neglect the mass of the bottom quark although it may be included in principle. For the electroweak theory we use the R_ξ-gauge. The gauge parameter ξ will not be fixed in the following.

The total contribution of fig.1 to the decay width between Barbieri-Remiddi (BR) wave functions (which are our zero order solutions of the BS-equation) becomes:

$$\Gamma_1 = \Gamma_0 - 2\Gamma_0\sigma_n^2 + \frac{e^2 m}{16\pi s^2}[(1 + 2\frac{M^2}{m^2})(1 - \frac{M^2}{m^2}) + 2\rho(m^2,\xi)]\langle\frac{p^2 - m^2}{m^2}\rangle, \quad (2)$$

$$\rho(p^2,\xi) = \frac{1-\xi}{2\xi}(1 - \frac{M^2}{2p^2}\frac{1+\xi}{\xi}) \quad , \qquad \sigma_n = \frac{2\alpha_s}{3n} \quad (3)$$

where Γ_0 is the well known zero order result which is twice the decay width of a free top quark. It proves sufficient to evaluate the graphs shown in fig. 2 to leading order. We obtain ($\xi > M^2/m^2$ and $m^2 > M^2$):

$$\Gamma_2 = \frac{e^2}{16\pi s^2}\langle\frac{4\pi\bar{\alpha}}{q^2}\rangle\left[(1 + \frac{m^2}{2M^2})(1 - \frac{M^2}{m^2})(1 + 3\frac{M^2}{m^2}) + 4\rho(m^2,\xi)\right] \quad (4)$$

We note already here that with the help of the Schrödinger equation the gauge dependent terms proportional $\rho(m^2,\xi)$ exactly cancel in (2) plus (4)!

To $O(\alpha_s^2\Gamma_0)$ also the terms of second order BS-perturbation theory contribute ($h_0^\Sigma = H^\Sigma|_{P_0=M_{n,0}}, h_1 = \partial H/\partial P_0|_{P_0=M_{n,0}}$)

$$\Gamma_3 = -2\mathrm{Im}[\langle\langle h_0^{QCD}g_1 h_0^\Sigma\rangle\rangle + \langle\langle h_0^\Sigma g_1 h_0^{QCD}\rangle\rangle] \quad (5)$$

$$\Gamma_4 = -2\mathrm{Im}[\langle\langle h_0^{QCD}\rangle\rangle\langle\langle h_1^\Sigma\rangle\rangle + \langle\langle h_0^\Sigma\rangle\rangle\langle\langle h_1^{QCD}\rangle\rangle] \quad (6)$$

where h_0^{QCD} denotes some QCD perturbation and $\langle\langle ..\rangle\rangle$ means between BR wave functions. g_1 is the reduced Green function for the state under consideration. The first contribution to Γ_3 is shown in fig. 3. It is sufficient to evaluate (5) and (6) to leading nonrelativistic order [6]:

$$\Gamma_3 = \frac{\Gamma_0}{2}\langle h_0^{QCD}\rangle\langle\frac{1}{\omega}\rangle \quad (7)$$

$$\Gamma_4 = -\frac{\Gamma_0}{2}\langle h_0^{QCD}\rangle\langle\frac{1}{\omega}\rangle - 2[\mathrm{Im}\langle\langle h_0^\Sigma\rangle\rangle]\langle\langle\frac{\partial}{\partial P_0}K_{BR}\big|_{P_0=M_{n,0}}\rangle\rangle. \quad (8)$$

By adding Γ_1 (2) to Γ_2 (4) *all* gauge dependent terms are found to cancel. Moreover a striking feature of the sum $\Gamma_1 + \Gamma_2$ is that even the *gauge-independent* terms precisely sum to zero as well, which is by no means "natural" in view of the explicit expressions. In a similar manner also in the sum $\Gamma_3 + \Gamma_4$ large cancellations take place so that finally

$$\Gamma_{boundstate} = \Gamma_0(1 - \frac{1}{2}\sigma_n^2 + O(\alpha_s^3)). \tag{9}$$

Recalling that $1 - \sigma_n^2/2 \approx (1 - \langle \vec{k}^2/m^2 \rangle)^{1/2}$, the residual effect may be interpreted as an approximation of the γ-factor for the time dilatation for the weakly bound top. The origin of that term, the relativistically generalized Coulomb-kernel K_{BR} gives some credit to this interpretation. For the gauge dependent terms in the sum $\Gamma_1 + \Gamma_2$ one can show that the usual QED-like Ward identity, which is valid here within the expectation value, is responsible for their cancellation.

A reader familiar with phenomenological calculations may wonder why no running coupling in α_s or even a more "realistic" potential involving a confining piece has been used. The simple answer is that any generalization of this type would completely ruin the strictly perturbative approach and, as a consequence, also the built-in gauge-independence advocated here. Any further correction like e.g. the gluon vacuum polarization (cf. h_0^{QCD} above) must be accounted for as a separate perturbation and must under no circumstances be mixed with higher order leading logs in perturbation theory, as summarized in a running α_s. Our result provides a theoretical basis for including only time-dilatation effects in Γ in explicit computations of the Green function for toponium near threshold [3].

One of the authors (W.M) thanks the organizers for financial support. This work is supported in part by the Austrian Science Foundation (FWF) in project P10063-PHY within the framework of the EEC- Program "Human Capital and Mobility", Network "Physics at High Energy Colliders", contract CHRX-CT93-0357 (DG 12 COMA).

1. V.S. Fadin, V.A. Khoze, Sov.J. Nucl.Phys. **48** (1988) 309.
2. Y. Sumino, K. Fujii, H. Hagiwara, H. Murayama, C.-K. Ng, Phys.Rev. **D47** (1993) 56.
3. M. Jeżabek, J.H. Kühn, "The Top Width, a Theoretical Update" TTP-93-4; M. Jeżabek, T.Teubner, Z.Phys. **C59** (1993) 669.
4. H. Überall, Phys.Rev. **119** (1960) 365.
5. W. Kummer and W. Mödritsch, "Rigorous QCD-Potential for the $t\bar{t}$-System at Threshold", TUW–94–14.
6. W. Mödritsch, W.Kummer "Relativistic and Gauge Independent Off-Shell Corrections to the Toponium Decay Width", TUW–94–06

CHARMONIUM DECAYS WITH BOTH RELATIVISTIC AND QCD RADIATIVE CORRECTIONS AND THE DETERMINATION OF $\alpha_s(m_c)$

Kuang-Ta CHAO, Han-Wen HUANG, Jing-Hua LIU, Yu-Quan LIU, Jian TANG

Department of Physics, Peking University, Beijing 100871, P.R.China

ABSTRACT

We estimate the decay rates of $\eta_c \longrightarrow 2\gamma$, $J/\psi \longrightarrow e^+e^-$, and $J/\psi \longrightarrow 3g$ by taking into account both relativistic and QCD radiative corrections. The decay amplitudes are derived in the Bethe-Salpeter formalism. We find that the relativistic correction to the ratio $R \equiv \Gamma(\eta_c \longrightarrow 2\gamma)/\Gamma(J/\psi \longrightarrow e^+e^-)$ is negative and tends to compensate the positive contribution from the QCD radiative correction. For $J/\psi \longrightarrow 3g$ we find the relativistic correction is indeed very large. By using the experimental data for the ratio $R_g \equiv \frac{\Gamma(J/\psi \longrightarrow 3g)}{\Gamma(J/\psi \longrightarrow e^+e^-)} \approx 10$, and the calculated widths, we obtain $\alpha_s(m_c) = 0.29 \pm 0.02$, which is consistent with the well determined QCD scale parameter $\Lambda_{\overline{MS}}^{(4)} \approx 200 MeV$.

Both QCD radiative corrections and relativistic corrections are important for charmonium decays (see, e.g., ref.1). The decay rates of many processes are subject to substantial relativictic corrections. In particular, the determination of $\alpha_s(m_c)$, the strong coupling constant at the charm quark mass scale, depends rather crucially on the relativistic corrections.

We first consider $\eta_c \longrightarrow 2\gamma$ and $J/\psi \longrightarrow e^+e^-$ decays[2]. These two processes proceed via the $c\bar{c}$ annihilation. In the Bethe-Salpeter (BS) formalism the annilination matrix elements can be written as follows

$$\langle 0 \mid \overline{Q}\Gamma Q \mid P \rangle = \int d^4q Tr\left[\Gamma \chi_P(q)\right], \tag{1}$$

where $\mid P \rangle$ represents a $Q\bar{Q}$ bound state, $P(q)$ is the total (relative) momentum of the $Q\bar{Q}$, $\chi_P(q)$ is its four dimensional BS wave function, Γ is the interaction vertex. If Γ is independent of q^0, Eq. (1) can be written as

$$\langle 0 \mid \overline{Q}\Gamma Q \mid P \rangle = \int d^3q Tr[\Gamma(\vec{q})\Phi_{\vec{P}}(\vec{q})], \tag{2}$$

where $\Phi_{\vec{P}}(\vec{q}) = \int dq^0 \chi_P(q)$ is the three dimensional BS wave function of the annihilated meson. In the BS formalism the wave functions can be expressed as follows (in the meson rest frame)

$$\Phi_{\vec{P}}^{0^-}(\vec{q}) = \Lambda_+^1(\vec{q})\gamma^0(1+\gamma^0)\gamma_5\gamma^0\Lambda_-^2(-\vec{q})\varphi(\vec{q}),$$

$$\Phi_{\vec{P}}^{1^-}(\vec{q}) = \Lambda_+^1(\vec{q})\gamma^0(1+\gamma^0)\hat{e}\gamma^0\Lambda_-^2(-\vec{q})f(\vec{q}), \tag{3}$$

where $\Phi_{\underset{P}{0^-}}(\vec{q})$, and $\Phi_{\underset{P}{1^-}}(\vec{q})$ represent the three dimensional wave functions of 0^- and 1^- mesons respectively.

For process $\eta_c \longrightarrow 2\gamma$ with the photon momenta and polarizations q_1, ϵ_1 and q_2, ϵ_2, the amplitude can be written as

$$T = \langle 0 \mid \bar{c}\Gamma_{\mu\nu}(q)c \mid \eta_c \rangle \epsilon_1^\mu(\lambda_1)\epsilon_2^\nu(\lambda_2) + \langle 0 \mid \bar{c}\Gamma'_{\mu\nu}(q)c \mid \eta_c \rangle \epsilon_2^\mu(\lambda_2)\epsilon_1^\nu(\lambda_1), \qquad (4)$$

where $\Gamma_{\mu\nu}(q) = \gamma_\mu \dfrac{e^2 e_Q^2}{\not{p}-m}\gamma_\nu$, $\Gamma'_{\mu\nu}(q) = \gamma_\mu \dfrac{e^2 e_Q^2}{\not{p'}-m}\gamma_\nu$, $p_1(p_2)$ is the quark (antiquark) momentum, $p = p_1 - q_1$, $p' = p_1 - q_2$, m and M represent the masses of c quark and η_c meson respectively, $e_Q = \frac{2}{3}$ for $Q = c$. Since $p_1^0 + p_2^0 = M$, it is reasonable to take $p_1^0 = p_2^0 = \frac{M}{2}$. Therefore, $p^0 = \frac{1}{2}M - q_1^0 = 0$, $p'^0 = \frac{1}{2}M - q_2^0 = 0$, the amplitude T becomes independent of q^0. Employing Eqs. (2) and (3), we get

$$T = B\epsilon^{\rho\sigma\mu\nu}q_{1\rho}q_{2\sigma}\epsilon_{1\mu}(\lambda_1)\epsilon_{2\nu}(\lambda_2)e^2 e_Q^2 + B'\epsilon^{\rho\sigma\mu\nu}q_{1\rho}q_{2\sigma}\epsilon_{1\nu}(\lambda_1)\epsilon_{2\mu}(\lambda_2)e^2 e_Q^2, \qquad (5)$$

where

$$B = B',$$

$$B = i\frac{2m}{M}\int d\vec{q}\, \frac{\sqrt{\vec{q}^2+m^2}+m}{\left(\vec{q}^2+m^2\right)\left(\vec{q}^2+\vec{q}_1^2+m^2-2\vec{q}\cdot\vec{q}_1\right)}\varphi(\vec{q}). \qquad (6)$$

Using $q_1 \cdot \epsilon_1 = 0$ and $q_2 \cdot \epsilon_2 = 0$, it is easy to get the decay width

$$\Gamma(\eta_c \to 2\gamma) = 3M^3\pi\alpha^2 e_Q^4 |B|^2. \qquad (7)$$

For process $J/\psi \longrightarrow e^+e^-$ with the electron (positron) momentum $k_1(k_2)$ and helicity $r_1(r_2)$, $\Gamma = -ie\gamma_\mu$, the amplitude can be written as

$$T = e^2 e_Q\langle 0 \mid \bar{c}\gamma_\mu c \mid J/\psi \rangle \bar{u}_{r_1}(k_1)\gamma^\mu v_{r_2}(k_2)\frac{1}{M^2}. \qquad (8)$$

Define the decay constant f_V by

$$f_V M e_\mu \equiv \langle 0 \mid \bar{c}\gamma_\mu c \mid J/\psi \rangle = \int d\vec{q}\, Tr[\gamma_\mu \Phi_{\vec{P}}(\vec{q})], \qquad (9)$$

where e_μ is the polarization vector of J/ψ meson. Then with (3) we find

$$f_V = \frac{2\sqrt{3}}{M}\int d\vec{q}\, (\frac{m+E}{E} - \frac{\vec{q}^2}{3E^2})f(\vec{q}), \qquad (10)$$

where $E = \sqrt{\vec{q}^2+m^2}$. Then it is easy to get the decay width

$$\Gamma(J/\psi \to e^+e^-) = \frac{4}{3}\pi\alpha^2 e_Q^2 f_V^2/M. \qquad (11)$$

308

Comparing (7) with (11), and then including also the QCD radiative corrections[1], we will get the ratio

$$R \equiv \frac{\Gamma(\eta_c \rightarrow 2\gamma)}{\Gamma(J/\psi \rightarrow e^+e^-)} = \frac{9}{4} M_{\eta_c}^3 M_{J/\psi} e_Q^2 \frac{|B|^2}{f_V^2} (1 + 1.96 \frac{\alpha_s(m_c)}{\pi}). \tag{12}$$

Expanding B (6) and f_V (10) in terms of $\vec{q}^2/m^2$, to the first order we have

$$B = \frac{2i}{Mm^2} \int d\vec{q}\, \varphi(\vec{q})(1 - \frac{5}{4}\frac{\vec{q}^2}{m^2}),$$

$$f_V = -\frac{4\sqrt{3}}{M} \int d\vec{q}\, f(\vec{q})(1 - \frac{5}{12}\frac{\vec{q}^2}{m^2}). \tag{13}$$

From (13) we can clearly see that the effects of relativistic kinematics are to suppress the decay widths. From (13) we also see that B is very sensitive to the c quark mass $m_c(\propto \frac{1}{m^2})$ so that R is also sensitive to $m_c(\propto \frac{1}{m^4})$. This result is similar to the QCD sum rule analyses[3].

To calculate the decay widths of these two processes, we employ QCD-motivated interquark potentials, i.e., a linear confinement potential plus the one gluon exchange potential to obtain the meson wave functions by solving the BS equation (see J.Tang, J.H.Liu, and K.T.Chao in ref.2). Substituting the obtained BS wave functions (with $m_c = 1.5GeV$) into (6), (7), and (10), (11), and (12), we get[2]

$$R = 0.95, \tag{14}$$

$$\Gamma(J/\psi \longrightarrow e^+e^-) = 5.45keV, \quad \Gamma(\eta_c \longrightarrow 2\gamma) = 5.2keV. \tag{15}$$

Our results are in agreement with the experimental value of $\Gamma(J/\psi \longrightarrow e^+e^-) = 5.36 \pm 0.29keV$[4] and the $CLEO$ data[5] $\Gamma(\eta_c \longrightarrow 2\gamma) = (5.9^{+2.1}_{-1.8} \pm 1.9)keV$, but slightly smaller than the $L3$ data[6] $\Gamma(\eta_c \longrightarrow 2\gamma) = (8.0 \pm 2.3 \pm 2.4)keV$. They are also consistent with the QCD sum rule result[3]. We have also used a smaller charm quark mass $m_c = 1.4GeV$ and get $R = 1.12$ and $\Gamma(\eta_c \longrightarrow 2\gamma) = 6.0keV$. Here in above calculations the value of $\alpha_s(m_c)$ in the QCD radiative correction factor is taken to be 0.29[1].

We have also calculated the relativistic correction to the hadronic decay $J/\psi \rightarrow 3g$ in a similar way[7]. We find with both relativistic and QCD radiative corrections the decay width is given by

$$\Gamma(J/\psi \rightarrow 3g) = \frac{640(\pi^2 - 9)\alpha_s^3}{81M^3}(1 - 3.7\frac{\alpha_s}{\pi})g_V^2, \tag{16}$$

where to the first order relativistic correction[7]

$$g_V \approx \int d\vec{q}\, f(\vec{q})[1 - (\frac{36 - \frac{7}{3}\pi^2}{8(\pi^2 - 9)} + \frac{5}{6})\frac{\vec{q}^2}{m^2}]$$

$$\approx \int d\vec{q}\, f(\vec{q})[1 + 2.7\frac{\vec{q}^2}{m^2}]^{-1}. \tag{17}$$

Here the wave function integral in the first equation of (17) has to be regulated as shown in the second equation of (17), otherwise it would be divergent for potentials like the Coulomb potential which are singular at the origin. Comparing (17) with (13), we see that the suppression due to relativistic correction for $J/\psi \to 3g$ is much more severe than for $J/\psi \to e^+e^-$. This result then rules out the conjecture that the relativistic correction to $J/\psi \to 3g$ may be neglibibly small[8]. Using the experimental data[4]

$$R_g \equiv \frac{\Gamma(J/\psi \to 3g)}{\Gamma(J/\psi \to e^+e^-)} \approx 10, \tag{18}$$

and calculated widths (16),(17), and (10),(11), we find

$$\alpha_s(m_c) = 0.29 \pm 0.02, \tag{19}$$

as campared with the value without relativistic corrections (but with QCD radiative corrections)

$$\alpha_s^0(m_c) = 0.19 \pm 0.02. \tag{20}$$

Clearly, it is the strong suppression due to the relativistic correction to $J/\psi \to 3g$ that enhances the value of $\alpha_s(m_c)$ and then makes it consistent with the well determined QCD scale parameter $\Lambda_{\overline{MS}}^{(4)} \approx 200 MeV$.

1. W. Kwong, P. B. Mackenzie, R. Rosenfeld, and J. L. Rosner, *Phys. Rev.* **D37** (1988) 3210, and references therein.
2. J. Tang, H. W. Huang, J. H. Liu, and K. T. Chao, PUTP-94-12; see also J. Tang, J. H. Liu, and K. T. Chao, *Phys. Rev.* **D** (to appear) and references therein.
3. L. J. Reinders, H. Rubinstein and S. Yazaki, *Phys. Rep.* **127** (1985) 1; *Phys. Lett.* **B113** (1982) 411; R. Kirschner and A. Schiller, *Z. Phys.* **C16** (1982) 141.
4. Particle Data Group, *Phys. Rev.* **D45** (1992) No.11.
5. *CLEO* Collaboration, *Phys. Lett.* **B243** (1990) 169.
6. *L3* Collaboration, *Phys. Lett.* **B318** (1993) 575.
7. K. T. Chao, H. W. Huang, and Y. Q. Liu, PUTP-94-20.
8. M. Consoli and J. H. Field, *Phys. Rev.* **D49** (1994) 1293; M.Beyer *et al.*, *Z. Phys.* **C55** (1992) 307.

MASS FORMULA FOR LIGHT MESONS

CLAUDE SEMAY *

Université de Mons-Hainaut, 19 Avenue Maistriau, B-7000 Mons, Belgium

ABSTRACT

A mass formula for light mesons is constructed. It is shown that a good fit can be obtained by assuming that the light quarks are ultrarelativistic particles confined by a linear potential, and that the short-range part of the quark interaction stems from instanton effects.

1. Introduction

Blask *et al.*[1], have developed a nonrelativistic quark model which describes quite well all mesons (including η and η') and all baryons composed of u, d or s quarks. The long-range part of their interaction is the usual linear confinement potential, but their short-range part is a pairing force stemming from instanton effects. This force presents the peculiarity to act only on quark-antiquark state with zero spin and zero angular momentum. The main problem of this model, and more generally of all nonrelativistic models, is that the velocity of a light quark inside a meson is not small compared with the speed of light. This makes the interpretation of the parameters of such models questionable.

It is thus interesting to use the interaction proposed by Blask *et al.* in a relativistic calculation of light meson spectra. Such a work was partly achieved in a previous work[2] where the masses of the spin-triplet light mesons were calculated in the framework of the relativistic two-body Dirac equation. In this model, which contains only two parameters besides quark masses, the leading Regge trajectories are well described. These results are encouraging, and it would be interesting to study the spin-singlet light mesons by turning on the instanton forces.

The purpose of this work is to test the relevance of such proposals by constructing a meson mass formula directly inspired by a relativistic model of a confined quark-antiquark pair interacting by instanton forces. Note that our mass formula can be reinterpreted in the framework of the spectrum-generating algebra[3].

2. Model

In agreement with several relativistic quark models[4], a good mass formula for mesons, in which the instanton forces do not act ($L \neq 0$ or $S \neq 0$), is given by (v is the radial quantum number)

$$M^2(i,j;v,L,S,J) = (M_0^2)_{ij} + A_{ij}v + B_{ij}L + C_{ij}S + D_{ij}J \tag{1}$$

*Chercheur qualifié F.N.R.S.

with parameters $(M_0^2)_{ij} = (M_{ij})^2 + eM_{ij}$ and $X_{ij} = x + x'M_{ij}$ (X denoting A, B, C or D). These parameters have a flavor dependence with the quantity $M_{ij} = M_i + M_j$, where M_i is the "constituent mass" of the quark i. This constituent mass takes into account the current mass and some dynamical effects due to the confinement and the relativistic motion. For simplicity, all parameters are assumed to be a linear function of M_{ij}.

In order to describe the pseudoscalar nonet π, K, η and η', it is necessary to introduce in the mass formula above the contribution of the instantons forces. They act only for $L = S = J = 0$ configuration since they vanish for $S = 1$ states, and since the interaction is a contact one. The best result is obtained when we assume that these forces act only for $v = 0$ mesons. The instanton interaction maximazes the flavor mixing. Consequently, in the flavor basis ($|1\rangle$, $|0\rangle$, $|s\bar{s}\rangle$) with $|1\rangle = (|u\bar{u}\rangle - |d\bar{d}\rangle)/\sqrt{2}$ and $|0\rangle = (|u\bar{u}\rangle + |d\bar{d}\rangle)/\sqrt{2}$, the instanton interaction matrix $\langle M'^2 \rangle$ has the form

$$\langle M'^2 \rangle = \begin{pmatrix} -h & 0 & 0 \\ 0 & h & \sqrt{2}h \\ 0 & \sqrt{2}h & 0 \end{pmatrix} \tag{2}$$

where the parameter h is positive. Using crossing symmetry, the matrix elements for strange pseudoscalar mesons are then given by (n stands for u or d quarks)

$$\langle n\bar{s} | M'^2 | n\bar{s} \rangle = \langle s\bar{n} | M'^2 | s\bar{n} \rangle = -h \tag{3}$$

As the formalism adopted here is relativistic, the contribution of the instanton forces M'^2 is added to the M^2 formula.

Collecting results above, the square mass formula for all light mesons is given by

$$\langle q_i\bar{q}_j; v, L, S, J | M^2 | q_{i'}\bar{q}_{j'}; v', L', S', J' \rangle$$
$$= \delta_{ii'}\delta_{jj'}\delta_{vv'}\delta_{LL'}\delta_{SS'}\delta_{JJ'} \Big\{ (M_{ij})^2 + eM_{ij} + (a + a'M_{ij})v + (b + b'M_{ij})L$$
$$+ (c + c'M_{ij})S + (d + d'M_{ij})J \Big\} + \langle M'^2 \rangle_{ij,i'j'} \tag{4}$$

with

$$\langle M'^2 \rangle_{ij,i'j'} = h\delta_{v0}\delta_{L0}\delta_{S0}\delta_{J0}\delta_{v'0}\delta_{L'0}\delta_{S'0}\delta_{J'0} \begin{cases} (-1)\delta_{ii'}\delta_{jj'} & \text{if } i \neq j \\ (1 - \delta_{ii'})\delta_{i'j'} & \text{if } i = j \end{cases} \tag{5}$$

3. Results

The parameters of the mass formula are found by minimizing a χ^2-function based on the square masses of a set of 36 well-established mesons selected from the 1992 compilation of masses of the Particle Data Group (PDG)[5]. The best results is obtained with the following set of parameters: $M_n = 0.319$ GeV, $M_s = 0.498$ GeV, $a = 0.848$ GeV^2, $a' = 0.855$ GeV, $b = 0.784$ GeV^2, $b' = 0.518$ GeV, $c = 0.188$ GeV^2,

$c' = 0.058$ GeV, $d = 0.194$ GeV^2, $d' = -0.248$ GeV, $e = -0.135$ GeV, $h = 0.302$ GeV^2.

The agreement between theory and experiment is generally good. Nevertheless, some peculiar points must be discussed: i) Our model suggests that the lowest $^3P_0$ isovector state is the not-well-established resonance $a_0(1320)$, which implies that the $a_0(980)$ is not a $q\bar{q}$ meson. Actually, experimental considerations[5] and theoretical works [6,7] suggest that the $a_0(980)$ is a $K\bar{K}$ molecule. The situation is similar in the ω family, where the $f_0(1240)$ is preferred to the $f_0(975)$ as the lowest $^3P_0$ isoscalar $n\bar{n}$ state. The inclusion of the states $a_0(980)$ and $f_0(975)$ in the selected set of mesons considerably worsens the fit. ii) A good assignment for all K mesons of the PDG tables cannot be found within our model. A possible reason is the existence of a large mixing between $S = 0$ and $S = 1$ states, which is not taken into account in our mass formula. iii) Our model fails to reproduce correctly the mass of the η'. The theoretical value found is 1.087 GeV, while the experimental value is 0.958 GeV. Note that, in the nonrelativistic model of Blask *et al.*, the value found is 1.116 GeV. It is possible that a more sophisticated parametrization of the instanton contribution (Eq. 5) improves our fit but it is not the purpose of this work.

A comparison with relativistic dynamical models[4] suggests that: i) light quarks can be treated as ultrarelativistic particles confined by a linear potential. ii) the confinement potential is mainly scalar but must contain other Lorentz structures.

A mass formula has been constructed for the mesons composed of u, d or s quarks. It provides a good fit for all these mesons, except for some K mesons and the η' meson, and it shows that the short-range part of the quark interaction can be described by instanton effects, which take into account both scattering and annihilation mechanisms. The model can be used as a guide in more sophisticated approaches.

4. References

1. W.H. Blask *et al.* *Z. Phys.* **A337** (1990)327.
2. C. Semay and R. Ceuleneer, *Phys. Rev.* **D48** (1993) 4361.
3. F. Iachello, Nimai C. Mukhopadhyay and L. Zhang, *Phys. Rev.* **D44** (1991) 898.
4. C. Semay, *J. Phys.* **G20** (1994) 689.
5. Particle Data Group, K. Hikasa *et al.*, *Phys. Rev.* **D45** (1992) S1.
6. J. Weinstein and N. Isgur, *Phys. Rev.* **D41**, (1990) 2236.
7. B. Silvestre-Brac and C. Semay, *Z. Phys.* **C59** (1993) 457.

MESON DECAYS

LUCA GAMBERALE

Dipartimento di Fisica, Università di Milano, via Celoria 16, 20133 Milano, Italy

ABSTRACT

Starting from the knowledge of the meson eigenstates at rest of the QCD hamiltonian derived in the framework of ACD it is possible to calculate the 3-meson vertex adopting a perturbative strategy. Results are in good agreement with experiment.

1. The vacuum structure of QCD and the resulting hamiltonian

The generally accepted theory of QCD is based on the notion of *Asymptotic Freedom* (AF), which is considered the theoretical basis for the applicability of the perturbative expansion at high energy scales. This idea is based on the assumption that the perturbative vacuum, i.e. the quantum state on which we construct the perturbation theory, is a good approximation of the true vacuum, i.e. the quantum state of minimum energy.

The study of the minimum of the QCD energy density performed with variational techniques, initiated by J.K.Savvidy in 1977 [1], continued by Nielsen and Olesen [2] and later by G.Preparata [3,4,5], has shown that this assumption is completely wrong. The outcome of this work is that there exist quantum states of the gauge field (Savvidy states) that have an energy density lower than that of the perturbative vacuum by a quantity diverging like Λ^4 (ultraviolet cut-off). This means that at any scale the perturbative vacuum is not a good approximation of the true vacuum thus implying that the standard perturbative procedure is theoretically unfounded.

The probable structure of the true QCD vacuum stems from the study of the magnetic condensation of the Savvidy states and is supplied by the *chromomagnetic liquid* (CML) [6]. The CML is a collection of needle-shaped domains uniformly distributed in all points and all directions of the space in which the chromomagnetic field assumes huge values (proportional to Λ^2) and in which the color charges are confined. In this way the color charges placed in a given needle follow an independent dynamics with respect to those placed in different needles, and as a result in each needle the dynamics is 1+1-dimensional (like in QCD_2, which is a naturally confined theory). The reason why it is not possible to observe the existence of the needles is that they rotate: we can only observe an average over directions.

In this scenario it has been possible to construct an effective theory of QCD (Anisotropic Chromo Dynamics, ACD) [6,7,8] in an "enlarged" space-time in which to each point of the usual space-time is associated a Lorentz-invariant pseudosphere in order to include the "notion" of the needle-shaped domains.

In ACD the hamiltonian can be written in the following form [6,7,8]:

$$H_{int} = H_{Kin} + H^{(0)} + H',$$ (1)

where H_{Kin} is the kinetic hamiltonian and $H^{(0)}$ has the following characteristics:

1. - it preserves the number of quarks, antiquarks and gluons separately

2. - it is the only infra-red singular term of the hamiltonian. It contains the confining term.

On the other hand the H' term:

1. - preserves the difference of the number of quarks and antiquarks; it is a pair-creation term and is responsible of the transitions between hadronic states

2. - is regular in the coordinate space

3. - can be treated as a perturbation.

The structure of $H^{(0)}$ and H' allows us to adopt a perturbative strategy which consists in solving the eigenvalue problem for the operator $H_{Kin} + H^{(0)}$ then making a perturbation with H'. The diagonalization of $H_{Kin} + H^{(0)}$ determines successfully the meson spectrum and the meson wave functions ϕ_n at rest in terms of six parameters: 5 quark masses and the string tension [9].

2. The 3-meson vertex

In this section I discuss the method I have adopted in the calculation of transition amplitudes between hadronic states. To focus our ideas let us consider the matrix element

$$< M_2, \vec{p}_2, M_3, \vec{p}_3 \mid S \mid M_1, \vec{p}_1 >$$ (2)

which represents the transition amplitude of meson 1 with momentum $\vec{p}_1$ into mesons 2 and 3 with momentum $\vec{p}_2$ and $\vec{p}_3$ respectively. To lowest order in perturbation expansion of the S matrix the amplitude can be written:

$$M = 2 \int d^3\vec{x} d^3\vec{x}' \left[\int \frac{d^3\vec{k}}{(2\pi)^3} \Phi_2^*(\vec{p}, \vec{k} - \frac{\vec{p}}{2}) \Phi_1(\vec{p}, \vec{k} - \frac{\vec{p}}{2}) e^{i\vec{k}\cdot\vec{x}} \right] \cdot$$ (3)

$$\cdot \left[\int \frac{d^3\vec{q}}{(2\pi)^3} \Phi_3^*(\vec{0}, \vec{q}) e^{-i\vec{q}\cdot\vec{x}'} \right] V(\vec{x}, \vec{x}')$$ (4)

where Φ_1, Φ_2, Φ_3 are the wave functions of the boosted mesons whose expression is given in Ref. [10] and where $\vec{p}$ is directed along the 3-axis and whose modulus satisfies the kinematical condition

$$\sqrt{p^2 + M_1^2} = \sqrt{p^2 + M_2^2} + M_3.$$ (5)

$V(\vec{x}, \vec{x}')$ is the creation potential in the coordinate space and in the case of equal quark masses reads

$$V(\vec{x}, \vec{x}') = \frac{\mu^2}{2\pi^2\sqrt{3}}\delta^2(\hat{x} - \hat{x}')\frac{v(mr, mr')}{rr'}\left[-i1^{(1)}\vec{\sigma}^{(2)}1^{(3)} + \vec{\sigma}^{(1)} \times \vec{\sigma}^{(2)}1^{(3)}\right]\cdot\hat{x} \qquad (6)$$

where m is the quark mass and

$$v(\rho, \rho') = \int_0^{+\infty} dt[K_0(\rho - t)K_0(\rho' - t) - K_0(\rho + t)K_0(\rho' + t)]. \qquad (7)$$

For the $\rho \to \pi\pi$ decay the computed transition amplitude is

$$M_{\rho\to\pi\pi} = 4.1\text{GeV} \qquad (8)$$

corresponding to a coupling constant

$$g_{\rho\to\pi\pi} = 5.72 \qquad (9)$$

to be compared with the experimental value

$$g_{\rho\to\pi\pi}^{\text{exp}} = 6.01, \qquad (10)$$

a very encouraging result!

3. Conclusions

The success of the $\rho \to \pi\pi$ calculation is another piece of evidence of the goodness of the theory since now developed. The calculation will be soon applied to the calculation of $K^* \to K\pi$, $\phi \to KK$. It is also possible to calculate Vector Meson Dominance contributions that, in conjunction with the electromagnetic matrix element shown in section 4, will allow us to compute $V \to P\gamma$ processes.

1. J.K.Savvidy *Physics Letters* **B71**, (1977), 71.
2. N.K.Nielsen and P.Olesen *Nuclear Physics* **B134**, (1978), 376.
3. M.Consoli and G.Preparata *Physics Letters* **B154**, (1985), 411.
4. G.Preparata *Il Nuovo Cimento* **A96**, (1986), 366.
5. L.Gamberale, G.Preparata and S.S.Xue *Il Nuovo Cimento* **A105**, (1992), 309.
6. G.Preparata *Il Nuovo Cimento* **A103**, (1990), 1073.
7. J.L.Basdevant and G.Preparata *Il Nuovo Cimento* **A67**, (1982), 19.
8. J.L.Basdevant P.Colangelo and G.Preparata *Il Nuovo Cimento* **A71**, (1982), 445.
9. see M.Scorletti's talk at this conference
10. L.Cosmai, M.Pellicoro and G.Preparata *Nuclear Physics* **B228**, (1983), 31.

SECTION C

Quark pair creation effects, mixing, decays;
exotic states, glueballs;
heavy quark effective theory.

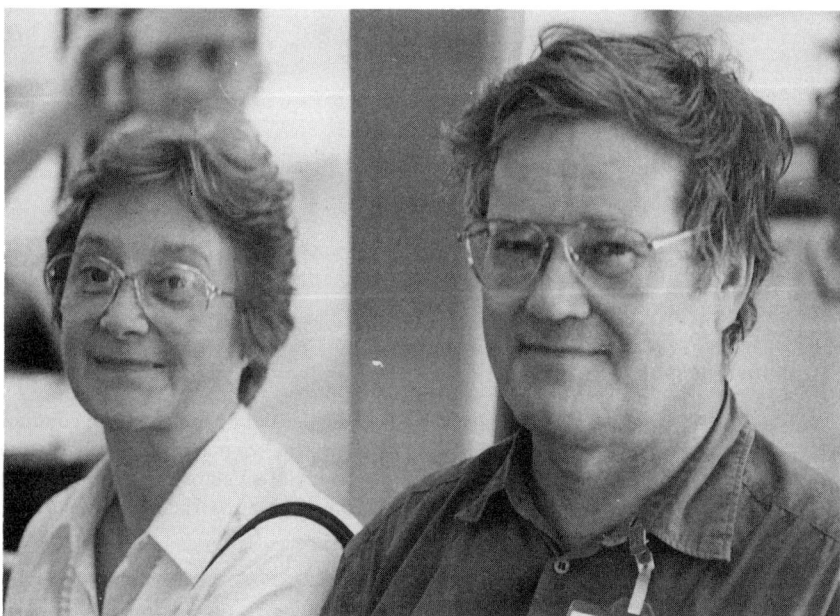

HEAVY BARYONS: CURRENT, PION AND PHOTON TRANSITIONS

J.G. KÖRNER*

and

J. LANDGRAF

Institut für Physik, Johannes Gutenberg-Universität
Staudinger Weg 7, D-55099 Mainz, Germany

ABSTRACT

We discuss the structure of current-induced bottom baryon to charm baryon transitions, and the structure of pion and photon transitions between heavy charm or bottom baryons in the Heavy Quark Symmetry limit as $m_Q \to \infty$. Our discussion involves the ground state s-wave heavy baryons as well as the excited p-wave heavy baryon states.

1. Introduction

The Heavy Quark Effective Theory (HQET) formulated in 1990[1] is so well known by now that it no longer needs an extensive introduction. The HQET provides a systematic expansion of QCD in terms of inverse powers of the heavy quark mass. The leading term in this expansion gives rise to a new spin and flavour symmetry at equal velocities, termed Heavy Quark Symmetry. Corrections to the Heavy Quark Symmetry limit can be classified and evaluated order by order in $1/m_Q$ by considering the contributions of the nonleading terms in the effective HQET fields and the HQET Langrangian. We might mention that there exist different formulations of HQET which differ from one another starting at $O(1/m_Q^2)^{2,3}$. They can be transformed into each other by appropriate field redefinitions [4]. Among these, the Foldy-Wouthuysen type HQET introduced in [3] is closest to the original quantum mechanical version of the Foldy-Wouthuysen transformation introduced in the 50's to deal with recoil corrections when calculating e.g. properties of the hydrogen atom.

In this review we will only be concerned with the leading order term in the HQET expansion, i.e. in the Heavy Quark Symmetry limit. In this limit the dynamics of the light and heavy constituents decouple and the calculation of the Heavy Quark Symmetry predictions for transition amplitudes essentially amounts to an angular momentum coupling exercise, however involved it may be. It is then not surprising that such calculations have been done before even before the conceptual foundations of HQET had been laid down in 1990. For example, the Heavy Quark Symmetry

*Supported in part by the BMFT, FRG, under contract 06MZ730

structure of current-induced charm baryon transitions had been written down as early as 1976[5]. Also, and this will be explicated in this talk, the technical tools needed to do the requisite angular momentum coupling calculations already existed for quite some time in the form of the Wigner 6-j-symbol calculus developed and used extensively by atomic and nuclear physicists working on problems closely related to Heavy Quark Symmetry calculations.

At present most of the attention of experimentalists and theoreticians working in heavy quark physics is directed towards the application of HQET in the meson sector where data is starting to become quite abundant. This data will be supplemented in the not-too-distant future by corresponding data on heavy baryon decays and there will be a need to analyze this data in terms of HQET as applied to heavy baryons. This will be the subject of the present review. We shall discuss the structure of flavour changing bottom baryon to charm baryon decays, as well as pion and photon transition between heavy baryons with the same flavour, all in the Heavy Quark Symmetry limit as $m_Q \to \infty$. Our discussion involves both ground state heavy baryons and their p-wave excitations. We present our results in the form of building blocks listed in tabular form. The building blocks may then easily be assembled to obtain the Heavy Quark Symmetry predictions for the current, pion and photon transition amplitudes of heavy baryons.

2. Classification of s- and p-Wave Heavy Baryon States

A heavy baryon is made up of a light diquark system (qq) and a heavy quark Q. The light diquark system has bosonic quantum numbers j^P with the total angular momentum $j = 0, 1, 2 \ldots$ and parity $P = \pm 1$. To each diquark system with spin-parity j^P there is a degenerate heavy baryon doublet with $J^P = (j \pm 1/2)^P$ ($j = 0$ is an exception). It is important to realize that the Heavy Quark Symmetry structure of the heavy baryon states is entirely determinated by the spin-parity j^P of the light diquark system.

From our experience with light baryons and light mesons we know that one can get a reasonable description of the light particle spectrum in the constituent quark model picture. This is particularly true for the enumeration of states, their spins and their parities. As much as we know up to now, gluon degrees of freedom do not seem to contribute to the particle spectrum. It is thus quite natural to try the same constituent approach to enumerate the light diquark states, their spins and their parities. From the spin degrees of freedom of the two light quarks one obtains a spin 0 and a spin 1 state. The total orbital state of the diquark system is characterized by two angular degrees of freedom which we take to be the two independent relative momenta $k = \frac{1}{2}(p_1 - p_2)$ and $K = \frac{1}{2}(p_1 + p_2 - 2p_3)$ that can be formed from the two light quark momenta p_1 and p_2 and the heavy quark momentum p_3. The k-orbital momentum describes relative orbital exitations of the two quarks, and the K-orbital momentum describes orbital excitations of the center of mass of the two light quarks relative to the heavy quark. The (k, K) basis is quite convenient for two reasons. First, Copley, Isgur and Karl[6] have found that the (k, K) basis diagonalizes the Hamiltonian, when harmonic interquark forces are used. Second, it allows one to classify the diquark states in terms of $SU(2N_f) \otimes O(3)$ representations.

Let us do just this for the two flavour case $N_f = 2$ for the s- and p-wave states.

a) s-wave ground state $\sim \square\square_{10} \otimes \begin{smallmatrix}\square\\\square\end{smallmatrix}_1$

where the spin-flavour content of the $SU(4)$ diquark representation $\underline{10}$ can be determined by looking at how the $\underline{10}$ representation decomposes under the $SU(2)_{spin} \otimes SU(2)_{flavour}$ subgroup. One has

$$\square\square_{10} = \left(\begin{smallmatrix}\square\\\square\end{smallmatrix}_1 \otimes \begin{smallmatrix}\square\\\square\end{smallmatrix}_1 + \square\square_3 \otimes \square\square_3 \right) \tag{1}$$

When coupling in the heavy quark one finally has the particle content

$$[q_1 q_2] : 0^+ \quad \rightarrow \quad \tfrac{1}{2}^+ \quad \Lambda_Q$$

$$\{q_1 q_2\} : 1^+ \quad \Big\langle \quad \begin{array}{l} \tfrac{1}{2}^+ \quad \Sigma_Q \\ \tfrac{3}{2}^+ \quad \Sigma_Q^* \end{array} \tag{2}$$

b) p-wave ($l_k = 0$, $l_K = 1$) $\sim \square\square_{10} \otimes \boxed{K}_3$: The $\underline{10}$ representation decomposes as before. The spin 0 and spin 1 pieces of the $\underline{10}$ couple with $l_K = 1$ to give $j^P = 1^-$ and $j^P = 0^-, 1^-, 2^-$, respectively. One has the particle content

$$[q_1 q_2] : 1^- \quad \Big\langle \quad \begin{array}{l} \tfrac{1}{2}^- \\ \tfrac{3}{2}^- \end{array} \Big\} \quad \Lambda_{QK1}^{**}$$

$$\{q_1 q_2\} : 0^- \quad \rightarrow \quad \tfrac{1}{2}^- \quad \Sigma_{QK0}^{**}$$

$$1^- \quad \Big\langle \quad \begin{array}{l} \tfrac{1}{2}^- \\ \tfrac{3}{2}^- \end{array} \Big\} \quad \Sigma_{QK1}^{**}$$

$$2^- \quad \Big\langle \quad \begin{array}{l} \tfrac{3}{2}^- \\ \tfrac{5}{2}^- \end{array} \Big\} \quad \Sigma_{QK2}^{**} . \tag{3}$$

c) p-wave ($l_k = 1$, $l_K = 0$) $\sim \begin{smallmatrix}\square\\\square\end{smallmatrix}_6 \otimes \boxed{K}_3$: The diquark representation $\underline{6}$ of $SU(4)$ decomposes under the $SU(2)_{spin} \otimes SU(2)_{flavour}$ subgroup as

$$\begin{smallmatrix}\square\\\square\end{smallmatrix}_6 = \left(\begin{smallmatrix}\square\\\square\end{smallmatrix}_1 \otimes \square\square_3 + \square\square_3 \otimes \begin{smallmatrix}\square\\\square\end{smallmatrix}_1 \right) . \tag{4}$$

After coupling with the k-orbital angular momentum $l_k = 1$ and the heavy quark spin $s = 1/2$, the particle content can then be determined to be

$$[q_1 q_2] : 0^- \quad \rightarrow \quad \tfrac{1}{2}^- \qquad \Lambda^{**}_{Qk0}$$

$$1^- \quad \Big\langle \quad \left. \begin{array}{c} \tfrac{1}{2}^- \\ \tfrac{3}{2}^- \end{array} \right\} \qquad \Lambda^{**}_{Qk1}$$

$$2^- \quad \Big\langle \quad \left. \begin{array}{c} \tfrac{3}{2}^- \\ \tfrac{5}{2}^- \end{array} \right\} \qquad \Lambda^{**}_{Qk2}$$

$$\{q_1 q_2\} : 1^- \quad \Big\langle \quad \left. \begin{array}{c} \tfrac{1}{2}^- \\ \tfrac{3}{2}^- \end{array} \right\} \qquad \Sigma^{**}_{Qk1} \tag{5}$$

One thus has altogether seven Λ-type p-wave states and seven Σ-type p-wave states. The analysis can easily be extended to the case $SU(6)\otimes 0(3)$ bringing in the strangeness quark in addition.

Let us mention that, in the charm sector the states $\Lambda_c(2285)$ and $\Sigma_c(2453)$ are well established while there is first evidence for the $\Sigma_c^*(2510)$ state. Recently two excited states $\Lambda_c^{**}(2593)$ and $\Lambda_c^{**}(2627)$ have been seen which very likely correspond to the two p-wave states making up the $\Lambda_{cK_1}^{**}$ Heavy Quark Symmetry doublet. The charm-strangeness states $\Xi_c(2470)$ and $\Omega_c(2720)$ have been seen and first evidence was presented for the $\tfrac{1}{2}^+$ $\Xi_c'(2570)$ state with the flavour configuration $c\{sq\}$. In the bottom sector the $\Lambda_b(5640)$ has made its way into the Particle Data Booklet listing while some indirect evidence has been presented for the $\Xi_b(5800)$.

3. Heavy Baryon Spin Wave Functions

We are now in the position to write down the covariant spin wave functions of the heavy baryons. These will be needed to derive the predictions of Heavy Quark Symmetry for current, pion and photon transitions. One has a light diquark system with J^P quantum numbers j^P which couples with a heavy quark with $J^P = 1/2^+$ to form a degenerate pair of states with $J^P = (j \pm 1/2)^P$. The coupling goes according to the scheme

$$
\begin{array}{ccccc}
 & & & \nearrow & (j - 1/2)^P \\
j^P & \otimes & 1/2^+ & & \\
 & & & \searrow & (j + 1/2)^P \\
\text{light diquark} & & \text{heavy quark} & & \text{heavy baryon} \\
\varphi^{\mu_1 \cdots \mu_j} & & & & u^{(\mu_1)\mu_2 \cdots \mu_j}
\end{array}
\tag{6}
$$

where $\varphi^{\mu_1 \cdots \mu_j}$ stands for the spin wave function of the light diquark with spin j and where $u^{\mu_1 \mu_2 \cdots \mu_j}$ and $u^{\mu_2 \cdots \mu_j}$ are the Rarita-Schwinger spinor wave functions of the degenerate pair of heavy baryons with $J = j + 1/2$ and $J = j - 1/2$, respectively.

The case $j = 0$ is special. In this case there is a coupling only to one heavy baryon state with $J = 1/2$ as e.g. for the Λ-type baryon ground state Λ_Q.

It is then straightforward to write down the covariant spin wave functions $\Psi_{\alpha\beta\gamma}$ of the Heavy Quark Symmetry baryon doublets. One has

$$\Psi_{\alpha\beta\gamma} = (\varphi_{\mu_1\cdots\mu_j})_{\alpha\beta} \left\{ \begin{array}{c} M_{\gamma\gamma'}^{\mu_1} u_{\gamma'}^{\mu_2\cdots\mu_j} \\ M_{\gamma\gamma'} u_{\gamma'}^{\mu_1\cdots\mu_j} \end{array} \right\} \tag{7}$$

where we have explicitly written out the Dirac spinor indices α, β and γ. The spinor indices α and β refer to the light quark system and the index γ refers to the heavy quark. Dropping the spinor indices one has

$$\Psi = \varphi_{\mu_1\cdots\mu_j} \psi^{\mu_1\cdots\mu_j} \tag{8}$$

where the "superfield" heavy baryon wave function $\psi^{\mu_1\cdots\mu_j}$ stands for the two spin wave functions $\{j-1/2, j+1/2\}$ as indicated in Eq.(7). What remains to be done is to determine the matrices M^{μ_1} and M in Eq.(7). They can be worked out by noting that the heavy baryon spin wave function has to satisfy the mass-shell condition on the heavy quark spinor index, i.e.

$$\not{v}\psi^{\mu_1\cdots\mu_j} = \psi^{\mu_1\cdots\mu_j} \tag{9}$$

Eq.(9) is solved by

$$\begin{array}{rcl} M^{\mu_1} & = & N_j \gamma_\perp^{\mu_1} \gamma_5 \\ M & = & N_j' \cdot \mathbb{1} \end{array} \tag{10}$$

where the transverse gamma matrix $\gamma_\perp^\mu$ is defined by $\gamma_\perp^\mu = \gamma^\mu - \not{v}v^\mu$ and where v^μ is the four-velocity of the heavy baryon $v^\mu = p^\mu/M$. The normalization of the coupling matrices M^μ and M is fixed by the normalization condition

$$\bar{\psi}^{\mu_1\cdots\mu_j} \psi_{\mu_1\cdots\mu_j} = (-1)^{J-1/2} 2M \tag{11}$$

which gives $N_0 = 0$, $N_1 = \sqrt{1/3}$, $N_2 = \sqrt{1/10}$ and $N_0' = 0$, $N_1' = N_2' = 1$ for the s- and p-wave cases discussed in this review. There is an implicit understanding that the set of tensor indices "$\mu_1\cdots\mu_j$" is always completely symmetrized, traceless with regard to any pair of indices and transverse to the line of flight in every index. This is annoted explicitly in Table 1 and 2 where the s- and p-wave heavy baryon wave functions are listed. For example, the notation $\{\mu_1^\perp \mu_2^\perp\}_0$ implies symmetrization, tracelessness and transversity of the two tensor indices $\mu_1\mu_2$ as specified above. For the sake of completeness we have also included the light-side spin wave functions as given in a constituent quark model approach for the light side, where $\hat{\chi}^0 = \frac{1}{2\sqrt{2}}(\not{v}+1)C$ and $\hat{\chi}^{1\mu} = \frac{1}{2\sqrt{2}}(\not{v}+1)\gamma_\perp^\mu C$ with C the charge conjugation matrix.

Table 1. Spin wave functions (s.w.f.) of Λ-type s- and p-wave heavy baryons. Light-side spin wave functions are constituent spin wave functions.

	light side s.w.f. $\hat{\phi}^{\mu_1\cdots\mu_j}$	j^P	heavy side s.w.f. $\psi_{\mu_1\cdots\mu_j}$	J^P
Λ_Q	$\hat{\chi}$	0^+	u	$\frac{1}{2}^+$
$\{\Lambda_{QK1}\}$	$\hat{\chi}^0 K_\perp^{\mu_1}$	1^-	$\frac{1}{\sqrt{3}}\gamma_{\mu_1}^\perp \gamma_5 u$	$\frac{1}{2}^-$
			u_{μ_1}	$\frac{3}{2}^-$
Λ_{Qk0}	$\frac{1}{\sqrt{3}}\hat{\chi}^1 \cdot k_\perp$	0^-	u	$\frac{1}{2}^-$
$\{\Lambda_{Qk1}\}$	$\frac{i}{\sqrt{2}}\varepsilon(\mu_1\hat{\chi}^1 k_\perp v)$	1^-	$\frac{1}{\sqrt{3}}\gamma_{\mu_1}^\perp \gamma_5 u$	$\frac{1}{2}^-$
			u_{μ_1}	$\frac{3}{2}^-$
$\{\Lambda_{Qk2}\}$	$\frac{1}{2}\{\hat{\chi}^{1,\mu_1} k_\perp^{\mu_2}\}_0$	2^-	$\frac{1}{\sqrt{10}}\gamma_{\{\mu_1}^\perp \gamma_5 u_{\mu_2}\}_0$	$\frac{3}{2}^-$
			$u_{\mu_1\mu_2}$	$\frac{5}{2}^-$

Table 2. Spin wave functions (s.w.f.) of Σ-type s- and p-wave heavy baryons.

	light side s.w.f. $\hat{\phi}^{\mu_1\cdots\mu_j}$	j^P	heavy side s.w.f. $\psi_{\mu_1\cdots\mu_j}$	J^P
$\{\Sigma_Q\}$	$\hat{\chi}^{1\mu_1}$	1^+	$\frac{1}{\sqrt{3}}\gamma_{\mu_1}^\perp \gamma_5 u$	$\frac{1}{2}^+$
			u_{μ_1}	$\frac{3}{2}^+$
$\{\Sigma_{Qk1}\}$	$\hat{\chi}^0 k_\perp^{\mu_1}$	1^-	$\frac{1}{\sqrt{3}}\gamma_{\mu_1}^\perp \gamma_5 u$	$\frac{1}{2}^-$
			u_{μ_1}	$\frac{3}{2}^-$
Σ_{QK0}	$\frac{1}{\sqrt{3}}\hat{\chi}^1 \cdot K_\perp$	0^-	u	$\frac{1}{2}^-$
$\{\Sigma_{QK1}\}$	$\frac{i}{\sqrt{2}}\varepsilon(\mu_1\hat{\chi}^1 K_\perp v)$	1^-	$\frac{1}{\sqrt{3}}\gamma_{\mu_1}^\perp \gamma_5 u$	$\frac{1}{2}^-$
			u_{μ_1}	$\frac{3}{2}^-$
$\{\Sigma_{QK2}\}$	$\frac{1}{2}\{\hat{\chi}^{1,\mu_1} K_\perp^{\mu_2}\}_0$	2^-	$\frac{1}{\sqrt{10}}\gamma_{\{\mu_1}^\perp \gamma_5 u_{\mu_2}\}_0$	$\frac{3}{2}^-$
			$u_{\mu_1\mu_2}$	$\frac{5}{2}^-$

4. Generic Picture of Current, Pion and Photon Transitions

In Fig. 1 we have drawn the generic diagrams that describe $b \to c$ current transitions, and $c \to c$ pion and photon transitions between heavy baryons in the Heavy Quark Symmetry limit. The heavy-side and light-side transitions occur completely independent of each other (they "factorize") except for the requirement that the heavy side and the light side have the same velocity in the initial and final state, respectively, which are also the velocities of the initial and final heavy baryons. The $b \to c$ current transition induced by the flavour-spinor matrix Γ is hard and accordingly there is a change of velocities $v_1 \to v_2$, whereas there is no velocity change in the pion and photon transitions. The heavy-side transitions are completely specified whereas the light-side transitions $j_1^{P_1} \to j_2^{P_2}$, $j_1^{P_1} \to j_2^{P_2} + \pi$ and $j_1^{P_1} \to j_2^{P_2} + \gamma$ are described by a number of form factors or coupling factors which parametrize the light-side transitions. The pion and the photon couple only to the light side. In the case of the pion this is due to its flavour content. In the case of the photon the coupling of the photon to the heavy side involves a spin flip which is down by $1/m_Q$ and thus the photon couples only to the light side in the Heavy Quark Symmetry limit.

Referring to Fig. 1 we are now in the position to write down the generic expressions for the current, pion and photon transitions according to the spin-flavour flow depicted in Fig. 1. One has

current transitions:

$$\bar{\psi}_2^{\nu_1 \cdots \nu_{j_2}} \Gamma \psi^{\mu_1 \cdots \mu_{j_1}} \left(\sum_{i=1}^{N} f_i(\omega) t^i_{\nu_1 \cdots \nu_{j_2}; \mu_1 \cdots \mu_{j_1}} \right) \tag{12}$$

$$
\begin{aligned}
n_1 \cdot n_2 &= 1 & N &= j_{\min} + 1 \\
n_1 \cdot n_2 &= -1 & N &= j_{\min}
\end{aligned}
$$

pion transitions:

$$\bar{\psi}_2^{\nu_1 \cdots \nu_{j_2}} \psi^{\mu_1 \cdots \mu_{j_1}} \left(\sum_{i=1}^{N} f_i^\pi t^i_{\nu_1 \cdots \nu_{j_2}; \mu_1 \cdots \mu_{j_1}} \right) \tag{13}$$

$$
\begin{aligned}
n_1 \cdot n_2 &= 1 & N &= j_{\min} \\
n_1 \cdot n_2 &= -1 & N &= j_{\min} + 1
\end{aligned}
$$

photon transitions:

$$\bar{\psi}_2^{\nu_1 \cdots \nu_{j_2}} \psi^{\mu_1 \cdots \mu_{j_1}} \left(\sum_{i=1}^{N} f_i^\gamma t^i_{\nu_1 \cdots \nu_{j_2}; \mu_1 \cdots \mu_{j_1}} \right) \tag{14}$$

$$
\begin{aligned}
j_1 &= j_2 & N &= 2j_1 \\
j_1 &\neq j_2 & N &= 2j_{\min} + 1
\end{aligned}
$$

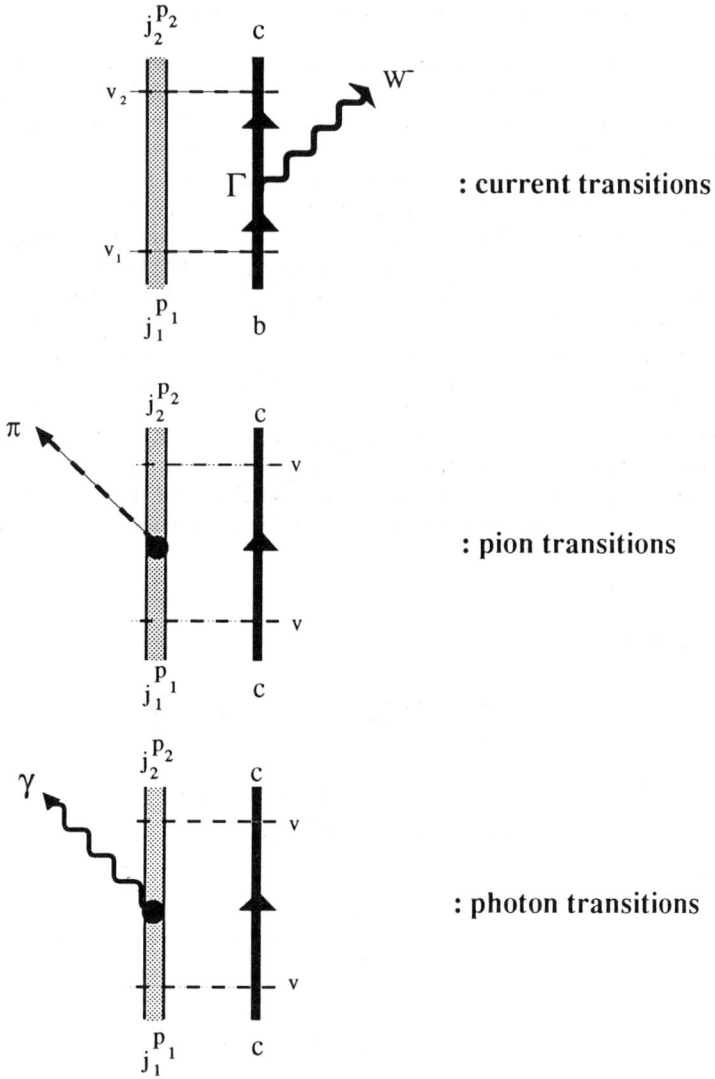

Fig. 1. Generic picture of bottom to charm current transitions, and pion and photon transitions in the charm sector in the Heavy Quark Symmetry limit $m_Q \to \infty$

where the $\psi^{\mu_1\cdots\mu_j}$ are the heavy baryon spin wave functions introduced in Sec. 3.

In each of the above cases we have also given the result of counting the number N of independent form factors or coupling factors. These are easy to count by using either helicity amplitude counting or LS partial wave amplitude counting. In the case of current and pion transitions the counting involves the normalities of the light-side diquarks which is defined by $n = (-1)^j P$.

All three coupling expressions can also be written down in terms of Wigner's 6-j symbols as will be discussed later for the pion and photon transitions. The generic expressions Eq.(12), Eq.(13) and Eq.(14) completely determine the heavy quark symmetry structure of the current, pion and photon transitions. What remains to be done is to write down independent sets of covariant coupling tensors. This will be done in the next section.

5. Coupling Structure of Current, Pion and Photon Transitions

What remains to be done is to tabulate explicit expressions for the tensors $t^i_{\nu_1\cdots\nu_{j_2};\mu_1\cdots\mu_{j_1}}$ describing the light-side transitions. We shall treat the current, pion and photon transitions in turn and shall work out some sample transitions in order to familiarize the reader with the use of the tables.

Current transitions:

The tensors $t^i_{\nu_1\cdots\nu_{j_2};\mu_1\cdots\mu_{j_1}}$ have to be build from the vectors $v_1^{\nu_i}$ and $v_2^{\mu_i}$, the metric tensors $g_{\mu_i\mu_k}$ and, depending on parity, from the Levi-Civita object $\varepsilon(\mu_i\nu_k v_1 v_2) := \varepsilon_{\mu_i\nu_k\alpha\beta}v_1^\alpha v_2^\beta$. The relevant tensors of interest have been listed in Table 3. For the $1^+ \rightarrow 1^+$ and $1^+ \rightarrow 2^-$ transitions the number of linearly independent tensors is $N = 2$. We have chosen to diagonalize the light-side transition in terms of partial-wave amplitudes with definite partial wave L_V. Table 3 also contains some normalization information in that all the amplitudes square up to $(\omega^2 - 1)^{L_V}$.

As an application consider the current induced transitions $\Lambda_b \rightarrow \Lambda_c, \Lambda_c^{**}$. They involve the following diquark transitions:

$$
\begin{aligned}
\Lambda_b &\rightarrow \Lambda_c &&: &0^+ &\rightarrow 0^+ &&\text{allowed} &&(N = 1)\\
&\rightarrow \{\Lambda^{**}_{cK1}\} &&: &&\rightarrow 1^- &&\text{allowed} &&(N = 1)\\
&\rightarrow \Lambda^{**}_{ck0} &&: &&\rightarrow 0^- &&\text{forbidden}\\
&\rightarrow \{\Lambda^{**}_{ck1}\} &&: &&\rightarrow 1^- &&\text{allowed} \cdot &&(N = 1)\\
&\rightarrow \{\Lambda^{**}_{ck1}\} &&: &&\rightarrow 2^- &&\text{forbidden}
\end{aligned}
\tag{15}
$$

where the curly bracket notation indicates that the particle symbol stands for a Heavy Quark Symmetry doublet. It is noteworthy that Heavy Quark Symmetry predicts that

Table 3. Tensor structure of diquark transitions. Partial wave L_V is the partial wave of the transition $j_1^{P_1} \to j_2^{P_2} + 0^+$. Sign of the product of normalities $n_1 \cdot n_2$ determines the number N of independent transitions or Isgur-Wise functions.

diquark transition	partial wave		covariant coupling
$j_1^{P_1} \to j_2^{P_2}$	L_V	$n_1 \cdot n_2$	$t^i_{\mu_1 \dots \mu_{j_1}; \nu_1 \dots \nu_{j_2}}$
$0^+ \to 0^+$	0	$+1$	1
$0^+ \to 0^-$	forbidden	-1	-
1^-	1	$+1$	$v_{1\mu}$
2^-	forbidden	-1	-
$1^+ \to 1^+$	0	$+1$	$\frac{1}{\sqrt{3}}\left(g_{\mu\nu} - \frac{1}{\omega+1}v_{1\mu}v_{2\nu}\right)$
	2	$+1$	$\frac{1}{\sqrt{6}}\left((\omega^2-1)g_{\mu\nu} - (\omega+2)v_{1\mu}v_{2\nu}\right)$
$1^+ \to 0^-$	1	$+1$	$v_{2\mu}$
1^-	1	-1	$\frac{1}{\sqrt{2}}\varepsilon(\mu_1\nu_1 v_1 v_2)$
2^-	1	$+1$	$\sqrt{\frac{3}{5}}\left(v_{1\mu_1}g_{\mu_2\nu_1} - \frac{1}{\omega+1}v_{1\mu_1}v_{1\mu_2}v_{2\nu_1}\right)$
	3	$+1$	$\frac{1}{\sqrt{10}}\left(2(\omega^2-1)v_{1\mu_1}g_{\mu_2\ nu_1}\right.$
			$\left. -(2\omega+3)v_{1\mu_1}v_{1\mu_2}v_{2\nu_1}\right)$

only four of the possible seven transitions to the p-wave states are allowed. Using the relevant entries in Tables 1 and 3 one can then write down the Heavy Quark Symmetry structure of the full transition amplitudes. One has

i) $\Lambda_b \to \Lambda_c$ $1/2^+ \to 1/2^+$

$$M^\lambda = \bar{u}_2 \Gamma^\lambda u_1 f^{(0)}(\omega) \qquad f^{(0)}(1) = 1 \tag{16}$$

ii) $\Lambda_b \to \{\Lambda_{cK1}^{**}\}$ $1/2^+ \to \left\{ {1/2^- \atop 3/2^-} \right\}$

$$M^\lambda = \left\{ {-\frac{1}{\sqrt{3}} \bar{u}_2 \gamma_5 \gamma_{\perp}{}^\mu_2 \atop \bar{u}_2^\mu} \right\} \Gamma^\lambda u_1 f_1^{(1)}(\omega) v_{1\mu} \tag{17}$$

iii) $\Lambda_b \to \{\Lambda_{ck1}^{**}\}$ $1/2^+ \to \left\{ {1/2^- \atop 3/2^-} \right\}$

$$\tag{18}$$

the same as case ii) with $f_1^{(1)}(\omega) \to f_2^{(1)}(\omega)$

In the case of a left-chiral current transition as in the Standard Model one has $\Gamma^\lambda = \gamma^\lambda(1 - \gamma_5)$. For the elastic transition $\Lambda_b \to \Lambda_c$ one has the normalization condition $f^{(0)}(\omega = 1) = 1$ at equal velocities $v_1 = v_2$ due to the normalization of the diquark states, whereas there is no such normalization condition for the transition form factors $f_1^{(1)}(\omega)$ and $f_2^{(1)}(\omega)$ at zero recoil $\omega = 1$.

One can then easily calculate the contributions of Λ_c, Λ_{cK1}^{**} and Λ_{ck1} to the Bjørken sum rule[7] by squaring the relevant transition amplitudes and, in the case of Λ_{cK1}^{**} and Λ_{ck1}^{**}, by summing over the contributions of the two degenerate partners in the respective doublets. One obtains

$$1 = |f^{(0)}(\omega)|^2 + (\omega^2 - 1)(|f_1^{(1)}(\omega)|^2 + |f_2^{(1)}(\omega)|^2 + \ldots) \tag{19}$$

It is noteworthy that the transition $\Lambda_b \to \Lambda_{ck1}^{**}$ is predicted to be zero when the light-side transition is calculated in a constituent quark model approach with $SU(2N_f) \times O(3)$ symmetry. The reason is that the light-side transition involves a $s = 0$ to a $s = 1$ quark spin transition which is zero in the constituent picture. This would then imply that there are altogether only two nonzero transitions to the seven Λ_c^{**} p-wave states, which, in the light of the Bjørken sum rule, would imply that the quasi-elastic transition $\Lambda_b \to \Lambda_c$ constitutes a large fraction of the total inclusive semileptonic $\Lambda_b \to X_c$ decay rate.

Pion transitions:

The $(j_1 + j_2)$ rank tensors $t^i_{\nu_1 \cdots \nu_{j_2}; \mu_1 \cdots \mu_{j_1}}$ describing the light-side transitions $j_1^{P_1} \to j_2^{P_2} + \pi$ have to be composed from the building blocks $g_{\perp\mu\nu} = g_{\mu\nu} - v_\mu v_\nu$, $p_{\perp\mu} = p_\mu - p \cdot v \, v_\mu$

and, depending on parity, from the Levi-Civita tensor $\varepsilon(\mu_i\nu_k p\,v)$. In the case when there are two independent transitions we have diagonalized the light-side transition by going to the LS-basis. The tensors and amplitudes are now labelled by the partial wave l_π of the pion emission process. Again, we have introduced some normalization information in Table 4. The normalization of the partial wave amplitudes f_{l_π} is such that a given partial wave amplitude f_{l_π} contributes as $|f_{l_\pi}|^2|\vec{p}|^{2l_\pi}$ to the spin-summed square of the diquark transition amplitude. As an example we write down the pion transition amplitudes for the ground state to ground state transition $\{\Sigma_c\} \to \Lambda_c + \pi$. Using Tables 1 and 2 for the heavy-side baryon wave functions and Table 4 for the $1^+ \to 0^+ + \pi$ light-side pion transition one has

$$M^\pi = \bar{u}_2(v) \left\{ \begin{array}{c} \frac{1}{\sqrt{3}}\gamma_1^\mu \gamma_5 u_1(v) \\ u^\mu(v) \end{array} \right\} f_p p_\mu^\perp \tag{20}$$

Calculating the decay rate in the degeneracy limit $M_{\Sigma_c^*} = M_{\Sigma_c} = M_1$ one finds

$$\Gamma_{\Sigma_c^* \to \Lambda_c + \pi} = \Gamma_{\Sigma_c \to \Lambda_c + \pi} = \frac{1}{6\pi}\frac{M_2}{M_1}\,|\,f_p\,|^2|\,\vec{p}\,|^3 \tag{21}$$

That the decay rates from degenerate doublet partners into a singlet state are equal is a general result. This general result is much easier to derive in the $6\text{-}j$ symbol approach than in the covariant approach used so far. Looking again at the pion transition in Fig. 1 one sees that one has to perform altogether three angular couplings. They are

i) $j_1{}^{P_1} \otimes 1/2^+ \Rightarrow J_1{}^{P_1}$

ii) $j_2{}^{P_2} \otimes 1/2^+ \Rightarrow J_2{}^{P_2}$

iii) $J_2{}^{P_2} \otimes L_\pi \Rightarrow J_1{}^{P_1}$

$$\tag{22}$$

where $L_\pi = l_\pi$ is the orbital momentum of the pion and $J_1{}^{P_1}$ and $J_2{}^{P_2}$ denote the J^P quantum numbers of the initial and final baryons. The heavy quark has $1/2^+$ quantum numbers. This is a coupling problem well-known from atomic and nuclear physics and the problem is solved by Wigner's $6\text{-}j$ symbol calculus. One finds

$$M^\pi(J_1 J_1^z \to J_2 J_2^z + L_\pi m) = M_{L_\pi}(-1)^{L_\pi + j_2 + 1/2 + J}(2j_1 + 1)^{1/2}(2J_2 + 1)^{1/2}$$
$$\left\{ \begin{array}{ccc} j_2 & j_1 & L_\pi \\ J & J_2 & 1/2 \end{array} \right\} \langle Lm J_2 J_2^z \mid J_1 J_1^z \rangle. \tag{23}$$

where $\left\{ \begin{array}{ccc} j_2 & j & L_\pi \\ J_1 & J_2 & 1/2 \end{array} \right\}$ is Wigner's $6\text{-}j$ symbol and $\langle L_\pi M J_2 J_2^z | J_1 J_1^z \rangle$ is the Clebsch-Gordan coefficient coupling L_π and J_2 to J_1. M_{L_π} is the reduced amplitude of the transition and is proportional to f_{l_π}. Then by using the standard orthogonality

Table 4. Tensor structure of pion couplings to diquark states. The pion is in a definite orbital state l_π. Tensor structure of transitions with $\left(j_1^{P_1}, j_2^{P_2}\right) \to \left(j_1^{-P_1}, j_2^{-P_2}\right) \to \left(j_2^{P_2}, j_1^{P_1}\right) \to \left(j_2^{-P_2}, j_1^{-P_1}\right)$ are identical and are not always listed here.

diquark transition $j_1^{P_1} \to j_2^{P_2} + \pi$	orbital wave l_π	covariant coupling $t^i_{\mu_1 \dots \mu_{j_1}; \nu_1 \dots \nu_{j_2}}$
$0^+ \to 0^+ + \pi$	forbidden	-
$1^+ \to 0^+ + \pi$	1	$p^\perp_{\mu_1}$
$1^+ + \pi$	1	$\frac{1}{\sqrt{2}}\varepsilon(\mu_1\nu_1 pv)$
$0^- \to 0^+ + \pi$	0	1 (scalar)
$1^+ + \pi$	forbidden	-
$0^- + \pi$	forbidden	-
$1^- \to 0^+ + \pi$	forbidden	-
$1^+ + \pi$	0	$\frac{1}{\sqrt{3}}g^\perp_{\mu_1\nu_1}$
	2	$\sqrt{\frac{3}{2}}(p^\perp_{\mu_1}p^\perp_{\nu_1} - \frac{1}{3}p^2_\perp g^\perp_{\mu_1\nu_1})$
$0^- + \pi$	1	$p^\perp_{\mu_1}$
$1^- + \pi$	1	$\frac{1}{\sqrt{2}}\varepsilon(\mu_1\nu_1 pv)$
$2^- \to 0^+ + \pi$	2	$\sqrt{\frac{3}{2}}p^\perp_{\mu_1}p^\perp_{\mu_2}$
$1^+ + \pi$	2	$p^\perp_{\mu_2}\varepsilon(\mu_1\nu_1 pv)$
$0^- + \pi$	forbidden	-
$1^- + \pi$	1	$\sqrt{\frac{3}{5}}g^\perp_{\mu_1\nu_1}p^\perp_{\mu_2}$
	3	$\sqrt{\frac{5}{2}}\{p^\perp_{\mu_1}p^\perp_{\mu_2}p^\perp_{\nu_1} - \frac{1}{5}(p^2_\perp g^\perp_{\mu_1\mu_2}p^\perp_{\nu_3}+\text{cycl.}(\mu_1\mu_2\nu_1))\}$
$2^- + \pi$	1	$\sqrt{\frac{2}{5}}g^\perp_{\mu_1\nu_1}\varepsilon(\mu_2\nu_2 pv)$
	3	$\sqrt{\frac{2}{5}}(p^\perp_{\mu_1}p^\perp_{\nu_1} - \frac{1}{5}g_{\mu_1\nu_1}p^2_\perp)\varepsilon(\mu_2\nu_2 pv)$

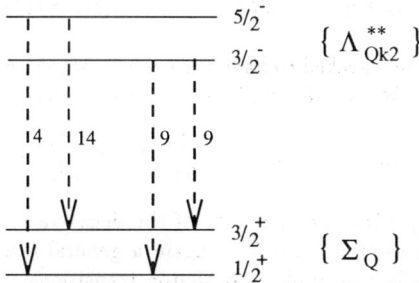

Fig. 2. One-pion transition strengths for the transitions $\{\Lambda^{**}_{QK2}\} \to \{\Sigma_Q\} + \pi$. Degeneracy levels are split for illustrative purposes.

relation for the 6-j symbols one immediately concludes that the pion decay rates from degenerate doublet partners into a singlet state (or vice versa) are equal.[8,9]

Similarily one can calculate the doublet to doublet transition rates for e.g. $\{\Lambda^{**}_{Qk2}\} \rightarrow \{\Sigma_Q\} + \pi$. The rates are in the ratios 4 : 14 : 9 : 9 as represented in Fig. 2. This result can easily be calculated using the 6-j formula Eq.(23) but involves infinitely more labour in the covariant approach. Also, the result "$4 + 14 = 9 + 9$" for doublet to doublet one-pion transitions is a general result which again can easily be derived using the 6-j approach.[8,9]

Photon transitions:

In the photon transition case $j_1^{P_1} \rightarrow j_2^{P_2} + \gamma$ one has to use the field strength tensor $F_{\alpha\beta} = k_\alpha \varepsilon_\beta - k_\beta \varepsilon_\alpha$ or, depending on parity, its dual $\tilde{F}_{\alpha\beta} = \frac{1}{2}\varepsilon_{\alpha\beta\gamma\delta}F^{\gamma\delta}$ in order to guarantee a gauge invariant coupling of the photon to the light side. As in the pion transition case further building blocks for the diquark transition tensor are the metric tensor, the velocity v_α and the photon momentum k_μ. The diquark photon transitions listed in Table 5 are labelled by the total angular momentum J_γ of the photon (spin of the photon plus its orbital angular momentum). The amplitudes are normalized such that the spin summed square of a given diquark transition amplitude f^{J_γ} is $|f^{J_\gamma}|^2 |\vec{k}|^{2J_\gamma+1}$. It is then an easy matter to derive the Heavy Quark Symmetry structure of photon transitions between heavy baryon states using Tables 1, 2 and 5. As an illustration we write down the amplitude for the ground state transition $\{\Sigma_Q\} \rightarrow \Lambda_Q + \gamma$. One has

$$M^\gamma = \bar{u}_2 \left\{ \begin{array}{c} \frac{1}{\sqrt{3}}\gamma_\perp^{\mu_1}\gamma_5 u_1 \\ u_1^{\mu_1} \end{array} \right\} \frac{1}{\sqrt{2}} f^{M1} \tilde{F}_{\alpha\beta} g^\alpha_{\mu_1} v^\beta \tag{24}$$

Using standard $\varepsilon_{\alpha\beta\gamma\delta}$-tensor identities one obtains

$$\Sigma_c \rightarrow \Lambda_c + \gamma : \qquad M^\gamma = i\frac{1}{\sqrt{6}} f^{M1} \bar{u}_2 \slashed{k} \slashed{\varepsilon}^* u_1 \tag{25}$$

$$\Sigma_c^* \rightarrow \Lambda_c + \gamma : \qquad M^\gamma = \frac{1}{\sqrt{2}} f^{M1} \bar{u}_2 \varepsilon(\mu_1 v k \varepsilon^*) u_1^{\mu_1} \tag{26}$$

The transition (26) can be checked to have the correct $M1$ coupling structure. In the degeneracy limit $M_{\Sigma_c^*} = M_{\Sigma_c} = M_1$ one finds the rate expressions

$$\Gamma_{\Sigma_c \rightarrow \Lambda_c + \gamma} = \Gamma_{\Sigma_c^* \rightarrow \Lambda_c + \gamma} = \frac{1}{6\pi} |f^{M1}|^2 \frac{M_2}{M_1} |\vec{k}|^3 \tag{27}$$

where $|\vec{k}| = (M_1^2 - M_2^2)/2M_1$. The equality of the decay rates of heavy quark symmetry partners into the ground state Λ_c is again a general result that can easily be derived in the 6-j formalism as applied to photon transitions. In order to derive the relevant decay formula in the 6-j approach one again compounds the three angular momentum couplings[8,10]

Table 5. Tensor structure of photon couplings to diquark states. Photon is in definite multipole state EJ (electric) MJ (magnetic). Sign of the product of naturalities determines whether coupling is to field strength tensor $F_{\alpha\beta}$ $(n_1 \cdot n_2 = +1)$ or to its dual $\tilde{F}_{\alpha\beta}$ $(n_1 \cdot n_2 = -1)$. Tensor structure of transitions with $(j_1^{P_1}, j_2^{P_2}) \rightarrow (j_1^{-P_1}, j_2^{-P_2}) \rightarrow (j_2^{P_2} j_1^{P_1}) \rightarrow (j_2^{-P_2} j_1^{-P_1})$ are identical and are not always listed separately.

diquark transition $j_1^{P_1} \rightarrow j_2^{P_2} + \gamma$	multipoles	$n_1 n_2$	covariant coupling $t^i_{\mu_1 \ldots \mu_{j_1}; \nu_1 \ldots \nu_{j_2}}$
$0^+ \rightarrow 0^+ + \gamma$	forbidden	$+1$	
$1^+ \rightarrow 0^+ + \gamma$	M1	-1	$\frac{1}{\sqrt{2}} \tilde{F}_{\alpha\beta} g^\alpha_{\mu_1} v^\beta$
$1^+ + \gamma$	M1	$+1$	$\frac{1}{2} F_{\alpha\beta} g^\alpha_{\mu_1} g^\beta_{\nu_1}$
	E2	$+1$	$\frac{1}{2} F_{\alpha\beta}(2k_{\mu_1} g^\alpha_{\nu_1} v^\beta + k \cdot v g^\alpha_{\mu_1} g^\beta_{\nu_1})$
$0^- \rightarrow 0^+ + \gamma$	forbidden	-1	
$1^- \rightarrow 0^+ + \gamma$	E1	$+1$	$\frac{1}{\sqrt{2}} F_{\alpha\beta} g^\alpha_{\mu_1} v^\beta$
$1^+ + \gamma$	E1	-1	$\frac{1}{2} \tilde{F}_{\alpha\beta} g^\alpha_{\mu_1} g^\beta_{\nu_1}$
	M2	-1	$\frac{1}{2} \tilde{F}_{\alpha\beta}(2k_{\mu_1} g^\alpha_{\nu_1} v^\beta + k \cdot v g^\alpha_{\mu_1} g^\beta_{\nu_1})$
$2^- \rightarrow 0^+ + \gamma$	M2	-1	$\tilde{F}_{\alpha\beta} k_{\mu_1} g^\alpha_{\mu_2} v^\beta$
$1^+ + \gamma$	E1	$+1$	$\sqrt{\frac{3}{10}} F_{\alpha\beta} g^\alpha_{\mu_1} g_{\mu_2 \nu_1} v^\beta$
	M2	$+1$	$\sqrt{\frac{1}{6}} F_{\alpha\beta}(v \cdot k g_{\mu_2 \nu_1} g^\alpha_{\mu_1} v^\beta + 2k_{\mu_2} g^\alpha_{\mu_1} g^\beta_{\nu_1})$
	E3	$+1$	$\sqrt{\frac{1}{30}} F_{\alpha\beta}((v \cdot k)^2 g_{\mu_2 \nu_1} g^\alpha_{\mu_1} v^\beta$ $+ \frac{5}{4} v \cdot k k_{\mu_2} g^\alpha_{\mu_1} g^\beta_{\nu_1}$ $+ \frac{15}{4} v^\beta k_{\mu_2}(k_{\nu_1} g^\alpha_{\mu_1} + k_{\mu_1} g^\alpha_{\nu_1}))$
$2^- \rightarrow 0^- + \gamma$	E2	$+1$	$F_{\alpha\beta} k_{\mu_1} g^\alpha_{\mu_2} v^\beta$
$1^- + \gamma$	M1	-1	$\sqrt{\frac{3}{10}} \tilde{F}_{\alpha\beta} g^\alpha_{\mu_1} g_{\mu_2 \nu_1} v^\beta$
	E2	-1	$\sqrt{\frac{1}{6}} \tilde{F}_{\alpha\beta}(v \cdot k g_{\mu_2 \nu_1} g^\alpha_{\mu_1} v^\beta + 2k_{\mu_2} g^\alpha_{\mu_1} g^\beta_{\nu_1})$
	M3	-1	$\sqrt{\frac{1}{30}} \tilde{F}_{\alpha\beta}((v \cdot k)^2 g_{\mu_2 \nu_1} g^\alpha_{\mu_1} v^\beta$ $+ \frac{5}{4} v \cdot k k_{\mu_2} g^\alpha_{\mu_1} g^\beta_{\nu_1}$ $+ \frac{15}{4} v^\beta k_{\mu_2}(k_{\nu_1} g^\alpha_{\mu_1} + k_{\mu_1} g^\alpha_{\nu_1}))$
$2^- + \gamma$	M1	$+1$	$\sqrt{\frac{1}{5}} F_{\alpha\beta} g_{\mu_1 \nu_1} g^\alpha_{\mu_2} g^\beta_{\nu_2}$
	E2	$+1$	$\sqrt{\frac{3}{7}} F_{\alpha\beta} g_{\mu_1 \nu_1}(2k_{\mu_2} v^\beta g^\alpha_{\nu_2} + v \cdot k g^\alpha_{\mu_2} g^\beta_{\nu_2})$
	M3	$+1$	$\sqrt{\frac{3}{10}} F_{\alpha\beta} g^\alpha_{\mu_2} g^\beta_{\nu_2}((v \cdot k)^2 g_{\mu_1 \nu_1} + \frac{5}{2} k_{\mu_1} k_{\nu_1})$
	E4	$+1$	$\sqrt{\frac{1}{14}} F_{\alpha\beta}(2k_{\mu_2} v^\beta g^\alpha_{\nu_2}$ $+ v \cdot k g^\alpha_{\mu_2} g^\beta_{\nu_2})((v \cdot k)^2 g_{\mu_1 \nu_1}$ $+ \frac{7}{2} k_{\mu_1} k_{\nu_1})$

i) $\qquad j_1{}^{P_1} \otimes 1/2^+ \Rightarrow J_1{}^{P_1}$

ii) $\qquad j_2{}^{P_2} \otimes 1/2^+ \Rightarrow J_2{}^{P_2}$

iii) $\qquad J_2{}^{P_2} \otimes J_\gamma \Rightarrow J_1{}^{P_1}$

$$(28)$$

where the notation is identical to the pion case treated before except for the replacement $L_\pi \to J_\gamma$ and where J_γ is the total angular momentum of the photon. The heavy baryon photon transition amplitude may then be written as

$$M^\gamma(J_1 J_1^z \to J_2 J_2^z + J_\gamma m) = M_{J_\gamma}(-1)^{J_\gamma+j_2+\frac{1}{2}+J_1}(2j_1+1)^{\frac{1}{2}}(2J_2+1)^{\frac{1}{2}}$$
$$\begin{Bmatrix} J_\gamma & j_2 & j_1 \\ \frac{1}{2} & J_1 & J_2 \end{Bmatrix} \langle J_\gamma m J_2 J_2^z \mid J_1 J_1^z \rangle \qquad (29)$$

The reduced matrix elements M_{J_γ} correspond to the multipole amplitudes f^{J_γ} as e.g. in Eq.(24). Again, using orthogonality relations for the 6-j symbols, one can deduce that the one-photon rates of doublet partners into singlet states are equal. Similarily there is a sum rule for photon transitions between doublets as discussed for the pion transitions.[8,10]

6. Summary and Conclusion

We have provided a comprehensive set of formulas that allow one to work out the predictions of Heavy Quark Symmetry for current, pion and photon transitions in the baryon sector. We have chosen to present the material in a form which emphasizes the similarities between the three different types of transitions. The formulation is general and easily extends to transitions involving higher orbital excitations. The coupling of the various angular momentum involved in the transitions has been done using conventional covariant techniques, and, in the case of pion and photon transitions, also by 6-j coupling methods. Although we have chosen to express our results for the pion and photon transitions in terms of transition amplitudes the formulas can easily be transcribed to the language of chiral and gauge invariant[9,10] effective Lagrangians. In conclusion one may state that we are certainly looking forward to analyze the forthcoming wealth of data on heavy baryon decays to see how Heavy Quark Symmetry is at work.

1. N.Isgur and M.B.Wise, Phys.Lett. **B232** (1989) 113; **B237** (1990) 527; E.Eichten and B.Hill, Phys.Lett. **B234** (1990) 511; B.Grinstein, Nucl.Phys. **B339** (1990) 253; H.Georgi, Phys.Lett. **B240** (1990) 447; F.Hussain, J.G.Körner, K.Schilcher, G.Thompson and J.L.Wu, Phys.Lett. **B249** (1990) 295
2. A.Falk, H.Georgi, B.Grinstein and M.B.Wise, Nucl.Phys. **B343** (1990) 1; T.Mannel, W.Roberts and Z.Ryzak, Nucl.Phys. **B368** (1992) 204

3. J.G.Körner and G.Thompson, Phys.Lett. **264** (1991) 185
4. S.Balk, J.G.Körner and D.Pirjol, Nucl.Phys. **B428** (1994) 499
5. A.de Rujula, H.Georgi and S.L.Glashow, Phys.Rev.Lett. **37** (1976) 785;
   J.G.Körner and M.Kuroda, Phys.Lett. **B67** (1977) 455;
   P.M.Stevenson, Phys.Rev. **D18** (1978) 4063
6. L.A.Copley, N.Isgur and G.Karl, Phys.Rev. **D20** (1979) 768
7. J.D.Bjørken, I.Dunietz and J.Taron, Nucl.Phys. **B371** (1992) 111
8. J.G.Körner, M.Krämer and D.Pirjol,
   Progr.Part.Nucl.Phys., **Vol33** (1994) 787
9. A.Ilakovac, U.Kilian, J.G.Körner and J.Landgraf, to be published
10. J.G.Körner and J.Landgraf, to be published

GLUONIC EXCITATIONS IN HEAVY MESONS AND
THEIR DECAYS BY FLUX-TUBE BREAKING

PHILIP PAGE*

Theoretical Physics, University of Oxford, 1 Keble Road, Oxford OX1 3NP, UK

ABSTRACT

We have examined decays of low-lying gluonic excitations of mesons (hybrids) by chromoelectric flux-tube breaking. An analytical calculation of non-relativistic flux-tube model decay amplitudes is performed in an harmonic oscillator approximation. Specific decay signatures of all J^{PC} charmonium hybrids are identified and the widths predicted. We introduce a new selection rule which can be used to understand the systematics of numerical decay calculations.

1. Mesons with an excited gluonic field

We define *charmonium hybrids* as charm-anticharm bound systems (mesons) with an excitation of the gluonic degree of freedom. Unless otherwise stated, we just refer to these systems as *"hybrids"* from now on. They represent new confined states of QCD beyond the quark model, and hence an *important* experimental test of QCD. Interesting features of hybrids are :

- Their uniqueness as a bound systems with both "valent" fermions and bosons.
- Hybrids often have *exotic quantum numbers* not found in the quark model, e.g. $J^{PC} = 0^{+-}, 1^{-+}$ and 2^{+-} for the lowest lying hybrids in the flux-tube model[1], which facilitate easier experimental detection of non-quark model states.
- Hybrids are expected to have *masses* of 4.1 ± 0.4 GeV (literature average)[†]
- Some hybrids are believed to be very *stable* (i.e. having small widths). For hybrids below the $D^{**}D$ threshold the flux tube model predicts very small widths (decays into DD, D^*D and D^*D^* are almost forbidden). Hence we are specifically interested in hybrid decay widths to $D^{**}D$ above threshold.

The reason for our interest in *charmonium* hybrids derives from the expectation[1,2] that their masses are better defined than is the case for their light quark counterparts and, due to the smaller amount of phase space available in the corresponding decay channels, their widths are smaller. Bottomonium hybrids are expected to be more difficult to produce than charmonium hybrids. Experimentally, the situation is

---

*Based on work done in collaboration with Frank Close[5].

[†]A mass of ≈ 4.3 GeV has been estimated with a numerical simulation in the flux-tube model[7]. The fact that these masses are well-defined means that we can make firm experimental predictions.

(i) *Light quark hybrids* : There has recently been a claim of a broad $J^{PC} = 1^{-+}$ exotic resonance with isospin one by the AGS at Brookhaven[3], with mass in the range 1.6 – 2.2 GeV. They studied pion-nucleus inelastic scattering, and found evidence for the above resonance decaying into $f_1(1285)\pi^- \to K^+\pi^0\pi^-$, a process which is predicted to be important (with a width of 50 MeV) in the flux tube model[2]. The current data suffers from low statistics, making its discovery an ambiguous claim and its existence in need of independent corroboration.

(ii) *Charmonium hybrids* : Beijing e^+e^- annihilation experiments may soon have high enough luminosity to produce $J^{PC} = 1^{--}$ hybrids above the DD threshold.

(iii) *Bottomonium hybrids* : Detection in e^+e^- annihilation of a $J^{PC} = 1^{--}$ hybrid above the BB threshold at the SLAC B-factory is a possibility.

2. The flux-tube model decay amplitude for *hybrid* $\to D^{**}D$

In the Isgur-Paton non-relativistic flux-tube model of QCD[1] the *gluonic field* of a hybrid is represented by *beads*, which are connected to each other and the quarks at the ends via a non-relativistic string. The gluonic field is excited, giving rise to an excited adiabatic potential between the quarks, which enables the study of hybrids without further experimental input.

The decay amplitude[2,4] of $A \to BC$ by pair creation is similar to the $^3P_0$-model amplitude (where A is the initial meson decaying into mesons B and C). There is an additional *overlap* of the string of A to break at the pair creation position into the strings of B and C, though. The model also predicts an *overlap* for *hybrid* $\to BC$. This prohibits pair creation on the hybrid q$\bar{q}$-axis.

We perform an *analytical calculation*[5] of the decay amplitude *hybrid* $\to D^{**}D$ by assuming the outgoing D^{**} and D wave functions to be L=1 and L=0 S.H.O. wave functions with inverse radii $\beta_{D^{**}}$ and β_D respectively. The initial hybrid wave function is proportional to $r^\delta D(\Omega) \exp(-\beta_{hybrid}^2 r^2/2)$, where $D(\Omega)$ is a Wigner rotation function and $0 < \delta < 1$.

The calculation is performed for the case $\beta_{D^{**}} = \beta_D \equiv \beta$. The main reason for this simplification is that *(a)* the D^{**}, D^* and D are expected to have similar β's, and that *(b)* the systematics of earlier numerical calculations for light quarks can be understood in this limit.

The only free parameter in the model is the *lattice size* (or longitudinal distance between beads), which is related to the overall normalization of decays[4]. All other parameters have previously been estimated in the context of the model[1].

For decays of *ordinary mesons into ordinary mesons*, it is generally the case that the $^3P_0$ and the flux-tube models coincide in the limit where the string tension vanishes. It is possible to make a stronger statement : *The $^3P_0$ and flux-tube models coincide when $\beta_B = \beta_C$ even if the string tension is non-zero*[5]. This result at least holds for two final state L=0 mesons *and* for final state L=0 and L=1 mesons. This also explains the systematics of earlier numerical calculations[4].

For decays of *hybrid mesons into ordinary mesons*, the flux-tube model predictions are much more distinctive. When $\beta_B = \beta_C$ the hybrid decay width to two L=0 mesons

336

is zero (i.e. $hybrid \to DD, D^*D$, and D^*D^* is forbidden). Hence the first non-zero decay width is to L=1 and L=0 mesons. This explains our interest in $hybrid \to D^{**}D$.

When $\beta_B = \beta_C$ there is also an important *selection rule* operating in the moving frame of the initial $q\bar{q}$-pair for $hybrid \to D^{**}D$: The one unit of angular momentum of the hybrid around the $q\bar{q}$-axis is exactly absorbed by the component of the outgoing D^{**} angular momentum along the $q\bar{q}$-axis.

The dominant decay modes of $hybrid \to D^{**}D$ are displayed in Table 1. The eight lowest lying hybrids in the model are assumed to have masses of 4.30 GeV or 4.35 GeV (and in addition a small hyperfine splitting[6], which affects phase space appreciably). They lie around the $D^{**}D$ threshold of ≈ 4.3 GeV. The magnitudes of the decays are normalized to ordinary meson decays[4]. All resonances are approximated to be narrow. We assume the D, D^{**}_{2++}, D^{**}_{1+L} (low mass), D^{**}_{0++} and D^{**}_{1+H} (high mass) have masses 1.87, 2.46, 2.42, 2.40 and 2.45 GeV, respectively. Also $\beta_{hybrid} = 0.35$ GeV, $\beta = 0.37$ GeV, $\delta = 0.62$ and the $D^{**}_{1+L} / D^{**}_{1+H}$ mixing is $41°$.

Table 1. Dominant widths in MeV for $hybrid\ (c\bar{c}g) \to D^{**}D$ for various J^{PC} in partial wave L, both for 4.30 GeV hybrids (Γ_1) and 4.35 GeV hybrids (Γ_2).

$c\bar{c}g$	D^{**}	L	Γ_1	Γ_2	$c\bar{c}g$	D^{**}	L	Γ_1	Γ_2	$c\bar{c}g$	D^{**}	L	Γ_1	Γ_2
2^{-+}	2^{++}	S	0	120	2^{+-}	2^{++}	P	0	40	1^{++}	2^{++}	P	0	40
	1^{+L}	D	.02	10		1^{+L}	P	.3	10		1^{+L}	P	6	70
	0^{++}	D	.9	8		1^{+H}	P	0	40		1^{+H}	P	0	50
1^{-+}	1^{+L}	S	20	30	1^{+-}	2^{++}	P	0	20	0^{+-}	1^{+L}	P	30	150
		D	1	40		1^{+L}	P	20	120		1^{+H}	P	0	90
	1^{+H}	S	0	300		0^{++}	P	50	140	1^{--}	1^{+L}	S	60	150
0^{-+}	0^{++}	S	-	400		1^{+H}	P	0	30		1^{+H}	S	0	170

3. Conclusions and Acknowledgements

The flux-tube model predicts a surprising *stability of gluonic excitations* in heavy mesons. Above the $D^{**}D$ threshold hybrids decay preferentially into $D^{**}D$ (with a width that is driven by the available phase space), instead of DD, D^*D, D^*D^*, perhaps explaining why hybrids have not yet been found experimentally.

I would like to thank Ted Barnes and Jack Paton for consultation.

4. References

1. N. Isgur, J. Paton, *Phys. Rev.* **D31** (1985) 2910.
2. N. Isgur, R. Kokoski and J. Paton, *Phys. Rev. Lett.* **54** (1985) 869.
3. J.H. Lee *et al., Phys. Lett.* **B323** (1994) 227.
4. R. Kokoski, N. Isgur, *Phys. Rev.* **D35** (1987) 907.
5. F.E. Close, P.R. Page, *in preperation*.
6. J. Merlin, J. Paton, *Phys. Rev.* **D35** (1987) 1668.
7. T. Barnes, F.E. Close, E. Swanson, *private communication*.

HEAVY MESON DYNAMICS
IN A QCD RELATIVISTIC POTENTIAL MODEL

FULVIA DE FAZIO

Dipartimento di Fisica dell'Università di Bari,
Istituto Nazionale di Fisica Nucleare, Sezione di Bari,
Via Amendola 173, 70126, Bari, Italy

ABSTRACT

We use a QCD relativistic potential model to compute the strong coupling constant g appearing in the effective Lagrangian which describes the interaction of 0^- and 1^- $\bar{q}Q$ states with soft pions in the limit $m_Q \to \infty$. We compare our results with other approaches; in particular, in the non relativistic limit, we are able to reproduce the constituent quark model result: $g = 1$, while the inclusion of relativistic effects due to the light quark gives $g = \frac{1}{3}$, in agreement with QCD sum rules. We also estimate heavy meson radiative decay rates, with results in agreement with available experimental data.

1. The Strong Coupling Constant $g_{D^* D\pi}$

The decay $D^{*+} \to D^0 \pi^+$ is described in terms of a strong coupling constant $g_{D^* D\pi}$ defined by: $< D^0(k)\pi^+(q)|D^{*+}(p, \epsilon) > = g_{D^* D\pi} \, \epsilon^\mu q_\mu$. CLEO collaboration measurement [1]: $BR(D^{*+} \to D^0 \pi^+) \simeq 68.1 \pm 1.0 \pm 1.3\%$ and the upper bound[2] : $\Gamma(D^{*+}) < 131 \, KeV$ provide us with the constraint : $g_{D^* D\pi} < 20.6$.

The interest in the evaluation of this coupling constant is manifold. The form factor $F_1(q^2)$ describing the semileptonic decay $B \to \pi \ell \nu$ is believed to be dominated by the B^* pole, so that its value at maximum transferred momentum is proportional to $g_{B^* B\pi}$, which is related to $g_{D^* D\pi}$ by [3]: $g_{P^* P\pi} = 2 \, m_P \, g/f_\pi$, where $P(P^*)$ is a 0^- (1^-) heavy meson of mass m_P, and g is independent of m_P. Besides, g appears in the effective Lagrangian describing the interaction between heavy mesons and light Nambu-Goldstone bosons [4,5,6].

In non relativistic quark models [6,7] $g \simeq 1$, while recent QCD sum rules [8] and HQET [9,10] analyses give: $g \simeq 0.2 - 0.4$. We wish to show that the inclusion of the relativistic effects in the bound state can lower the value $g = 1$, providing an explanation of the discrepancy between the different approaches.

We shall obtain g in the framework of a QCD relativistic potential model [11]. In this model the $\bar{q}Q$ heavy states D and D^* are described in terms of the creation operators of the constituent quarks and of a meson wave function ψ, normalized according to: $\frac{1}{(2\pi)^3} \int d\vec{k}|\psi|^2 = 2\sqrt{m_D^2 + \vec{p}^2}$, where $\vec{p}$ is the meson momentum and $\vec{k}$ is the quark relative momentum. ψ satisfies a Salpeter equation*which includes

---

*This equation arises from the bound-state Bethe Salpeter equation by considering the instantaneous approximation and restricting the Fock space to the $\bar{q}Q$ pairs [12].

relativistic effects in the quark kinematics. In this equation the interquark potential V coincides, in the meson rest frame, with the Richardson potential, with a linear behaviour for large distances, reproducing QCD confinement, and a coulombic shape for small distances, with the further assumption that V is constant near the origin in order to avoid unphysical singularities. The values of the quark masses, obtained by fits to meson masses, are: $m_u = m_d = 38 \; MeV$, $m_s = 115 \; MeV$, $m_c = 1452 \; MeV$, $m_b = 4890 \; MeV$.

To evaluate $g_{D^* D \pi}$ let us consider the matrix element of the axial current A_μ between the states D^* and D:

$$< D^0(k)|A_\mu|D^{*+}(p, \epsilon) > = -i\{ \; \epsilon_\mu \; (m_{D^*} + m_D)A_1(q^2) - \frac{\epsilon \cdot q}{m_D + m_{D^*}}(p+k)_\mu A_2(q^2)$$
$$- \frac{\epsilon \cdot q}{q^2}2m_{D^*}q_\mu[A_3(q^2) - A_0(q^2)]\} , \qquad (1)$$

where $2m_{D^*}A_3 = (m_D + m_{D^*})A_1 + (m_{D^*} - m_D)A_2 \quad (q = p - k)$.

Taking the derivative of A_μ, we can link the l.h.s. of Eq. (1) to the matrix element of the pseudoscalar current between the same states, which is supposed to be dominated by the π^+ pole. Performing the overlap of the states D, D^* and the axial current, we have in the chiral limit: $g_{D^* D \pi} = \frac{2m_{D^*}}{f_\pi} A_0(0)$, and finally:

$$g = A_0(0) = \frac{1}{4m_D} \int_0^\infty dk |\tilde{u}(k)|^2 \frac{E_q + m_q}{E_q} \left[1 - \frac{k^2}{3(E_q + m_q)^2} \right] , \qquad (2)$$

where $E_q = \sqrt{k^2 + m_q^2}$ and $\tilde{u}(k) = \frac{k \, \psi(k)}{\sqrt{2\pi}}$. The wave functions $\tilde{u}(k)$ come from a numerical solution of the Salpeter equation. We obtain [11]: $A_0(0) = 0.4$ (D case); $A_0(0) = 0.39$ (B case), giving so: $g_{D^* D \pi} = 12.3$ and $g_{B^* B \pi} = 31.7$. We can observe that we have only 2% deviation from the scaling result $g_{D^* D \pi}/g_{B^* B \pi} = m_D/m_B$.

It is interesting to notice that, in the non relativistic limit, i.e. $E_q \simeq m_q \gg k$, we obtain: $g = \frac{1}{2m_D} \int_0^\infty dk |\tilde{u}(k)|^2 = 1$, reproducing the constituent quark model result. On the other hand, in the limit $m_q \to 0$, $m_Q \to \infty$, the result is: $g = 1/3$, in agreement with the QCD sum rules determination.

2. Radiative Heavy Meson Decays

The evaluation of radiative decay rates involves the knowledge of the matrix element of the electromagnetic current $J_\mu^{e.m.}$ between the states D^* and D:

$$< D^+(k)|J_\mu^{e.m.}|D^{*+}(p, \epsilon) > = \left(\frac{e_Q}{\Lambda_Q} + \frac{e_q}{\Lambda_q} \right) \epsilon_{\mu\nu\alpha\beta} \; \epsilon^\nu k^\alpha p^\beta , \qquad (3)$$

where e_Q (e_q) is the heavy (light) quark electric charge. In the framework of the relativistic QCD model we are able to evaluate the "effective" masses Λ_q, Λ_Q by computing the overlap of meson states and the electromagnetic current. We find

Table 1. Heavy meson radiative decay rates

Decay rate/ BR	theory	experiment
$\Gamma(D^{*+})$	46.21 KeV	$< 131 \ KeV$
$BR(D^{*+} \to D^+\pi^0)$	31.3%	$30.8 \pm 0.4 \pm 0.8\%$
$BR(D^{*+} \to D^0\pi^+)$	67.7%	$68.1 \pm 1.0 \pm 1.3\%$
$BR(D^{*+} \to D^+\gamma)$	1.0%	$1.1 \pm 1.4 \pm 1.6\%$
$\Gamma(D^{*0})$	41.6 KeV	
$BR(D^{*0} \to D^0\pi^0)$	50.0%	$63.6 \pm 2.3 \pm 3.3\%$
$BR(D^{*0} \to D^0\gamma)$	50.0%	$36.4 \pm 2.3 \pm 3.3\%$
$\Gamma(D_s^*) = \Gamma(D_s^* \to D_s\gamma)$	0.382 KeV	
$\Gamma(B^{*-}) = \Gamma(B^{*-} \to B^-\gamma)$	0.243 KeV	
$\Gamma(B^{*0}) = \Gamma(B^{*0} \to B^0\gamma)$	$9.2 \ 10^{-2} \ KeV$	
$\Gamma(B_s^*) = \Gamma(B_s^* \to B_s\gamma)$	$8.0 \ 10^{-2} \ KeV$	

$\Lambda_c \simeq m_c = 1.57 \ GeV$, $\Lambda_b \simeq m_b = 4.95 \ GeV$ and $\Lambda_q = 0.48 \ GeV \gg m_q$. The B.R. are reported in Table I [11] together with the available experimental data.

We conclude that the inclusion of relativistic effects in a QCD potential model has allowed us to explain the discrepancy among different models in the evaluation of the heavy meson coupling constant with soft pions. Moreover, the same model, applied to heavy meson radiative transitions, gives results in agreement with experimental data.

I thank P. Colangelo and G. Nardulli for their precious collaboration.

3. References

1. CLEO Collaboration, F. Butler et al., *Phys.Rev.Lett.* **69** (1992) 2041.
2. ACCMOR Collaboration, S. Barlag et al., *Phys.Lett.* **B278** (1992) 480.
3. T. N. Pham, *Phys. Rev.* **D25** (1982) 2955;
   S. Nussinov and W. Wetzel, *Phys.Rev.* **D36** (1987) 139.
4. M. B. Wise, *Phys. Rev.* **D45** (1992) R2188.
5. G. Burdman and J. F. Donoghue, *Phys. Lett.* **B280** (1992) 287.
6. T.-M. Yan et al., *Phys.Rev.* **D46** (1992) 1148.
7. N. Isgur and M. B. Wise, *Phys.Rev.* **D41** (1990) 151.
8. P. Colangelo, A. Deandrea, N. Di Bartolomeo, F. Feruglio, R. Gatto and G. Nardulli, preprint UGVA-DPT 1994/06-856, BARI-TH/94-171.
9. R. Casalbuoni et al., *Phys.Lett.* **B299** (1993) 139.
10. P. Colangelo, F. De Fazio and G. Nardulli, *Phys.Lett.* **B316** (1993) 555.
11. P. Colangelo, F. De Fazio and G. Nardulli, *Phys.Lett.* **B334** (1994) 175.
12. P. Colangelo, G. Nardulli and M. Pietroni, *Phys.Rev.* **D43** (1991) 3002.

HADRONS OF ARBITRARY SPIN

AND

HEAVY QUARK SYMMETRY

F. HUSSAIN and G. THOMPSON

International Centre for Theoretical Physics
Trieste, 34100, Italy

ABSTRACT

In this short talk we illustrate the use of the LSZ reduction theorem and interpolating fields in the heavy quark effective theory to investigate the structure of the Bethe-Salpeter amplitude for heavy hadrons in the Heavy Quark Effective Theory. We show how a simple form of this amplitude, for arbitrary meson and baryon resonances, used extensively in heavy hadron decay calculations, follows naturally upto $O(1/M)$ from these field theoretic considerations.

1. Interpolating Fields and Heavy Quark Mass Limit

In field theory a variety of interpolating fields can be used to represent a bound state [1]. Consider as an example a vector (1^-) heavy meson with $Q\bar{q}$ quantum numbers. Using the equations of motion, $(i\not{D} - m)\psi = 0$, we can show [2] that the following two interpolating fields, for a vector meson, with momentum P and mass M,

$$\phi_\mu^1(x) = Z_3^{1/2}\frac{1}{V_1}T\bar{\psi}_q(x)\gamma_\mu^\perp\psi_Q(x),$$

$$\phi_\mu^2(x) = -Z_3^{1/2}\frac{1}{V_2}\frac{i\partial_\nu}{M}T\bar{\psi}_q(x)\gamma^\nu\gamma_\mu^\perp\psi_Q(x) \tag{1}$$

are equal in the heavy quark limit, $m_Q \to \infty$, $m_Q + m_q = M$, alongwith the important relation between the normalisation constants $V_1 = V_2$. The neglected term is proportional to $\bar{\psi}_q\frac{\vec{D}_\mu^\perp}{M}\psi_Q$. Here $\gamma_\mu^\perp = \gamma_\mu - \not{v}v_\mu$ and $v = \frac{P}{M}$. The superscript $\perp$ on a vector $A_\mu^\perp$ indicates that it is orthogonal to the velocity, v, of the heavy meson, i.e. $v.A^\perp = 0$.

This result turns out to be a general feature of the heavy quark limit. In fact one can show that given a particular interpolating field $N_1T\bar{\psi}_q(x)\Gamma\psi_Q$ for an arbitrary resonance, it is equal to the interpolating field $-N_2\frac{i\partial_\nu}{M}T\bar{\psi}_q(x)\gamma^\nu\Gamma\psi_Q$ upto terms of $O(\vec{D}^\perp/M)$, alongwith the relation between the normalisation constants, $N_1 = N_2$, in the heavy quark limit mass limit. Here Γ is some Dirac matrix, possibly with derivatives. The terms neglected are exactly the kind of terms dropped in deriving the heavy quark effective theory (HQET) [3] at leading order. Physically this means that we consider all momenta transverse to the line of flight of the meson to be small.

2. Consequences for Bethe-Salpeter Amplitudes

We now discuss the consequences of the equality of interpolating fields for matrix elements. Consider the Bethe-Salpeter (B-S) amplitude for heavy vector meson with momentum P

$$M_\alpha{}^\beta(x_1, x_2) = \langle 0|T\psi_{Q\alpha}(x_1)G(x_1, x_2)\bar{\psi}^\beta(x_2)|P\rangle. \tag{2}$$

Here $G(x_1, x_2)$ is some colour matrix to make the Bethe-Salpeter amplitude gauge invariant. Now use the reduction theorem to write the B-S amplitude as

$$M_\alpha{}^\beta(x_1, x_2)$$
$$= \lim_{P^2 \to M^2}(P^2 - M^2)\int d^4y\, e^{-iP\cdot y}\langle 0|T\psi_{Q\alpha}(x_1)G(x_1, x_2)\bar{\psi}^\beta(x_2)\phi^\dagger(y)|0\rangle. \tag{3}$$

where $\phi^\dagger = \epsilon^\mu \phi_\mu^\dagger$ with $\phi_\mu = \phi_\mu^1$ or ϕ_μ^2.

When one uses the second of the interpolating fields, we can integrate by parts and shift the derivative in the interpolating fields onto the exponential with the result that $\frac{i\partial_\mu}{M}$ becomes ψ. It is then easy to see that the requirement that the use of any one of the alternative interpolating fields should give the same answers in the reduction theorem, alongwith the heavy quark limit, leads to the condition

$$\begin{aligned}\bar{\psi}_Q(y)\psi &= \bar{\psi}_Q \\ \psi\psi(y) &= -\psi(y)\end{aligned} \tag{4}$$

within the reduction formula, in the mass shell limit. In other words, in the rest frame of the on-shell heavy *meson*, only the quark part of the heavy quark field and only the antiquark part of the light quark field contribute to the matrix element. It is obvious that this physical requirement will be enforced by taking the interpolating fields for the heavy vector meson to be

$$\phi_\mu(x) = \frac{1}{2}(\phi_\mu^1(x) + \phi_\mu^2(x)). \tag{5}$$

Thus in Eq. (3), the interpolating field is now to be understood as the above combination.

As is well known by now [4], one can go to the zeroth order HQET by simply replacing the QCD heavy fields by the zeroth order heavy quark effective fields. Thus we shall consider the heavy quark fields, ψ_q, to be the corresponding fields appearing in the effective theory. One can now decouple these fields [4] through the transformation

$$\psi_Q(x) = W\left[\begin{matrix}x\\v\end{matrix}\right]Q(x), \tag{6}$$

where

$$W\left[\begin{matrix}x\\v\end{matrix}\right] = Pexp\left[ig\int_{-\infty}^{v\cdot x}ds\, A\cdot v\right] \tag{7}$$

is a path-ordered exponential, wherein the path is a straight line from $-\infty$ to x along the v-direction.

Because the transformed fields, Q and $\bar{Q}$ are now decoupled from the gluons, the right hand side of eq. (3) factorises and we can now use the heavy quark propagator

$$\langle 0|Q_\alpha(x_1)\bar{Q}^\gamma(y)|0\rangle = \int d^4p \left[\frac{e^{-ip\cdot(x_1-y)}}{\not{v}\cdot p - m_Q}\right]^\gamma_\alpha .$$

(8)

Putting everything together we arrive at the final form of the B-S amplitude for the heavy meson

$$M_\alpha{}^\beta(x_1,x_2) = \chi_\alpha{}^\delta(v)A_\delta{}^\beta(x_1,x_2),$$

(9)

with

$$A_\delta{}^\beta(x_1,x_2) = \lim_{P^2\to M^2} Z_3^{1/2}\frac{(P^2-M^2)}{V_1} .$$

$$\int d^4y d^4p e^{-iP\cdot y}\frac{e^{-ip\cdot(x_1-y)}}{v\cdot p - m_Q}\langle 0|\bar{\psi}_q^\beta(x_2)W\begin{bmatrix}x_1\\v\end{bmatrix}G(x_1,x_2)W\begin{bmatrix}y\\v\end{bmatrix}^{-1}\psi_{q\delta}(y)|0\rangle .$$

(10)

with

$$\chi_\alpha{}^\delta(v) = \frac{1}{2}[(1+\not{v})\not{q}]_\alpha{}^\delta .$$

(11)

The B-S amplitude in this form now satisfies the Bargmann-Wigner equation on the 'heavy' label α leading to all the nice results of the heavy quark symmetry.

It is easy now to generalise this result to the pseudoscalar case and to arbitrary orbital resonances and also to heavy baryons of arbitrary spin.

3. Acknowledgements

We would like to thank John Strathdee for very useful discussions.

4. References

1. D. Lurie, *Particles and Fields*, Interscience Publishers, John Wiley and Sons, Inc., N.Y., 1968.
2. F. Hussain and G. Thompson, *Phys. Lett.* **B355** (1994) 205.
3. E. Eichten and F. Feinberg, *Phys. Rev.* **D23** (1981) 2724; G.P. Lepage, B.A. Thacker, *Nucl. Phys.* **B** (Proc. Suppl.) (1988) 199; E. Eichten , ibid., 170; E. Eichten and F. Feinberg, *Phys. Rev. Lett.* **43** (9179) 1205; W.E. Caswell and G.P. Lepage, *Phys. Lett.* **167B** (1986) 437; M.B. Voloshin and M.A. Shifman, *Sov. J. Nucl. Phys.* **47** (3) (1988) 511; H. Georgi, *Phys. Lett.* **B240** (1990) 447.
4. J.G. Körner and G. Thompson, *Phys. Lett.* **B264** (1991) 185.

THE QUARK-RESONANCE MODEL AND THE ANOMALOUS
MAGNETIC MOMENT OF THE MUON

ELISABETTA PALLANTE

I.N.F.N., Laboratori Nazionali di Frascati, Via E. Fermi,
00044 Frascati (Rome) ITALY.

ABSTRACT

An effective model for low energy hadronic interactions is formulated arising form the bosonization of a Nambu-Jona Lasinio type Lagrangian including all chiral invariant multiquark effective interactions. The bosonized Lagrangian includes non-renormalizable quark-meson vertices which are next-to-leading in the low energy inverse cutoff expansion. Next-to-leading corrections and gluonic corrections in the present framework are calculated for the specific case of the hadronic vacuum polarization contribution to the anomalous magnetic moment of the muon.

1. Introduction

In the framework of effective fermion models *à la* Nambu-Jona Lasinio for low energy hadronic interactions, the Quark-Resonance model[1] can be thought of as a generalized NJL model (see[2] for a recent formulation). While the original NJL model only includes the lowest dimensional non-renormalizable four fermion interaction terms, the Quark-Resonance (QR) Lagrangian results from the bosonization of the infinite tower of chiral invariant multiquark effective interactions ordered by an expansion in inverse powers of the ultraviolet cutoff Λ_χ ($\simeq 1$ GeV). The addition of higher dimensional multifermion interactions with increasing powers of derivatives takes into account in a perturbative way the nonlocality of the effective low energy action.

In section 2 the quark-resonance Lagrangian is constructed up to including $1/\Lambda_\chi^2$ terms, which are next-to-leading in the inverse cutoff expansion and leading in the $1/N_c$ expansion. All meson SU(3) flavour octet quantum numbers are included: the pseudoscalar mesons π, K, η, the vector, axial, scalar and pseudoscalar resonances. Two parameters of the leading vector resonance chiral effective Lagrangian are derived, including $(Q^2/\Lambda_\chi^2)\ln(\Lambda_\chi^2/Q^2)$ corrections to the leading logarithmic ENJL contribution: the coupling of the vector resonance to the external vector current and the vector mass. They enter the calculation of the hadronic vacuum polarization contribution to the muon $g - 2$, which is studied in section 3.

2. The quark-resonance model

The effective action of the quark-resonance model, coming from the bosonization of the most general NJL model, in the *constituent* quark base and in the presence of external vector, axial, scalar and pseudoscalar sources is given by[1]:

$$e^{-\Gamma_{eff}[R;v,a,s,p]} = \frac{1}{Z}\int \mathcal{D}G_\mu \exp\left(-\int d^4x \frac{1}{4}G^{(a)}_{\mu\nu}G^{(a)\mu\nu}\right)e^{-f[R]}\int \mathcal{D}Q\mathcal{D}\bar{Q}$$

$$\exp\left[\int d^4x\left(\bar{Q}\gamma^\mu(\partial_\mu + iG_\mu)Q + \sum_0^\infty \left(\frac{1}{\Lambda_\chi}\right)^n \bar{Q}RQ\right)\right], \quad (1)$$

where the functional $f[R]$ contains the terms with auxiliary boson fields which are not coupled to fermions. The most general structure of the R operator can be represented by: $R = \beta(\Lambda_\chi)\times\{\gamma_{Dirac}\}\times\{W_\mu^+, W_\mu^-, \tilde{H}\}\times\{\nabla_\mu^n, (\nabla_\mu^{CT})^n\}$, where the generic coupling $\beta(\Lambda_\chi)$ is not deducible from symmetry principles. ∇_μ and ∇_μ^{CT} are the covariant derivatives of the *constituent* quarks which contain the vector and axial currents with the pseudoscalar meson field. The set $\{W_\mu^+, W_\mu^-, \tilde{H}\}$ contains all possible fields introduced by the bosonization which can couple to the *constituent* quark bilinears and which can be identified with the physical degrees of freedom of the low energy effective theory: the vector field W_μ^+, the axial-vector field W_μ^- and the scalar field $\tilde{H}$. The QR Lagrangian at leading order in the $1/\Lambda_\chi$ expansion and in the $1/N_c$ expansion, in the *constituent* quark base, coincides with the bosonization of the Extended NJL model[2]. Higher order terms up to $1/\Lambda_\chi^2$ are all the quark-meson bilinears which are locally chiral invariant and with some additional *caveats* which guarantee the equivalence between the *current* and the *constituent* quark-meson Lagrangian. At order $\frac{1}{\Lambda_\chi}$ there are no invariants. At order $\frac{1}{\Lambda_\chi^2}$ we distinguish four classes according to the types of interactions among resonances: the totally derivative terms, the Vector set, the Scalar set and the mixed Scalar-Vector set. The latter contributes to the vector Lagrangian parameters proportionally to M_Q, where M_Q is the VEV of the scalar field $\tilde{H}$. Corrections of order M_Q^2/Λ_χ^2 can affect the very low energy behaviour of the 2-point vector Green's function and can be relevant for the hadronic vacuum polarization contribution to the muon $g - 2$.

$1/\Lambda_\chi^2$ terms generate two types of corrections to the parameters of the effective meson Lagrangian:

- next-to-leading power corrections to the leading logarithms (NPLL): (Q^2/Λ_χ^2) $\ln(\Lambda_\chi^2/Q^2)$,

- next-to-leading power corrections: $Q^2/\Lambda_\chi^2 \cdot 1$.

It is crucial for the predictivity of the model the fact that NPLL corrections arise always from a finite class of $1/\Lambda_\chi^2$ terms, while genuine power corrections arise from an infinite tower of higher dimensional terms. NPLL corrections have been explicitly derived[1] for the two-point vector Green's function[1,3] $\Pi_V^1(Q^2) = f_V^2(Q^2)M_V^2(Q^2)/Q^2 + M_V^2(Q^2)$, written in terms of the running vector-photon coupling $f_V(Q^2)$ and the running vector mass $M_V(Q^2)$. The two parameters can be parametrized in terms of two new coefficients up to NPLL order: $\beta_\Gamma^1, \beta_V^1$. Their numerical value has been determined via a fit[1] of the derivative of the 2-point vector Green's function with a modelization of the experimental data on $e^+e^- \rightarrow hadrons$ in the isovector channel $(I = 1, J = 1)$ and in the intermediate energy region $0.5 \leq Q \leq 0.9$ GeV. The best values of the two free coefficients are $\beta_\Gamma^1 = -0.75 \pm 0.01$, $\beta_V^1 = -0.79 \pm 0.01$. The difference between the leading order prediction and the next-to-leading order (i.e. including up to NPLL corrections) prediction of the vector 2-point function[1] reaches a 30% at 0.7 GeV.

3. The hadronic vacuum polarization contribution to the muon $g - 2$

The quark-resonance model is a good tool to give a theoretical estimate of the hadronic vacuum polarization contribution to the muon $g - 2$ and to understand the size of higher order corrections to the leading order prediction. The interest to the calculation of this quantity is strongly motivated by the necessity of reaching a highly accurate theoretical prediction of it. The most recent numerical estimates, obtained from the best fit of the $e^+e^- \rightarrow hadrons$ total cross section, give the following values[4,5,6]: $7.07(.066)(.17) \cdot 10^{-8}, 6.84(.11) \cdot 10^{-8}, 7.100(.105)(.49) \cdot 10^{-8}$, where the first error is statistical and the second one is systematic. a_μ^h is related to the renormalized hadronic photon self-energy $\Pi_R^h(Q^2)$ through the following integral[7,8]

$$a_\mu^h = \frac{\alpha}{\pi} \int_0^1 dx(1 - x) \left[-e^2 \, \Pi_R^h\left(\frac{x^2}{1 - x} m_\mu^2 \right) \right]. \tag{2}$$

$\Pi_R^h(Q^2)$ is given in terms of the vector two point function $\Pi_V^1(Q^2)^{1,3}$: $\Pi_R^h(Q^2) = \sum_{i=u,d,s} Q_i^2 (\Pi_V^1(Q^2) - \Pi_V^1(0))$, where $Q_i = (2/3, -1/3, -1/3)$ are the charges of the SU(3) flavour quarks u,d,s and the renormalized photon self-energy satisfies the constraint $\Pi_R^h(0) = 0$. Because the typical momenta of the off-shell photons are of the order of the squared muon mass $(Q^2 \sim .01 GeV^2)$ the integral is dominated by the low energy contribution to $\Pi_R^h(Q^2)$. Using a first derivative approximation (i.e. $\Pi_R^h(Q^2) \sim Q^2 \frac{d\Pi_R^h}{dQ^2}(0)$ the ENJL model gives the following prediction:

$$\frac{d\Pi_V^1}{dQ^2}(0) = -\frac{N_c}{16\pi^2} \frac{1}{M_Q^2} \frac{4}{15} \left[e^{-\frac{M_Q^2}{\Lambda_\chi^2}} + \frac{5}{6} \frac{1 - g_A}{g_A} \Gamma(0, \frac{M_Q^2}{\Lambda_\chi^2}) \right], \tag{3}$$

where g_A is the mixing parameter between the axial-vector and the pseudoscalar mesons. With the values of the best fit 1 of ref. [2] $M_Q = 265 MeV$, $\Lambda_\chi = 1.165 GeV$ and $g_A = 0.61$ we obtain for a_μ^h the value $a_\mu^h = 8.66 \cdot 10^{-8}$. Adding the contribution from chiral loops due to π, K exchanges equal to[7] $(0.71 \pm 0.07) \cdot 10^{-8}$ we obtain $a_\mu(had) = a_\mu(ENJL) + a_\mu(\chi loops) = 9.37 \cdot 10^{-8}$, which is a rather high value compared to the phenomenological estimates. The first derivative approximation is not enough accurate for the prediction of $g - 2$.

The evaluation of the full dispersive integral (2) requires the knowledge of the long distance (ld) plus the short distance (sd) behaviour of $\Pi_R^h(Q^2)$. We can do our best performing the matching between the ld prediction coming from the effective theory and the sd prediction coming from perturbative QCD. A value of $a_\mu^h(Fig.2a) =$

$6.7 \cdot 10^{-8}$ has been obtained in the ENJL framework[7] (i.e. the leading order) with a best fitted matching point $\hat{x} \simeq 0.91$ which corresponds to an euclidean $\hat{Q}^2 = \hat{x}^2/(1 - \hat{x})m_\mu^2 \simeq (320)^2 \ MeV^2$. The value obtained for a_μ^h is quite better than the first approximation value $8.66 \cdot 10^{-8}$.

To see how "extra" corrections modify the ENJL prediction is sufficient to study the integral (2) over the ld part in the range $\int_0^{\hat{x}}$ for different values of $\hat{x}$ corresponding to an equivalent value of $\hat{Q}^2 = (0.3)^2, (0.5)^2, (0.8)^2 \ GeV^2$.

NPLL corrections proportional to Q^2 in the QR model lead to a vector two-point function which decreases faster in Q^2 then the ENJL prediction. The dispersive integral gives for a_μ^h in the ENJL approximation the values 6.8×10^{-8} with $\hat{Q}^2 = (0.5)^2$ GeV^2 and 7.1×10^{-8} with $\hat{Q}^2 = (0.8)^2$ GeV^2, while in the QR model gives 6.9×10^{-8} and 7.3×10^{-8} respectively[9].

The correction induced by higher order terms proportional to Q^2 and which are relevant in the intermediate Q^2 region does not exceed three percent. This proves that a_μ^h is practically only sensitive to perturbative corrections which modify the very low Q^2 region, i.e. $Q^2 \leq (500 MeV)^2$.

Gluonic corrections can be parametrized following ref. [5]. The leading contribution in the $1/N_c$ expansion involves only one unknown parameter g which is related to the lowest dimensional gluon vacuum condensate[9]. They modify the ENJL vector two-point function for all Q^2. The corrections to the ENJL (i.e. leading order) prediction of a_μ^h are 10% for g=0.5 and 15% for g=0.25. They decrease a_μ^h. A better determination of non-gluonic contributions to a_μ^h can be used to constrain the value of the g parameter in low energy effective fermion models.

1. E. Pallante and R. Petronzio, "Quark-Resonance model", preprint ROM2F93/37, to be published in *Z. Ph.* **C**.
2. J. Bijnens, C. Bruno and E. de Rafael, *Nucl. Phys.* **B390** (1993) 501.
3. J. Bijnens, E. de Rafael and H. Zheng, Preprint CERN-TH 6924/93, CTP-93/P2917, NORDITA-93/43 N,P.
4. T. Kinoshita, B. Nižić and Y. Okamoto, *Phys. Rev.* **D31** (1985) 2108.
5. J.A. Casas, C. Lopez and F.J. Ynduráin, *Phys. Rev.* **D32** (1985) 736.
6. L.M. Kurdadze et al., *Sov. J. Nucl. Phys.* **40** (1984) 286.
7. E. de Rafael, *Phys. Lett.* **B322** (1994) 239.
8. B.E. Lautrup, A. Peterman and E. de Rafael, *Phys. Rep.* **3C** (1972) 193.
9. E. Pallante, "The hadronic vacuum polarization contribution to the muon $g-2$ in the quark-resonance model", prep. LNF 94/031 (P), ROM2F 94/12, to be published in *Phys. Lett.* **B**.

EVIDENCE ON $qq\bar{q}\bar{q}$ HADRON SPECTRUM

FROM $\gamma\gamma \to$ *vector meson vector meson* REACTIONS ?

MITJA ROSINA

Faculty of Natural Sciences and Technology, and Jožef Stefan Institute,
University of Ljubljana, Jadranska 19, P.O.B.64, 61111 Ljubljana, Slovenia

and

BORUT BAJC

Jožef Stefan Institute, Jamova 39, p.p. 100, 61111 Ljubljana, Slovenia

ABSTRACT

The $\gamma\gamma \to \rho^o\rho^o \to 4\pi$ reaction shows a broad "resonance" at 1.5 GeV with no counterpart in the $\rho^+\rho^-$ channel. This $(J^P, J_z) = (2^+, 2)$, $I = 0$ and 2 resonance is considered as a candidate for a $qq\bar{q}\bar{q}$ state. We show, however, that it can also be explained by potential scattering of $\rho^o\rho^o$ via the $\sigma-$ exchange.

1. The Motivation

After many successes of quark models to describe a single meson or baryon, their predictive power in the two-hadron sector remains questionable. A fair description of the nucleon-nucleon interaction was accompanied with the prediction of a rich dibaryon spectrum which has never been observed. It is interesting to see how different quark models perform in the two-meson sector, as compared to mesonic models. We are also motivated by our experimental colleagues in Ljubljana (ARGUS coll.[1,2]).

2. The Experimental "Puzzle"

The $\gamma\gamma \to \rho\rho$ reaction [1,2] has a large and broad peak near the nominal threshold in the $\rho^o\rho^o$, (2^+2) channel and a much smaller cross section in other $\rho^o\rho^o$ channels as well as in the $\rho^+\rho^-$ channel. Since so far the quark models [3] covered more features simultaneously than the effective mesonic models [4], the opinion prevails that explicit quark degrees of freedom (a $qq\bar{q}\bar{q}$ molecule) are essential. Here we challenge this view.

3. The Hypothesis

The peak near the nominal threshold and its strong spin dependence are reminiscent of the p-n scattering at low energy. There the S=0 channel has a very high cross section while in the S=1 chanel it is much lower, because in the S=0 state the potential well contains almost exactly 1/4 wavelength of the relative wavefunction.

We hypothesize that a similar trick of nature is played also in $\rho\rho$ scattering. We assume that both photons are converted in ρ via vector dominance, and both ρ then interact by a scalar isoscalar potential which is just strong enough to (almost) bind them for one spin orientation and fails to do so for the others. Also, an isoscalar potential cannot exchange charges and does not lead to the $\rho^+\rho^-$ final state.

4. A Toy Potential Model

To illustrate this idea, we choose a Yukawa-type σ-exchange potential with the same parameters as in the Bonn potential for the nucleon-nucleon system [5] ($g^2/4\pi = 7.07$, $m_\sigma = 0.55$ GeV), but multiplying it with a factor $(\frac{2}{3})^2$ (two quarks rather than three at each vertex). Solving the nonrelativistic Schrödinger equation we showed that with a very plausible hard core ($r_c = 0.17$ fm) a weakly bound or antibound state at $E \sim 0$ can be obtained. If the potential is assumed to be slightly spin dependent one spin channel will have a state close to zero and other spin channels will miss it. This demonstrates that the proposed hypothesis can be realized.

5. A Relativistic Potential Model

Our starting Lagrangian is gauge invariant and respects vector dominance:

$$
\begin{aligned}
\mathcal{L} = & -\frac{1}{4}(\partial_\mu B_\nu - \partial_\nu B_\mu)^2 - \frac{1}{4}(\partial_\mu \rho_\nu - \partial_\nu \rho_\mu)^2 + \frac{1}{2}m_\rho^2 \rho_\mu^2 + \frac{1}{2}(\partial_\mu \sigma)^2 - \frac{1}{2}m_s^2 \sigma^2 \\
& -\frac{1}{2}m_\rho^2[\rho_\mu^2 - (\rho_\mu - \frac{e}{g}B_\mu)^2] + \frac{g_s}{2}\sigma(\partial_\mu \rho_\nu - \partial_\nu \rho_\mu)^2 \,.
\end{aligned}
\tag{1}
$$

The amplitude was calculated with the Bethe-Salpeter equation in the leading order of e/g with the σ exchange kernel given from Eq. (1). After the Blankenbecler-Sugar reduction [6] and the partial wave expansion [7] the integral equations for different channels were solved with the matrix inversion method [8], taking 16 points for the magnitude of the relative three-momentum from 0 to the cutoff Λ. Assuming that the effect of different off-shellness of the final ρ is small, we got the cross section by weighing the amplitude squared with Breit-Wigner factors.

We choose for the vector dominance factor $e/g = 0.036$, which is consistent with the value obtained within a larger model [9] from the ρ and $a_1(1260)$ decay widths. The mass in the two-body propagator was chosen $m = 0.692$ GeV instead of $m = m_\rho = 0.77$ GeV, since the ρ meson is broad. The results proved to be relatively insensitive to the choice of the σ mass, so that we will report only the case $m_s = 0.5$ GeV.

The results are shown in fig. 1, for different choices of the paramaters Λ and g_s together with the experimental values. A combination of the two parameters in a reasonable range exists, which reproduces quite well the experimental results.

Some models [3] predict, that the enhancement in the $(2^+, 2)$ channel is a universal feature of $\gamma\gamma \rightarrow 2\,vector\,mesons$. Experimentally, there are too few events to judge. Due to the sensitivity to potential parameters in our models we predict that such enhancement is very difficult to appear in more than one case.

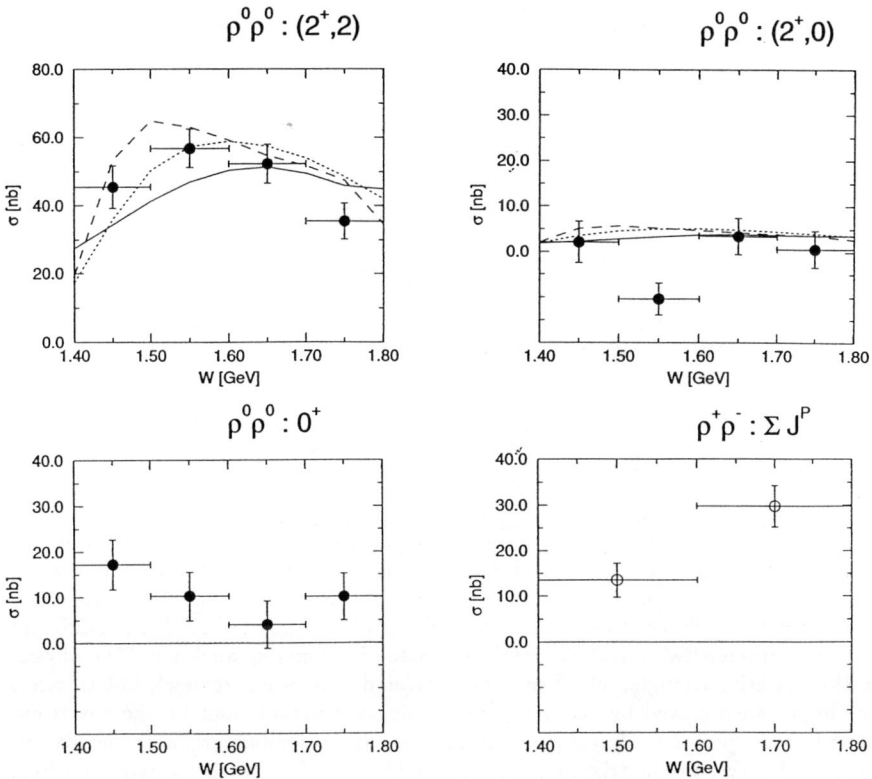

Figure 1: Cross sections for various channels in the relativistic potential model with scalar exchange: — ($\Lambda = 2.0$ GeV, $g_s = 12.0$ GeV^{-1}); $\cdots$ ($\Lambda = 2.2$ GeV, $g_s = 10.4$ GeV^{-1}); - - ($\Lambda = 2.4$ GeV, $g_s = 9.13$ GeV^{-1}); experiments: • ARGUS [1], ∘ ARGUS [2].

Work supported by the Ministry of Science and Technology of Slovenia.

1. ARGUS Coll., H. Albrecht et al., *Z. Phys.* **C50** (1991) 1.
2. ARGUS Coll., H. Albrecht et al., *Phys. Lett.* **B267** (1991) 535.
3. N. N. Achasov, S. A. Devyanin and G. N. Shestakov, *Z. Phys* **C27** (1985) 99.
4. B. Moussallam, *Z. Phys.* **C39** (1988) 535.
5. R. Machleidt, K. Holinde and Ch. Elster, *Phys. Rep.* **149** (1987) 1.
6. R. Blankenbecler and R. Sugar, *Phys. Rev.* **142** (1966) 1051.
7. M. Jacob and G.C. Wick, *Ann. of Phys.* **7** (1959) 404.
8. G. E. Brown, A. D. Jackson and T. T. S. Kuo, *Nucl. Phys.* **A133** (1969) 481.
9. B. W. Lee and H. T. Nieh, *Phys. Rev.* **166** (1968) 1507; S. Gasiorowicz and D. A. Geffen, *Rev. Mod. Phys.* **41** (1969) 531.

The impact of f_B on the Quark-Mixing

DECIO COCOLICCHIO

Dipartimento di Matematica, Università della Basilicata, Potenza
Via N. Sauro 85, 85100 Potenza, Italy

Istituto Nazionale di Fisica Nucleare, Sezione di Milano
Via G. Celoria 16, 20133 Milano, Italy

ABSTRACT

Survey of the current status on some of the issues in quark mixing after the recent experimental determination of the top quark mass m_t. Weak decays of hadrons seem to provide a rather detailed knowledge of the flavour sector of the standard model and also serve as a probe of that part of the strong interactions phenomenology which is less understood: the confinement of quarks and gluons into hadrons. This intricate interplay between weak and strong interactions represents one of the most attractive and challenging questions in the recent years.

The decay constants of the heavy pseudoscalar mesons represent a measure of the strength of the Quark-antiquark attraction inside the bound state and are of great interest for the study of quark mixing and CP-violation. The aim of this contribution is not merely the description of the quark mixing in the standard model, the different sensible parametrizations, the numerical update of the mixing angles and the physics of the unitarity triangle, which may be recovered in a recent review[1], but to stress the importance played by the heavy meson decay constants and by the deviations of the weak form factors $f_\pm(q^2=0)$ from unity to understand reliably the flavour mixing. In fact, the matrix element $|V_{us}|=0.22/|f_+^{K\to\pi}(0)|$ has been extracted from the available data on the $K_{\ell_3}^{+,0}$ decays, $|V_{cd}|=0.21/|f_+^{D\to\pi}(0)|$ derives from the results of the $\sigma(\nu N \to cX)$ compared with the D_{ℓ_3} experimental data, $|V_{cs}|=0.55/|f_+^{D\to K}(0)|$ from D_{ℓ_3} and $|V_{cb}|=0.04/|f_+^{B\to D}(0)|$ from a model dependent analysis of the inclusive lepton spectrum in semileptonic B decays. On the other side, the f_π and f_K constants result important scale setting parameters in chiral symmetry expansions whereas heavier meson decay constants f_D and f_B are essential parameters for weak decays and oscillation studies. Furthermore, the neutral B-meson decay constant f_B severely constrains the $(|V_{tb}|, |V_{td}|)$ values and it seems crucial especially to account for the splitting between the mass eigenstates which induces the B_d^0-$\overline{B_d^0}$ mixing

$$x_d = \frac{G_F^2}{6\pi^2}m_W^2 f_B^2 B_B m_B \eta_{QCD}^B S(x_t)|V_{tb}|^2|V_{td}|^2\tau_B \qquad (1)$$

which is proportional to f_B^2. Irrespectively of the recent CDF evidence for an heavy top quark of near 170 GeV/c^2, a best fit analysis of the elements in the quark mixing matrix cannot constrain considerably the allowed region of the product $f_B\sqrt{B}$, the bag-constant B_K and in general the hadronic uncertainties governed by the quark-meson duality.

Table 1. Some theoretical predictions for the B meson decay constant.

Method	Group	Value of f_B (in MeV)
Heavy Quark	Harvard	
		190 ± 50
Effective Theory	German	
	ITEP	115 ± 15
QCD Sum Rules	Hadronic	$135 - 185$
	Spectral	211 ± 36
	APE Collab.	$290 \pm 15 \pm 45$
Lattice	BNL Group	187 ± 37
	UKQCD Group	$155 - 242$
	non-relativistic	110
Potential	semi-relativistic	163
	relativistic	140

Treating f_B as unknown, the experimental constraints on the value of the semileptonic B-width, on ε_K (the K^0-$\bar{K}^0$ mixing parameter), on x_d (the observable of B^0-$\bar{B}^0$ mixing) as well as on the present knowledge of other charged current data yield[2,3] a distinctive feature of two disconnected best fit solutions for the quark mixing, in dependence of the sign of the CP violating phase. This ambiguity is translated into uncertainties for the range of the quark mixing parameters V_{Qq} which could be impressively depicted and become transparent considering the unitarity relation:

$$V_{ud}V_{ub}^* + V_{cd}V_{cb}^* + V_{td}V_{tb}^* = 0 \qquad (2)$$

in the (ρ, η) plane of the Wolfenstein parametrization. It is worth noting that we recover as the most likely solutions for the B-meson decay constant, the values $f_B = (224 \pm 28)$ MeV for a positive value of ρ and $f_B = (125 \pm 14)$ MeV for a negative one, respectively. The forecasts for the asymmetries occurring in $\bar{b} \to \bar{u}u\bar{d}$ processes like the decay $B^0 \to \pi^+\pi^-$ and in the $\bar{b} \to \bar{c}c\bar{s}$ case as in the channel $B^0 \to \psi K_S$ are

$$\hat{\mathcal{A}}(B_d^0 \to \pi^+\pi^-) = -\sin(2\alpha)\frac{x_d}{1+x_d^2} \simeq \begin{cases} 0.400 & \text{for } \alpha \simeq 120° \\ -0.253 & \text{for } \alpha \simeq 17° \end{cases} \qquad (3)$$

$$\hat{\mathcal{A}}(B_d^0 \to \psi K_S) = -\sin(2\beta)\frac{x_d}{1+x_d^2} \simeq \begin{cases} -0.278 & \text{for } \beta \simeq 19° \\ -0.107 & \text{for } \beta \simeq 7° \end{cases} \qquad (4)$$

where the upper expectation is linked with a positive value of ρ (larger f_B) and the lower one with the negative ρ solution (smaller f_B). In the lack of any experimental information, considerable effort has been devoted to the theoretical predictions of f_B and its dependence on m_B. Notwithstanding the extensive literature on the subject, the current status for f_B is rather confusing, with a plethora of theoretical predictions not always in mutual agreement even if obtained within the same method (see Table I).

Most of them seem to predict the larger value of almost 300 MeV favored in recent lattice calculations even if some approach proposes like what has been obtained earlier in non-relativistic potential models. A reliable calculation of the decay constants might be carried out from first principles along with the successful approach given by the QCD sum rules. Anyway, in this case the selection of the input parameters (quark masses, $\alpha_S(\mu)$, vacuum condensates, ...) becomes so dramatic, that relatively large differences cannot be avoided. However it seems that a progressive convergence is going on towards a relatively high value for f_B of the order of near 250 MeV. A different conclusion in general pertains for the quark potentials models, which do not seem to reproduce a reliable high value of f_B.

Methods based on the dynamical chiral symmetry breaking, on the other hand, might play a crucial role to face this disappointing situation. In this approach, various models for evaluating f_B have been proposed. They can be classified in two main categories based on the asymptotic and the infrared behaviour of QCD. For large momenta, the following asymptotic solution for the quark self-energy emerges in Landau gauge: $\Sigma(p^2) \simeq 4m/p^2$ up to a logarithm. Similar solution has been successfully used by Pagels and Stokar and many others to calculate the pion decay constant and the electromagnetic form factors of the pion. In small momenta, the nonperturbative instanton contributions to the binding kernel of the meson is expected to dominate. The singular effective gluon propagator might then be rewritten consistently with the form of the quark propagator to give a realistic solution which can break chiral symmetry dynamically. Actually, both approaches do not seem to reproduce a reliable high value for f_B which requires massive constituents. So far, the relativistically covariant formulation of f_B, and analogous decay constants, must be connected to extensions of equations and techniques with explicit quark-mass terms breaking the global flavour chiral symmetry. The method is based on the Jackiw-Johnson sum rule[4] which let us express f_B^2 in terms of the properties of the fermion propagator and the axial-vector vertex[5]. An alternative to this sum-rule derivation, can relate the decay constant to the regularized bound-state wave function[6].

Only an accurate detection of f_B would help us to discriminate among these different theoretical models. The cleanest experimental determination of this constant will eventually come from the purely leptonic decay $B \to \tau\nu$. Even though its branching ratio is expected[7] of the order of $1.1 \cdot 10^{-4}$, its identification does not seem to be affected of large backgrounds.

References

1. D. Cocolicchio, "*CP–asymmetries in B decays*", Proceedings *Advanced Study Conference on Heavy Flavours*, (Editions Frontieres, 1994), p. 367.
2. D. Cocolicchio and J. R. Cudell, *Physics Letters* **B245** (1990) 591.
3. A. Acuto and D. Cocolicchio, *Physical Review* **D47** (1993) 1501.
4. R. Jackiw and J. Johnson, *Physical Review* **D8** (1973) 2386.
5. A. Barducci et al., *Physical Review* **D38** (1988) 238.
6. D. Cocolicchio, "*f_B in dynamical symmetry breaking*", to appear.
7. D. Cocolicchio and L. Maiani, *Physics Letters* **B291** (1992) 155.

SCALAR MESON DECAYS:
$K\overline{K}$ AND $\pi\pi$ THRESHOLD EXPANSION PARAMETERS

Leonard LEŚNIAK

Department of Theoretical Physics,
Henryk Niewodniczański Institute of Nuclear Physics,
PL 31-342 Kraków, Poland

ABSTRACT

Coupled channel analysis of the scalar mesons $f_0(975)$ and $f_0(1400)$ decaying into the $\pi\pi$ and $K\overline{K}$ pairs has been done. The spin 0 and isospin 0 scattering amplitudes have been calculated in the energy range between the $\pi\pi$ threshold and about 1400 MeV. Threshold expansion of the $K\overline{K}$ and $\pi\pi$ amplitudes has been performed. We have evaluated three real parameters of the $\pi\pi$ effective range expansion and three complex parameters of the $K\overline{K}$ threshold expansion of the phase shifts and the inelasticity. M-matrix parameters have also been obtained. Good agreement with near threshold $\pi\pi$ data has been found. New $K\overline{K}$ data are needed in order to understand better the nature of scalar mesons.

Decay channels of two scalar mesons $f_0(975)$ and $f_0(1400)$ are coupled together. Recently [1] we have done a coupled channel analysis of the scalar- isoscalar scattering amplitudes solving a system of Lippmann-Schwinger equations with separable $\pi\pi$ and $K\overline{K}$ interactions. Experimentally the low energy $\pi\pi$ interactions are not very precisely known but the $K\overline{K}$ interactions are essentially unknown. Two scalar mesons $f_0(975)$ and $a_0(980)$ lie very close to the $K\overline{K}$ threshold and are often interpreted[2-4] as the $K\overline{K}$ unstable bound states (deuteronlike or molecular states). Since different interpretations of the scalar mesons exist in the literature [5-7] a determination of the $K\overline{K}$ threshold parameters is crucial for understanding the nature of scalar mesons. This necessity has also been stressed by N. Törnqvist [8].

We use the effective range expansion in the $\pi\pi$ and $K\overline{K}$ channels:

$$k\cot\delta = \frac{1}{a} + \frac{1}{2}rk^2 + vk^4 + O(k^6), \tag{1}$$

where δ is the scattering phase shift, k is the meson relative momentum, a is the scattering length, r is the effective range of the interaction and the parameter v can be related to the shape of the intermeson potentials.

Above the $K\overline{K}$ threshold we define the complex $K\overline{K}$ phase shift $\delta = \delta_K + i\rho$, where δ_K is the real $K\overline{K}$ phase shift and ρ is related to the inelasticity parameter $\eta = exp(-2\rho)$. In the $K\overline{K}$ channel the expansion (1) can still be valid if the parameters a, r and v are complex.

Table 1. Low momentum parameters of the $\pi\pi$ scalar, $I = 0$ scattering (the average pion mass $m_\pi = 137.27$ MeV)

Set No	$a_\pi(m_\pi^{-1})$	$r_\pi(m_\pi^{-1})$	$v_\pi(m_\pi^{-3})$
1	0.172 ± 0.008	-8.60	3.28
2	0.174 ± 0.008	-8.51	3.25

The effective range parameters are given in tables 1 and 2 for two sets of experimental data analysed in [1]. These data sets differ qualitatively in a vicinity of the $K\overline{K}$ threshold: the $K\overline{K}$ phase shifts tend to decrease at threshold for the set 1 and increase for the set 2. The model [1] describes better the data set 1 than the set 2.

Table 2. Low momentum parameters of the $K\overline{K}$ scalar, $I = 0$ scattering

Set No	a_K fm	r_K fm	v_K fm^3	R_K fm	V_K fm^3
1	$-1.73 + i\,0.59$	$-0.057 + i\,0.032$	$0.016 - i\,0.0044$	0.38	-0.66
2	$-1.58 + i\,0.61$	$-0.352 + i\,0.043$	$0.028 - i\,0.0057$	0.20	-0.83

In table 2 we have introduced two additional real parameters R_K and V_K satisfying the following expression valid for real δ_K:

$$k \cot\delta_K = \frac{1}{Re\, a_K} + \frac{1}{2} R_K k^2 + V_K k^4 + O(k^6). \tag{2}$$

These parameters are not independent on a_K, r_K and v_K but have been introduced for a comparison with the $\pi\pi$ channel.

In a recent analysis of the $\pi N \longrightarrow \pi\pi N$ data D. Počanić et al. [9] have obtained the $\pi\pi$ scattering length $a = (0.177 \pm 0.006)\, m_\pi^{-1}$ which is in a very good agreement with the predicted values seen in table 1. Using the chiral perturbation theory Gasser and Leutwyler [10] have got a value $(0.20 \pm 0.01)\, m_\pi^{-1}$ while the values calculated by Roberts et al. [11] are $0.16\, m_\pi^{-1}$ or $0.17\, m_\pi^{-1}$.

The $\pi\pi$ effective range is not well determined experimentally. Following the analyses of Rosselet et al. [12] and Belkov and Buniatov [13] we have derived the values $r_\pi = (-1.4 \pm 3.7)\, m_\pi^{-1}$ and $(-8.1 \pm 5.3)\, m_\pi^{-1}$, respectively. Within the Weinberg approach [14] $r_\pi = -8.48\, m_\pi^{-1}$.

The experimental information about the $K\overline{K}$ threshold parameters is very scarce. As seen in table 2 the real part of the $K\overline{K}$ scattering length is negative and large. This last result is confirmed by the experimental study of Wetzel et al. [15].

For a full description of the two coupled $\pi\pi$ and $K\overline{K}$ channels (including the $K\overline{K} \longrightarrow \pi\pi$ annihilation process) one can introduce a real and symmetric matrix M related to the scattering matrix T by

$$M = T^{-1} + i\,\hat{k}, \tag{3}$$

where $\hat{k}$ is a diagonal 2×2 matrix of the $K\overline{K}$ and $\pi\pi$ momenta in the center–of–mass system.

At the $K\overline{K}$ threshold the M–matrix elements can be expanded as

$$M_{ij} = A_{ij} + \tfrac{1}{2}B_{ij}k_1^2 + C_{ij}k_1^4 + O(k_1^6), \tag{4}$$

where A_{ij}, B_{ij} and C_{ij} are real coefficients and k_1 is the $K\overline{K}$ momentum (i, j=1,2). Every threshold parameter in two channels introduced in eq. (1) can be related to a set of the M_{ij} expansion parameters.

The coefficients A_{ij}, B_{ij} and C_{ij} are shown in table 3 for the data set 1.

Table 3. M–matrix expansion parameters at the $K\overline{K}$ threshold

reaction channel	i j	A_{ij} fm^{-1}	B_{ij} fm	C_{ij} fm^3
$K\overline{K}$	1 1	-0.483	-8.10×10^{-2}	1.83×10^{-2}
$\pi\pi$	2 2	0.476	-1.58×10^{-1}	1.43×10^{-3}
$K\overline{K} \longleftrightarrow \pi\pi$	1 2	0.669	-1.57×10^{-2}	5.93×10^{-3}

The $K\overline{K}$ threshold parameters given in tables 2 and 3 can be useful in future analyses of experimental data obtained at new accelerators like CEBAF, COSY and DAΦNE.

This work has been partially supported by the Polish Committee for Scientific Research (grant No 2 0198 9101) and by Maria Skłodowska–Curie Fund II (No PAA/NSF–94–158).

References

1. R. Kamiński, L. Leśniak and J. P. Maillet, *Orsay report* IPNO/TH 93–31, to appear in *Phys. Rev.* **D**.
2. J. Weinstein and N. Isgur, *Phys. Rev.* **D41** (1990) 2236.
3. T. Barnes, *Phys.Lett.* **165B** (1985) 434 and *report* ORNL-CCIP-94-08.
4. F. Cannata, J. P. Dedonder and L. Leśniak, *Z. Phys.* **A 334** (1989) 457.
5. D. Morgan and M. R. Pennington, *Phys. Rev.* **D48** (1993) 1185.
6. B. S. Zou and D. V. Bugg, *Phys. Rev.* **D48** (1993) 3948.
7. F. E. Close et al., *Phys. Lett.* **B319** (1994) 291.
8. N. A. Törnqvist, *Helsinki report* HU-SEFT R 1994-03.
9. D. Počanić et al., *Phys. Rev. Lett.* **72** (1994) 1156.
10. J. Gasser and H. Leutwyler, *Phys. Lett.* **B1**, (1985) 325.
11. C. D. Roberts et al., *Phys. Rev.* **D49**, (1994) 125.
12. L. Rosselet et al., *Phys. Rev.* **D15** (1977) 574.
13. A. A. Belkov and C. A. Buniatov, *Particles & Nuclei*, **13** (1982) 5.
14. S. Weinberg, *Phys. Rev. Lett.* **17** (1966) 616.
15. W. W. Wetzel et al., *Nucl. Phys.* **B115** (1976) 208.

EFFECTS OF THE ONE-GLUON ANNIHILATION PROCESS

ON LIGHT DIQUONIA

B. SILVESTRE-BRAC

Institut des Sciences Nucléaires,53 Avenue des Martyrs
F-38026 Grenoble-Cedex, FRANCE

and

C. SEMAY

Université de Mons-Hainaut,19 Avenue Maistriau
B-7000 Mons, BELGIQUE

ABSTRACT

The spectra of all $q^2\bar{q}^2$ systems composed of u, d or s quarks are investigated in the framework of the nonrelativistic quark model using a Hamiltonian which takes into account the one-gluon annihilation process. The shift in energy due to this effect is studied as well as the mixing of different flavour states. The four-body problem is solved by the variational method in harmonic oscillator bases including all configurations up to 6 quanta. The contribution of the one-gluon annihilation process to the structures of light diquonia is found quite weak.

1. Introduction

In recent papers [1,2], we performed systematic studies of systems composed of two quarks and two antiquarks, which are called diquonia. These works were made in the framework of the nonrelativistic quark model (NRQM) using the potential proposed by Bhaduri *et al.* [3]. Some interesting candidates could be stable under strong interaction, but they all contain heavy quarks and should be difficult to be produced presently. Nevertheless, we have found that some neutral diquonia of type $nq\bar{n}\bar{q}$ with $q \neq n$ (n denotes either the u or the d quark) could appear as narrow resonances or even as stable particles; but reliable conclusions about these systems require a correct treatment of the annihilation effects.

In most potentials used in literature, as the Bhaduri's one, only scattering diagrams are considered; the one-gluon exchange gives rise to the short-range part while the multi-gluon exchanges are the source of the linear confinement. Nevertheless, it is well-known that annihilation effects can be very large between light quarks and antiquarks. For instance, if one uses Bhaduri's potential the masses of the mesons η and π are the same, which is very far from the experimental situation. In this case an annihilation mechanism including many gluons is necessary. Moreover, the one-gluon annihilation process, which does not exist in mesons, occurs in diquonia, and can thus reinforce the contributions of the other annihilation mechanisms in these systems.

The effect of one-gluon annihilation is always repulsive, but correlatively it induces mixing in the flavor sectors, which should lead to attraction. A good description of the neutral diquonia, and in particular of $ns\bar{n}\bar{s}$ states, which are very promising candidates for bound or resonant diquonia [1,2] implies to perform a complete and detailed calculation of the one-gluon annihilation effects. This is the subject of this contribution in the sector of (u, d, s) quarks where the effect is the most important.

2. The model and results

In NRQM, we must solve the Schrödinger equation with a total hamiltonian H_A which is the sum of nonrelativistic kinetic energy K and a potential V composed of two-body terms $V_{ij}(r)$. In the spirit of this paper, the potential itself is the sum of a scattering contribution V_S and a one-gluon annihilation term V_A. For our calculations we chose for V_S the potential proposed by Bhaduri; it contains a central part of type "coulomb + linear" and a hyperfine term of short-range Yukawa type. We denote $H_S = K + V_S$, so that the total hamiltonian is $H_A = H_S + V_A$ and the effect of the annihilation contribution is made clear by comparing the values resulting from H_A and H_S.

The expression for V_A is obtained by the usual nonrelativistic reduction of the corresponding Feynman diagram. Taken as such, it contains an irrealistic $\delta(\mathbf{r})$ form factor; we smooth it with the same Yukawa factor that was used for the equivalent hyperfine term of V_S. Consequently the V_A term looks like :

$$V_{Aij}(r) = \frac{9}{8} \, g \frac{\kappa}{mm'} \, P_{ij}^{\text{gluon}} \, \frac{\exp(r/r_0)}{rr_0^2} \tag{1}$$

In this expression m is the mass of the quark for the destroyed pair, m' the mass of the quark for the created pair; κ and r_0 are the same parameters which appear in the hyperfine term (their values are given in [3]). P_{ij}^{gluon} is a projector which imposes to the quark pair to possess the same isospin, spin and color quantum numbers than the gluon. The statistical value g takes the values 1, $\sqrt{2}$, 2 according to the fact that the transition between the initial and final states can contain respectively 0, 1 or 2 $n\bar{n}$ pairs.

The wave function Ψ is expanded on basis functions Ψ_r which are the tensor products of the basis states for the color, isospin, spin and space degrees of freedom. The space function is itself expanded on a harmonic oscillator (HO) basis for each of the three jacobi coordinates corresponding to a diquark-antidiquark coupling (see [1]). The diquark and antidiquark isospins are no longer good quantum numbers, as it was the case when V_A was absent; only the total isospin I remains a good quantum number.

The presence of V_A allows the coupling between sectors of different flavors. To simplify the notation we will denote by $|0\rangle$ and $|1\rangle$ states in which the (nn) diquark is coupled to isospin 0 and 1 respectively. In the same way $|D\rangle$ is the (ns) diquark and $|S\rangle$ is the (ss) diquark.

Just as an example, let us take states with isospin $I=0$. They can be obtained by coupling a $|0\rangle$ diquark with the same corresponding antidiquark. The resulting flavor function is denoted [00]. But we have also the three other possibilities [11], [DD] and [SS]. So we have in that case four different flavor sectors, and the projection of the wave function on each of these sectors is expanded on basis states including color, spin and space. If one uses H_S only, these flavor subspaces are disconnected and one can diagonalize the hamiltonian matrix separately for each of them. The technics for evaluating the matrix elements of H_S employs Brody-Moshinsky coefficient and has been described in detail in [1]. The effect of V_A is first to add new contributions in each flavor sector, and second to make a coupling between them. The calculation of the new matrix elements is more involved; it needs to change the set of Jacobi coordinates towards a meson-meson coupling and to evaluate integrals with HO functions of different size. We will not enter into details here, but all these technical difficulties have been solved. For isospin $I=1/2$ we have only three different flavor spaces namely [0D], [1D] and [DS], while for $I=1$ we have again four sectors namely [11], [DD], [01] and [10]. States with isospins $I=3/2$ and $I=2$ are insensitive to V_A and have not been considered here.

All the basis states have been included up to a number of quanta $N=6$; this represents about 1000 basis states when the flavor coupling is effective. The HO size is determined by minimizing the ground state energy. In addition to the energy, we have also calculated the proportion of the wave function in each flavor space which gives strong indications on the importance of V_A. For each state we have also computed the charge conjugaison quantum number C which must be equal to $+1$ or -1; in the case of diquonia, the expression for this quantity is not at all trivial and the good resulting values are a tremendous check for our numerical codes.

For each isospin sector, we have studied the 3 spin possibilities $S=0$, 1 or 2 and mainly states with zero orbital momentum. Let us briefly comment our results. Whatever the state under consideration, the effect of V_A is always repulsive; it is rather moderate since the shift in energy does not exceeds 120 MeV in the most dramatic case. All the states which were degenerate previously, are now well separated. Correlatively the flavor mixing is always weak, since the proportion within a given sector is always greater than 90 %; thus it is quite easy to "follow" a state when the annihilation term is switched on.

In summary, the effect of the one-gluon annihilation, although palpable, is not determining. What would be the effect of multi-gluon annihilation ? Such a study is in progress.

3. References

1. B. Silvestre-Brac and C. Semay, Z. Phys. C **57**, 273 (1993).
2. B. Silvestre-Brac and C. Semay, Z. Phys. C **59**, 457 (1993).
3. R.K. Bhaduri, L.E. Cohler and Y. Nogami, Il Nuovo Cimento A **65**, 376 (1981).

SECTION D

Phenomenology and experiments;
miscellanea

GRIBOV CONFINEMENT AND SCALAR MESONS

Y. DOKSHITZER

Department of Theoretical Physics, Lund University
Solvegatan 14A, 223 62 Lund, Sweden

ABSTRACT

Phenomenological aspects of the Gribov light quark confinement theory and the notion of the infrared regular effective QCD coupling were discussed.

QCD INTERCONNECTION EFFECTS IN HADRONIC W^+W^- AND $t\bar{t}$ EVENTS

VALERY A. KHOZE

Department of Physics, University of Durham,
Durham, DH1 3LE, England

ABSTRACT

In the events of the type $e^+e^- \to W^+W^- \to 4$ jets, $e^+e^- \to t\bar{t} \to bW^+\bar{b}W^-$, particle production could depend in a non-trivial way on the kinematics of the process. It is shown that QCD interference effects are negligible for energetic perturbative emission, but soft perturbative gluons and non-perturbative fragmentation could induce colour correlations. Possible consequences for LEP 2 and NLC events are briefly addressed.

In high-energy physics, "tomorrow belongs" to the detailed study of heavy unstable particles (W bosons, top quarks, SUSY particles, ...). An important aim of future experiments is the precise determination of their parameters, primarily masses. This requires a detailed understanding of production and decay mechanisms (including interference effects) and, in particular, of the effects arising from the large width of many of these objects, $\Gamma \sim O(1 \text{ GeV})$. This talk is concerned with the QCD interconnection phenomena that may occur when two unstable particles decay and hadronize close to each other in space and time. The word 'interconnection' is here introduced to cover those aspects of final-state particle production that are not dictated by the separate decays of unstable objects, but can only be understood in terms of the joint action of the two. (For further details and a long list of references see Refs.[1,2].)

Let us start from hadronic W^+W^- events. QCD interferences between W^+ and W^- undermine the traditional meaning of a W mass in the process $e^+e^- \to W^+W^- \to q_1\bar{q}_2q_3\bar{q}_4$. Specifically, it is not possible to subdivide the final state into two groups of particles, one of which is produced by the $q_1\bar{q}_2$ system of the W^+ decay and the other by the $q_3\bar{q}_4$ system of the W^- decay: some particles originate from the collective action of the two systems. Since a determination of the W mass is one of the main objectives of LEP 2, it is important to understand how large the ambiguities can be. A statistical error of 55 MeV per experiment is expected, so the precision of the theoretical predictions should match or exceed this accuracy. A complete description of interference effects is not possible since non-perturbative QCD is not well understood. The concept of colour interconnection/rearrangement is therefore useful to quantify effects. In a rearrangement two original colour singlets (such as $q_1\bar{q}_2$ and $q_3\bar{q}_4$) are transmuted into two new ones (such as $q_1\bar{q}_4$ and $q_3\bar{q}_2$). Subsequently each singlet system is assumed to hadronize independently according to the standard algorithms. Depending on whether a reconnection has occurred or not, the hadronic final state is then going to be somewhat different. For a detailed understanding of QCD interconnection effects one needs to examine the space-time picture of the process. It was shown in [1] that interference is negligibly small for energetic perturbative gluon emission. Firstly, the W^+ and W^- decay at separate times after production, which leads to large relative phases for radiation off the two constituents of a rearranged system, and a corresponding dampening of the QCD cascades. Secondly, within the perturbative scenario the colour transmutation appears only in order α_s^2 and is colour-suppressed. It was concluded that only a few low-energy particles could be affected. In order to understand the interconnection effects occuring at the hadronization stage, the standard Lund fragmentation model has been considerably extended and several alternative models for the space-time structure of the fragmentation process have been developed [1]. Comparing different models with the no-reconnection scenario, it turns out that interconnection effects are very small. The change in the averaged charged multiplicity is at the level of a per cent or less, and similar statements hold for rapidity distributions, thrust distributions and so on. The total contribution to the systematic error on the W mass reconstruction may be as large as 40 MeV. This is good news. Otherwise, LEP 2 would not have significant advantages in the measurements of M_W over hadronic machines where the accuracy is steadily improving (the current combined results give $M_W = 80.23 \pm 0.18$ GeV and with the increase in statistics at the Tevatron further improvements are expected).

Clearly, colour rearrangement effects are interesting in their own right, for instance, as a new probe of the non-perturbative QCD dynamics. However, the standard measures considered in [1] seem to be below the experimental precision one may expect at LEP 2. A more optimistic conclusion has been reached in Ref. [3], where some specific ways to disentangle colour reconnection phenomena were proposed. But personally I still believe that one will need good luck in order to establish the nature and size of the QCD rearrangement effects in real-life experiments.

We turn now to the process $e^+e^- \to t\bar{t} \to bW^+\bar{b}W^-$ [2]. We assume that the W's decay leptonically, so the colour flow is generated only by the t and b quarks. Further, we restrict ourselves to the region a few GeV above the $t\bar{t}$ threshold. The interplay of several particle production sources is reminiscent of the colour rearrangement effects we have studied for $e^+e^- \to W^+W^-$ but there are important differences. From the onset, W^+W^- events consist of two separate colour singlets, $q_1\bar{q}_2$ and $q_3\bar{q}_4$, so that there is no logical imperative of an interplay between the two. Something extra has to happen to induce a colour rearrangement to $q_1\bar{q}_4$ and $q_3\bar{q}_2$ singlets, such as a perturbative exchange of gluons or a non-perturbative string overlap. This introduces a sizeable dependence on the space-time picture, i.e. on how far separated the W^+ and W^- decay vertices are. Except in the unlikely case that top-flavoured hadrons would have time to form, the process $e^+e^- \to t\bar{t} \to bW^+\bar{b}W^-$ only involves one colour singlet. Therefore an interplay is here inevitable, while a colour rearrangement of the above kind is impossible.

One of the main objectives of a Next Linear e^+e^- Collider will be to determine the m_t with an accuracy of about 300 MeV. One method is to reconstruct the top invariant mass event by event, another is to measure the top momentum distribution. In either case, the colour flow restructuring could introduce the potentiality for a systematic bias in the m_t determination. In Ref. [2] we concentrated on the possible manifestations of the QCD interconnection effects in the distribution of the particle flow in the final state. As a specific example, we studied the multiplicity of leptonic top decays as a function of the angle between the b and $\bar{b}$ jets. The main conclusions of our study are:

- The interconnection phenomena should be readily visible in the variation of the average multiplicity as a function of the angle between the b and $\bar{b}$.

- A more detailed test is obtained by splitting the particle content in momentum bins. The high-momentum particles are mainly associated with the $\widehat{tb}$ and $\widehat{\bar{t}\bar{b}}$ dipoles and follow the b and $\bar{b}$ directions, while the low-momentum ones are sensitive to the influence of the $\widehat{b\bar{b}}$ dipole.

- A correct description of the event shapes in top decay, combined with sensible reconstruction algorithms, should give errors on the top mass that are significantly less than 100 MeV.

The possibility of interference interconnection effects in the $t\bar{t}$ production is surely not restricted to the events studied here. One could discuss also hadronic W decays and/or the interferences with beam jets in $pp/p\bar{p} \rightarrow t\bar{t}$ events. The problem with these processes is that there are too many other uncertainties. At the moment, the main uncertainties come from the modelling of the non-perturbative fragmentation. However, the main conclusion that the QCD interconnection does not induce any sizeable restructuring of the final particle flows should remain valid for all of these cases.

1. T. Sjöstrand and V.A. Khoze, *Z. Phys.* **C62** (1994) 281; *Phys. Rev. Lett.* **72** (1994) 28.
2. V.A. Khoze and T. Sjöstrand, *Phys. Lett.* **B328** (1994) 466.
3. G. Güstafson and J. Häkkinen, Lund preprint, LU TP 94-9.

HIGH PRECISION CHARMONIUM SPECTROSCOPY

IN ANTIPROTON-PROTON ANNIHILATIONS

NADIA PASTRONE

Istituto Nazionale di Fisica Nucleare, sez. Torino,
Via P. Giuria 1, 10125 Torino, Italy

for the E760 Collaboration

ABSTRACT

Charmonium spectroscopy has been successfully studied by experiment 760, at Fermilab $\bar{p}$ Accumulator, resulting in data of unprecedented accuracy. Charmonium states are resonantly produced from $\bar{p}p$ annihilations and masses, widths and branching ratios are precisely measured studying their electromagnetic decays. The main results of the experiment are presented.

1. Introduction

The Fermilab experiment 760 was mainly dedicated to the study of charmonium spectroscopy in $\bar{p}p$ annihilations, exploiting a technique that was first pioneered by experiment R704 at CERN-ISR. In $\bar{p}p$ annihilations all charmonium states can be directly formed and measured without being limited by detector resolution, while in e^+e^- experiments this is possible only for $J^{PC} = 1^{--}$ states.

In E760 a beam of antiprotons circulating in the Accumulator ring collides with an internal hydrogen gas-jet target, providing high luminosity (up to 10^{31} cm^{-2}s^{-1}) and a pointlike annihilation source.

The masses and widths of the $(\bar{c}c)$ states are determined directly from the antiproton beam energy by measuring the excitation curves of the resonances as the energy of the beam is changed in small steps. The resolution is limited only by the accuracy of the $\bar{p}$ beam. The product $B_{R \to \bar{p}p} \times B_{R \to out}$ is also measured when detector efficiency and acceptance are known.

The beam was decelerated from 8.9 GeV to an energy just above the resonance and then the scan was performed by steps of between 170-500 keV (center of mass energy).

A direct measurement of sub-MeV widths for the charmonium states is allowed since the beam was cooled to $(\Delta p/p)_{rms} \approx 2 \times 10^{-4}$, resulting in a spread in the center of mass energy of 240 keV. The absolute calibration of the beam energy was obtained using the mass of the ψ', known to ± 100 keV/c^2 from measurements at e^+e^- colliders. This contributes a systematic error of only 33 keV/c^2 to the measurement of J/ψ mass ($M_{J/\psi} = 3096.87 \pm 0.03 \pm 0.03$ MeV/c^2).

Due to the presence of a large background of non resonant $\bar{p}p$ strong interactions, the rare events from direct formation of charmonium resonances (1 part in 10^6 of the total cross section) were selected by detecting their decays into e^+e^- or, depending on the state quantum numbers, into $\gamma\gamma$. For higher $(c\bar{c})$ excitations that do not decay directly into a two body electromagnetic final state, it is possible to obtain a strong signature by detecting their inclusive decay to J/ψ or η_c, which in turn decay to e^+e^- or $\gamma\gamma$ respectively. The detector [3], with a large acceptance (2π in azimuth and from 2^0 to 70^0 in polar angle) was optimized for the identification of e.m. final states from charmonium decay. Trigger hodoscopes and tracking were completed by a Cherenkov counter to tag electrons and an e.m. calorimeter to identify electrons and photons and measure their energies and directions. Luminosity was monitored measuring the yield of recoil protons from forward $\bar{p}p$ elastic scattering.

E760 collected data for an integrated luminosity of ≈ 30 pb^{-1}, during a total period of nine months in 1990-1991.

2. Experimental results

2.1. Measurement of the parameters of J/ψ, ψ', χ_1 and χ_2

The resonance formation was easily identified in the E760 detector, measuring the invariant mass of the electron-positron pair, m_{ee}, in the decay channels:

$J/\psi \to e^+e^-$; $\qquad \psi' \to e^+e^-$; $\qquad \psi' \to J/\psi + X$; $\qquad \chi_{1,2} \to J/\psi + \gamma$

By selecting these decay modes an almost background free sample can be obtained.

Total widths of $(99\pm12\pm6)$ keV and $(306\pm36\pm16)$ keV were measured respectively for J/ψ and ψ' states, from the study of the line shapes [3].

The analysis of the excitation curves of χ_1 and χ_2, shown in fig.1a together with a typical beam profile, yielded the results[1] reported in the table (Γ_{χ_1} measured for the first time). Fig.1b compares the existing measurements [?] of M_{χ_1} and M_{χ_2}.

	M_R (MeV/c^2)	Γ_R (MeV)	$\Gamma(R \to \bar{p}p) \times B(R \to (e^+e^-)_{J/\psi}\gamma)$ (eV)	$\Gamma(R \to \bar{p}p)$ (eV)
χ_1	$3510.53 \pm 0.04 \pm 0.12$	$0.88 \pm 0.11 \pm 0.08$	$1.29 \pm 0.09 \pm 0.13$	$69 \pm 9 \pm 10$
χ_2	$3556.15 \pm 0.07 \pm 0.12$	$1.98 \pm 0.17 \pm 0.07$	$1.67 \pm 0.09 \pm 0.12$	$180 \pm 16 \pm 26$

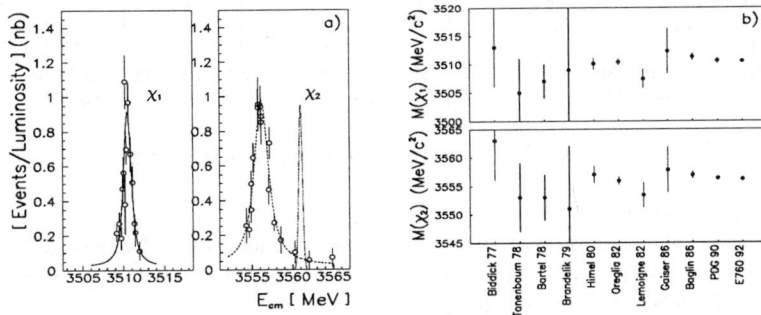

Figure 1: Excitation curves of χ_1 and χ_2 (a) and comparison of results on $M_{\chi_{1,2}}$

2.2. Search for the singlet state $h_c(^1P_1)$

The observation of $^1P_1$, never seen in e^+e^- experiments, is important because a comparison of its mass with the center of gravity of the three triplet ^3P states:

$$m_{og} = \frac{\sum_j (2J+1) m_{\chi_j}}{\sum_j (2J+1)} = 3525.27 \pm 0.12 \; MeV \quad \text{(using E760 } M_{\chi_1} \text{ and } M_{\chi_2}\text{)}$$

provides a measurement of the deviation of the vector part of the $q\bar{q}$ interaction from pure one gluon exchange. The process $\bar{p}p \to h_c$ is forbidden by helicity conservation rules in massless QCD. Moreover the width is expected to be narrow (< 1 MeV).

E760 searched for the h_c near m_{cog} (500 keV steps), collecting an integrated luminosity of 16 pb^{-1} and focusing on the decays:

$$h_c \to \eta_c \gamma \to \gamma\gamma \qquad h_c \to J\psi\pi^0 \to e^+e^-\pi^0 \qquad h_c \to J\psi 2\pi \to e^+e^- 2\pi$$

No evidence of the $\eta_c\gamma$ decay channel was seen, owing to the very small branching ratio $\eta_c \to \gamma\gamma$. The J/ψ in the other two final states provides a strong signature.

A structure in the $J/\psi\pi^0$ channel can be seen in fig.2a, interpreted as the first evidence of the $^1P_1$ state of charmonium [2,6]. From the study of the excitation curve a value for the mass, $M_{h_c} = (3526.2 \pm 0.15 \pm 0.2) \; MeV/c^2$ and an upper limit on the width, $\Gamma_{h_c} \leq 1.1 MeV$ (90%C.L.) were derived. For values of Γ_{h_c} in the range 0.5 - 1.0 MeV, the product of the branching fractions, $B_{h_c \to \bar{p}p} \times B_{h_c \to J\psi\pi^0}$ ranges from $(1.7 \pm 0.4) \times 10^{-7}$ to $(2.3 \pm 0.6) \times 10^{-7}$. The probability that the resonant signal is a fluctuation of the flat continuum is $\leq 1/400$.

No events were found that could fit the states $J/\psi\pi^0\pi^0$ or $J/\psi\pi^+\pi^-$; this sets a limit to the ratio $B_{h_c \to J/\psi 2\pi}/B_{h_c \to J\psi\pi^0} < 0.18$ at 90% confidence level.

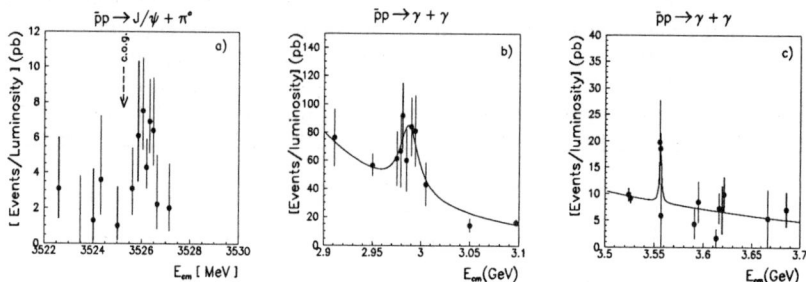

Figure 2: Measured cross sections vs center-of-mass energy

2.3. The $\gamma\gamma$ final states

The study of the reaction $\bar{p}p \to (c\bar{c})_{C_{even}} \to \gamma\gamma$ is difficult since decay branching fractions to $\gamma\gamma$ are small and backgrounds from all-neutral final states of $\bar{p}p$ interactions, especially $\pi^0\pi^0$ and $\pi^0\gamma$, are quite large.

A scan (3.56 pb^{-1}) was performed in the η_c region where a structure is seen around 2990 MeV (fig.2b) above a background of comparable magnitude [7]. To optimize signal to background ratio it was necessary to restrict the angular acceptance to $|\cos(\theta_\gamma^*)| \leq 0.25$, where θ_γ^* is the $\gamma\bar{p}$ angle in the center-of-mass system. A fit to a background cross section (parametrized with a power law) plus a Breit-Wigner line shape, gave a mass $M_{\eta_c} = 2987.5^{+3.0}_{-2.8}$ MeV/c^2 (to be compared to the world average[5] of 2978.8 ± 1.9 MeV/c^2), a total width $\Gamma_{\eta_c} = 23.7^{+11.1}_{-7.0}$ MeV and a product of $B_{\eta_c \to \bar{p}p} \times B_{\eta_c \to \gamma\gamma} = (35.4^{+8.0}_{-7.2} \pm 2.0) \times 10^{-8}$. The partial decay width to $\gamma\gamma$ ($\Gamma_{\gamma\gamma} = 7.0^{+2.9}_{-2.0} \pm 2.3$ keV) is compared to results from other experiments [5] in fig.3a.

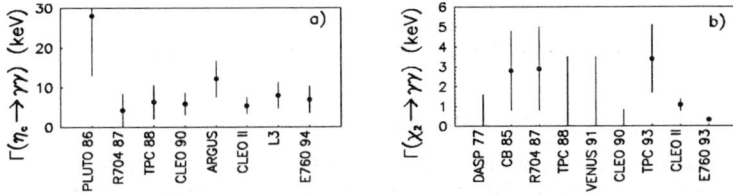

Figure 3: $\Gamma_{\eta_c \to \gamma\gamma}$ (a) and $\Gamma_{\chi_2 \to \gamma\gamma}$ (b): comparison of existing results

Data were also collected in the region $3.5 \leq E_{cm} \leq 3.7$ GeV, where 1P and 2S resonance are formed (fig. 2c with events selected cutting at $\mid cos(\theta^*_\gamma) \mid \leq 0.4$).
A four σ enhancement above the background level was observed at the $\chi_2(3556)$ [4]. This event excess, fitted with a Breit-Wigner distribution of the χ_2 mass and width, gives for the product of branching ratios $B_{\chi_2 \to \bar{p}p} \times B_{\chi_2 \to \gamma\gamma} = (1.60 \pm 0.39 \pm 0.16) \times 10^{-8}$. The measured partial decay width to $\gamma\gamma$, $\Gamma_{\gamma\gamma} = 0.32 \pm 0.08 \pm 0.05$ keV, is compared to other recent results [5] in fig.3b.

A more systematic and higher sensitivity search of the η'_c must be performed in the future, since no evidence of an enhancement was seen in the range of energy spanned.

3. Conclusions

Considerable progress has been achieved in the field of charmonium spectroscopy with the study of resonant formation in $\bar{p}p$ annihilations [6].

During the next period of fixed target mode at Fermilab, experiment 835 will take data with an istantaneus luminosity increased by a factor five. A further step in accuracy and sensitivity is forseen to provide new measurements of η_c, χ_0, h_c, η'_c states and to search for yet undetected narrow states of charmonium as $^3D_2$, $^1D_2$ or bound states $(D\bar{D}^*)$, $(D^*\bar{D}^*)$ near the thresholds.

1. T. A. Armstrong et al., *Nucl. Phys.* **B373** (1992) 35.
2. T. A. Armstrong et al., *Phys. Rev. Lett.* **69** (1992) 2337.
3. T. A. Armstrong et al., *Phys. Rev.* **D47** (1993) 772.
4. T. A. Armstrong et al., *Phys. Rev. Lett.* **70** (1993) 2988.
5. Particle Data Group, *Phys. Rev.* **D50** (1994) and references therein.
6. R. Cester and P. A. Rapidis, *Ann. Rev. Part. Sci.* **44** (1994) 329.
7. T. A. Armstrong et al., *Phys. Rev.* in preparation.

RECENT RESULTS OF
THE CERN HYPERON BEAM EXPERIMENT WA89

THE WA89 COLLABORATION *

represented by

Corinne BERAT

ISN, 53 Avenue des Martyrs, 38026 Grenoble Cedex, France

ABSTRACT

Recent results from the WA89 collaboration, the CERN hyperon beam experiment, are presented. Measurements of the x_f and $p_\perp$ dependence of the production cross-section for charmed (non-strange) baryons have been obtained. A leading particle effect is observed, as already seen with charmed mesons. Results concerning the charmed strange baryons Ξ_c^+, Ξ_c^0 are also quoted together with the first evidence for the symmetric state $\Xi_c^{+'}$.

1. Introduction

The WA89 fixed target experiment [1] is dedicated to the study of charmed strange baryons (production, decay properties and observation of yet unobserved baryons belonging to the SU(4) multiplets) and to the search for possible exotic multiquark states. It uses the high energy charged hyperon beam (330 GeV/c) at the CERN SPS : the Σ^- projectiles are expected to enhance the production of final states containing a strange quark, in order to obtain charmed strange baryons. The experiment is using the Omega spectrometer in the CERN West Hall. The full setup has been described elsewhere [2,3]. The present results emerged from the analysis of 100 Million interaction triggers recorded during the first data taking in 1991.

2. Charmed (Non Strange) Baryons

The lowest mass charmed baryon Λ_c^+ has been studied in the $pK^-\pi^+$ decay channel. It was used to search for the Σ_c baryons ($\Sigma_c \to \Lambda_c^+\pi$). Σ_c^0 's have been reconstructed (figure 1), after selecting events within ± 10 MeV/c^2 around the Λ_c^+ mass. When applying the same method to the isospin partner Σ_c^{++}, no peak is visible, although the overall acceptance is the same. This observation is quantified by an upper limit on the ratio of the particle yields:

$$N(\Sigma_c^{++})/N(\Sigma_c^0) < 0.52 \qquad (\text{at } 90\text{ \% CL}) \qquad \text{for } x_f > 0.2$$

We interpret this as the leading particle effect, which is supposed to favour the forward production of those particles which have quarks in common with the projectile. The same effect has also been observed in the D meson production [4].

*CERN - Genoa Univ./INFN - Grenoble ISN/IN2P3 - Heidelberg MPI f. Kernphysik - Heidelberg Univ. - Mainz Univ. Inst. f. Kernphysik - Moscow Lebedev Phys. Inst.

Figure 2 shows the first determination of the x_f- and $p_\perp^2$-distributions of the Σ_c^0. The distributions are fitted with the parametrisation $(1 - x_f)^n\, e^{-bp_\perp^2}$ commonly used to describe the x_f- and $p_\perp^2$-behaviour of double differential cross-sections.

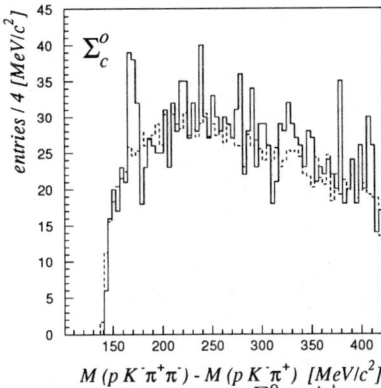

Fig. 1. Solid line: signal of $\Sigma_c^0 \to \Lambda_c^+ \pi^-$. Dashed line: distribution from Λ_c^+ sidebands.

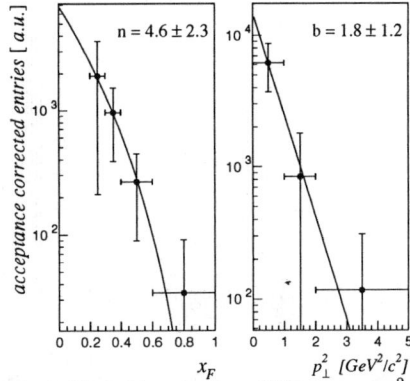

Fig. 2. First determination of the x_f- and $p_\perp^2$-distributions of Σ_c^0 (statistical errors only).

Similar distributions are obtained for Λ_c^+ and D^- (not reported here), but the present statistics do not yet allow conclusive comparisons between the spectral shapes for the different particles Λ_c^+, Σ_c^0 and D^-. For the same reason (low statistics) a detailed investigation of the leading particle effects in terms of x_f-distributions is not possible. However, none of the observed x_f-distributions shows any rise towards high x_f, which indicates that there is probably no strong diffractive component.

3. Charmed Strange Baryons

Up to now, the charmed strange baryons Ξ_c^+ and Ξ_c^0 have been reconstructed in two decay channels $\Lambda K^-\pi^+\pi^+$, $\Xi^-\pi^+\pi^+$ and $\Lambda K^-\pi^+$, $\Xi^-\pi^+$, respectively. It is the first observation of the $\Lambda K^-\pi^+$ decay mode of the Ξ_c^0.

The statistically most significant decay mode $\Lambda K^-\pi^+\pi^+$ is used to determine the Ξ_c^+ lifetime [5]:

$$\tau\,(\Xi_c^+) = 0.32\,^{+0.08}_{-0.06}(stat.),$$

and this measurement is in agreement with the lifetime hierarchy:

$$\tau\,(\Xi_c^+) > \tau(\Lambda_c^+) > \tau\,(\Xi_c^0)$$

as predicted by Guberina et al.[7]

Using the available decay mode of the Ξ_c hyperons as a base, we looked for the yet unobserved $\Xi_c{}'$ multiplet partners. Their states are symmetric in the two light quarks. The chromomagnetic hyperfine interaction leads to a $\Xi_c{}'$ - Ξ_c mass splitting, for which most predictions give a value around 90 to 110 MeV/c^2, involving a radiative decay of the $\Xi_c{}'$. In figure 3 the mass splitting spectra for the two observed decay modes of the Ξ_c^+ are presented. An enhancement at a mass difference of 95 MeV/c^2 is observed: it is interpreted as the first experimental evidence for the $\Xi_c^{+'}$ state [6].

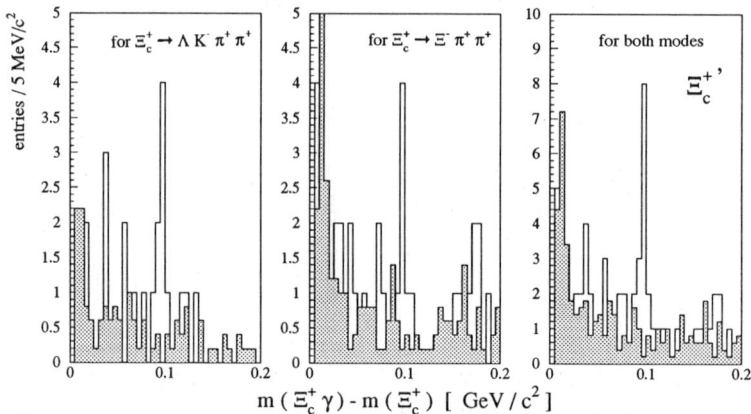

Fig. 3. $\Xi_c^{+'} \to \Xi_c^+ \gamma$ yield, shown for two Ξ_c^+ decay modes and for both modes. The hatched distributions show the scaled background from sidebands of the Ξ_c^+ mass distributions.

4. Outlook

The analysis of the 1991 data has demonstrated the ability of the WA89 experiment to reconstruct ground and excited states of charmed baryons. In 1993, the amount of recorded interactions was doubled with an improved setup. A large increase in charm statistics is thus expected from the 1993 data analysis, as well as from the 1994 current data taking. This should allow investigating the production and decay properties of charmed strange baryons with the same accuracy as currently available for Λ_c^+ and in the charmed meson sector.

5. References

1. A. Forino et al., *Proposal CERN/SPSC/87-43*
2. M. I. Adamovich et al., *Preprint CERN-PPE/94-86*
3. A. Simon, Talk Presented at the XXIX. Rencontres de Moriond, *QCD and High Energy Hadronic Interactions*, Méribel, March 19-26, 1994
4. S. Brons, *Doctoral thesis*, Heidelberg University, 1994 and references therein.
5. F. Dropmann, *Doctoral thesis*, Heidelberg University, 1994 and references therein.
6. B. Volkemer, *Doctoral thesis*, Mainz University, in preparation.
7. B. Guberina et al., *Z. Phys. C* **33** (1986) 297.

CONDENSATES AND CONFINEMENT

JÓZEF M. NAMYSŁOWSKI *

Institute of Theoretical Physics, Warsaw University
Hoża 69, PL-00-681 Warsaw, Poland

ABSTRACT

Nonzero quark and gluon condensates are responsible for confinement of quarks
and gluons. By confinement we mean: i) the absence of any asymptotic states for
either quarks, or gluons, ii) the absence of any continuum spectrum for partons,
iii) the absence of any colour-full parton bound systems, and iv) the presence of
hadrons which are colour-less bound states of quarks (antiquarks) and gluons.

Phenomenologically, the Local Parton-Hadron Duality of Dokshitzer and Troyan[1]
supports the presence of physical singularities in three-point vertex functions (quark-
gluon-quark, and triple-gluon). By Slavnov-Taylor identities the same physical singularities must appear[2] in the inverse of quark propagator, and in the inverse of
transverse gluon propagator. Just algebraically, these singularities in inverses of
propagators lead to the absence[2] of quark and gluon asymptotic states. These physical singularities are also supported by the existence of self-consistent solutions of
Dyson-Schwinger equations[3] which are exact solutions on the one-loop level, at these
singularities.

Both quark and gluon propagators are of the Wheeler type[2], i.e. they are real
propagators, equal to the avarage of the retarded and advanced propagators. Nonzero
values of quark and gluon condensates[1] insure that quark and gluon propagators correspond to short living quark and gluon states[3]. Time scales of these states are
inversly proportional to inverses of mass scales of condensates, and are equal to fractions of fermi. The absence of any continuum partonic states is guaranteed[2] by: 1)
the Wheeler character of the relative motion propagator, and 2) the vanishing of the
scalar product of the Wightman-Garding relative momentum and the hadron momentum. Both of these properties are due to the nonzero values of quark and gluon
condensates.

The nonzero values of quark and gluon condensates[1] are characterized by nonzero
mass scales χ and ν connected with these condensates by following matrix elements
of composite operators in the physical vacuum

$$-\chi^3 \equiv <| \sqrt{\alpha} : \overline{\psi}\psi :|>, \quad \nu^4 \equiv <| \frac{\alpha}{\pi} : G^a_{\mu\nu}G^{\mu\nu}_a :|> . \tag{1}$$

For u and d, s, c and b, and t quarks χ is equal, approximately, to the following
fractions of GeV: 1/4, 1/5, 1/10, and 1/50, respectively. The gluon condensate mass
scale ν is known[1] to be around 1/3 GeV.

---

*E-mail jmn@fuw.edu.pl

The nonzero values of χ and ν allow for the presence of the following physical singularities[2] in inverses of quark and gluon propagators

$$S^{-1}(p) = \not{p} - m + i\epsilon + \frac{\chi^3/M}{\not{p} - M + i\epsilon}, \qquad D_T^{-1}(p) = p^2 + i\epsilon + \frac{\nu^4}{p^2 + i\epsilon}, \qquad (2)$$

where m is the current quark mass, and M is the mass of corresponding pseudoscalar mesons: π-meson for u, d; K-meson for s; D-meson for c; and B-meson for b.

It is trivial to invert algebraically inverses of propagators in Eq.(2), and the result for the gluon transverse propagator is particularly easy

$$D_T(p^2) = \frac{p^2}{p^4 + \nu^4} = \frac{1}{2}\left(\frac{1}{p^2 + i\nu^2} + \frac{1}{p^2 - i\nu^2}\right). \qquad (3)$$

This equation demonstrates the absence of any real momentum pole for gluon, and therefore the absence of any asymptotic gluon state. In Eq.(3) there are two complex conjugate poles in the variable p^2, showing the real character of the Wheeler gluon propagator. The presence of these poles is not in conflict either with causality, or with unitarity, or with analyticity[2].

The presence of physical poles in Eq.(2), and the numerical values of the condensate mass scales χ, and ν, can be verified[2] by reproducing these mass scales from vacuum-to-vacuum transitions obtained by closing up in the position space, respectively, quark, and gluon lines, represented by nonperturbative propagators. For numerical stability of such calculations anomalous dimensions of propagators must be included. Then, the mass scales χ and ν are independent of a huge variation of an arbitrary mass scale μ, used as a normalization point in the renormalization group equation solutions. The normalization mass scale μ is varied[2] between 1 GeV and 100 000 GeV.

Nonzero mass scales χ and ν are responsible for the exponential damping of the Fourier transform of propagators. The time-dependence of the zero three-momentum gluon propagator is given[2] by the expression

$$\text{Fourier tr.}\left[\frac{-1}{p^2 + i\epsilon + \nu^4/(p^2 + i\epsilon)}\right]_{\vec{p}=0} = \frac{-1}{2\nu}e^{-\nu|\tau|/\sqrt{2}}\sin\left(\frac{\pi}{4} - \frac{\nu|\tau|}{\sqrt{2}}\right)$$

$$= \frac{-1}{2\sqrt{2}\nu}e^{-\nu|\tau|/\sqrt{2}}\left[\cos\left(\frac{\nu\tau}{\sqrt{2}}\right) - \sin\left(\frac{\nu|\tau|}{\sqrt{2}}\right)\right]. \qquad (4)$$

This expression has the time scale of the short living gluon state of the value $\tau = \sqrt{2}/\nu$. It is between 0.6 and 0.8 fm, for ν between 1/3 and 1/2 GeV.

The quark propagator, calculated algebraically from Eq.(2), is[2]

$$S(\not{p}) \equiv \frac{Z(\not{p} + \rho)}{p^2 - \rho^2} + \frac{Z^*(\not{p} + \rho^*)}{p^2 - \rho^{*2}}, \qquad (5)$$

where complex numbers Z and ρ are defined by the mass scales: m, M, and χ as follows

$$\text{Re}Z = \frac{1}{2}, \qquad \text{Im}Z = \frac{(M-m)/4}{\sqrt{\chi^3/M - \frac{1}{4}(M-m)^2}},$$

$$\text{Re}\rho = \frac{1}{2}(m+M), \qquad \text{Im}\rho = \sqrt{\chi^3/M - \frac{1}{4}(M-m)^2}. \tag{6}$$

The Wheeler relative motion propagator in quark-antiquark system is[2]

$$\frac{Z^2(\not p_1 + \rho)(\not p_2 + \rho)}{(p_1^2 - \rho^2)(p_2^2 - \rho^2)}\, \delta\left[\frac{qP(M^2 - \overline{P}^2)}{(p_1^2 - \rho^2)(p_2^2 - \rho^2)}\right]$$

$$+\frac{Z^{*2}(\not p_1 + \rho^*)(\not p_2 + \rho^*)}{(p_1^2 - \rho^{*2})(p_2^2 - \rho^{*2})}\, \delta\left[\frac{qP(M^2 - \overline{P}^{*2})}{(p_1^2 - \rho^{*2})(p_2^2 - \rho^{*2})}\right]. \tag{7}$$

From this expression for the Wheeler relative-motion propagator it follows immediately the definition of a momentum-dependent constituent quark mass $\mathcal{M}(q^2)$

$$\mathcal{M}(q^2) = \sqrt{\sqrt{(mM + \chi^3/M)^2 - q^2(m^2 + M^2 - 2\chi^3/M) + q^4} + q^2}. \tag{8}$$

For zero value of the Wightman-Garding relative momentum q in the quark-antiquark system the constituent quark mass takes the maximal value $\mathcal{M}(0)$, which is: 0.33 GeV for u and d quarks, 0.4 GeV for s quark, 1.7 GeV for c quark, and approximately 5 GeV for b quark. The explicit expression for the maximal value of the constituent quark mass is

$$\mathcal{M}(0) = (mM + \chi^3/M)^{1/2}. \tag{9}$$

For very large values of the relative momentum q the mass $\mathcal{M}(q^2)$ tends to zero, irrespective of flavour.

1. References

1. M. A. Shifman, A. I. Vainshtein and V. I. Zakharov, *Nucl.Phys.* **B147**(1979) 385,488; Yu. L. Dokshitzer and S. I. Troyan, *Proc. XIX Winter School of the LNPI*, **1** (1984) 14.
2. J. M. Namysłowski, in *Particles and Fields*, Proc. First German- Polish Symposium, ed.'s H. D. Doebner, M. Pawłowski and R. Raczka, World Scientific, Singapore 1994, p. 155; J. M. Namysłowski, "Nonperturbative QCD", IFT, Warsaw preprint IFT/10/94.
3. M. Stingl, *Phys. Rev.* **D34** (1986) 3863; H. Habel, U. Konning, H. G. Reusch, M. Stingl and S. Wigard, *Z.Phys.* **A336**(1990)423,435.

QUARK CONFINEMENT, INTERQUARK POTENTIAL AND DIRAC EQUATION

A.B.PESTOV

Bogoliubov Laboratory of Theoretical Physics, Joint Institute for Nuclear Research
141980 Dubna, Moscow region, Russia

ABSTRACT

It is substantiated from physical, geometrical and group-theoretical points of view that in the framework of QCD it is not only necessary but also possible to modify the Dirac equation so that the correspondence principle holds valid. The wave equation for describing the dynamics of quarks is proposed. It is shown that the interquark potential expresses the Coulomb law for the quarks and, in fact, coincides with the well–known Cornell potential.

1. Preliminaries

As it is well-known, QCD is conceptually a simple theory. Its structure is solely determined by symmetry principles. However, the connection between so important phenomena as confinement and quark–lepton symmetry, on the one hand, and the first principles of QCD, on the other one, are absent. There is only one natural way to solve this problem: one needs to modify the Dirac equation in view of unusual quark properties. Here we consider this step from different view points.

2. Material Point and Rigid Body

In the 30's it has been emphasized by Casimir that from the physical point of view the notion of a rigid body is as fundamental as the notion of a material point is. Now we have a possibility to put forward the main idea: a quark is an elementary particle with the essential properties of a rotating rigid body or a top. What is the problem? For an electron we have the fundamental sequence: geometrical point, material point and electron. So, we should also have the sequence: geometrical point, rigid body and quark. The sense of these correspondences is quite transparent but the latter sequence cannot be realized in the framework of the special theory of relativity because a rigid body cannot be considered as a fundamental concept in this theory.

3. Geometrical Framework

Nevertheless, there is a second important idea that the sequence: geometrical point, rigid body, quark can be realized in a new geometrical framework. The result is that if for an electron the underlying space-time manifold is a 4-plane E_4, then for a quark the underlying space-time manifold is a one sheeted hyperboloid H_4 which can be represented as a hyperquadric

$$(x^0)^2 - (x^1)^2 - (x^2)^2 - (x^3)^2 - (x^4)^2 = -a^2,$$

in the 5-dimensional Minkowski space–time with the coordinates x^0, x^1, x^2, x^3, x^4. Here a is the radius of H_4, which can be interpreted as the radius of the region of confinement.

4. Underlying symmetry group

Let us consider one very important and general property of E_4 and H_4. For E_4 there exists a 4–parameter group of transformations that can transform any point of E_4 to any a priori given point. It is the group of translations. The space-time manifold H_4 admits a 4-parameter group of that type as well. It means from the physical point of view that if for electrons translation is of fundamental significance, then for quarks it is rotation that can be represented as a motion in the space-time manifold H_4. In coordinates x^0, x^1, x^2, x^3, x^4 the operators analogous to the operators of translations $T_a = \partial/\partial x^a$, $a = 0, 1, 2, 3$, have the form

$$R_0 = (a + x_0{}^2/a)\, \partial/\partial x^0 + (x_0/a) \sum_{i=1}^{4} x^i \partial/\partial x^i,$$

$$R_1 = x^2\, \partial/\partial x^3 - x^3\, \partial/\partial x^2 + x^1\, \partial/\partial x^4 - x^4\, \partial/\partial x^1,$$

$$R_2 = x^3\, \partial/\partial x^1 - x^1\, \partial/\partial x^3 + x^2\, \partial/\partial x^4 - x^4\, \partial/\partial x^2,$$

$$R_3 = x^1\, \partial/\partial x^2 - x^2\, \partial/\partial x^1 + x^3\, \partial/\partial x^4 - x^4\, \partial/\partial x^3.$$

5. Hamiltonian

The Hamiltonian for the electrons has as usual the following form:

$$H_e = c(\alpha, \mathbf{P}_e) + \rho_3 m_e c^2,$$

where $\mathbf{P}_e = i\hbar \nabla$. For the quarks we have

$$H_q = c(\alpha, \mathbf{P}_q) + \rho_3 m_q c^2,$$

where

$$\mathbf{P}_q = \frac{i\hbar}{a}(\mathbf{r} \times \nabla + \frac{a^2 - r^2}{2a}\nabla + \frac{\mathbf{r}}{a}(\mathbf{r}, \nabla)).$$

Here $\mathbf{P}_q = i\hbar/a\,\mathbf{R}$, $\mathbf{R} = (R_1, R_2, R_3)$ and R_1, R_2, R_3 are expressed through the usual coordinates. For the eigenvalues E of H_q we have

$$E^2 = m_q^2 c^4 + n^2 \frac{c^2 \hbar^2}{a^2},$$

where n is integer. In the classical limit, when $n \to \infty$ and $a \gg \hbar/m_q c$, we obtain

$$E = m_q c^2 + \frac{L^2}{2I},$$

where $L = n\hbar$ and $I = m_q a^2$. The latter relation is consistent with the classical formula for the energy of a spherical top.

6. Coulomb law

With the help of the Hamiltonian H_q one can derive the Maxwell equations for quarks. As it is known, the Coulomb potential for electrons can be derived as a solution of the equations of electrostatics invariant under the group of Euclidean motions including rotations and translations. We can look for the Coulomb potential for quarks in an analogous manner. The solution can be represented in the following form:

$$\varphi(r) = q(\frac{1}{r} - \frac{r}{a^2} + \frac{1}{a}),$$

where q is the quark charge. Thus, the interquark potential expresses the Coulomb law for the quarks and, in fact, coincides with the well–known Cornell potential that was first very successfully used by the Cornell group [1] for the description of the charmonium. But in our case, it contains the quark charge and size of the confinement region a as free parameters. So, the analysis of the charmonium and bottomium in the new framework is a very important problem.

7. Quark–lepton symmetry

At large a, when $a \to \infty$ from the theory of quarks we derive the theory of electrons but with electrons evidently deconfined, because in this case the region of confinement is the Euclidean space. Thus, the symmetry between quarks and leptons has a natural explanation. One can also say that the physical meaning of the phenomenon called confinement consists in that quarks possess properties of a top.

Since the key ideas have been put forward, the problem is to develop in detail the classical and quantum mechanics of quarks, and it is quite possible that at the end we shall have the full and consistent theory of quarks.

1. E. Eichten, H. Gottfried, T. Kinoshita, K. D. Lane, T.–M. Yan, *Phys. Rev.* **D21** (1980) 203.
2. A.B. Pestov, *Hadronic Journal Suppl.* **8** (1993) 99.

POSTER SESSION

CLASSICAL DYNAMICS
OF ROTATING RELATIVISTIC STRING WITH MASSIVE ENDS:
THE REGGE TRAJECTORIES AND QUARK MASSES

B. M. BARBASHOV

Bogoliubov Laboratory of Theoretical Physics, Joint Institute for Nuclear Research
Dubna, SU-141980, Russia

ABSTRACT

Dynamic equations in the theory of the relativistic string with point spinless masses at the ends are formulated in terms of geometric invariants of world trajectories of the massive ends of the string (curvature k_i and torsion $\kappa_i(\tau)$ of the trajectories). With these characteristics we reproduce the string world surface up to its position in the Minkowski space E_2^1. The torsions $\kappa_i(\tau)$, $i = 1, 2$ obey a system of second order differential equations with delay arguments describing the retardation effects of the interaction of masses through the strings, k_i being constants. The constant torsions are investigated in detail. In this case the string world sheet is a helicoid in E_2^1. A nonlinear relation (the Regge trajectory) between the angular momentum of the system, J, and the mass squared, M^2, is derived. For given meson masses (M) and spins (J), the masses of quarks are calculated.

A NEW APPROACH TO $SU(3)$ SYMMETRY

OLE L. TRINHAMMER

Rungsted Gymnasium, Stadion Alle 14
DK-2960 Rungsted Kyst, Denmark

ABSTRACT

Baryons are described as stationary states in the Lie group SU(3). Confinement reflects the compactness of the configuration space. Asymptotic freedom is a high energy approximation of the dynamics. The predictions are based on a Hamiltonian inherited from QCD. The spectrum tends to predict realistically the observed neutral flavour baryons. The neutron-proton mass splitting agrees with the experimental value at the sub per mille level. Two neutral resonances to test the approach are predicted in the open charm system around 4400 MeV. It is suggested that QCD be considered an approximation.

HADRON PHENOMENOLOGY FROM Q-DEFORMED POINCARÉ ALGEBRA

Lauro Tomio [*], Jishnu Dey [†], Paulo Leal Ferreira
Instituto de Física Teórica - UNESP, 01405-900, São Paulo, Brasil

and

Rajkumar Roy Choudhury
Indian Statistical Institute, 203 B. T. Road, Calcutta 700 035, India

We find that a κ-deformed Poincaré algebra, defined by Lukierski et al.[1], provides an universal fitting of the experimental Regge trajectories (See ref. 2, for more details). We predict a flattening of the energy spectrum as a function of angular momentum (J), implying that many higher J states are densely packed in an energy region of say 13 to 20 GeV. With the slope parameter α' fixed, we obtain the deformation parameter ϵ by minimizing the total deviation, $\delta_T = \sum_{trajectory} \delta$, for the deformed trajectories, from the corresponding experimental ones. Our results significantly improves the overall fitting of 27 known Regge trajectories, for α' in the range $(0.6 - 0.9) GeV^{-2}$. Furthermore, allowing the slope to vary when minimizing the deviation, our best fit gives $\alpha' = 0.72 GeV^{-2}$ and $\epsilon = 0.14$ fm. [2]

1. J. Lukierski, A. Nowicki and H. Ruegg, *Phys.Lett.* **B 293** (1992) 344.
2. J. Dey, P. L. Ferreira, L. Tomio and R. R. Choudhury, *Phys. Lett.* **B 331** (1994) 355.

THREE PARTICLE RELATIVISTIC DYNAMICS AND APPLICATIONS [*]

Lauro Tomio and Sadhan K. Adhikari

Instituto de Física Teórica, UNESP, São Paulo, Brasil

By analysing the different dynamic contributions to low energy, we employ four types of three-dimensional three-particle relativistic equations and five sets of trinucleon potential models [1]. The relativistic correction to binding energy may vary a lot and even change sign depending on the relativistic formulation employed. The deviations of these observables from those obtained in nonrelativistic models follow the general universal trend of deviations introduced by off- and on-shell variations of two- and three- nucleon potentials in a nonrelativistic model calculation. Consequently, it will be difficult to separate unambiguously the effect of off- and on- shell variations of two- and three- nucleon potentials on low-energy three nucleon observables from the effect of relativistic dynamics.

1. S. K. Adhikari, T. Frederico, and L. Tomio, *Ann. of Phys.* (1994); S. K. Adhikari and L. Tomio, *Ann. of Phys.* (1994) (to appear).

---

[*]Supported partially by CNPq of Brasil.
[†]Supported by FAPESP of Brasil and in part by DST grant no. SP/S2/K04/93, Govt. of India. Permanent address : 1/10 Prince Golam Md. Road, Calcutta 700 026, India

PHASE TRANSITION IN MESONIC BAGS AND LATTICE THEORY *

Lauro Tomio, Carlos F. Araujo Jr., Jishnu Dey†
Instituto de Física Teórica, UNESP
01405-900, São Paulo, Brasil

and

Mira Dey
Department of Physics, Lady Brabourne College, Calcutta, India

By looking for physical quantities which may reflect the phase change in QCD, the sound velocity[1] (v_s) was shown in lattice theory to become zero at the transition point[2]. We find that a simple model, like the bag, where the dynamics is simple and clear, is able to reproduce the lattice result for the masses, at finite temperature[3]. The crossing of zero of v_s suggests that the transition point $T = T_v$ depends on the nature of the meson: the rise of the pion mass with T is much weaker than that of the meson rho in agreement with lattice result. Also, different transition points seem to be indicated by particle emission in heavy ion reactions. For the pion, v_s becomes zero at about $T_v = 133$ MeV, whereas for the ρ the temperature T_v is about 149 MeV.

1. R. V. Gavai and A. Gocksch, *Phys. Rev.* **D 33** (1986) R614.
2. S.Huang, K. J. M. Moriarty, E. Myers and J. Potvin, *Zeits. f. Phys.* **C 50** (1991) 221.
3. C.F. Araujo Jr., J. Dey, L. Tomio and M. Dey, *Mod. Phys. Lett.* **A** (1994) (to appear).

*Supports received: CNPq of Brasil (LT), CAPES of Brasil (CFA), FAPESP of Brasil (JD) and DST grant no. SP/S2/K04/93, Govt. of India. (JD and MD)

†Permanent address : 1/10 Prince Golam Md. Road, Calcutta 700 026, India